Cohen-Macaulay Representations

Mathematical
Surveys
and
Monographs

Volume 181

Cohen-Macaulay Representations

Graham J. Leuschke
Roger Wiegand

American Mathematical Society
Providence, Rhode Island

EDITORIAL COMMITTEE

Ralph L. Cohen, Chair Michael A. Singer
Jordan S. Ellenberg Benjamin Sudakov
Michael I. Weinstein

2010 *Mathematics Subject Classification.* Primary 13C14, 16G50;
Secondary 13B40, 13C05, 13C60, 13H10, 16G10, 16G60, 16G70.

For additional information and updates on this book, visit
www.ams.org/bookpages/surv-181

Library of Congress Cataloging-in-Publication Data

Leuschke, Graham J., 1973–
  Cohen-Macaulay representations / Graham J. Leuschke, Roger Wiegand.
    p. cm. — (Mathematical surveys and monographs ; v. 181)
  Includes bibliographical references and index.
  ISBN 978-0-8218-7581-0 (alk. paper)
  1. Cohen-Macaulay modules. 2. Representations of rings (Algebra) I. Wiegand, Roger, 1943–
II. Title.

QA251.3.L48 2012
512'.44—dc23
                                                          2012002344

Copying and reprinting. Individual readers of this publication, and nonprofit libraries acting for them, are permitted to make fair use of the material, such as to copy a chapter for use in teaching or research. Permission is granted to quote brief passages from this publication in reviews, provided the customary acknowledgment of the source is given.

Republication, systematic copying, or multiple reproduction of any material in this publication is permitted only under license from the American Mathematical Society. Requests for such permission should be addressed to the Acquisitions Department, American Mathematical Society, 201 Charles Street, Providence, Rhode Island 02904-2294 USA. Requests can also be made by e-mail to reprint-permission@ams.org.

© 2012 by the American Mathematical Society. All rights reserved.
The American Mathematical Society retains all rights
except those granted to the United States Government.
Printed in the United States of America.

∞ The paper used in this book is acid-free and falls within the guidelines
established to ensure permanence and durability.
Visit the AMS home page at http://www.ams.org/

10 9 8 7 6 5 4 3 2 1   17 16 15 14 13 12

To Conor, David and Andrea, and
to our students, past, present and future.

Contents

Preface	xi
Chapter 1. The Krull-Remak-Schmidt Theorem	**1**
§1. KRS in an additive category	1
§2. KRS over Henselian rings	6
§3. R-modules vs. $\widehat{R}$-modules	7
§4. Exercises	9
Chapter 2. Semigroups of Modules	**13**
§1. Krull monoids	14
§2. Realization in dimension one	17
§3. Realization in dimension two	23
§4. Flat local homomorphisms	26
§5. Exercises	27
Chapter 3. Dimension Zero	**29**
§1. Artinian rings with finite Cohen-Macaulay type	29
§2. Artinian pairs	32
§3. Exercises	39
Chapter 4. Dimension One	**41**
§1. Necessity of the Drozd-Roĭter conditions	42
§2. Sufficiency of the Drozd-Roĭter conditions	45
§3. ADE singularities	50
§4. The analytically ramified case	52
§5. Multiplicity two	54
§6. Ranks of indecomposable MCM modules	56
§7. Why MCM modules?	57
§8. Exercises	58
Chapter 5. Invariant Theory	**61**
§1. The skew group ring	61
§2. The endomorphism algebra	65

§3.	Group representations and the McKay-Gabriel quiver	71
§4.	Exercises	77

Chapter 6. Kleinian Singularities and Finite CM Type — 81
§1.	Invariant rings in dimension two	81
§2.	Kleinian singularities	83
§3.	McKay-Gabriel quivers of the Kleinian singularities	91
§4.	Geometric McKay correspondence	97
§5.	Exercises	106

Chapter 7. Isolated Singularities and Dimension Two — 109
§1.	Miyata's theorem	109
§2.	Isolated singularities	113
§3.	Two-dimensional CM rings of finite CM type	117
§4.	Exercises	120

Chapter 8. The Double Branched Cover — 123
§1.	Matrix factorizations	123
§2.	The double branched cover	128
§3.	Knörrer's periodicity	133
§4.	Exercises	139

Chapter 9. Hypersurfaces with Finite CM Type — 141
§1.	Simple singularities	141
§2.	Hypersurfaces in good characteristics	144
§3.	Gorenstein rings of finite CM type	150
§4.	Matrix factorizations for the Kleinian singularities	151
§5.	Bad characteristics	159
§6.	Exercises	160

Chapter 10. Ascent and Descent — 163
§1.	Descent	163
§2.	Ascent to the completion	164
§3.	Ascent along separable field extensions	170
§4.	Equicharacteristic Gorenstein singularities	172
§5.	Exercises	173

Chapter 11. Auslander-Buchweitz Theory — 175
§1.	Canonical modules	175
§2.	MCM approximations and FID hulls	179
§3.	Numerical invariants	189

§4.	The index and applications to finite CM type	194
§5.	Exercises	200

Chapter 12. Totally Reflexive Modules — 203
§1.	Stable Hom and Auslander transpose	203
§2.	Complete resolutions	207
§3.	Totally reflexive modules	209
§4.	Exercises	215

Chapter 13. Auslander-Reiten Theory — 217
§1.	AR sequences	217
§2.	AR quivers	224
§3.	Examples	228
§4.	Exercises	238

Chapter 14. Countable Cohen-Macaulay Type — 241
§1.	Structure	241
§2.	Burban-Drozd triples	244
§3.	Hypersurfaces of countable CM type	252
§4.	Other examples	260
§5.	Exercises	263

Chapter 15. The Brauer-Thrall Conjectures — 267
§1.	The Harada-Sai lemma	268
§2.	Faithful systems of parameters	270
§3.	Proof of Brauer-Thrall I	276
§4.	Brauer-Thrall II	280
§5.	Exercises	285

Chapter 16. Finite CM Type in Higher Dimensions — 287
§1.	Two examples	287
§2.	Classification for homogeneous CM rings	293
§3.	Exercises	295

Chapter 17. Bounded CM Type — 297
§1.	Hypersurface rings	297
§2.	Dimension one	299
§3.	Descent in dimension one	303
§4.	Exercises	307

Appendix A. Basics and Background — 309

§1.	Depth, syzygies, and Serre's conditions	309
§2.	Multiplicity and rank	313
§3.	Henselian rings	317
Appendix B.	Ramification Theory	321
§1.	Unramified homomorphisms	321
§2.	Purity of the branch locus	326
§3.	Galois extensions	335
Bibliography		341
Index		355

Preface

This book is about the representation theory of commutative local rings, specifically the study of maximal Cohen-Macaulay modules over Cohen-Macaulay local rings.

The guiding principle of representation theory, broadly speaking, is that we can understand an algebraic structure by studying the sets upon which it acts. Classically, this meant understanding finite groups by studying the vector spaces they act upon; the powerful tools of linear algebra can then be brought to bear, revealing information about the group that was otherwise hidden. In other branches of representation theory, such as the study of finite-dimensional associative algebras, sophisticated technical machinery has been built to investigate the properties of modules, and how restrictions on modules over a ring restrict the structure of the ring.

The representation theory of maximal Cohen-Macaulay modules began in the late 1970s and grew quickly, inspired by three other areas of algebra. Spectacular successes in the representation theory of finite-dimensional algebras during the 1960s and 70s set the standard for what one might hope for from a representation theory. In particular, this period saw: P. Gabriel's introduction of the representations of quivers and his theorem that a quiver has finite representation type if and only if it is a disjoint union of ADE Coxeter-Dynkin diagrams; M. Auslander's influential Queen Mary notes applying his work on functor categories to representation theory; Auslander and I. Reiten's foundational work on AR sequences; and key insights from the Kiev school, particularly Y. Drozd, L. A. Nazarova, and A. V. Roĭter. All these advances continued the work on finite representation type begun in the 1940s and 50s by T. Nakayama, R. Brauer, R. Thrall, and J. P. Jans. Secondly, the study of lattices over orders, a part of integral representation theory, blossomed in the late 1960s. Restricting attention to lattices rather than arbitrary modules allowed a rich theory to develop. In particular, the work of Drozd-Roĭter and H. Jacobinski around this time introduced the conditions we call "the Drozd-Roĭter conditions" classifying commutative orders with only a finite number of non-isomorphic indecomposable lattices. Finally, M. Hochster's study of the homological conjectures emphasized the importance of the maximal Cohen-Macaulay condition (even for non-finitely generated modules). The equality of the geometric invariant of dimension with the arithmetic one of depth makes this class of modules easy to work with, simultaneously ensuring that they faithfully reflect the structure of the ring.

The main focus of this book is on the problem of classifying Cohen-Macaulay local rings having only a finite number of indecomposable maximal Cohen-Macaulay modules, that is, having *finite CM type*. Notice that we wrote "the problem," rather than "the solution." Indeed, there is no complete classification to date. There are many partial results, however, including complete classifications in dimensions zero and one, a characterization in dimension two under some mild assumptions, and a complete understanding of the hypersurface singularities with this property. The tools developed to obtain these classifications have many applications to other problems as well, in addition to their inherent beauty. In particular there are applications to the study of other representation types, including countable type and bounded type.

This is not the first book about the representation theory of Cohen-Macaulay modules over Cohen-Macaulay local rings. The text [**Yos90**] by Y. Yoshino is a fantastic book and an invaluable resource, and has inspired us both on countless occasions. It has been the canonical reference for the subject for twenty years. In those years, however, there have been many advances. To give just two examples, we mention C. Huneke and Leuschke's elementary proof in 2002 of Auslander's theorem that finite CM type implies isolated singularity, and R. Wiegand's 2000 verification of F.-O. Schreyer's conjecture that finite CM type ascends to and descends from the completion. These developments alone might justify a new exposition. Furthermore, there are many facets of the subject not covered in Yoshino's book, some of which we are qualified to describe. Thus this book might be considered simultaneously an updated edition of [**Yos90**], a companion volume, and an alternative.

In addition to telling the basic story of finite CM type, our choice of material is guided by a number of themes.

(i) For a homomorphism of local rings $R \longrightarrow S$, which maximal Cohen-Macaulay modules over S "come from" R is a basic question. It is especially important when $S = \widehat{R}$, the completion of R, for then the Krull-Remak-Schmidt uniqueness theorem holds for direct-sum decompositions of $\widehat{R}$-modules.

(ii) The failure of the Krull-Remak-Schmidt theorem is often more interesting than its success. We can often quantify exactly how badly it fails.

(iii) A certain amount of non-commutativity can be useful even in pure commutative algebra. In particular, the endomorphism ring of a module, while technically a non-commutative ring, should be a standard object of consideration in commutative algebra.

(iv) An abstract, categorical point of view is not always a good thing in and of itself. We tend to be stubbornly concrete, emphasizing explicit constructions over universal properties.

The main material of the book is divided into 17 chapters. The first chapter contains some vital background information on the Krull-Remak-Schmidt Theorem, which we view as a version of the Fundamental Theorem of Arithmetic for modules, and on the relationship between modules over a local ring R and over its completion $\widehat{R}$. Chapter 2 is devoted to an analysis of exactly how badly the Krull-Remak-Schmidt Theorem can fail. Nothing here is specifically about Cohen-Macaulay rings or maximal Cohen-Macaulay modules.

Chapters 3 and 4 contain the classification theorems for Cohen-Macaulay local rings of finite CM type in dimensions zero and one. Here essentially everything is known. In particular Chapter 3 introduces an auxiliary representation-theoretic problem, the Artinian pair, which is then used in Chapter 4 to solve the problem of finite CM type over one-dimensional rings via the conductor-square construction.

The two-dimensional Cohen-Macaulay local rings of finite CM type are at a focal point in our telling of the theory, with connections to algebraic geometry, invariant theory, group representations, solid geometry, representations of quivers, and other areas, by way of the *McKay correspondence*. Chapter 5 sets the stage for this material, introducing (in arbitrary dimension) the necessary invariant theory and results of Auslander relating a ring of invariants to the associated skew group ring. These results are applied in Chapter 6 to show that two-dimensional rings of invariants have finite CM type. In particular this applies to the Kleinian singularities, also known as Du Val singularities, rational double points, or ADE hypersurface singularities. We also describe some aspects of the McKay correspondence, including the geometric results due to M. Artin and J.-L. Verdier. Finally Chapter 7 gives the full classification of complete local two-dimensional $\mathbb{C}$-algebras of finite CM type. This chapter also includes Auslander's theorem mentioned earlier that finite CM type implies isolated singularity.

In dimensions higher than two, our understanding of finite CM type is imperfect. We do, however, understand the Gorenstein case more or less completely. By a result of J. Herzog, a complete Gorenstein local ring of finite CM type is a hypersurface ring; these are completely classified in the equicharacteristic case. This classification is detailed in Chapter 9, including the theorem of R.-O. Buchweitz, G.-M. Greuel, and Schreyer which states that if a complete equicharacteristic hypersurface singularity over an algebraically closed field has finite CM type, then it is a simple singularity in the sense of V. I. Arnol'd. We also write down the matrix factorizations for the indecomposable MCM modules over the Kleinian singularities, from which the matrix factorizations in arbitrary dimension can be obtained. Our proof of the Buchweitz-Greuel-Schreyer result is by reduction to dimension two via the double branched cover construction and H. Knörrer's periodicity theorem. Chapter 8 contains these background results, after a brief presentation of the theory of matrix factorizations.

Chapter 10 addresses the critical questions of ascent and descent of finite CM type along ring extensions, particularly between a Cohen-Macaulay local ring and its completion, as well as passage to a local ring with a larger residue field. This allows us to extend the classification theorem for hypersurface singularities of finite CM type to non-algebraically closed fields.

Chapters 11 and 13 describe two powerful tools in the study of maximal Cohen-Macaulay modules over Cohen-Macaulay rings: MCM approximations and Auslander-Reiten sequences. We are not aware of another complete, concise and explicit treatment of Auslander and Buchweitz's theory of MCM approximations and hulls of finite injective dimension, which we believe deserves to be better known. The theory of Auslander-Reiten sequences and quivers, of course, is essential. Chapter 12 establishes some homological tools and introduces totally reflexive modules, whose homological behavior over general local rings mimics that of MCM modules over Gorenstein rings.

The last four chapters consider other representation types, namely countable and bounded CM type, and finite CM type in higher dimensions. Chapter 14 uses recent results of I. Burban and Drozd, based on a modification of the conductor-square construction, to prove Buchweitz-Greuel-Schreyer's classification of the hypersurface singularities with countable CM type. It also proves certain structural results for rings of countable CM type, due to Huneke and Leuschke. Chapter 15 contains a proof of the first Brauer-Thrall conjecture, that an excellent isolated singularity with bounded CM type necessarily has finite CM type. Our presentation follows the original proofs of E. Dieterich and Yoshino. The Brauer-Thrall theorem is then used, in Chapter 16, to prove that two three-dimensional examples have finite CM type. We also quote the theorem of D. Eisenbud and Herzog which classifies the homogeneous rings of finite CM type; in particular, their result says that there are no examples in dimension $\geqslant 3$ other than the ones we have described in the text. Finally, in Chapter 17, we consider the rings of bounded but infinite CM type. It happens that for hypersurface rings they are precisely the same as the rings of countable but infinite CM type. We also classify the one-dimensional rings of bounded CM type.

We include two Appendices. In Appendix A, we gather for ease of reference some basic definitions and results of commutative algebra that are prerequisites for the book. Appendix B, on the other hand, contains material that we require from ramification theory that is not generally covered in a general commutative algebra course. It includes the basics on unramified and étale homomorphisms, Henselian rings, ramification of prime ideals, and purity of the branch locus. We make essential use of these concepts, but they are peripheral to the main material of the book.

The knowledgeable reader will have noticed significant overlap between the topics mentioned above and those covered by Yoshino in his book [**Yos90**]. To a certain extent this is unavoidable; the basics of the area are what they are, and any book on Cohen-Macaulay representation types will mention them. However, the reader should be aware that our guiding principles are quite different from Yoshino's, and consequently there are few topics on which our presentation parallels that in [**Yos90**]. When it does, it is generally because both books follow the original presentation of Auslander, Auslander-Reiten, or Yoshino.

Early versions of this book have been used for advanced graduate courses at the University of Nebraska in Fall 2007 and at Syracuse University in Fall 2010. In each case, the students had had at least one full-semester course in commutative algebra at the level of Matsumura's book [**Mat89**]. A few more advanced topics are needed from time to time, such as the basics of group representations and character theory, properties of canonical modules and Gorenstein rings, Cohen's structure theory for complete local rings, the Artin-Rees Lemma, and the material on multiplicity and Serre's conditions in the Appendix. Many of these can be taken on faith at first encounter, or covered as extra topics.

The core of the book, Chapters 3 through 9, is already more material than could comfortably be covered in a semester course. One remedy would be to streamline the material, restricting to the case of complete local rings with algebraically closed residue fields of characteristic zero. One might also skip or sketch some of the more tangential material. We regard the following as essential: Chapter 3 (omitting most of the proof of Theorem 3.7); the first three sections of Chapter 4; Chapter 5; Chapter 6 (omitting the proof of Theorem 6.11, the calculations in §3, and §4); Chapters 7 and 8; and the first two sections of Chapter 9. Chapters 2 and 10 can each stand alone as optional topics, while the thread beginning with Chapters 11 and 13, continuing through Chapters 15 and 17 could serve as the basis of a completely separate course (though some knowledge of the first half of the book would be necessary to make sense of Chapters 14 and 16).

At the end of each chapter is a short section of exercises of varying difficulty, over 120 in all. Some are independent problems, while others ask the solver to fill in details of proofs omitted from the body of the text.

We gratefully acknowledge the many, many people and organizations whose support we enjoyed while writing this book. Our students at Nebraska and Syracuse endured early drafts of the text, and helped us improve it; thanks to Tom Bleier, Jesse Burke, Ela Çelikbas, Olgur Çelikbas, Justin Devries, Kos Diveris, Christina Eubanks-Turner, Inês Bonacho dos Anjos Henriques, Nick Imholte, Brian Johnson, Micah Leamer, Laura Lynch, Matt Mastroeni, Lori McDonnell, Sean Mead-Gluchacki, Livia Miller, Frank Moore, Terri Moore, Hamid Rahmati, Silvia Saccon, and Mark Yerrington. GJL was supported by National Science Foundation grants DMS-0556181 and DMS-0902119 during this project, and RW by a grant from the National Security Agency. The CIRM at Luminy hosted us for a highly productive and enjoyable week in June 2010. Each of us visited the other several times over the years, and enjoyed the hospitality of each other's home department, for which we thank UNL and SU, respectively.

GRAHAM J. LEUSCHKE
  Syracuse University
  `gjleusch@math.syr.edu`
ROGER WIEGAND
  University of Nebraska-Lincoln
  `rwiegand@math.unl.edu`
Syracuse and Lincoln, December 2011

CHAPTER 1

The Krull-Remak-Schmidt Theorem

In this chapter we will prove the Krull-Remak-Schmidt uniqueness theorem for direct-sum decompositions of finitely generated modules over complete local rings. The first such theorem, in the context of finite groups, was stated by Wedderburn [**Wed09**]: Let G be a finite group with two direct-product decompositions $G = H_1 \times \cdots \times H_m$ and $G = K_1 \times \cdots \times K_n$, where each H_i and each K_j is indecomposable. Then $m = n$, and, after renumbering, $H_i \cong K_i$ for each i. In 1911 Remak [**Rem11**] gave a complete proof, and actually proved more: H_i and K_i are *centrally isomorphic*, that is, there are isomorphisms $f_i \colon H_i \longrightarrow K_i$ such that $x^{-1}f(x)$ is in the center of G for each $x \in H_i$, $i = 1, \ldots, m$. These results were extended to groups with operators satisfying the ascending and descending chain conditions by Krull [**Kru25**] and Schmidt [**Sch29**]. In 1950 Azumaya [**Azu50**] proved an analogous result for possibly infinite direct sums of modules, with the assumption that the endomorphism ring of each factor is local in the non-commutative sense.

§1. KRS in an additive category

Looking ahead to an application in Chapter 3, we will clutter things up slightly by working in an additive category, rather than a category of modules. An *additive* category is a category $\mathcal{A}$ with 0-object such that (i) $\mathrm{Hom}_{\mathcal{A}}(M_1, M_2)$ is an abelian group for each pair M_1, M_2 of objects, (ii) composition is bilinear, and (iii) every finite set of objects has a biproduct. A *biproduct* of $M_1, \ldots, M_m$ consists of an object M together with maps $u_i \colon M_i \longrightarrow M$ and $p_i \colon M \longrightarrow M_i$, $i = 1, \ldots, m$, such that $p_i u_j = \delta_{ij}$ and $u_1 p_1 + \cdots + u_m p_m = 1_M$. We denote the biproduct by $M_1 \oplus \cdots \oplus M_m$.

We will need an additional condition on our additive category, that idempotents split (cf. [**Bas68**, Chapter I, §3, p. 19]). Given an object M and an idempotent $e \in \mathrm{End}_{\mathcal{A}}(M)$, we say that e *splits* provided there is a factorization $M \xrightarrow{p} K \xrightarrow{u} M$ such that $e = up$ and $pu = 1_K$.

The reader is probably familiar with the notion of an *abelian* category, that is, an additive category in which every map has a kernel and a cokernel, and in which every monomorphism (respectively epimorphism) is a kernel (respectively cokernel). Over any ring R the category R-Mod of all left R-modules is abelian; if R is left Noetherian, then the category R-mod of finitely generated left R-modules is abelian. It is easy to see that idempotents split in an abelian category. Indeed, suppose $e\colon M \longrightarrow M$ is an idempotent, and let $u\colon K \longrightarrow M$ be the kernel of $1_M - e$. Since $(1_M - e)e = 0$, the map e factors through u; that is, there is a map $p\colon M \longrightarrow K$ satisfying $up = e$. Then $upu = eu = eu + (1_M - e)u = u = u1_K$. Since u is a monomorphism (as kernels are always monomorphisms), it follows that $pu = 1_K$.

A non-zero object M in the additive category $\mathcal{A}$ is said to be *decomposable* if there exist non-zero objects M_1 and M_2 such that $M \cong M_1 \oplus M_2$. Otherwise, M is *indecomposable*. We leave the proof of the next result as an exercise:

1.1. PROPOSITION. *Let M be a non-zero object in an additive category $\mathcal{A}$, and let $E = \mathrm{End}_{\mathcal{A}}(M)$.*

   (i) *If 0 and 1 are the only idempotents of E, then M is indecomposable.*
   (ii) *Conversely, suppose $e = e^2 \in E$, with $e \neq 0, 1$. If both e and $1 - e$ split, then M is decomposable.* □

We say that the Krull-Remak-Schmidt Theorem (KRS for short) holds in the additive category $\mathcal{A}$ provided

   (i) every object in $\mathcal{A}$ is a biproduct of indecomposable objects, and
   (ii) if $M_1 \oplus \cdots \oplus M_m \cong N_1 \oplus \cdots \oplus N_n$, with each M_i and each N_j an indecomposable object in $\mathcal{A}$, then $m = n$ and, after renumbering, $M_i \cong N_i$ for each i.

It is easy to see that every Noetherian object is a biproduct of finitely many indecomposable objects (cf. Exercise 1.19), but there are easy examples to show that (ii) can fail. For perhaps the simplest example, let $R = k[x, y]$, the polynomial ring in two variables over a field. Letting $\mathfrak{m} = Rx + Ry$ and $\mathfrak{n} = R(x - 1) + Ry$, we get a short exact sequence
$$0 \longrightarrow \mathfrak{m} \cap \mathfrak{n} \longrightarrow \mathfrak{m} \oplus \mathfrak{n} \longrightarrow R \longrightarrow 0,$$
since $\mathfrak{m} + \mathfrak{n} = R$. This splits, so $\mathfrak{m} \oplus \mathfrak{n} \cong R \oplus (\mathfrak{m} \cap \mathfrak{n})$. Since neither $\mathfrak{m}$ nor $\mathfrak{n}$ is isomorphic to R as an R-module, KRS fails for finitely generated R-modules.

Alternatively, let D be a Dedekind domain with a non-principal ideal I. We have an isomorphism (see Exercise 1.20)

$$(1.1) \qquad R \oplus R \cong I \oplus I^{-1},$$

and of course all of the summands in (1.1) are indecomposable.

These examples indicate that KRS is likely to fail for modules over rings that aren't local. It can fail even for finitely generated modules over local rings. An example due to Swan is in Evans's paper [**Eva73**]. In Chapter 2 we will see just how badly it can fail. Azumaya [**Azu48**] observed that the crucial property for guaranteeing KRS is that the endomorphism rings of the summands be local in the non-commutative sense. To avoid a conflict of jargon, we define a ring Λ (not necessarily commutative) to be *nc-local* provided $\Lambda/\mathcal{J}(\Lambda)$ is a division ring, where $\mathcal{J}(-)$ denotes the Jacobson radical. Equivalently (cf. Exercise 1.21) $\Lambda \neq \{0\}$ and $\mathcal{J}(\Lambda)$ is exactly the set of non-units of Λ. It is clear from Proposition 1.1 that any object with nc-local endomorphism ring must be indecomposable.

We'll model our proof of KRS after the proof of unique factorization in the integers, by showing that an object with nc-local endomorphism ring behaves like a prime element in an integral domain. We'll even use the same notation, writing "$M \mid N$", for objects M and N, to indicate that there is an object Z such that $N \cong M \oplus Z$. Our inductive proof depends on direct-sum cancellation ((ii) below), analogous to the fact that $mz = my \implies z = y$ for non-zero elements m, z, y in an integral domain. Later in the chapter (Corollary 1.16) we'll prove cancellation for arbitrary finitely generated modules over a local ring, but for now we'll prove only that objects with nc-local endomorphism rings can be cancelled.

1.2. LEMMA. *Let $\mathcal{A}$ be an additive category in which idempotents split. Let M, X, Y, and Z be objects of $\mathcal{A}$, let $E = \mathrm{End}_{\mathcal{A}}(M)$, and assume that E is nc-local.*

 (i) *If $M \mid X \oplus Y$, then $M \mid X$ or $M \mid Y$ ("primelike").*
 (ii) *If $M \oplus Z \cong M \oplus Y$, then $Z \cong Y$ ("cancellation").*

PROOF. We'll prove (i) and (ii) simultaneously. In (i) we have an object Z such that $M \oplus Z \cong X \oplus Y$. In the proof of (ii) we set $X = M$ and again get an isomorphism $M \oplus Z \cong X \oplus Y$. Put $J = \mathcal{J}(E)$, the set of non-units of E.

Choose reciprocal isomorphisms $\varphi \colon M \oplus Z \longrightarrow X \oplus Y$ and $\psi \colon X \oplus Y \longrightarrow M \oplus Z$. Write

$$\varphi = \begin{bmatrix} \alpha & \beta \\ \gamma & \delta \end{bmatrix} \quad \text{and} \quad \psi = \begin{bmatrix} \mu & \nu \\ \sigma & \tau \end{bmatrix},$$

where $\alpha \colon M \longrightarrow X$, $\beta \colon Z \longrightarrow X$, $\gamma \colon M \longrightarrow Y$, $\delta \colon Z \longrightarrow Y$, $\mu \colon X \longrightarrow M$, $\nu \colon Y \longrightarrow M$, $\sigma \colon X \longrightarrow Z$ and $\tau \colon Y \longrightarrow Z$. Since $\psi \varphi = 1_{M \oplus Z} = \begin{bmatrix} 1_M & 0 \\ 0 & 1_Z \end{bmatrix}$, we have $\mu \alpha + \nu \gamma = 1_M$. Therefore, as E is local, either $\mu \alpha$

or $\nu\gamma$ must be outside J and hence an automorphism of M. Assuming that $\mu\alpha$ is an automorphism, we will produce an object W and maps

$$M \xrightarrow{u} X \xrightarrow{p} M \qquad W \xrightarrow{v} X \xrightarrow{q} W$$

satisfying $pu = 1_M$, $pv = 0$, $qv = 1_W$, $qu = 0$, and $up + vq = 1_X$. This will show that $X = M \oplus W$. (Similarly, the assumption that $\nu\gamma$ is an isomorphism forces M to be a direct summand of Y.)

Letting $u = \alpha$, $p = (\mu\alpha)^{-1}\mu$ and $e = up \in \text{End}_{\mathcal{A}}(X)$, we have $pu = 1_M$ and $e^2 = e$. By hypothesis, the idempotent $1 - e$ splits, so we can write $1 - e = vq$, where $X \xrightarrow{q} W \xrightarrow{v} X$ and $qv = 1_W$. From $e = up$ and $1 - e = vq$, we get the equation $up + vq = 1_X$. Moreover, $qu = (qv)(qu)(pu) = q(vq)(up)u = q(1-e)eu = 0$; similarly, $pv = pupvqv = pe(1-e)v = 0$. We have verified all of the required equations, so $X = M \oplus W$. This proves (i).

To prove (ii) we assume that $X = M$. Suppose first that α is a unit of E. We use α to diagonalize φ:

$$\begin{bmatrix} 1 & 0 \\ -\gamma\alpha^{-1} & 1 \end{bmatrix} \begin{bmatrix} \alpha & \beta \\ \gamma & \delta \end{bmatrix} \begin{bmatrix} 1 & -\alpha^{-1}\beta \\ 0 & 1 \end{bmatrix} = \begin{bmatrix} \alpha & 0 \\ 0 & -\gamma\alpha^{-1}\beta + \delta \end{bmatrix}$$

Since all the matrices on the left are invertible, so must be the one on the right, and it follows that $-\gamma\alpha^{-1}\beta + \delta \colon Z \longrightarrow Y$ is an isomorphism.

Suppose, on the other hand, that $\alpha \in J$. Then $\nu\gamma \notin J$ (as $\mu\alpha + \nu\gamma = 1_M$), and it follows that $\alpha + \nu\gamma \notin J$. We define a new map

$$\psi' = \begin{bmatrix} 1_M & \nu \\ \sigma & \tau \end{bmatrix} \colon M \oplus Y \longrightarrow M \oplus Z\,,$$

which we claim is an isomorphism. Assuming the claim, we can diagonalize ψ' as we did φ, obtaining, in the lower-right corner, an isomorphism from Y onto Z, and finishing the proof. To prove the claim, we use the equation $\psi\varphi = 1_{M \oplus Z}$ to get

$$\psi'\varphi = \begin{bmatrix} \alpha + \nu\gamma & \beta + \nu\gamma \\ 0 & 1_Z \end{bmatrix}.$$

As $\alpha + \nu\gamma$ is an automorphism of M, $\psi'\varphi$ is clearly an automorphism of $M \oplus Z$. Therefore $\psi' = (\psi'\varphi)\varphi^{-1}$ is an isomorphism. $\square$

1.3. THEOREM (Krull-Remak-Schmidt). *Let $\mathcal{A}$ be an additive category in which every idempotent splits. Let $M_1, \ldots, M_m$ and $N_1, \ldots, N_n$ be indecomposable objects of $\mathcal{A}$, with $M_1 \oplus \cdots \oplus M_m \cong N_1 \oplus \cdots \oplus N_n$. Assume that $\text{End}_{\mathcal{A}}(M_i)$ is nc-local for each i. Then $m = n$ and, after renumbering, $M_i \cong N_i$ for each i.*

PROOF. We use induction on m, the case $m = 1$ being trivial. Assuming $m \geqslant 2$, we see that $M_m \mid N_1 \oplus \cdots \oplus N_n$. By (i) of Lemma 1.2, $M_m \mid N_j$ for some j; by renumbering, we may assume that $M_m \mid N_n$. Since N_n is indecomposable and $M_m \neq 0$, we must have $M_m \cong N_n$. Now (ii) of Lemma 1.2 implies that $M_1 \oplus \cdots \oplus M_{m-1} \cong N_1 \oplus \cdots \oplus N_{n-1}$, and the inductive hypothesis completes the proof. □

Azumaya actually proved the *infinite* version of Theorem 1.3: If $\bigoplus_{i \in I} M_i \cong \bigoplus_{j \in J} N_j$ and the endomorphism ring of each M_i is nc-local, and each N_j is indecomposable, then there is a bijection $\sigma \colon I \longrightarrow J$ such that $M_i \cong N_{\sigma(i)}$ for each i. (Cf. [**Azu48**], or see [**Fac98**, Chapter 2].)

We want to find some situations where indecomposables automatically have nc-local endomorphism rings. It is well known that idempotents lift modulo any nil ideal. A typical proof of this fact actually yields the following stronger result, which we will use in the next section.

1.4. PROPOSITION. *Let I be a two-sided ideal of a (possibly non-commutative) ring Λ, and let e be an idempotent of Λ/I. Given any positive integer n, there is an element $x \in \Lambda$ such that $x + I = e$ and $x \equiv x^2 \pmod{I^n}$.*

PROOF. Start with an arbitrary element $u \in \Lambda$ such that $u + I = e$, and let $v = 1 - u$. In the binomial expansion of $(u+v)^{2n-1}$, let x be the sum of the first n terms: $x = u^{2n-1} + \cdots + \binom{2n-1}{n-1} u^n v^{n-1}$. Putting $y = 1 - x$ (the other half of the expansion), we see that $x - x^2 = xy \in \Lambda(uv)^n \Lambda$. Since $uv = u(1-u) \in I$, we have $x - x^2 \in I^n$. □

Here is an easy consequence, which will be needed in Chapter 3:

1.5. COROLLARY. *Let M be an indecomposable object in an additive category $\mathcal{A}$. Assume that idempotents split in $\mathcal{A}$. If $E := \mathrm{End}_\mathcal{A}(M)$ is left or right Artinian, then E is nc-local.*

PROOF. Since M is indecomposable, E has no non-trivial idempotents. Since $\mathcal{J}(E)$ is nilpotent, Proposition 1.4 implies that $E/\mathcal{J}(E)$ has no idempotents either. It follows easily from the Wedderburn-Artin Theorem [**Lam91**, (3.5)] that $E/\mathcal{J}(E)$ is a division ring, whence nc-local. □

1.6. COROLLARY. *Let R be a commutative Artinian ring. Then KRS holds in the category of finitely generated R-modules.*

PROOF. Let M be an indecomposable finitely generated R-module. By Exercise 1.22 $\mathrm{End}_R(M)$ is finitely generated as an R-module and therefore is a left (and right) Artinian ring. Now apply Corollary 1.5 and Theorem 1.3. □

§2. KRS over Henselian rings

We now proceed to prove KRS for finitely generated modules over complete and, more generally, Henselian local rings. Here we define a local ring $(R, \mathfrak{m}, k)$ to be *Henselian* provided, for every module-finite R-algebra Λ (not necessarily commutative), each idempotent of $\Lambda/\mathcal{J}(\Lambda)$ lifts to an idempotent of Λ. For the classical definition of "Henselian" in terms of factorization of polynomials, and for other equivalent conditions, see Theorem A.30.

1.7. LEMMA. *Let R be a commutative ring and Λ a module-finite R-algebra (not necessarily commutative). Then $\Lambda \mathcal{J}(R) \subseteq \mathcal{J}(\Lambda)$.*

PROOF. Let $f \in \Lambda \mathcal{J}(R)$. We want to show that $\Lambda(1 - \lambda f) = \Lambda$ for every $\lambda \in \Lambda$. Clearly $\Lambda(1 - \lambda f) + \Lambda \mathcal{J}(R) = \Lambda$, and now NAK completes the proof. □

1.8. THEOREM. *Let $(R, \mathfrak{m}, k)$ be a Henselian local ring, and let M be an indecomposable finitely generated R-module. Then $\mathrm{End}_R(M)$ is nc-local. In particular, KRS holds for the category of finitely generated modules over a Henselian local ring.*

PROOF. Let $E = \mathrm{End}_R(M)$ and $J = \mathcal{J}(E)$. Since E is a module-finite R-algebra (cf. Exercise 1.22), Lemma 1.7 implies that $\mathfrak{m}E \subseteq J$ and hence that E/J is a finite-dimensional k-algebra. It follows that E/J is semisimple Artinian. Moreover, since E has no non-trivial idempotents, neither does E/J. By the Wedderburn-Artin Theorem [**Lam91**, (3.5)], E/J is a division ring. □

1.9. COROLLARY (Hensel's Lemma). *Let $(R, \mathfrak{m}, k)$ be a complete local ring. Then R is Henselian.*

PROOF. Let Λ be a module-finite R-algebra, and put $J = \mathcal{J}(\Lambda)$. Again, $\mathfrak{m}\Lambda \subseteq J$, and $J/\mathfrak{m}\Lambda$ is a nilpotent ideal of $\Lambda/\mathfrak{m}\Lambda$ (since $\Lambda/\mathfrak{m}\Lambda$ is Artinian). By Proposition 1.4 we can lift each idempotent of Λ/J to an idempotent of $\Lambda/\mathfrak{m}\Lambda$. Therefore it will suffice to show that every idempotent e of $\Lambda/\mathfrak{m}\Lambda$ lifts to an idempotent of Λ. Using Proposition 1.4, we can choose, for each positive integer n, an element $x_n \in \Lambda$ such that $x_n + \mathfrak{m}\Lambda = e$ and $x_n \equiv x_n^2 \pmod{\mathfrak{m}^n \Lambda}$. (Of course $\mathfrak{m}^n \Lambda = (\mathfrak{m}\Lambda)^n$.) We claim that (x_n) is a Cauchy sequence for the $\mathfrak{m}\Lambda$-adic topology on Λ. To see this, let n be an arbitrary positive integer. Given any $m \geqslant n$, put $z = x_m + x_n - 2x_m x_n$. Then $z \equiv z^2 \pmod{\mathfrak{m}^n \Lambda}$. Also, since

$x_m \equiv x_n \pmod{\mathfrak{m}\Lambda}$, we see that $z \equiv 0 \pmod{\mathfrak{m}\Lambda}$, so $1 - z$ is a unit of Λ. Since $z(1 - z) \in \mathfrak{m}^n\Lambda$, it follows that $z \in \mathfrak{m}^n\Lambda$. Thus we have

$$x_m + x_n \equiv 2x_m x_n, \qquad x_m \equiv x_m^2, \qquad x_n \equiv x_n^2 \pmod{\mathfrak{m}^n\Lambda}.$$

Multiplying the first congruence, in turn, by x_m and by x_n, we learn that $x_m \equiv x_m x_n \equiv x_n \pmod{\mathfrak{m}^n\Lambda}$. If, now, $\ell \geqslant n$ and $m \geqslant n$, we see that $x_\ell \equiv x_m \pmod{\mathfrak{m}^n\Lambda}$. This verifies the claim. Since Λ is $\mathfrak{m}\Lambda$-adically complete (cf. Exercise 1.24), we let x be the limit of the sequence (x_n) and check that x is an idempotent lifting e. $\square$

1.10. COROLLARY. *KRS holds for finitely generated modules over complete local rings.* $\square$

Henselian local rings are *almost* characterized as those having the Krull-Remak-Schmidt property. Indeed, a theorem due to Evans states that a local ring R is Henselian if and only if for every module-finite commutative local R-algebra A the finitely generated A-modules satisfy KRS [**Eva73**].

§3. R-modules vs. $\widehat{R}$-modules

A major theme in this book is the study of direct-sum decompositions over local rings that are not necessarily complete. Here we record a few results that will allow us to use KRS over the completion $\widehat{R}$ to understand R-modules.

We begin with a result due to Guralnick [**Gur85**, Theorem A] on lifting homomorphisms modulo high powers of the maximal ideal of a local ring. Given finitely generated modules M and N over a local ring $(R, \mathfrak{m})$, we define a *lifting number* for the pair (M, N) to be a non-negative integer e satisfying the following property: For each positive integer f and each R-homomorphism $\xi \colon M/\mathfrak{m}^{e+f}M \longrightarrow N/\mathfrak{m}^{e+f}N$, there exists $\sigma \in \mathrm{Hom}_R(M, N)$ such that σ and ξ induce the same homomorphism $M/\mathfrak{m}^f M \longrightarrow N/\mathfrak{m}^f N$. (Thus the outer and bottom squares in the diagram below both commute, though the top square may not.)

$$\begin{array}{ccc} M & \xrightarrow{\sigma} & N \\ \downarrow & & \downarrow \\ M/\mathfrak{m}^{e+f}M & \xrightarrow{\xi} & N/\mathfrak{m}^{e+f}N \\ \downarrow & & \downarrow \\ M/\mathfrak{m}^f M & \xrightarrow[\overline{\xi}=\overline{\sigma}]{} & N/\mathfrak{m}^f N \end{array}$$

For example, 0 is a lifting number for the pair (M, N) if M is free.

1.11. LEMMA. *Every pair (M, N) of modules over a local ring $(R, \mathfrak{m})$ has a lifting number.*

PROOF. Choose exact sequences
$$F_1 \xrightarrow{\alpha} F_0 \longrightarrow M \longrightarrow 0,$$
$$G_1 \xrightarrow{\beta} G_0 \longrightarrow N \longrightarrow 0,$$
where F_i and G_i are finite-rank free R-modules. Define an R-homomorphism $\Phi \colon \operatorname{Hom}_R(F_0, G_0) \times \operatorname{Hom}_R(F_1, G_1) \longrightarrow \operatorname{Hom}_R(F_1, G_0)$ by $\Phi(\mu, \nu) = \mu\alpha - \beta\nu$. Applying the Artin-Rees Lemma to the submodule $\operatorname{im}(\Phi)$ of $\operatorname{Hom}_R(F_1, G_0)$, we get a positive integer e such that

(1.2) $\quad \operatorname{im}(\Phi) \cap \mathfrak{m}^{e+f} \operatorname{Hom}_R(F_1, G_0) \subseteq \mathfrak{m}^f \operatorname{im}(\Phi) \quad$ for each $f > 0$.

Suppose now that $f > 0$ and $\xi \colon M/\mathfrak{m}^{e+f}M \longrightarrow N/\mathfrak{m}^{e+f}N$ is an R-homomorphism. We can lift ξ to homomorphisms $\overline{\mu_0}$ and $\overline{\nu_0}$ making the following diagram commute.

(1.3)
$$\begin{array}{ccccccc}
F_1/\mathfrak{m}^{e+f}F_1 & \xrightarrow{\overline{\alpha}} & F_0/\mathfrak{m}^{e+f}F_0 & \longrightarrow & M/\mathfrak{m}^{e+f}M & \longrightarrow & 0 \\
{\scriptstyle \overline{\nu_0}}\downarrow & & {\scriptstyle \overline{\mu_0}}\downarrow & & \downarrow {\scriptstyle \xi} & & \\
G_1/\mathfrak{m}^{e+f}G_1 & \xrightarrow[\overline{\beta}]{} & G_0/\mathfrak{m}^{e+f}G_0 & \longrightarrow & N/\mathfrak{m}^{e+f}N & \longrightarrow & 0
\end{array}$$

Now lift $\overline{\mu_0}$ and $\overline{\nu_0}$ to homomorphisms $\mu_0 \in \operatorname{Hom}_R(F_0, G_0)$ and $\nu_0 \in \operatorname{Hom}_R(F_1, G_1)$. The commutativity of (1.3) implies that the image of $\Phi(\mu_0, \nu_0) \colon F_1 \longrightarrow G_0$ lies in $\mathfrak{m}^{e+f}G_0$. Choosing bases for F_1 and G_0, we see that the matrix representing $\Phi(\mu_0, \nu_0)$ has entries in $\mathfrak{m}^{e+f}$, so that $\Phi(\mu_0, \nu_0) \in \mathfrak{m}^{e+f} \operatorname{Hom}_R(F_1, G_0)$. By (1.2), $\Phi(\mu_0, \nu_0) \in \mathfrak{m}^f \operatorname{im}(\Phi) = \Phi(\mathfrak{m}^f(\operatorname{Hom}_R(F_0, G_0) \times \operatorname{Hom}_R(F_1, G_1)))$. Choose a pair of maps $(\mu_1, \nu_1) \in \mathfrak{m}^f(\operatorname{Hom}_R(F_0, G_0) \times \operatorname{Hom}_R(F_1, G_1))$ with $\Phi(\mu_1, \nu_1) = \Phi(\mu_0, \nu_0)$, and set $(\mu, \nu) = (\mu_0, \nu_0) - (\mu_1, \nu_1)$. Then $\Phi(\mu, \nu) = 0$, so μ induces an R-homomorphism $\sigma \colon M \longrightarrow N$. Since μ and μ_0 agree modulo $\mathfrak{m}^f$, it follows that σ and ξ induce the same map $M/\mathfrak{m}^f M \longrightarrow N/\mathfrak{m}^f N$. □

1.12. LEMMA. *If e is a lifting number for (M, N) and $e' \geqslant e$, then e' is also a lifting number for (M, N).*

PROOF. Let f' be a positive integer, and let
$$\xi \colon M/\mathfrak{m}^{e'+f'}M \longrightarrow N/\mathfrak{m}^{e'+f'}N$$
be an R-homomorphism. Put $f = f' + e' - e$. Since $e' + f' = e + f$ and e is a lifting number, there is a homomorphism $\sigma \colon M \longrightarrow N$ such that σ and ξ induce the same homomorphism $M/\mathfrak{m}^f M \longrightarrow N/\mathfrak{m}^f N$. Now

$f \geqslant f'$, and it follows that σ and ξ induce the same homomorphism $M/\mathfrak{m}^{f'}M \longrightarrow N/\mathfrak{m}^{f'}N$. □

We denote $e(M, N)$ the smallest lifting number for the pair (M, N).

1.13. THEOREM (Guralnick). *Let $(R, \mathfrak{m})$ be a local ring, and let M and N be finitely generated R-modules. If $M/\mathfrak{m}^{r+1}M \mid N/\mathfrak{m}^{r+1}N$ for some $r \geqslant \max\{e(M, N),\ e(N, M)\}$, then $M \mid N$.*

PROOF. Choose reciprocal homomorphisms

$$\xi\colon M/\mathfrak{m}^{r+1}M \longrightarrow N/\mathfrak{m}^{r+1}N \quad \text{and} \quad \eta\colon N/\mathfrak{m}^{r+1}N \longrightarrow M/\mathfrak{m}^{r+1}M$$

such that $\eta\xi = 1_{M/\mathfrak{m}^{r+1}M}$. Since r is a lifting number (Lemma 1.12), there exist R-homomorphisms $\sigma\colon M \longrightarrow N$ and $\tau\colon N \longrightarrow M$ such that σ agrees with ξ and τ agrees with η modulo $\mathfrak{m}$. By NAK, $\tau\sigma\colon M \longrightarrow M$ is surjective and therefore, by Exercise 1.27, an automorphism. It follows that $M \mid N$. □

1.14. COROLLARY. *Let $(R, \mathfrak{m})$ be a local ring and M, N finitely generated R-modules. If $M/\mathfrak{m}^n M \cong N/\mathfrak{m}^n N$ for all $n \gg 0$, then $M \cong N$.*

PROOF. By Theorem 1.13, $M \mid N$ and $N \mid M$. In particular, we have surjections $N \xrightarrow{\alpha} M$ and $M \xrightarrow{\beta} N$. Then $\beta\alpha$ is a surjective endomorphism of N and therefore is an automorphism (cf. Exercise 1.27). It follows that α is one-to-one and therefore an isomorphism. □

1.15. COROLLARY. *Let $(R, \mathfrak{m})$ be a local ring and $(\widehat{R}, \widehat{\mathfrak{m}})$ its $\mathfrak{m}$-adic completion. Let M and N be finitely generated R-modules.*
  (i) If $\widehat{R} \otimes_R M \mid \widehat{R} \otimes_R N$, then $M \mid N$.
  (ii) If $\widehat{R} \otimes_R M \cong \widehat{R} \otimes_R N$, then $M \cong N$. □

1.16. COROLLARY. *Let M, N and P be finitely generated modules over a local ring $(R, \mathfrak{m})$. If $P \oplus M \cong P \oplus N$, then $M \cong N$.*

PROOF. We have $(\widehat{R} \otimes_R P) \oplus (\widehat{R} \otimes_R M) \cong (\widehat{R} \otimes_R P) \oplus (\widehat{R} \otimes_R N)$. Using KRS for complete rings (Corollary 1.9) we see that $\widehat{R} \otimes_R M \cong \widehat{R} \otimes_R N$. Now apply Corollary 1.15. □

§4. Exercises

1.17. EXERCISE. Prove Proposition 1.1: For a non-zero object M in an additive category $\mathcal{A}$, and $E = \text{End}_{\mathcal{A}}(M)$, if 0 and 1 are the only idempotents of E, then M is indecomposable. Conversely, suppose $e = e^2 \in E$, with $e \neq 0, 1$. If both e and $1 - e$ split, then M is decomposable.

1.18. EXERCISE. Let M be an object in an additive category. Show that every direct-sum (i.e., coproduct) decomposition $M = M_1 \oplus M_2$ has a biproduct structure.

1.19. EXERCISE. Let M be an object in an additive category.
  (i) Suppose that M has either the ascending chain condition or the descending chain condition on direct summands. Prove that M has an indecomposable direct summand.
  (ii) Prove that M is a direct sum (biproduct) of finitely many indecomposable objects.

1.20. EXERCISE. Prove Steinitz's Theorem ([**Ste11**]): Let I and J be non-zero fractional ideals of a Dedekind domain D. Then $I \oplus J \cong D \oplus IJ$.

1.21. EXERCISE. Let Λ be a ring with $1 \neq 0$. Prove that the following conditions are equivalent:
  (i) Λ is nc-local.
  (ii) $\mathcal{J}(\Lambda)$ is the set of non-units of Λ.
  (iii) The set of non-units of Λ is closed under addition.
(Warning: In a non-commutative ring one can have non-units x and y such that $xy = 1$.)

1.22. EXERCISE. Let M and N be finitely generated modules over a commutative Noetherian ring R. Prove that $\mathrm{Hom}_R(M, N)$ is finitely generated as an R-module.

1.23. EXERCISE. Let $(R, \mathfrak{m})$ be a Henselian local ring and X, Y, M finitely generated R-modules. Let $\alpha \colon X \longrightarrow M$ and $\beta \colon Y \longrightarrow M$ be homomorphisms which are not split surjections. Prove that $[\alpha\ \beta] \colon X \oplus Y \longrightarrow M$ is not a split surjection.

1.24. EXERCISE. Let M be a finitely generated module over a complete local ring $(R, \mathfrak{m})$. Show that M is complete for the topology defined by the submodules $\mathfrak{m}^n M, n \geqslant 1$.

1.25. EXERCISE. Prove Fitting's Lemma: Let Λ be any ring and M a Λ-module of finite length n. If $f \in \mathrm{End}_\Lambda(M)$, then $M = \ker(f^n) \oplus f^n(M)$. Conclude that if M is indecomposable then every non-unit of $\mathrm{End}_\Lambda(M)$ is nilpotent.

1.26. EXERCISE. Use Exercise 1.21 and Fitting's Lemma from the exercise above to prove that the endomorphism ring of any indecomposable finite-length module is nc-local. Thus, over any ring R, KRS holds for the category of left R-modules of finite length. (Be careful:

You're in a non-commutative setting, where the sum of two nilpotents might be a unit! If you get stuck, consult [**Fac98**, Lemma 2.21].)

1.27. EXERCISE. Let M be a Noetherian left Λ-module, and let $f \in \operatorname{End}_\Lambda(M)$.
  (i) If f is surjective, prove that f is an automorphism of M. (Consider the ascending chain of submodules $\ker(f^n)$.)
  (ii) If f is surjective and $f^2 = f$, prove that $f = 1_M$.

CHAPTER 2

Semigroups of Modules

In this chapter we analyze the different ways in which a finitely generated module over a local ring can be decomposed as a direct sum of indecomposable modules. Put another way, we are interested in exactly how badly KRS uniqueness can fail.

Our main result depends on a technical lemma, which provides indecomposable modules of varying ranks at the minimal prime ideals of a certain one-dimensional local ring. The proof of this lemma is left as an exercise, with hints directed at a similar argument in the next chapter.

Given a ring A, choose a set $V(A)$ of representatives for the isomorphism classes $[M]$ of finitely generated left A-modules. We make $V(A)$ into an additive semigroup in the obvious way: $[M] + [N] = [M \oplus N]$. This monoid encodes information about the direct-sum decompositions of finitely generated A-modules. (In what follows, we use the terms "semigroup" and "monoid" interchangeably.)

2.1. DEFINITION. For a finitely generated left A-module M, we denote by $\operatorname{add}(M)$ or $\operatorname{add}_A(M)$ the full subcategory of A-mod consisting of finitely generated modules that are isomorphic to direct summands of direct sums of copies of M. Also, $+(M)$ is the subsemigroup of $V(A)$ consisting of representatives of the isomorphism classes in $\operatorname{add}(M)$.

In the special case where R is a complete local ring, it follows from KRS (Corollary 1.9) that $V(R)$ is a *free monoid*, that is, $V(R) \cong \mathbb{N}_0^{(I)}$, where $\mathbb{N}_0$ is the additive semigroup of non-negative integers and the index set I is the set of atoms of $V(R)$, that is, the set of representatives for the indecomposable finitely generated R-modules. Furthermore, if M is a finitely generated R-module, then $+(M)$ is free as well.

For a general local ring R, the semigroup $V(R)$ is naturally a subsemigroup of $V(\widehat{R})$ by Corollary 1.15, and similarly $+(M)$ is a subsemigroup of $+(\widehat{M})$ for an R-module M. This forces various structural restrictions on which semigroups can arise as $V(R)$ for a local ring R, or as $+(M)$ for a finitely generated R-module M. In short, $+(M)$ must be a *finitely generated semigroup*. In §1 we detail these restrictions, and in the rest of the chapter we prove two realization theorems, which show that every finitely generated Krull monoid can be realized in the form $+(M)$ for a suitable local ring R and MCM R-module M. Both these theorems actually realize a semigroup Λ together with a given embedding $\Lambda \subseteq \mathbb{N}_0^{(n)}$. The first construction (Theorem 2.12) gives a one-dimensional domain R and a finitely generated torsion-free module M realizing an *expanded* subsemigroup Λ as $+(M)$, while the second (Theorem 2.17) gives a two-dimensional unique factorization domain R and a finitely generated reflexive module M realizing Λ as $+(M)$, assuming only that Λ is a *full* subsemigroup of $\mathbb{N}_0^{(t)}$. (See Proposition 2.4 for the terminology.)

§1. Krull monoids

In this section, we let $(R, \mathfrak{m}, k)$ be a local ring with completion $(\widehat{R}, \widehat{\mathfrak{m}}, k)$. Let $V(R)$ and $V(\widehat{R})$ denote the semigroups, with respect to direct sum, of finitely generated modules over R and $\widehat{R}$, respectively. We write all our semigroups additively, though we will keep the "multiplicative" notation inspired by direct sums, $x \mid y$, meaning that there exists z such that $x + z = y$. We write 0 for the neutral element $[0]$ corresponding to the zero module.

There is a natural homomorphism of semigroups

$$j \colon V(R) \longrightarrow V(\widehat{R})$$

taking $[M]$ to $[\widehat{R} \otimes_R M]$. This homomorphism is injective by Corollary 1.15, so we consider $V(R)$ as a subsemigroup of $V(R)$. It follows that $V(R)$ is *cancellative*: if $x + z = y + z$ for $x, y, z \in V(R)$, then $x = y$. Since in this chapter we will deal only with local rings, all of our semigroups will be tacitly assumed to be cancellative. We also see that $V(R)$ is *reduced*, i.e. $x + y = 0$ implies $x = y = 0$.

The semigroup homomorphism $j \colon V(R) \longrightarrow V(\widehat{R})$ actually satisfies a much stronger condition than injectivity. A *divisor homomorphism* is a semigroup homomorphism $j \colon \Lambda \longrightarrow \Lambda'$ such that $j(x) \mid j(y)$ implies $x \mid y$ for all x and y in Λ. Corollary 1.15 says that $j \colon V(R) \longrightarrow V(\widehat{R})$ is a divisor homomorphism. Similarly, if M is a finitely generated R-module, the map $+(M) \hookrightarrow +(\widehat{M})$ is a divisor homomorphism. A reassuring consequence is that a finitely generated module over a local ring has only finitely many direct-sum decompositions. (Cf. (1.1), which shows that this fails over a Dedekind domain with infinite class group.) To be precise, let us say that two direct-sum decompositions $M \cong M_1 \oplus \cdots \oplus M_m$ and $M \cong N_1 \oplus \cdots \oplus N_n$ are *equivalent* provided $m = n$ and, after a permutation, $M_i \cong N_i$ for each i. (We do not require that the summands be indecomposable.) The next theorem appears as Theorem 1.1 in [**Wie99**], with a slightly non-commutative proof. We will give a commutative proof here.

2.2. THEOREM. *Let $(R, \mathfrak{m})$ be a local ring, and let M be a finitely generated R-module. Then there are only finitely many isomorphism classes of indecomposable modules in $\mathrm{add}_R(M)$. In particular, M has, up to equivalence, only finitely many direct sum decompositions.*

PROOF. Let $\widehat{R}$ be the $\mathfrak{m}$-adic completion of R, and write $\widehat{R} \otimes_R M = V_1^{(n_1)} \oplus \cdots \oplus V_t^{(n_t)}$, where each V_i is an indecomposable $\widehat{R}$-module and each $n_i > 0$. If $L \in \mathrm{add}(M)$, then $\widehat{R} \otimes_R L \cong V_1^{(a_1)} \oplus \cdots \oplus V_t^{(a_t)}$ for suitable non-negative integers a_i; moreover, the integers a_i are uniquely determined by the isomorphism class $[L]$, by Corollary 1.9. Thus we have a well-defined map $j: +(M) \longrightarrow \mathbb{N}_0^t$, taking $[L]$ to $(a_1, \ldots, a_t)$. Moreover, this map is one-to-one, by faithfully flat descent (Corollary 1.15).

If $[L] \in +(M)$ and $j([L])$ is a minimal non-zero element of $j(+(M))$, then L is clearly indecomposable. Conversely, if $[L] \in \mathrm{add}(M)$ and L is indecomposable, we claim that $j([L])$ is a minimal non-zero element of $j(+(M))$. For, suppose that $j([X]) < j([L])$, where $[X] \in +(M)$ is non-zero. Then $\widehat{R} \otimes_R X \mid \widehat{R} \otimes_R L$, so $X \mid L$ by Corollary 1.15. But $X \neq 0$ and $X \not\cong L$ (else $j([X]) = j([L])$), and we have a contradiction to the indecomposability of L.

By Dickson's Lemma (Exercise 2.20), $j(+(M))$ contains only finitely many minimal non-zero elements, and, by what we have just shown, $\mathrm{add}(M)$ has only finitely many isomorphism classes of indecomposable modules.

For the last statement, let $n = \mu_R(M)$, the number of elements in a minimal generating set for M, and let $\{N_1, \ldots, N_t\}$ be a complete set of representatives for the isomorphism classes of direct summands of M. Any direct summand of M is isomorphic to $N_1^{(r_1)} \oplus \cdots \oplus N_t^{(r_t)}$, where each r_i is non-negative and $r_1 + \cdots + r_t \leqslant n$. It follows that there are, up to isomorphism, only finitely many direct summands of M. Let $\{L_1, \ldots, L_s\}$ be a set of representatives for the non-zero direct summands of M. Any direct-sum decomposition of M must have the form $M \cong L_1^{(u_1)} \oplus \cdots \oplus L_s^{(u_s)}$, with $u_1 + \cdots + u_s \leqslant n$, and it follows that there are only finitely many such decompositions. □

We will see in Example 2.13 that $\mathrm{add}(M)$ may contain indecomposable modules that do not occur as direct summands of M.

2.3. DEFINITION. A *Krull monoid* is a monoid that admits a divisor homomorphism into a free monoid.

Every finitely generated Krull monoid admits a divisor homomorphism into $\mathbb{N}_0^{(t)}$ for some positive integer t. Conversely, it follows easily from Dickson's Lemma (Exercise 2.20) that a monoid admitting a divisor homomorphism to $\mathbb{N}_0^{(t)}$ must be finitely generated.

Finitely generated Krull monoids are called *positive normal affine semigroups* in [**BH93**]. From [**BH93**, 6.1.10], we obtain the following characterization of these monoids:

2.4. PROPOSITION. *The following conditions on a semigroup Λ are equivalent:*
  (i) Λ *is a finitely generated Krull monoid.*
  (ii) $\Lambda \cong G \cap \mathbb{N}_0^{(t)}$ *for some positive integer t and some subgroup G of $\mathbb{Z}^{(t)}$. (That is, Λ is isomorphic to a* full *subsemigroup of $\mathbb{N}_0^{(t)}$.)*
  (iii) $\Lambda \cong W \cap \mathbb{N}_0^{(u)}$ *for some positive integer u and some $\mathbb{Q}$-subspace W of $\mathbb{Q}^{(n)}$. (That is, Λ is isomorphic to an* expanded *subsemigroup of $\mathbb{N}_0^{(u)}$.)*
  (iv) *There exist positive integers m and n, and an $m \times n$ matrix α over $\mathbb{Z}$, such that $\Lambda \cong \mathbb{N}^{(n)} \cap \ker(\alpha)$.* □

Observe that the descriptors "full" and "expanded" refer specifically to a given embedding of a semigroup into a free semigroup, while the definition of a Krull monoid is intrinsic. In addition, note that the group G and the vector space W are not mysterious; they are the group, respectively vector space, generated by Λ.

It's obvious that every expanded subsemigroup of $\mathbb{N}^{(t)}$ is also a full subsemigroup, but the converse can fail. For example, the subsemigroup

$$\Lambda = \left\{ \begin{bmatrix} x \\ y \end{bmatrix} \in \mathbb{N}_0^{(2)} \,\middle|\, x \equiv y \bmod 3 \right\}$$

of $\mathbb{N}_0^{(2)}$ is not the restriction to $\mathbb{N}_0^{(2)}$ of the kernel of a matrix, so is not expanded. However, Λ is isomorphic to

$$\Lambda' = \left\{ \begin{bmatrix} x \\ y \\ z \end{bmatrix} \in \mathbb{N}_0^{(3)} \,\middle|\, x + 2y = 3z \right\}.$$

As this example indicates, the number u of (iii) might be larger than the number t of (ii).

Condition (iv) says that a finitely generated Krull monoid can be regarded as the collection of non-negative integer solutions of a homogeneous system of linear equations. For this reason these monoids are sometimes called *Diophantine* monoids.

The key to understanding the monoids V(R) and +(M) is knowing which modules over the completion $\widehat{R}$ actually come from R-modules. Recall that if $R \longrightarrow S$ is a ring homomorphism, we say that an S-module N is *extended* (from R) provided there is an R-module M such that $N \cong S \otimes_R M$. In the two remaining sections, we will prove a pair of criteria—one in dimension one, and one in dimension two—for identifying which finitely generated modules over the completion $\widehat{R}$ of a local ring R are extended. In both cases, a key ingredient is that modules of finite length are *always* extended. We leave the proof of this fact as an exercise.

2.5. LEMMA. *Let R be a local ring with completion $\widehat{R}$, and let L be an $\widehat{R}$-module of finite length. Then L also has finite length as an R-module, and the natural map $L \longrightarrow \widehat{R} \otimes_R L$ is an isomorphism.* □

§2. Realization in dimension one

In the one-dimensional case, a beautiful result due to Levy and Odenthal [**LO96**] tells us exactly which $\widehat{R}$-modules are extended from R. See Corollary 2.8 below. First, we define for any one-dimensional local ring $(R, \mathfrak{m}, k)$ the *Artinian localization* K(R) by K(R) = $U^{-1}R$, where U is the complement of the union of the minimal prime ideals (the prime ideals distinct from $\mathfrak{m}$). If R is Cohen-Macaulay, then K(R) is just the total quotient ring {non-zerodivisors}^{-1}R. If R is not Cohen-Macaulay, then the natural map $R \longrightarrow$ K(R) is not injective.

2.6. PROPOSITION. *Let $(R, \mathfrak{m}, k)$ be a one-dimensional local ring, and let N be a finitely generated $\widehat{R}$-module. Then N is extended from R if and only if $K(\widehat{R}) \otimes_{\widehat{R}} N$ is extended from K(R).*

PROOF. To simplify notation, we set $K =$ K(R) and $L =$ K($\widehat{R}$). (Keep in mind, however, that these may not be fields.) If $\mathfrak{q}$ is a minimal prime ideal of $\widehat{R}$, then $\mathfrak{q} \cap R$ is a minimal prime ideal of R, since "going down" holds for flat extensions [**BH93**, Lemma A.9]. Therefore the inclusion $R \longrightarrow \widehat{R}$ induces a homomorphism $K \longrightarrow L$, and this homomorphism is faithfully flat, since the map Spec($\widehat{R}$) $\longrightarrow$ Spec(R) is surjective [**BH93**, Lemma A.10]. The "only if" direction is then clear from the identification $L \otimes_K K \otimes_R M \cong L \otimes_{\widehat{R}} \widehat{R} \otimes_R M$.

For the converse, let X be a finitely generated K-module such that $L \otimes_K X \cong L \otimes_{\widehat{R}} N$. Since K is a localization of R, there is a finitely generated R-module M such that $K \otimes_R M \cong X$. Since $L \otimes_{\widehat{R}} N \cong L \otimes_{\widehat{R}} (\widehat{R} \otimes_R M)$, there is a homomorphism $\varphi \colon N \longrightarrow \widehat{R} \otimes_R M$ inducing an isomorphism from $L \otimes_{\widehat{R}} N$ to $L \otimes_{\widehat{R}} (\widehat{R} \otimes_R M)$. Then the kernel U

and cokernel V of φ have finite length and therefore are extended by Lemma 2.5. Now we break the exact sequence
$$0 \longrightarrow U \longrightarrow N \longrightarrow S \otimes_R M \longrightarrow V \longrightarrow 0$$
into two short exact sequences:
$$0 \longrightarrow U \longrightarrow N \longrightarrow W \longrightarrow 0$$
$$0 \longrightarrow W \longrightarrow \widehat{R} \otimes_R M \longrightarrow V \longrightarrow 0\,.$$
Applying (ii) of Lemma 2.7 below to the second short exact sequence, we see that W is extended. Now we apply (i) of the lemma to the first short exact sequence, to conclude that N is extended. $\square$

2.7. LEMMA. *Let $(R, \mathfrak{m})$ be a local ring with completion $\widehat{R}$, and let*
$$0 \longrightarrow X \longrightarrow Y \longrightarrow Z \longrightarrow 0$$
be an exact sequence of finitely generated $\widehat{R}$-modules.

  (i) *Assume X and Z are extended. If $\operatorname{Ext}^1_{\widehat{R}}(Z, X)$ has finite length as an R-module (e.g. if Z is locally free on the punctured spectrum of $\widehat{R}$), then Y is extended.*

  (ii) *Assume Y and Z are extended. If $\operatorname{Hom}_{\widehat{R}}(Y, Z)$ has finite length as an R-module (e.g. if Z has finite length), then X is extended.*

  (iii) *Assume X and Y are extended. If $\operatorname{Hom}_{\widehat{R}}(X, Y)$ has finite length as an R-module (e.g. if X has finite length), then Z is extended.*

PROOF. For (i), write $X = \widehat{R} \otimes_R X_0$ and $Z = \widehat{R} \otimes_R Z_0$, where X_0 and Z_0 are finitely generated R-modules. The natural map
$$\widehat{R} \otimes_R \operatorname{Ext}^1_R(Z_0, X_0) \longrightarrow \operatorname{Ext}^1_{\widehat{R}}(Z, X)$$
is an isomorphism since Z_0 is finitely presented, and $\operatorname{Ext}^1_R(Z_0, X_0)$ has finite length by faithful flatness. Therefore the natural map
$$\operatorname{Ext}^1_R(Z_0, X_0) \longrightarrow \widehat{R} \otimes_R \operatorname{Ext}^1_R(Z_0, X_0)$$
is an isomorphism by Lemma 2.5. Combining the two isomorphisms, we see that the given exact sequence, when regarded as an element of $\operatorname{Ext}^1_{\widehat{R}}(Z, X)$, comes from a short exact sequence $0 \longrightarrow X_0 \longrightarrow Y_0 \longrightarrow Z_0 \longrightarrow 0$. Clearly, then, $\widehat{R} \otimes_R Y_0 \cong Y$.

To prove (ii), we write $Y = \widehat{R} \otimes_R Y_0$ and $Z = \widehat{R} \otimes_R Z_0$, where Y_0 and Z_0 are finitely generated R-modules. As in the proof of (i) we see that the natural map $\operatorname{Hom}_R(Y_0, Z_0) \longrightarrow \operatorname{Hom}_{\widehat{R}}(Y, Z)$ is an isomorphism. Therefore the given $\widehat{R}$-homomorphism $\beta\colon Y \longrightarrow Z$ comes from a homomorphism $\beta_0\colon Y_0 \longrightarrow Z_0$ in $\operatorname{Hom}_R(Y_0, Z_0)$. Clearly, then,

$X \cong \widehat{R} \otimes_R (\ker \beta_0)$. The proof of (iii) is essentially the same: Write $Y = \widehat{R} \otimes_R Y_0$ and $X = \widehat{R} \otimes_R X_0$; show that $\alpha \colon X \longrightarrow Y$ comes from some $\alpha_0 \in \operatorname{Hom}_R(X_0, Y_0)$, and deduce that $Z \cong \widehat{R} \otimes_R (\operatorname{cok} \alpha_0)$. □

2.8. COROLLARY (Levy-Odenthal). *Let $(R, \mathfrak{m}, k)$ be a local ring of dimension one for which the completion $\widehat{R}$ is reduced, and let N be a finitely generated $\widehat{R}$-module. Then N is extended from R if and only if*

$$\dim_{\widehat{R}_\mathfrak{p}}(N_\mathfrak{p}) = \dim_{\widehat{R}_\mathfrak{q}}(N_\mathfrak{q})$$

whenever $\mathfrak{p}$ and $\mathfrak{q}$ are minimal prime ideals of $\widehat{R}$ lying over the same prime ideal of R. In particular, if R is a domain, then N is extended if and only if N has constant rank. □

This gives us a strategy for producing strange direct-sum behavior:
(i) Find a one-dimensional domain R whose completion is reduced but has lots of minimal primes.
(ii) Build indecomposable modules with highly non-constant ranks over $\widehat{R}$.
(iii) Put them together in different ways to get constant-rank modules.

Suppose, to illustrate, that R is a domain whose completion $\widehat{R}$ has two minimal primes $\mathfrak{p}$ and $\mathfrak{q}$. Suppose we can build indecomposable $\widehat{R}$-modules U, V, W and X, with ranks $(\dim_{\widehat{R}_\mathfrak{p}}(-), \dim_{\widehat{R}_\mathfrak{q}}(-)) = (2,0)$, $(0,2)$, $(2,1)$, and $(1,2)$, respectively. Then $U \oplus V$ has constant rank $(2,2)$, so is extended; say, $U \oplus V \cong \widehat{M}$. Similarly, there are R-modules N, F and G such that $V \oplus W \oplus W \cong \widehat{N}$, $W \oplus X \cong \widehat{F}$, and $U \oplus X \oplus X \cong \widehat{G}$. Using KRS over $\widehat{R}$, we see easily that no non-zero proper direct summand of any of the modules $\widehat{M}$, $\widehat{N}$, $\widehat{F}$, $\widehat{G}$ has constant rank. It follows from Corollary 2.8 that M, N, F, and G are indecomposable, and of course no two of them are isomorphic since (again by KRS) their completions are pairwise non-isomorphic. Finally, we see that $M \oplus F \oplus F \cong N \oplus G$, since the two modules have isomorphic completions. Thus we easily obtain a mild violation of KRS uniqueness over R.

It's easy to accomplish (i), getting a one-dimensional domain with a lot of splitting but no ramification. In order to facilitate (ii), however, we want to ensure that each analytic branch has infinite Cohen-Macaulay type. The following construction from [**Wie01**, (2.3)] does the job nicely:

2.9. CONSTRUCTION (R. Wiegand). Fix a positive integer s, and let k be any field with $|k| \geqslant s$. Choose distinct elements $t_1, \ldots, t_s \in k$. Let

Σ be the complement of the union of the maximal ideals $(x - t_i)k[x]$, $i = 1, \ldots, s$. We define R by the pullback diagram

(2.1)
$$\begin{array}{ccc} R & \longrightarrow & \Sigma^{-1}k[x] \\ \downarrow & & \downarrow \pi \\ k & \longrightarrow & \dfrac{\Sigma^{-1}k[x]}{(x - t_1)^4 \cdots (x - t_s)^4} \end{array},$$

where π is the natural quotient map. Then R is a one-dimensional local domain, (2.1) is the conductor square for R (cf. Construction 4.1), and $\widehat{R}$ is reduced with exactly s minimal prime ideals. Indeed, we can rewrite the bottom line R_{art} as $k \hookrightarrow D_1 \times \cdots \times D_s$, where $D_i \cong k[x]/(x^4)$ for each i. The conductor square for the completion is then

$$\begin{array}{ccc} \widehat{R} & \longrightarrow & T_1 \times \cdots \times T_s \\ \downarrow & & \downarrow \pi \\ k & \longrightarrow & D_1 \times \cdots \times D_s \end{array},$$

where each T_i is isomorphic to $k[\![x]\!]$. (If $\text{char}(k) \neq 2, 3$, then R is the ring of rational functions $f \in k(T)$ such that $f(t_1) = \cdots = f(t_s) \neq \infty$ and the derivatives f', f'' and f''' vanish at each t_i.)

Let $\mathfrak{p}_1, \ldots, \mathfrak{p}_s$ be the minimal prime ideals of $\widehat{R}$. Define the *rank* of a finitely generated $\widehat{R}$-module N to be the s-tuple $(r_1, \ldots, r_s)$, where r_i is the dimension of $N_{\mathfrak{p}_i}$ as a vector space over $R_{\mathfrak{p}_i}$.

The next theorem [**Wie01**, (2.4)] says that even the case $s = 2$ of this example yields the pathology discussed after Corollary 2.8.

2.10. THEOREM. *Fix a positive integer s, and let R be the ring of Construction 2.9. Let $(r_1, \ldots, r_s)$ be any sequence of non-negative integers with not all the r_i equal to zero. Then $\widehat{R}$ has an indecomposable torsion-free module N with* $\text{rank}(N) = (r_1, \ldots, r_s)$.

PROOF. Set $P = T_1^{(r_1)} \times \cdots \times T_s^{(r_s)}$, a projective module over $\widehat{R} \cong T_1 \times \cdots \times T_s$. Lemma 2.11 below, a jazzed-up version of Theorem 3.7, yields an indecomposable $\widehat{R}_{\text{art}}$-module $V \hookrightarrow W$ with $W = D_1^{(r_1)} \times \cdots D_s^{(r_s)}$. Since $P/\mathfrak{c}P \cong W$, Construction 4.1 implies that there exists a torsion-free $\widehat{R}$-module M, namely, the pullback of P and V over W, such that $M_{\text{art}} = (V \hookrightarrow W)$. NAK implies that M is indecomposable, and the ranks of M at the minimal primes are precisely $(r_1, \ldots, r_s)$. $\square$

We leave the proof of the next lemma as a challenging exercise (Exercise 2.23).

2.11. LEMMA. *Let k be a field. Fix an integer $s \geqslant 1$, set $D_i = k[x]/(x^4)$ for $i = 1, \ldots, s$, and let $D = D_1 \times \cdots \times D_s$. Let $(r_1, \ldots, r_s)$ be an s-tuple of non-negative integers with at least one positive entry, and assume that $r_1 \geqslant r_i$ for every i. Then the Artinian pair $k \hookrightarrow D$ has an indecomposable module $V \hookrightarrow W$, where $W = D_1^{(r_1)} \times \cdots \times D_s^{(r_s)}$.* □

Recalling condition (iv) of Proposition 2.4, we say that the finitely generated Krull monoid Λ *can be defined by m equations* provided $\Lambda \cong \mathbb{N}_0^{(n)} \cap \ker(\alpha)$ for some n and some $m \times n$ integer matrix α. Given such an embedding of Λ in $\mathbb{N}_0^{(n)}$, we say a column vector $\lambda \in \Lambda$ is *strictly positive* provided each of its entries is a positive integer. By decreasing n (and removing some columns from α) if necessary, we can harmlessly assume, without changing m, that Λ contains a strictly positive element λ. Specifically, choose an element $\lambda \in \Lambda$ with the largest number of strictly positive coordinates, and throw away all the columns of α corresponding to zero entries of λ. If any element $\lambda' \in \Lambda$ had a non-zero entry in one of the deleted position, then $\lambda + \lambda'$ would have more positive entries than λ, a contradiction.

2.12. THEOREM. *Fix a non-negative integer m, and consider the ring R of Construction 2.9 with $s = m + 1$. Let Λ be a finitely generated Krull monoid defined by m equations and containing a strictly positive element λ. Then there exist a torsion-free R-module M and a commutative diagram*

$$\begin{array}{ccc} \Lambda & \hookrightarrow & \mathbb{N}_0^{(n)} \\ \varphi \downarrow & & \downarrow \psi \\ +(M) & \xrightarrow{j} & +(\widehat{R} \otimes_R M) \end{array}$$

in which

(i) j is the natural map taking $[N]$ to $[\widehat{R} \otimes_R N]$,
(ii) φ and ψ are semigroup isomorphisms, and
(iii) $\varphi(\lambda) = [M]$.

PROOF. We have $\Lambda = \mathbb{N}_0^{(n)} \cap \ker(\alpha)$, where $\alpha = [a_{ij}]$ is an $m \times n$ matrix over $\mathbb{Z}$. Choose a positive integer h such that $a_{ij} + h \geqslant 0$ for all i, j. For $j = 1, \ldots, n$, choose, using Theorem 2.10, a torsion-free $\widehat{R}$-module L_j such that $\mathrm{rank}(L_j) = (a_{1j} + h, \ldots, a_{mj} + h, h)$.

Given any column vector $\beta = [b_1, b_2, \ldots, b_n]^{\mathrm{tr}} \in \mathbb{N}_0^{(n)}$, put $N_\beta = L_1^{(b_1)} \oplus \cdots \oplus L_n^{(b_n)}$. The rank of N_β is

$$\left(\sum_{j=1}^n (a_{1j} + h)\, b_j, \ldots, \sum_{j=1}^n (a_{mj} + h)\, b_j, \left(\sum_{j=1}^n b_j \right) h \right).$$

Since R is a domain, Corollary 2.8 implies that N_β is in the image of $j\colon \mathrm{V}(R) \longrightarrow \mathrm{V}(\widehat{R})$ if and only if $\sum_{j=1}^n (a_{ij} + h) b_j = \left(\sum_{j=1}^n b_j \right) h$ for each i, that is, if and only if $\beta \in \mathbb{N}_0^{(n)} \cap \ker(\alpha) = \Lambda$. To complete the proof, we let M be the R-module (unique up to isomorphism) such that $\widehat{M} \cong N_\lambda$. □

This corollary makes it very easy to demonstrate spectacular failure of KRS uniqueness:

2.13. EXAMPLE. Let
$$\Lambda = \left\{ \begin{bmatrix} x \\ y \\ z \end{bmatrix} \in \mathbb{N}_0^{(3)} \;\Big|\; 72x + y = 73z \right\}.$$

This has three atoms (minimal non-zero elements), namely

$$\alpha = \begin{bmatrix} 1 \\ 1 \\ 1 \end{bmatrix}, \qquad \beta = \begin{bmatrix} 0 \\ 73 \\ 1 \end{bmatrix}, \qquad \gamma = \begin{bmatrix} 73 \\ 0 \\ 72 \end{bmatrix}.$$

Note that $73\alpha = \beta + \gamma$. Taking $s = 2$ in Construction 2.9, we get a local ring R and indecomposable R-modules A, B, C such that $A^{(t)}$ has only the obvious direct-sum decompositions for $t \leqslant 72$, but $A^{(73)} \cong B \oplus C$.

We define the *splitting number* $\mathrm{spl}(R)$ of a one-dimensional local ring R by

$$\mathrm{spl}(R) = \left| \mathrm{Spec}(\widehat{R}) \right| - |\mathrm{Spec}(R)|.$$

The splitting number of the ring R in Construction 2.9 is $s - 1$. Corollary 2.12 says that every finitely generated Krull monoid defined by m equations can be realized as $+(M)$ for some finitely generated module over a one-dimensional local ring (in fact, a domain essentially of finite type over $\mathbb{Q}$) with splitting number m. This is the best possible:

2.14. PROPOSITION. *Let M be a finitely generated module over a one-dimensional local ring R with splitting number m. The embedding $+(M) \hookrightarrow \mathrm{V}(\widehat{R})$ exhibits $+(M)$ as an expanded subsemigroup of the free semigroup $+(\widehat{R} \otimes_R M)$. Moreover, $+(M)$ is defined by m equations.*

PROOF. Write $\widehat{R}\otimes_R M = V_1^{(e_1)} \oplus \cdots \oplus V_n^{(e_n)}$, where the V_j are pairwise non-isomorphic indecomposable $\widehat{R}$-modules and the e_i are all positive. We have an embedding $+(M) \hookrightarrow \mathbb{N}_0^{(n)}$ taking $[N]$ to $[b_1, \ldots, b_n]^{\mathrm{tr}}$, where $\widehat{R}\otimes_R N \cong V_1^{(b_1)} \oplus \cdots \oplus V_n^{(b_n)}$, and we identify $+(M)$ with its image Λ in $\mathbb{N}_0^{(n)}$. Given a prime $\mathfrak{p} \in \mathrm{Spec}(R)$ with, say, t primes $\mathfrak{q}_1, \ldots, \mathfrak{q}_t$ lying over it, there are $t-1$ homogeneous linear equations on the b_j that say that $\widehat{N}$ has constant rank on the fiber over $\mathfrak{p}$ (cf. Corollary 2.8). Letting $\mathfrak{p}$ vary over $\mathrm{Spec}(R)$, we obtain exactly $m = \mathrm{spl}(R)$ equations that must be satisfied by elements of Λ. Conversely, if the b_j satisfy these equations, then $N := V_1^{(b_1)} \oplus \cdots \oplus V_n^{(b_n)}$ has constant rank on each fiber of $\mathrm{Spec}(\widehat{R}) \longrightarrow \mathrm{Spec}(R)$. By Corollary 2.8, N is extended from an R-module, say $N \cong \widehat{R}\otimes_R L$. Clearly $\widehat{R}\otimes_R L \mid \widehat{M^{(u)}}$ if u is large enough, and it follows from Proposition 2.19 that $L \in +(M)$, whence $[b_1, \ldots, b_n]^{\mathrm{tr}} \in \Lambda$. □

In [**Kat02**] Kattchee showed that, for each m, there is a finitely generated Krull monoid Λ that cannot be defined by m equations. Thus no single one-dimensional local ring can realize *every* finitely generated Krull monoid in the form $+(M)$ for a finitely generated module M.

§3. Realization in dimension two

Suppose we have a finitely generated Krull semigroup Λ and a full embedding $\Lambda \subseteq \mathbb{N}_0^{(t)}$, i.e. Λ is the intersection of $\mathbb{N}^{(t)}$ with a subgroup of $\mathbb{Z}^{(t)}$. By Proposition 2.14, we cannot realize this embedding in the form $+(M) \hookrightarrow +(\widehat{R}\otimes_R M)$ for a module M over a one-dimensional local ring R unless Λ is actually an *expanded* subsemigroup of $\mathbb{N}_0^{(t)}$, i.e. the intersection of $\mathbb{N}^{(t)}$ with a subspace of $\mathbb{Q}^{(t)}$. If, however, we go to a two-dimensional ring, then we can realize Λ as $+(M)$, though the ring that does the realizing is less tractable than the one-dimensional rings that realize expanded subsemigroups.

As in the last section, we need a criterion for an $\widehat{R}$-module to be extended from R. For general two-dimensional rings, we know of no such criterion, so we shall restrict to analytically normal domains. (A local domain $(R,\mathfrak{m})$ is *analytically normal* provided its completion $(\widehat{R},\widehat{\mathfrak{m}})$ is also a normal domain.)

We recall two facts from Bourbaki [**Bou98**, Chapter VII]. Firstly, over a Noetherian normal domain R one can assign to each finitely generated R-module M a *divisor class* $\mathrm{cl}(M) \in \mathrm{Cl}(R)$ in such a way that

(i) Taking divisor classes $\mathrm{cl}(-)$ is additive on exact sequences, and

(ii) if J is a fractional ideal of R, then $\mathrm{cl}(J)$ is the isomorphism class $[J^{**}]$ of the divisorial (i.e. reflexive) ideal J^{**}, where $-^*$ denotes the dual $\mathrm{Hom}_R(-, R)$.

Secondly, each finitely generated torsion-free module M over a Noetherian normal domain R has a "Bourbaki sequence," namely a short exact sequence

$$(2.2) \qquad 0 \longrightarrow F \longrightarrow M \longrightarrow J \longrightarrow 0$$

wherein F is a free R-module and J is an ideal of R.

The following criterion for a module to be extended is Proposition 3 of [**RWW99**] (cf. also [**Wes88**, (1.5)]).

2.15. PROPOSITION. *Let R be a two-dimensional local ring whose $\mathfrak{m}$-adic completion $\widehat{R}$ is a normal domain. Let N be a finitely generated torsion-free $\widehat{R}$-module. Then N is extended from R if and only if $\mathrm{cl}(N)$ is in the image of the natural homomorphism $\Phi \colon \mathrm{Cl}(R) \longrightarrow \mathrm{Cl}(\widehat{R})$.*

PROOF. Suppose $N \cong \widehat{R} \otimes_R M$. Then M is finitely generated and torsion-free, by faithfully flat descent. Choose a Bourbaki sequence of the form (2.2) for M; tensoring with $\widehat{R}$ and using the additivity of $\mathrm{cl}(-)$ on short exact sequences, we find

$$\mathrm{cl}(N) = \mathrm{cl}(\widehat{R} \otimes_R J) = [(\widehat{R} \otimes_R J)^{**}] = \Phi(\mathrm{cl}(J)).$$

For the converse, choose a Bourbaki sequence

$$0 \longrightarrow G \longrightarrow N \longrightarrow L \longrightarrow 0$$

over $\widehat{R}$, so that G is a free $\widehat{R}$-module and L is an ideal of $\widehat{R}$. Then $\mathrm{cl}(L) = \mathrm{cl}(N)$, and since $\mathrm{cl}(N)$ is in the image of Φ there is a divisorial ideal I of R such that $\widehat{R} \otimes_R I \cong L^{**}$. Set $V = L^{**}/L$. Then V has finite length and hence is extended by Lemma 2.5; it follows from Lemma 2.7(i) and the short exact sequence $0 \longrightarrow L \longrightarrow L^{**} \longrightarrow V \longrightarrow 0$ that L is extended. Moreover, $\widehat{R}_{\mathfrak{p}}$ is a discrete valuation ring for each height-one prime ideal $\mathfrak{p}$, so that $\mathrm{Ext}^1_{\widehat{R}}(L, G)$ has finite length. Now Lemma 2.7(ii) says that N is extended since G and L are. □

As in the last section, we need to guarantee that the complete ring $\widehat{R}$ has a sufficiently rich supply of MCM modules. This is [**Wie01**, Lemma 3.2].

2.16. LEMMA. *Let s be any positive integer. There is a complete local normal domain B, containing $\mathbb{C}$, such that $\dim(B) = 2$ and $\mathrm{Cl}(B)$ contains a copy of $(\mathbb{R}/\mathbb{Z})^{(s)}$.*

PROOF. Choose a positive integer d such that $(d-1)(d-2) \geqslant s$, and let V be a smooth projective plane curve of degree d over $\mathbb{C}$. Let A be the homogeneous coordinate ring of V for some embedding $V \hookrightarrow \mathbb{P}^2_\mathbb{C}$. Then A is a two-dimensional normal domain, by [**Har77**, Chap. II, Exercise 8.4(b)]. By [**Har77**, Appendix B, Sect. 5], $\text{Pic}^0(V) \cong D := (\mathbb{R}/\mathbb{Z})^{2g}$, where $g = \frac{1}{2}(d-1)(d-2)$, the genus of V. Here $\text{Pic}^0(V)$ is the kernel of the degree map $\text{Pic}(V) \longrightarrow \mathbb{Z}$, so $\text{Cl}(V) = \text{Pic}(V) = D \oplus \mathbb{Z}\sigma$, where σ is the class of a divisor of degree 1. There is a short exact sequence
$$0 \longrightarrow \mathbb{Z} \longrightarrow \text{Cl}(V) \longrightarrow \text{Cl}(A) \longrightarrow 0,$$
in which $1 \in \mathbb{Z}$ maps to the divisor class $\tau := [H \cdot V]$, where H is a line in $\mathbb{P}^2_\mathbb{C}$. (Cf. [**Har77**, Chap. II, Exercise 6.3].) Thus $\text{Cl}(A) \cong \text{Cl}(V)/\mathbb{Z}\tau$. Since τ has degree d, we see that $\tau - d\sigma \in D$. Choose an element $\delta \in D$ with $d\delta = \tau - d\sigma$. Recalling that $\text{Cl}(V) = \text{Pic}(V) = D \oplus \mathbb{Z}\sigma$, we define a surjection $f \colon \text{Cl}(V) \longrightarrow D \oplus \mathbb{Z}/(d)$ by sending $x \in D$ to $(x, 0)$ and σ to $(-\delta, 1 + (d))$. Then $\ker(f) = \mathbb{Z}\tau$, so $\text{Cl}(A) \cong D \oplus \mathbb{Z}/d\mathbb{Z}$.

Let $\mathfrak{P}$ be the irrelevant maximal ideal of A. By [**Har77**, Chap. II, Exercise 6.3(d)], $\text{Cl}(A_\mathfrak{P}) \cong \text{Cl}(A)$. The $\mathfrak{P}$-adic completion B of A is an integrally closed domain, by [**ZS75**, Chap. VIII, Sect. 13]. Moreover $\text{Cl}(A_\mathfrak{P}) \longrightarrow \text{Cl}(B)$ is injective by faithfully flat descent, so $\text{Cl}(B)$ contains a copy of $D = (\mathbb{R}/\mathbb{Z})^{(d-1)(d-2)}$, which, in turn, contains a copy of $(\mathbb{R}/\mathbb{Z})^{(s)}$. □

We now have everything we need to prove our realization theorem for full subsemigroups of $\mathbb{N}_0^{(t)}$.

2.17. THEOREM. *Let t be a positive integer, and let Λ be a full subsemigroup of $\mathbb{N}_0^{(t)}$. Assume that Λ contains a strictly positive element λ. Then there exist a two-dimensional local unique factorization domain R, a finitely generated reflexive (= MCM) R-module M, and a commutative diagram of semigroups*

$$\begin{array}{ccc} \Lambda & \hookrightarrow & \mathbb{N}_0^{(t)} \\ {\scriptstyle \varphi} \downarrow & & \downarrow {\scriptstyle \psi} \\ +(M) & \xrightarrow{j} & +(\widehat{R} \otimes_R M) \end{array}$$

in which

(i) *j is the natural map taking $[N]$ to $[\widehat{R} \otimes_R N]$,*
(ii) *φ and ψ are isomorphisms, and*
(iii) *$\varphi(\lambda) = [M]$.*

PROOF. Let G be the subgroup of $\mathbb{Z}^{(t)}$ generated by Λ, and write $\mathbb{Z}^{(t)}/G = C_1 \oplus \cdots \oplus C_s$, where each C_i is a cyclic group. Then $\mathbb{Z}^{(t)}/G$ can be embedded in $(\mathbb{R}/\mathbb{Z})^{(s)}$.

Let B be the complete local domain provided by Lemma 2.16. Since $\mathbb{Z}^{(t)}/G$ embeds in $\mathrm{Cl}(B)$, there is a group homomorphism $\varpi \colon \mathbb{Z}^{(t)} \longrightarrow \mathrm{Cl}(B)$ with $\ker(\varpi) = G$. Let $\{e_1, \ldots, e_t\}$ be the standard basis of $\mathbb{Z}^{(t)}$. For each $i \leqslant t$, write $\varpi(e_i) = [L_i]$, where L_i is a divisorial ideal of B representing the divisor class of $\varpi(e_i)$.

Next we use Heitmann's amazing theorem [**Hei93**], which implies that B is the completion of some local unique factorization domain R. For each element $m = (m_1, \ldots, m_t) \in \mathbb{N}_0^{(t)}$, we let $\psi(m)$ be the isomorphism class of the B-module $L_1^{(m_1)} \oplus \cdots \oplus L_t^{(m_t)}$. The divisor class of this module is $m_1[L_1] + \cdots + m_t[L_t] = \varpi(m_1, \ldots, m_t)$. By Proposition 2.15, the module $L_1^{(m_1)} \oplus \cdots \oplus L_t^{(m_t)}$ is the completion of an R-module if and only if its divisor class is trivial, that is, if and only if $m \in G \cap \mathbb{N}_0^{(t)}$. But $m \in G \cap \mathbb{N}_0^{(t)} = \Lambda$, since Λ is a full subsemigroup of $\mathbb{N}_0^{(t)}$. Therefore $L_1^{(m_1)} \oplus \cdots \oplus L_t^{(m_t)}$ is the completion of an R-module if and only if $m \in \Lambda$. If $m \in \Lambda$, we let $\varphi(m)$ be the isomorphism class of a module whose completion is isomorphic to $L_1^{(m_1)} \oplus \cdots \oplus L_t^{(m_t)}$. In particular, choosing a module M such that $[M] = \varphi(\lambda)$, we get the desired commutative diagram. □

§4. Flat local homomorphisms

Here we prove a generalization (Proposition 2.19) of the fact that $\mathrm{V}(R) \longrightarrow \mathrm{V}(\widehat{R})$ is a divisor homomorphism. This is [**HW09**, Theorem 1.3]. We begin with a general result that does not even require the ring to be local [**Wie98**, Theorem 1.1].

2.18. PROPOSITION. *Let $A \longrightarrow B$ be a faithfully flat homomorphism of commutative rings, and let U and V be finitely presented A-modules. Then $U \in \mathrm{add}_A V$ if and only if $B \otimes_A U \in \mathrm{add}_B(B \otimes_A V)$.*

PROOF. The "only if" direction is clear. For the converse, we may assume, by replacing V by a direct sum of copies of V, that $B \otimes_A U \mid B \otimes_A V$. Choose B-homomorphisms $B \otimes_A U \xrightarrow{\alpha} B \otimes_A V$ and $B \otimes_A V \xrightarrow{\beta} B \otimes_A U$ such that $\beta\alpha = 1_{B \otimes_A U}$. Since V is finitely presented and B is flat over A, the natural map $B \otimes_A \mathrm{Hom}_A(V, U) \longrightarrow \mathrm{Hom}_B(B \otimes_A V, B \otimes_A U)$ is an isomorphism. Therefore we can write $\beta = b_1 \otimes \sigma_1 + \cdots + b_r \otimes \sigma_r$, with $b_i \in B$ and $\sigma_i \in \mathrm{Hom}_A(V, U)$ for each i. Put $\sigma = [\sigma_1 \cdots \sigma_r] \colon V^{(r)} \longrightarrow U$. We will show that σ is a split surjection. Since

$$(1_B \otimes \sigma) \begin{bmatrix} b_1 \\ \vdots \\ b_r \end{bmatrix} \alpha = 1_{B \otimes_A U},$$

we see that $1_B \otimes \sigma \colon B \otimes_A V^{(r)} \longrightarrow B \otimes_A U$ is a split surjection. Therefore the induced map $(1_B \otimes \sigma)_* \colon \operatorname{Hom}_B(B \otimes_A U, B \otimes_A V^{(r)}) \longrightarrow \operatorname{Hom}_B(B \otimes_A U, B \otimes_A U)$ is surjective. Since U too is finitely presented, the vertical maps in the following commutative square are isomorphisms.
(2.3)

$$\begin{array}{ccc} B \otimes_A \operatorname{Hom}_A(U, V^{(r)}) & \xrightarrow{1_B \otimes \sigma_*} & B \otimes_A \operatorname{Hom}_A(U, U) \\ \cong \downarrow & & \downarrow \cong \\ \operatorname{Hom}_B(B \otimes_A U, B \otimes_A V^{(r)}) & \xrightarrow[(1_B \otimes \sigma)_*]{} & \operatorname{Hom}_B(B \otimes_A U, B \otimes_A U) \end{array}$$

Therefore $1_B \otimes_A \sigma_*$ is surjective as well. By faithful flatness, σ_* is surjective, and hence σ is a split surjection. $\square$

2.19. PROPOSITION. *Let $R \longrightarrow S$ be a flat local homomorphism of Noetherian local rings. Then the homomorphism $j \colon \mathrm{V}(R) \longrightarrow \mathrm{V}(S)$ taking $[M]$ to $[S \otimes_R M]$ is a divisor homomorphism.*

PROOF. Suppose M and N are finitely generated R-modules and that $S \otimes_R M \mid S \otimes_R N$. We want to show that $M \mid N$. By Theorem 1.13 it will be enough to show that $M/\mathfrak{m}^t M \mid N/\mathfrak{m}^t N$ for all $t \geqslant 1$. By passing to the flat local homomorphism $R/\mathfrak{m}^t \longrightarrow S/\mathfrak{m}^t S$, we may assume that R is Artinian and hence, by Corollary 1.6, that finitely generated modules satisfy KRS.

By Proposition 2.18, we know at least that $M \mid N^{(r)}$ for some $r \geqslant 1$. By Corollary 1.9 (or Theorem 1.3 and Corollary 1.5) M is uniquely a direct sum of indecomposable modules. If M itself is indecomposable, KRS immediately implies that $M \mid N$. An easy induction argument using direct-sum cancellation (Corollary 1.16) completes the proof (cf. Exercise 2.24). $\square$

§5. Exercises

2.20. EXERCISE. A subset C of a poset X is called a *clutter* (or *antichain*) provided no two elements of C are comparable. Consider the following property of a poset X: (†) X has the descending chain condition, and every clutter in X is finite. Prove that if X and Y both satisfy (†), then $X \times Y$ (with the product partial ordering defined by

$(x_1, y_1) \leq (x_2, y_2) \iff x_1 \leq x_2$ and $y_1 \leq y_2$) satisfies (†). Deduce Dickson's Lemma [**Dic13**]: Every clutter in $\mathbb{N}_0^{(t)}$ is finite.

2.21. EXERCISE. Prove the equivalence of conditions (i)–(iv) from Proposition 2.4.

2.22. EXERCISE. Prove Lemma 2.5.

2.23. EXERCISE ([**Wie01**, Lemma 2.2]). Prove the existence of the indecomposable $\widehat{R}_{\text{art}}$-module $V \hookrightarrow W$ in Lemma 2.11, as follows. Let $C = k^{(r_1)}$, viewed as column vectors. Define the "truncated diagonal" $\partial \colon C \longrightarrow W = D_1^{(r_1)} \times \cdots \times D_s^{(r_s)}$ by sending an element $[c_1, \ldots, c_{r_1}]^{\text{tr}}$ to the vector whose i^{th} entry is $[c_1, \ldots, c_{r_i}]^{\text{tr}}$. (Here we use $r_1 \geq r_i$ for all i.) Let V be the k-subspace of W consisting of all elements
$$\{\partial(u) + X\partial(v) + X^3\partial(Hv)\},$$
as u and v run over C, where $X = (x, 0, \ldots, 0)$ and H is the nilpotent Jordan block with 1 on the superdiagonal and 0 elsewhere.

(i) Prove that W is generated as a D-module by all elements of the form $\partial(u)$, $u \in C$, so that in particular $DV = W$. (Hint: it suffices to consider elements $w = (w_1, \ldots, w_s)$ with only one non-zero entry w_i, and such that $w_i \in D_i^{(r_i)}$ has only one non-zero entry, which is equal to 1.)

(ii) Prove that $V \hookrightarrow W$ is indecomposable along the same lines as the arguments in Chapter 4. (Hint: use the fact that $\{1, x, x^2, x^3\}$ is linearly independent over k. For additional inspiration, take a peek at the descending induction argument in Case 3.16 of the construction in the next chapter, with $\alpha = x$, $\beta = x^3$, and $t = 0$.)

2.24. EXERCISE. Complete the proof of Proposition 2.19.

CHAPTER 3

Dimension Zero

In this chapter we prove that the zero-dimensional commutative, Noetherian rings of finite representation type are exactly the Artinian principal ideal rings. We also introduce Artinian pairs, which will be used in the next chapter to classify the one-dimensional rings of finite Cohen-Macaulay type. The Drozd-Roĭter conditions (dr1) and (dr2) are shown to be necessary for finite representation type in Theorem 3.7, and in Theorem 3.23 we reduce the proof of their sufficiency to some special cases, where we can appeal to the matrix calculations of Green and Reiner.

Here are the main definitions of this book.

3.1. DEFINITION. Let $(R, \mathfrak{m})$ be a local ring of dimension d. A non-zero finitely generated R-module is *maximal Cohen-Macaulay (MCM)* if depth $M = d$. We say that R has *finite Cohen-Macaulay (CM) type* if there are, up to isomorphism, only a finite number of indecomposable maximal Cohen-Macaulay modules.

§1. Artinian rings with finite Cohen-Macaulay type

We'll say that an Artinian ring (possibly not local) has finite CM type provided it has only finitely many indecomposable finitely generated modules up to isomorphism. (Of course this causes no conflict in the local case.) To see that this condition forces R to be a principal ideal ring, and in several other constructions of indecomposable modules, we use the following result:

3.2. LEMMA. *Let R be any commutative ring, n a positive integer and H the nilpotent $n \times n$ Jordan block with 1's on the superdiagonal and 0's elsewhere. If α is an $n \times n$ matrix over R and $\alpha H = H \alpha$, then $\alpha \in R[H]$, that is, there are constants $r_i \in R$ such that $\alpha = r_0 I + r_1 H + \cdots + r_{n-1} H^{n-1}$.*

Note that α has r_0 on the main diagonal, r_1 on the first superdiagonal, and so on. Such matrices are often called "striped" in the literature.

PROOF. Let $\alpha = [a_{ij}]$. Left multiplication by H moves each row up one step and kills the bottom row, while right multiplication shifts each column to the right and kills the first column. The relation $\alpha H = H\alpha$ therefore yields the equations $a_{i,j-1} = a_{i+1,j}$ for $i,j = 1, \ldots, n$, with the convention that $a_{k\ell} = 0$ if $k = n+1$ or $\ell = 0$. These equations show (a) that each of the diagonals (of slope -1) is constant and (b) that $a_{21} = \cdots = a_{n1} = 0$. Combining (a) and (b), we see that α is upper triangular. Letting b_j be the constant on the diagonal $[a_{1,j+1} \ a_{2,j+2} \ \ldots \ a_{n-j,n}]$, for $0 \leqslant j \leqslant n-1$, we see that $\alpha = \sum_{j=0}^{n-1} b_j H^j$. □

When R is a field, there is a fancy proof: H is "cyclic" or "non-derogatory", that is, its characteristic and minimal polynomials coincide. The centralizer of a non-derogatory matrix B is always just $R[B]$ (see [**Jac75**, Corollary, p. 107]).

3.3. THEOREM. *Let R be a Noetherian ring. These are equivalent:*

(i) *R is an Artinian principal ideal ring.*
(ii) *R has only finitely many indecomposable finitely generated modules, up to isomorphism.*
(iii) *R is Artinian, and there is a bound on the number of generators required for indecomposable finitely generated R-modules.*

Under these conditions, the number of isomorphism classes of indecomposable finitely generated modules is exactly the length of R.

PROOF. Assuming (i), we will prove (ii) and verify the last statement. Since R is a product of finitely many local rings, we may assume that R is local, with maximal ideal $\mathfrak{m}$. As R is a principal ideal ring, the length ℓ of R is the least integer t such that $\mathfrak{m}^t = 0$. Since every finitely generated R-module is a direct sum of cyclic modules, the indecomposable modules are exactly the modules $R/\mathfrak{m}^t, 1 \leqslant t \leqslant \ell$.

To see that (ii) $\implies$ (iii), suppose R is not Artinian. Choose a maximal ideal $\mathfrak{m}$ of positive height. The ideals $\mathfrak{m}^t$, $t \geqslant 1$, then form a strictly descending chain of ideals (cf. Exercise 3.24). Therefore the R-modules $R/\mathfrak{m}^t$ are indecomposable and, since they have different annihilators, pairwise non-isomorphic, contradicting (ii).

To complete the proof, we show that (iii) $\implies$ (i). Again, we may assume that R is local with maximal ideal $\mathfrak{m}$. Supposing R is not a principal ideal ring, we will build, for every n, an indecomposable finitely generated R-module requiring exactly n generators. By passing to $R/\mathfrak{m}^2$, we may assume that $\mathfrak{m}^2 = 0$, so that now $\mathfrak{m}$ is a vector space over $k := R/\mathfrak{m}$. Choose two k-linearly independent elements $x, y \in \mathfrak{m}$.

Fix $n \geq 1$, let I be the $n \times n$ identity matrix, and let H be the $n \times n$ nilpotent Jordan block of Lemma 3.2. Put $\Psi = yI + xH$ and $M = \operatorname{cok}(\Psi)$. Since the entries of Ψ are in $\mathfrak{m}$, the R-module M needs exactly n generators.

To show that M is indecomposable, let $f = f^2 \in \operatorname{End}_R(M)$, and assume that $f \neq 1_M$. We will show that $f = 0$. There exist $n \times n$ matrices F and G over R making the following diagram commute.

$$\begin{array}{ccccccc} R^{(n)} & \xrightarrow{\Psi} & R^{(n)} & \longrightarrow & M & \longrightarrow & 0 \\ \downarrow G & & \downarrow F & & \downarrow f & & \\ R^{(n)} & \xrightarrow{\Psi} & R^{(n)} & \longrightarrow & M & \longrightarrow & 0 \end{array}$$

The equation $F\Psi = \Psi G$ yields $yF + xFH = yG + xHG$. Since x and y are linearly independent, we obtain, after reducing all entries of F, G and H modulo $\mathfrak{m}$, that $\overline{F} = \overline{G}$ and $\overline{F}\,\overline{H} = \overline{H}\,\overline{G}$. Therefore $\overline{F}$ and $\overline{H}$ commute, and by Lemma 3.2 $\overline{F}$ is an upper-triangular matrix with constant diagonal.

Now f is not surjective, by Exercise 1.27, and therefore neither is F. By NAK, $\overline{F}$ is not surjective, so $\overline{F}$ must be strictly upper triangular. But then $\overline{F}^n = 0$, and it follows that $\operatorname{im}(f) = \operatorname{im}(f^n) \subseteq \mathfrak{m}M$. Now NAK implies that $1 - f$ is surjective. Since $1 - f$ is idempotent, Exercise 1.27 implies that $f = 0$. □

This construction is far from new. See, for example, the papers of Higman [**Hig54**], Heller and Reiner [**HR61**], and Warfield [**War70**]. A similar construction arises in the classification of pairs of matrices up to simultaneous equivalence (see Dieudonné's discussion [**Die46**] of the work of Kronecker [**Kro74**] and Weierstrass [**Wei68**]). Essentially the same construction shows that certain higher-dimensional rings have unbounded CM type:

3.4. PROPOSITION. *Let $(S, \mathfrak{n}, k)$ be a CM local ring of dimension at least two, and let z be an indeterminate. Set $R = S[z]/(z^2)$. Then R has indecomposable MCM modules of arbitrarily large rank.*

PROOF. Fix $n \geq 2$, and let W be a free S-module of rank $2n$. Let I be the $n \times n$ identity matrix and H the $n \times n$ nilpotent Jordan block with 1 on the superdiagonal and 0 elsewhere. Let $\{x, y\}$ be part of a minimal generating set for the maximal ideal $\mathfrak{n}$ of S, and put $\Psi = yI + xH$. Finally, put $\Phi = \begin{bmatrix} 0 & \Psi \\ 0 & 0 \end{bmatrix}$. Noting that $\Phi^2 = 0$, we make W into an R-module by letting z act as $\Phi \colon W \longrightarrow W$. Then W is a MCM R-module, and one shows as in the proof of Theorem 3.3 that W is indecomposable over R. □

§2. Artinian pairs

Here we introduce the main computational tool for building indecomposable MCM modules over one-dimensional rings.

3.5. DEFINITION. An *Artinian pair* is a module-finite extension of commutative Artinian rings $(A \hookrightarrow B)$. Given an Artinian pair $\mathbf{A} = (A \hookrightarrow B)$, an $\mathbf{A}$-module is a pair $(V \hookrightarrow W)$, where W is a finitely generated projective B-module and V is an A-submodule of W with the property that $BV = W$. A morphism $(V_1 \hookrightarrow W_1) \longrightarrow (V_2 \hookrightarrow W_2)$ of $\mathbf{A}$-modules is a B-homomorphism from W_1 to W_2 that carries V_1 into V_2. We say that the $\mathbf{A}$-module $(V \hookrightarrow W)$ has *constant rank* n provided $W \cong B^{(n)}$.

With biproducts (direct sums) defined in the obvious way, we get an additive category $\mathbf{A}$-mod. To see that Theorem 1.3 applies in this context, we note first that the endomorphism ring of every $\mathbf{A}$-module is a module-finite A-algebra and therefore is left Artinian. Next, suppose ϵ is an idempotent endomorphism of an $\mathbf{A}$-module $\mathfrak{X} = (V \hookrightarrow W)$. Then $\mathfrak{Y} = (\epsilon(V) \hookrightarrow \epsilon(W))$ is also an $\mathbf{A}$-module. The projection $p \colon \mathfrak{X} \longrightarrow \mathfrak{Y}$ and inclusion $u \colon \mathfrak{Y} \hookrightarrow \mathfrak{X}$ give a factorization $\epsilon = up$, with $pu = 1_{\mathfrak{Y}}$. Thus idempotents split in $\mathbf{A}$-mod. Combining Theorem 1.3 and Corollary 1.5, we obtain the following:

3.6. THEOREM. *Let $\mathbf{A}$ be an Artinian pair, and let $\mathbf{M}_1, \ldots, \mathbf{M}_s$ and $\mathbf{N}_1, \ldots, \mathbf{N}_t$ be indecomposable $\mathbf{A}$-modules such that $\mathbf{M}_1 \oplus \cdots \oplus \mathbf{M}_s \cong \mathbf{N}_1 \oplus \cdots \oplus \mathbf{N}_t$. Then $s = t$, and, after renumbering, $\mathbf{M}_i \cong \mathbf{N}_i$ for each i.* □

We say $\mathbf{A}$ has *finite representation type* provided there are, up to isomorphism, only finitely many indecomposable $\mathbf{A}$-modules.

Our main result in this chapter is Theorem 3.7, which gives necessary conditions for an Artinian pair to have finite representation type. As we will see in the next chapter, these conditions are actually sufficient for finite representation type. The conditions were introduced by Drozd and Roĭter [**DR67**] in 1966, and we will refer to them as the Drozd-Roĭter conditions. (See the historical remarks in Section §2 of Chapter 4.)

3.7. THEOREM. *Let $\mathbf{A} = (A \hookrightarrow B)$ be an Artinian pair in which A is local, with maximal ideal $\mathfrak{m}$ and residue field k. Assume that at least one of the following conditions fails:*

(dr1) $\dim_k (B/\mathfrak{m}B) \leqslant 3$
(dr2) $\dim_k \left(\dfrac{\mathfrak{m}B + A}{\mathfrak{m}^2 B + A} \right) \leqslant 1$.

Let n be an arbitrary positive integer. Then there is an indecomposable $\mathbf{A}$-module of constant rank n. Moreover, if $|k|$ is infinite, there are at least $|k|$ pairwise non-isomorphic indecomposable $\mathbf{A}$-modules of rank n.

3.8. REMARK. If the field k is infinite then the number of isomorphism classes of $\mathbf{A}$-modules is *at most $|k|$*. To see this, note that there are, up to isomorphism, only countably many finitely generated projective B-modules W. Also, since any such W has finite length as an A-module, we see that $|W| \leqslant |k|$ and hence that W has at most $|k|$ A-submodules V. It follows that the number of possibilities for $(V \hookrightarrow W)$ is bounded by $\aleph_0 |k| = |k|$.

The proof of Theorem 3.7 involves a basic construction and a dreary analysis of the many cases that must be considered in order to implement the construction.

3.9. ASSUMPTIONS. Throughout the rest of this chapter, $\mathbf{A} = (A \hookrightarrow B)$ is an Artinian pair in which A is local, with maximal ideal $\mathfrak{m}$ and residue field k.

The next three results will allow us to pass to a more manageable Artinian pair $k \hookrightarrow D$, where D is a suitable finite-dimensional k-algebra. The proofs of the first two lemmas are exercises.

3.10. LEMMA. *Let C be a subring of B containing A. The functor $(V \hookrightarrow W) \rightsquigarrow (V \hookrightarrow B \otimes_C W)$ from $(A \hookrightarrow C)$-mod to $(A \hookrightarrow B)$-mod is faithful and full. The functor is injective on isomorphism classes and preserves indecomposability.* □

3.11. LEMMA. *For a nilpotent ideal I of B, let $\mathbf{E} = \left(\frac{A+I}{I} \hookrightarrow \frac{B}{I} \right)$. The functor $(V \hookrightarrow W) \rightsquigarrow \left(\frac{V+IW}{IW} \hookrightarrow \frac{W}{IW} \right)$, from $\mathbf{A}$-mod to $\mathbf{E}$-mod, is surjective on isomorphism classes and reflects indecomposable objects.* □

3.12. PROPOSITION. *Let $A \hookrightarrow B$ be an Artinian pair for which either (dr1) or (dr2) fails. There is a ring C between A and B such that, with $D = C/\mathfrak{m}C$, we have either*

  (i) $\dim_k(D) \geqslant 4$, or
  (ii) D contains elements α and β such that $\{1, \alpha, \beta\}$ is linearly independent over k and $\alpha^2 = \alpha\beta = \beta^2 = 0$.

PROOF. If (dr1) fails, we take $C = B$. Otherwise (dr2) fails, and we put $C = A + \mathfrak{m}B$. Since $\dim_k \left(\frac{\mathfrak{m}B+A}{\mathfrak{m}^2 B+A} \right) \geqslant 2$, we can choose elements $x, y \in \mathfrak{m}B$ such that the images of x and y in $\frac{\mathfrak{m}B+A}{\mathfrak{m}^2 B+A}$ are linearly independent. Since $D := C/\mathfrak{m}C$ maps onto $\frac{\mathfrak{m}B+A}{\mathfrak{m}^2 B+A}$, the images $\alpha, \beta \in D$

of x, y are linearly independent, and they obviously satisfy the required equations. □

Now let's begin the proof of Theorem 3.7. We have an Artinian pair $A \hookrightarrow B$, where $(A, \mathfrak{m}, k)$ is local and either (dr1) or (dr2) fails. We want to build indecomposable **A**-modules $V \hookrightarrow W$, with $W = B^{(n)}$. By Lemmas 3.10 and 3.11, we can pass to the Artinian pair $k \hookrightarrow D$ provided by Proposition 3.12. We fix a positive integer n. Our goal is to build an indecomposable $(k \hookrightarrow D)$-module $(V \hookrightarrow D^{(n)})$ and, if k is infinite, a family $\{(V_t \hookrightarrow D^{(n)})\}_{t \in T}$ of pairwise non-isomorphic indecomposable $(k \hookrightarrow D)$-modules, with $|T| = |k|$.

3.13. CONSTRUCTION. We describe a general construction, a modification of constructions found in [**DR67**, **Wie89**, **ÇWW95**]. Let n be a fixed positive integer, and suppose we have chosen $\alpha, \beta \in D$ with $\{1, \alpha, \beta\}$ linearly independent over k. Let I be the $n \times n$ identity matrix, and let H the $n \times n$ nilpotent Jordan block in Lemma 3.2. For $t \in k$, we consider the $n \times 2n$ matrix $\Psi_t = [I \mid \alpha I + \beta(tI + H)]$. Put $W = D^{(n)}$, viewed as columns, and let V_t be the k-subspace of W spanned by the columns of Ψ_t.

Suppose we have a morphism $(V_t \hookrightarrow W) \longrightarrow (V_u \hookrightarrow W)$, given by an $n \times n$ matrix φ over D. The requirement that $\varphi(V) \subseteq V$ says there is a $2n \times 2n$ matrix θ over k such that

(3.1) $$\varphi \Psi_t = \Psi_u \theta.$$

Write $\theta = \begin{bmatrix} A & B \\ P & Q \end{bmatrix}$, where A, B, P, Q are $n \times n$ blocks. Then (3.1) gives the following two equations:

(3.2) $$\varphi = A + \alpha P + \beta(uI + H)P$$
$$\alpha \varphi + \beta \varphi (tI + H) = B + \alpha Q + \beta(uI + H)Q.$$

Substituting the first equation into the second and combining terms, we get a mess:

(3.3)
$$-B + \alpha(A - Q) + \beta(tA - uQ + AH - HQ) + (\alpha + t\beta)(\alpha + u\beta)P$$
$$+ \alpha\beta(HP + PH) + \beta^2(HPH + tHP + uPH) = 0.$$

3.14. CASE. *D satisfies (ii). (There exist $\alpha, \beta \in D$ such that $\{1, \alpha, \beta\}$ is linearly independent and $\alpha^2 = \alpha\beta = \beta^2 = 0$.)*

From (3.3) and the linear independence of $\{1, \alpha, \beta\}$, we get the equations

(3.4) $\quad B = 0, \qquad A = Q, \qquad A((t-u)I + H) = HA.$

If φ is an isomorphism, we see from (3.2) that A has to be invertible. If, in addition, $t \neq u$, the third equation in (3.4) gives a contradiction, since the left side is invertible and the right side is not. Thus $(V_t \hookrightarrow W) \not\cong (V_u \hookrightarrow W)$ if $t \neq u$. To see that $(V_t \hookrightarrow W)$ is indecomposable, we take $u = t$ and suppose that φ, as above, is idempotent. Squaring the first equation in (3.2), and comparing "1" and "A" terms, we see that $A^2 = A$ and $P = AP + PA$. But equation (3.4) says that $AH = HA$, and it follows that A is in $k[H]$, which is a local ring. Therefore $A = 0$ or I, and either possibility forces $P = 0$. Thus $\varphi = 0$ or 1, as desired. Thus we may take $T = k$ in this case. □

3.15. ASSUMPTIONS. Having dealt with the case (ii), we assume from now on that (i) holds, that is, $\dim_k(D) \geqslant 4$.

3.16. CASE. *D has an element α such that $\{1, \alpha, \alpha^2\}$ is linearly independent.*

Choose any element $\beta \in D$ such that $\{1, \alpha, \beta, \alpha^2\}$ is linearly independent. We let E be the set of elements $t \in k$ for which $\{1, \alpha, \beta, (\alpha + t\beta)^2\}$ is linearly independent. Then E is non-empty (since it contains 0). Also, E is open in the Zariski topology on k and therefore is cofinite in k. Moreover, if $t \in E$, the set

$$E_t = \{u \in E \mid \{1, \alpha, \beta, (\alpha + t\beta)(\alpha + u\beta)\} \text{ is linearly independent}\}$$

is non-empty and cofinite in E. We will show that $(V_t \hookrightarrow W)$ is indecomposable for each $t \in E$, and that $(V_t \hookrightarrow W) \not\cong (V_u \hookrightarrow W)$ if t and u are distinct elements of E with $u \in E_t$. Assuming this has been done we can complete the proof in this case as follows: Define an equivalence relation $\sim$ on E by declaring that $t \sim u$ if and only if $(V_u \hookrightarrow D) \cong (V_t \hookrightarrow D)$, and let T be a set of representatives. Then $T \neq \emptyset$, and $(V_t \hookrightarrow W)$ is indecomposable for each $t \in T$. Moreover, each equivalence class is finite and E is cofinite in k. Therefore, if k is infinite, it follows that $|T| = |k|$.

Suppose $t \in E$ and $u \in E_t$ (possibly with $t = u$), and let $\varphi \colon (V_t \hookrightarrow W) \longrightarrow (V_u \hookrightarrow W)$ be a homomorphism. Keeping the notation of (3.1)–(3.3), one can show, by descending induction on i and j, that for all $i, j = 0, \ldots, n$ we have $H^i P H^j = 0$. (Cf. Exercise 3.30.) Therefore $P = 0$, and we again obtain equations (3.4). The rest of the proof proceeds exactly as in Case 3.14. □

The following lemma, whose proof is left as an exercise, is useful in treating the remaining case, when every element of D satisfies a monic quadratic equation over k:

3.17. LEMMA. *Let ℓ be a field, and let A be a finite-dimensional ℓ-algebra with $\dim_\ell(A) \geq 3$. Assume that $\{1, \alpha, \alpha^2\}$ is linearly dependent over ℓ for every $\alpha \in A$. Write $A = A_1 \times \cdots \times A_s$, where each A_i is local, with maximal ideal $\mathfrak{m}_i$. Let $\mathfrak{N} = \mathfrak{m}_1 \times \cdots \times \mathfrak{m}_s$, the nilradical of A.*

  (i) *If $x \in \mathfrak{N}$, then $x^2 = 0$.*
  (ii) *There are at least $|\ell|$ distinct rings between ℓ and A.*
  (iii) *If $s \geq 2$, then $A_i/\mathfrak{m}_i = \ell$ for each i.*
  (iv) *If $s \geq 3$ then $|\ell| = 2$.* □

3.18. ASSUMPTIONS. From now on, we assume that $\{1, \alpha, \alpha^2\}$ is linearly dependent over k for each $\alpha \in D$ (and that $\dim_k(D) \geq 4$). We write $D = D_1 \times \cdots \times D_s$, where each D_i is local, with maximal ideal $\mathfrak{m}_i$; we let $\mathfrak{N} = \mathfrak{m}_1 \times \cdots \times \mathfrak{m}_t$, the nilradical of D.

3.19. CASE. $\dim_k(\mathfrak{N}) \geq 2$.

Choose $\alpha, \beta \in \mathfrak{N}$ so that $\{1, \alpha, \beta\}$ is linearly independent. Then $\alpha^2 = \beta^2 = 0$ by Lemma 3.17. If $\{1, \alpha, \beta, \alpha\beta\}$ is linearly independent, we can use the mess (3.3) to complete the proof. Otherwise, we can write $\alpha\beta = a + b\alpha + c\beta$ with $a, b, c \in k$. Multiplying this equation first by α and then by β, we learn that $\alpha\beta = 0$, and we are in Case 3.14. □

3.20. ASSUMPTION. We assume from now on that $\dim_k(\mathfrak{N}) \leq 1$.

From Lemma 3.17 we see that s (the number of components) cannot be 2. Also, if $s = 3$, then, after renumbering if necessary, we have $\mathfrak{N} = \mathfrak{m}_1 \times 0 \times 0$ with $\mathfrak{m}_1 \neq 0$. Now put $\alpha = (x, 1, 0)$, where x is a nonzero element of $\mathfrak{m}_1$, and check that $\{1, \alpha, \alpha^2\}$ is linearly independent, contradicting Assumption 3.18. We have proved that either $s = 1$ or $s \geq 4$.

3.21. CASE. $s = 1$ *(D is local)*.

By Assumptions 3.15 and 3.20, $K := D/\mathfrak{N}$ must have degree at least three over k. On the other hand, Assumption 3.18 implies that each element of K has degree at most 2 over k. Therefore K/k is not separable, $\mathrm{char}(k) = 2$, $\alpha^2 \in k$ for each $\alpha \in K$, and $[K : k] \geq 4$. Now choose two elements $\alpha, \beta \in K$ such that $[k(\alpha, \beta) : k] = 4$. By Lemma 3.11 we can safely pass to the Artinian pair (k, K) and build our modules there; for compatibility with the notation in Construction 3.13, we rename K and call it D. Now we have $\alpha, \beta \in D$ such that $\{1, \alpha, \beta, \alpha\beta\}$ is linearly independent and both α^2 and β^2 are in k. If, now, $\varphi \colon (V_t \hookrightarrow W) \longrightarrow (V_u \hookrightarrow W)$ is a morphism, the mess (3.3) provides the following equations:

(3.5)
$$B = (\alpha^2 + tu\beta^2)P + \beta^2(HPH + tHP + uPH), \qquad A = Q,$$
$$A((t-u)I + H) = HA, \qquad (t+u)P + HP + PH = 0.$$

Suppose $t \neq u$. Then $t + u \neq 0$ (characteristic 2), and the fourth equation shows, via a descending induction argument as in Case 3.16, that $P = 0$. (Cf. Exercise 3.30.) Now the third equation shows, as in Case 3.14, that φ is not an isomorphism.

Now suppose $t = u$ and $\varphi^2 = \varphi$. Using the third and fourth equations of (3.5), the fact that char$(k) = 2$, and Lemma 3.2, we see that both A and P are in $k[H]$. In particular, A, P and H commute, and, since we are in characteristic two, we can square both sides of (3.2) painlessly. Equating φ and φ^2, we see that $P = 0$ and $A = A^2$. Since $k[H]$ is local, $A = 0$ or I. □

One case remains:

3.22. CASE. $s \geqslant 4$.

By Lemma 3.17, $|k| = 2$ and $D_i/\mathfrak{m}_i = k$ for each i. By Lemma 3.11 we can forget about the radical and assume that $D = k \times \cdots \times k$ (at least 4 components). Alas, this case does not yield to our general construction, but Dade's construction [**Dad63**] saves the day. (Dade works in greater generality, but the main idea is visible in the computation that follows. The key issue is that D has at least 4 components.)

Put $W = D^{(n)}$, and let V be the k-subspace of W consisting of all elements $(x, y, x+y, x+Hy, x, \ldots, x)$, where x and y range over $k^{(n)}$. (Again, H is the nilpotent Jordan block with 1's on the superdiagonal.) Clearly $DW = V$. To see that $(V \hookrightarrow W)$ is indecomposable, suppose φ is an endomorphism of $(V \hookrightarrow W)$, that is, a D-endomorphism of W carrying V into V. We write $\varphi = (\alpha, \beta, \gamma, \delta, \varepsilon_5, \ldots, \varepsilon_s)$, where each component is an $n \times n$ matrix over k. Since $\varphi((x, 0, x, x, x, \ldots, x))$ and $\varphi((0, y, y, Hy, 0, \ldots, 0))$ are in V, there are matrices σ, τ, ξ, η satisfying the following two equations for all $x \in k^{(n)}$:

$$(\alpha x,\ 0,\ \gamma x,\ \delta x,\ \varepsilon_5 x,\ \ldots,\ \varepsilon_s x) =$$
$$(\sigma x,\ \tau x,\ (\sigma + \tau)x,\ (\sigma + H\tau)x,\ \sigma x,\ \ldots,\ \sigma x);$$

$$(0,\ \beta y,\ \gamma y,\ \delta Hy,\ 0,\ \ldots,\ 0) =$$
$$(\xi y,\ \eta y,\ (\xi + \eta)y,\ (\xi + H\eta)y,\ \xi y,\ \ldots,\ \xi y).$$

The first equation shows that $\varphi = (\alpha, \alpha, \ldots, \alpha)$, and the second then shows that $\alpha H = H\alpha$. By Lemma 3.2 $\alpha \in k[H] \cong k[x]/(x^n)$, which is a local ring. If, now, $\varphi^2 = \varphi$, then $\alpha^2 = \alpha$, and hence $\alpha = 0$ or I_n. This shows that $(V \hookrightarrow W)$ is indecomposable and completes the proof of Theorem 3.7. □

We close the chapter with the following partial converse to Theorem 3.7. This statement, the sufficiency of the Drozd-Roĭter conditions, is due to Drozd-Roĭter [**DR67**] and Green-Reiner [**GR78**] in the special case where the residue field $A/\mathfrak{m}$ is finite. In this case they reduced to the situation where where $A/\mathfrak{m} \longrightarrow B/\mathfrak{n}$ is an isomorphism for each maximal ideal $\mathfrak{n}$ of B. In this situation they showed, via explicit matrix decompositions, that conditions (dr1) and (dr2) imply that **A** has finite representation type. These matrix decompositions depend only on the fact that the residue fields of B are all equal to k, and not on the fact that k is finite. The generalization stated here is due to R. Wiegand [**Wie89**] and depends crucially on the matrix decompositions in [**GR78**].

3.23. THEOREM. *Let* $\mathbf{A} = (A \hookrightarrow B)$ *be an Artinian pair in which A is local, with maximal ideal $\mathfrak{m}$ and residue field k. Assume that B is a principal ideal ring and either*

(i) *the field extension $k \hookrightarrow B/\mathfrak{n}$ is separable for every maximal ideal $\mathfrak{n}$ of B, or*
(ii) *B is reduced (hence a direct product of fields).*

If **A** *satisfies (dr1) and (dr2), then* **A** *has finite representation type.*

PROOF. As in [**GR78**] we will reduce to the case where the residue fields of B are all equal to k. By (dr1) B has at most three maximal ideals, and at most one of these has a residue field ℓ properly extending k. Moreover, $[\ell : k] \leqslant 3$. Assuming $\ell \neq k$, we choose a primitive element θ for ℓ/k, let $f \in A[T]$ be a monic polynomial reducing to the minimal polynomial for θ over k, and pass to the Artinian pair $\mathbf{A}' = (A' \hookrightarrow B')$, where $A' = A[T]/(f)$ and $B' = B \otimes_A A' = B[T]/(f)$. Each of the conditions (i), (ii) guarantees that B' is a principal ideal ring.

One checks that the Drozd-Roĭter conditions ascend to $\mathbf{A}'$, and finite representation type descends. (This is not difficult; the details are worked out in [**Wie89**].) If $k(\theta)/k$ is a separable, non-Galois extension of degree 3, then B' has a residue field that is separable of degree 2 over k, and we simply repeat the construction. Thus it suffices to prove the theorem in the case where each residue field of B is equal to k. For this case, we appeal to the matrix decompositions in [**GR78**], which work perfectly well over any field. □

§3. Exercises

3.24. EXERCISE. Let $\mathfrak{m}$ be a maximal ideal of a Noetherian ring R, and assume that $\mathfrak{m}$ is not a minimal prime ideal of R. Then $\{\mathfrak{m}^t \mid t \geqslant 1\}$ is an infinite strictly descending chain of ideals.

3.25. EXERCISE. Let $(R, \mathfrak{m}, k)$ be a commutative local Artinian ring, and assume k is infinite.
 (i) If $\mathcal{G}$ is a set of pairwise non-isomorphic finitely generated R-modules, prove that $|\mathcal{G}| \leqslant |k|$.
 (ii) Suppose R is not a principal ideal ring. Modify the proof of Theorem 3.3 to show that for each $n \geqslant 1$ there is a family $\mathcal{G}_n$ of pairwise non-isomorphic indecomposable modules, all requiring exactly n generators, with $|\mathcal{G}_n| = |k|$. (Hint: Given a unit t of R, let $\Psi_t = (y + tx)I + xH$. Show that an isomorphism between $\mathrm{cok}(\Psi_t)$ and $\mathrm{cok}(\Psi_u)$ forces t and u to be congruent modulo $\mathfrak{m}$.)

3.26. EXERCISE. Prove Lemmas 3.10 and 3.11.

3.27. EXERCISE. Prove Lemma 3.17. (For the second assertion, suppose there are fewer than $|\ell|$ intermediate rings. Mimic the proof of the primitive element theorem to show that $D = k[\alpha]$ for some α.)

3.28. EXERCISE. With E and E_t as in 3.16, prove that $|k - E| \leqslant 1$ and that $|E - E_t| \leqslant 1$.

3.29. EXERCISE. Let $\mathbf{A} = (A \hookrightarrow B)$ be an Artinian pair, and let C_1 and C_2 be distinct rings between A and B. Prove that the $\mathbf{A}$-modules $(C_1 \hookrightarrow B)$ and $(C_2 \hookrightarrow B)$ are not isomorphic.

3.30. EXERCISE. Work out the details of the descending induction arguments in Case 3.16 and Case 3.21. (In Case 3.16, assuming that $H^{i+1}\gamma H^j = 0$ and $H^i\gamma H^{j+1} = 0$, multiply the mess (3.3) by H^i on the left and H^j on the right. In Case 3.21, use the fourth equation in (3.5) and do the same thing.)

CHAPTER 4

Dimension One

In this chapter we give necessary and sufficient conditions for a one-dimensional local ring to have finite Cohen-Macaulay type. In the main case of interest, where the completion $\widehat{R}$ is reduced, these conditions are simply the liftings of the Drozd-Roĭter conditions (dr1) and (dr2) of Chapter 3. Necessity of these conditions follows easily from Theorem 3.7. To prove that they are sufficient, we will reduce the problem to consideration of some special cases, where we can appeal to the matrix decompositions of Green and Reiner [**GR78**] and, in characteristic two, Çimen [**Çim94, Çim98**].

Throughout this chapter $(R, \mathfrak{m}, k)$ is a one-dimensional local ring (with maximal ideal $\mathfrak{m}$ and residue field k). Let K denote the total quotient ring $\{\text{non-zerodivisors}\}^{-1}R$ and $\overline{R}$ the integral closure of R in K. If R is reduced (hence CM), then $\overline{R} = \overline{R/\mathfrak{p}_1} \times \cdots \times \overline{R/\mathfrak{p}_s}$, where the $\mathfrak{p}_i$ are the minimal prime ideals of R, and each ring $\overline{R/\mathfrak{p}_i}$ is a semilocal principal ideal domain.

When R is CM, a finitely generated R-module M is MCM if and only if it is torsion-free, that is, the torsion submodule is zero.

The main result in this chapter is Theorem 4.10, which states that a one-dimensional local ring $(R, \mathfrak{m}, k)$, with reduced completion, has finite CM type if and only if R satisfies the following two conditions:

(DR1) $\mu_R(\overline{R}) \leqslant 3$, and

(DR2) $\frac{\mathfrak{m}\overline{R}+R}{R}$ is a cyclic R-module.

The first condition just says that the multiplicity of R is at most three (cf. Theorem A.29). When the multiplicity is three we have to consider the second condition. One can check, for example, that $k[\![t^3, t^5]\!]$ satisfies (DR2) but that $k[\![t^3, t^7]\!]$ does not.

The case where the completion is not reduced is dealt with separately, in Theorem 4.16. In particular, we find (Corollary 4.17) that a one-dimensional local ring R has finite CM type if and only if its completion does. The analogous statement fails badly in higher dimension without some additional assumptions; cf. Chapter 10. Furthermore, Proposition 4.15 shows that if a one-dimensional CM local ring has finite CM type, then its completion is reduced; in particular R is an isolated singularity, which property will appear again in Chapter 7. We also treat the case of multiplicity two directly, without any reducedness assumption.

As a look ahead to later chapters, in §3 we discuss the alternative classification of finite CM type in dimension one due to Greuel and Knörrer in terms of the ADE hypersurface singularities.

§1. Necessity of the Drozd-Roĭter conditions

Looking ahead to Chapter 17, we work in a somewhat more general context than is strictly required for Theorem 4.10. In particular, we will not assume that R is reduced, and $\overline{R}$ will be replaced by a more general extension ring S. By a *finite birational extension* of R we mean a ring S between R and its total quotient ring K such that S is finitely generated as an R-module.

4.1. CONSTRUCTION. Let $(R, \mathfrak{m}, k)$ be a CM local ring of dimension one, and let S be a finite birational extension of R. Put $\mathfrak{c} = (R :_R S)$, the *conductor* of S into R. This is the largest common ideal of R and S. Set $A = R/\mathfrak{c}$ and $B = S/\mathfrak{c}$. Then the *conductor square of* $R \hookrightarrow S$

(4.1)
$$\begin{array}{ccc} R & \hookrightarrow & S \\ \downarrow & & \downarrow \pi \\ A & \hookrightarrow & B \end{array}$$

is a pullback diagram, that is, $R = \pi^{-1}(A)$. Since S is a module-finite extension of R contained in the total quotient ring K, the conductor contains a non-zerodivisor (clear denominators), so that the bottom line $\mathbf{A} = (A \hookrightarrow B)$ is an Artinian pair in the sense of Chapter 3.

Suppose that M is a MCM R-module. Then M is torsion-free, so the natural map $M \longrightarrow K \otimes_R M$ is injective. Let SM be the S-submodule of $K \otimes_R M$ generated by the image of M; equivalently, $SM = (S \otimes_R M)/\text{torsion}$. If we furthermore assume that SM is a projective S-module, then the inclusion $M/\mathfrak{c}M \hookrightarrow SM/\mathfrak{c}M$ gives a module over the Artinian pair $A \hookrightarrow B$.

In the special case where S is the integral closure $\overline{R}$, the situation clarifies. Since $\overline{R}$ is a direct product of semilocal principal ideal domains, and $\overline{R}M$ is torsion-free for any MCM R-module M, it follows that $\overline{R}M$ is $\overline{R}$-projective. Thus $M/\mathfrak{c}M \hookrightarrow \overline{R}M/\mathfrak{c}M$ is automatically a module over the Artinian pair $R/\mathfrak{c} \hookrightarrow \overline{R}/\mathfrak{c}$. We dignify this special case with the notation $R_{\text{art}} = (R/\mathfrak{c} \hookrightarrow \overline{R}/\mathfrak{c})$ and $M_{\text{art}} = (M/\mathfrak{c}M \hookrightarrow \overline{R}M/\mathfrak{c}M)$.

Now return to the case of an arbitrary finite birational extension $R \hookrightarrow S$, and let $V \hookrightarrow W$ be a module over the associated Artinian pair

$\mathbf{A} = (A \hookrightarrow B) = (R/\mathfrak{c} \hookrightarrow S/\mathfrak{c})$. Assume that there exists a finitely generated projective S-module P such that $W \cong P/\mathfrak{c}P$. (This is a real restriction; see the comments below.) We can then define an R-module M by a similar pullback diagram

(4.2)
$$\begin{array}{ccc} M & \hookrightarrow & P \\ \downarrow & & \downarrow \tau \\ V & \hookrightarrow & W \end{array}$$

so that $M = \tau^{-1}(V)$. Using the fact that $BV = W$, one can check that $SM = P$, so that in particular M is a MCM R-module. Moreover, $M/\mathfrak{c}M = V$ and $SM/\mathfrak{c}M = W$, so that two non-isomorphic $\mathbf{A}$-modules that are both liftable have non-isomorphic liftings.

If in particular $V \hookrightarrow W$ is an $\mathbf{A}$-module of constant rank, so that $W \cong B^{(n)}$ for some n, then there is clearly a projective S-module P such that $P/\mathfrak{c}P \cong W$, namely $P = S^{(n)}$. Furthermore, in this case M has constant rank n over R (see Definition A.27). It follows that every $\mathbf{A}$-module of constant rank lifts to a MCM R-module of constant rank. Moreover, every $\mathbf{A}$-module is a direct summand of one of constant rank, so is a direct summand of a module extended from R.

By analogy with the terminology "weakly liftable" of [**ADS93**], we say that a module $V \hookrightarrow W$ over the Artinian pair $\mathbf{A} = R/\mathfrak{c} \hookrightarrow S/\mathfrak{c}$ is *weakly extended* (from R) if there exists a MCM R-module M such that $V \hookrightarrow W$ is a direct summand of the $\mathbf{A}$-module $M/\mathfrak{c}M \hookrightarrow SM/\mathfrak{c}M$. The discussion above shows that every $\mathbf{A}$-module is weakly extended from R.

Now we lift the Drozd-Roĭter conditions up to the finite birational extension $R \hookrightarrow S$.

4.2. THEOREM. *Let $(R, \mathfrak{m}, k)$ be a local ring of dimension one, and let S be a finite birational extension of R. Assume that either*

(i) $\mu_R(S) \geqslant 4$, or
(ii) $\mu_R\left(\frac{\mathfrak{m}S+R}{R}\right) \geqslant 2$.

Then R has infinite Cohen-Macaulay type. Moreover, given an arbitrary positive integer n, there is an indecomposable MCM R-module M of constant rank n; if k is infinite, there are $|k|$ pairwise non-isomorphic indecomposable MCM R-modules of constant rank n.

PROOF. The hypotheses imply that either (dr1) or (dr2) of Theorem 3.7 fails for the Artinian pair $\mathbf{A} = (R/\mathfrak{c} \hookrightarrow S/\mathfrak{c})$. Therefore there exist indecomposable $\mathbf{A}$-modules of arbitrary constant rank n, in fact, $|k|$ of them if k is infinite. Each of these pulls back to R, so that there exist the same number of MCM R-modules of constant rank n for each $n \geqslant 1$. Furthermore these MCM modules are pairwise non-isomorphic. Finally, we must show that if $V \hookrightarrow W$ is indecomposable and M is a lifting to R, then M is indecomposable as well. Suppose $M \cong X \oplus Y$. Then $SM = SX \oplus SY$, and it follows that $(V \hookrightarrow W)$ is the direct sum of the **A**-modules $(X/\mathfrak{c}X \hookrightarrow SX/\mathfrak{c}X)$ and $(Y/\mathfrak{c}Y \hookrightarrow SY/\mathfrak{c}Y)$. Therefore either $X/\mathfrak{c}X = 0$ or $Y/\mathfrak{c}Y = 0$. By NAK, either $X = 0$ or $Y = 0$. □

The requisite extension S of Theorem 4.2 always exists if R is CM of multiplicity at least 4, as we now show.

4.3. PROPOSITION. *Let $(R, \mathfrak{m})$ be a one-dimensional CM local ring and set $e = \mathrm{e}(R)$, the multiplicity of R. (See Appendix A §2.) Then R has a finite birational extension S requiring e generators as an R-module.*

PROOF. Let K again be the total quotient ring of R. Let $S_n = (\mathfrak{m}^n :_K \mathfrak{m}^n)$ for $n \geqslant 1$, and put $S = \bigcup_n S_n$. To see that this works, we may harmlessly assume that k is infinite. (This is relatively standard, but see Theorem 10.14 for the details on extending the residue field.) Let $Rf \subseteq \mathfrak{m}$ be a minimal reduction of $\mathfrak{m}$. Choose n so large that

(a) $\mathfrak{m}^{i+1} = f\mathfrak{m}^i$ for $i \geqslant n$, and
(b) $\mu_R(\mathfrak{m}^i) = \mathrm{e}(R)$ for $i \geqslant n$.

Since f is a non-zerodivisor (as R is CM), it follows from (a) that $S = S_n$. We claim that $Sf^n = \mathfrak{m}^n$. We have $Sf^n = S_n f^n \subseteq \mathfrak{m}^n$. For the reverse inclusion, let $\alpha \in \mathfrak{m}^n$. Then $\frac{\alpha}{f^n}\mathfrak{m}^n \subseteq \frac{1}{f^n}\mathfrak{m}^{2n} = \frac{1}{f^n}f^n\mathfrak{m}^n = \mathfrak{m}^n$. This shows that $\frac{\alpha}{f^n} \in S_n$, and the claim follows. Therefore $S \cong \mathfrak{m}^n$ (as R-modules), and now (b) implies that $\mu(S) = \mathrm{e}(R)$. □

4.4. REMARK. Observe that the proof of the proposition above gives more than is claimed: for any one-dimensional CM local ring R and any ideal I of R containing a non-zerodivisor, there exists some $n \geqslant 1$ such that I^n is projective as a module over its endomorphism ring $S = \mathrm{End}_R(I^n)$, which is a finite birational extension of R. (Ideals projective over their endomorphism ring are called *stable* in [**Lip71**] and [**SV74**].) Since S is semilocal, I^n is isomorphic to S as an S-module, whence as an R-module. Furthermore, n may be taken to be the least integer such that $\mu_R(I^n)$ achieves its stable value. Sally and Vasconcelos show in [**SV74**, Theorem 2.5] that this n is at most $\max\{1, \mathrm{e}(R) - 1\}$, where $\mathrm{e}(R)$ is the multiplicity of R. This will be useful in Theorem 4.18 below.

4.5. REMARK. With R as in Theorem 4.2 and with k infinite, there are *at most* $|k|$ isomorphism classes of R-modules of constant rank. To

see this, we note that there are at most $|k|$ isomorphism classes of finite-length modules and that every module of finite length has cardinality at most $|k|$. Given an arbitrary MCM R-module M of constant rank n, one can build an exact sequence
$$0 \longrightarrow T \longrightarrow M \longrightarrow R^{(n)} \longrightarrow U \longrightarrow 0,$$
in which both T and U have finite length. Let W be the kernel of $R^{(n)} \longrightarrow U$ (and the cokernel of $T \longrightarrow M$). Since $|U| \leqslant |k|$, we see that $|\operatorname{Hom}_R(R^{(n)}, U)| \leqslant |k|$. Since there are at most $|k|$ possibilities for U, we see that there are at most $|k|^2 = |k|$ possibilities for W. Since there are at most $|k|$ possibilities for T, and since $\left|\operatorname{Ext}_R^1(W, T)\right| \leqslant |k|$, we see that there are at most $|k|$ possibilities for M.

§2. Sufficiency of the Drozd-Roĭter conditions

In this section we will prove, modulo the matrix calculations of Green and Reiner [**GR78**] and Çimen [**Çim94**, **Çim98**], that the Drozd-Roĭter conditions imply finite CM type. Recall that a local ring $(R, \mathfrak{m})$ is said to be *analytically unramified* provided its completion $\widehat{R}$ is reduced. The next result gives an equivalent condition [**Kru30**, **Nag58**] for one-dimensional CM local rings, namely finiteness of the integral closure.

4.6. THEOREM. *Let $(R, \mathfrak{m})$ be a local ring, and let $\overline{R}$ be the integral closure of R in its total quotient ring.*
  (i) *If R is analytically unramified, then $\overline{R}$ is finitely generated as an R-module.*
  (ii) *Suppose R is one-dimensional and CM. If $\overline{R}$ is finitely generated as an R-module then R is analytically unramified.*

PROOF. See [**Mat89**, p. 263] or [**HS06**, 4.6.2] for a proof of (i). With the assumptions in (ii), we'll show first that R is reduced. Suppose x is a non-zero nilpotent element of R and t a non-zerodivisor in $\mathfrak{m}$. Then
$$R\frac{x}{t} \subset R\frac{x}{t^2} \subset R\frac{x}{t^3} \subset \cdots$$
is an infinite strictly ascending chain of R-submodules of $\overline{R}$, contradicting finiteness of $\overline{R}$. Now assume that R is reduced and let $\mathfrak{p}_1 \ldots, \mathfrak{p}_s$ be the minimal prime ideals of R. There are inclusions
$$R \hookrightarrow R/\mathfrak{p}_1 \times \cdots \times R/\mathfrak{p}_s \hookrightarrow \overline{R/\mathfrak{p}_1} \times \cdots \times \overline{R/\mathfrak{p}_s} = \overline{R}.$$
Each of the rings $\overline{R/\mathfrak{p}_i}$ is a semilocal principal ideal domain. Since $\overline{R}$ is a finitely generated R-module, the $\mathfrak{m}$-adic completion of $\overline{R}$ is the

product of the completions of the localizations of the $\overline{R/\mathfrak{p}_i}$ at their maximal ideals. In particular, the $\mathfrak{m}$-adic completion of $\overline{R}$ is a direct product of discrete valuation rings. The flatness of $\widehat{R}$ implies that $\widehat{R}$ is contained in the $\mathfrak{m}$-adic completion of $\overline{R}$, hence is reduced. □

In the proof of part (ii) of the following proposition we encounter the subtlety mentioned in Construction 4.1: not every projective module over B is of the form $P/\mathfrak{c}P$ for a projective S-module. This is because $\overline{R}$ might not be a direct product of local rings. For example, the integral closure $\overline{R}$ of the ring $R = \mathbb{C}[x,y]_{(x,y)}/(y^2 - x^3 - x^2)$ has two maximal ideals (see Exercise 4.23), and so $\overline{R}/\mathfrak{c}$ is a direct product $B_1 \times B_2$ of two local rings. Obviously $B_1 \times 0$ does not come from a projective $\overline{R}$-module and hence cannot be the second component of an R_{art}-module of the form M_{art}. The reader may recognize that exactly the same phenomenon gives rise to modules over the completion $\widehat{R}$ that don't come from R-modules, as we saw in Chapter 2.

Recall that we use the notation $M_1 \mid M_2$, introduced in Chapter 1, to indicate that M_1 is isomorphic to a direct summand of M_2.

4.7. PROPOSITION. *Let $(R, \mathfrak{m}, k)$ be an analytically unramified local ring of dimension one, and assume $R \neq \overline{R}$. Let R_{art} be the Artinian pair $R/\mathfrak{c} \hookrightarrow \overline{R}/\mathfrak{c}$.*

  (i) *The functor $M \rightsquigarrow M_{\text{art}} = (M/\mathfrak{c}M \hookrightarrow \overline{R}M/\mathfrak{c}M)$, for M a MCM R-module, is injective on isomorphism classes.*
  (ii) *If M_1 and M_2 are MCM R-modules, then $M_1 \mid M_2$ if and only if $(M_1)_{\text{art}} \mid (M_2)_{\text{art}}$.*
  (iii) *The ring R has finite CM type if and only if the Artinian pair R_{art} has finite representation type.*

PROOF. (i) First observe that $M \rightsquigarrow M_{\text{art}}$ is indeed well-defined: since $\overline{R}$ is a direct product of principal ideal rings, $\overline{R}M$ is a projective $\overline{R}$-module, so $\overline{R}M/\mathfrak{c}M$ is a projective $\overline{R}/\mathfrak{c}$-module. Thus M_{art} is a module over R_{art}. Let M_1 and M_2 be MCM R-modules, and suppose that $(M_1)_{\text{art}} \cong (M_2)_{\text{art}}$. Write $(M_i)_{\text{art}} = (V_i \hookrightarrow W_i)$, and choose an isomorphism $\varphi: W_1 \longrightarrow W_2$ such that $\varphi(V_1) = V_2$. Since $\overline{R}M_1$ is $\overline{R}$-projective, we can lift φ to an $\overline{R}$-homomorphism $\psi: \overline{R}M_1 \longrightarrow \overline{R}M_2$

§2. SUFFICIENCY OF THE DROZD-ROĬTER CONDITIONS

carrying M_1 into M_2.

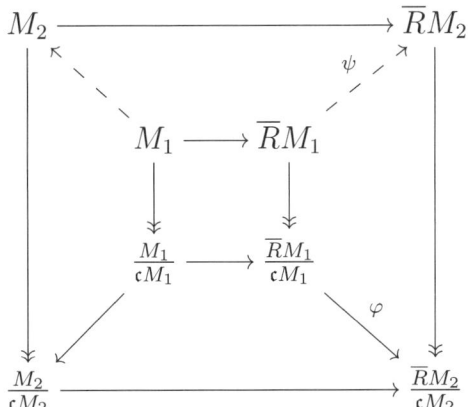

Since $\mathfrak{c} \subseteq \mathfrak{m}$, the induced R-homomorphism $M_1 \longrightarrow M_2$ is surjective, by NAK. (Here we need the assumption that $R \neq \overline{R}$.) Similarly, M_2 maps onto M_1, and it follows that $M_1 \cong M_2$ (see Exercise 4.26).

(ii) The "only if" direction is clear. For the converse, suppose there is an R_{art}-module $\mathfrak{X} = (V \hookrightarrow W)$ such that $(M_1)_{\text{art}} \oplus \mathfrak{X} \cong (M_2)_{\text{art}}$. Write $\overline{R} = D_1 \times \cdots \times D_s$, where each D_i is a semilocal principal ideal domain. Put $B_i = D_i/\mathfrak{c}D_i$, so that $\overline{R}/\mathfrak{c} = B_1 \times \cdots \times B_s$. Since $\overline{R}M_1$, and $\overline{R}M_2$ are projective $\overline{R}$-modules, there are non-negative integers e_i, f_i such that $\overline{R}M_1 \cong \prod_i D_i^{(e_i)}$ and $\overline{R}M_2 \cong \prod_i D_i^{(f_i)}$. Then $\overline{R}M_1/\mathfrak{c}M_1 \cong \prod_i B_i^{(e_i)}$, similarly $\overline{R}M_2/\mathfrak{c}M_2 \cong \prod_i B_i^{(f_i)}$, and $W = \prod_i B_i^{(f_i-e_i)}$. Letting $P = \prod_i D_i^{(f_i-e_i)}$, we see that $W \cong P/\mathfrak{c}P$. As discussed in Construction 4.1, it follows that there is a MCM R-module N such that $N_{\text{art}} \cong \mathfrak{X}$. We see from (i) that $M_1 \oplus N \cong M_2$.

(iii) Suppose R_{art} has finite representation type, and let $\mathbf{X}_1, \ldots, \mathbf{X}_t$ be a full set of representatives for the non-isomorphic indecomposable R_{art}-modules. Given a MCM R-module M, write $M_{\text{art}} \cong \mathbf{X}_1^{(n_1)} \oplus \cdots \oplus \mathbf{X}_t^{(n_t)}$, and put $j([M]) = (n_1, \ldots, n_t)$. By KRS (Theorem 3.6), j is a well-defined function from the set of isomorphism classes of MCM R-modules to $\mathbb{N}_0^{(t)}$, where $\mathbb{N}_0$ is the set of non-negative integers. Moreover, j is injective, by (i). Letting Σ be the image of j, we see, using (ii), that M is indecomposable if and only if $j[M]$ is a minimal non-zero element of Σ with respect to the product ordering. Dickson's Lemma (Exercise 2.20) says that every antichain (clutter) in $\mathbb{N}_0^{(t)}$ is finite. In particular, Σ has only finitely many minimal elements, and R has finite CM type.

We leave the proof of the converse (which will not be needed here) as an exercise. □

4.8. REMARK. It's worth observing that the proof of Proposition 4.7 uses KRS only over R_{art}, not over R (which is not assumed to be Henselian). In fact, since the completion $\widehat{R}$ is reduced, $(\widehat{R})_{\text{art}} = R_{\text{art}}$. Indeed, the bottom row $R/\mathfrak{c} \hookrightarrow \overline{R}/\mathfrak{c}$ of the conductor square for R is unaffected by completion since $\overline{R}/\mathfrak{c}$ has finite length. Therefore the $\mathfrak{m}$-adic completion of the conductor square for R is

$$(4.3) \qquad \begin{array}{ccc} \widehat{R} & \hookrightarrow & \widehat{R} \otimes_R \overline{R} \\ \downarrow & & \downarrow \widehat{\pi} \\ R/\mathfrak{c} & \hookrightarrow & \overline{R}/\mathfrak{c} \end{array}$$

which is still a pullback diagram by flatness of the completion. Note that $\widehat{R} \otimes_R \overline{R}$ is the integral closure of $\widehat{R}$. No non-zero ideal of $\overline{R}/\mathfrak{c}$ is contained in $R/\mathfrak{c}$, so $\ker \widehat{\pi}$ is the largest ideal of $\widehat{R} \otimes_R \overline{R}$ contained in $\widehat{R}$. Since also $\ker \widehat{\pi}$ contains a non-zerodivisor, $\ker \widehat{\pi}$ is the conductor for $\widehat{R}$ and (4.3) is the conductor square for $\widehat{R}$.

In particular, this shows that an analytically unramified local ring R of dimension one has finite CM type if and only if $\widehat{R}$ does. This is true as well in the case where $\widehat{R}$ is not reduced; see Corollary 4.17 below.

Returning now to sufficiency of the Drozd-Roĭter conditions, we will need the following observation from Bass's "ubiquity" paper [**Bas63**, (7.2)]:

4.9. LEMMA (Bass). *Let $(R, \mathfrak{m})$ be a one-dimensional Gorenstein local ring. Let M be a MCM R-module with no non-zero free direct summand. Let $E = \operatorname{End}_R(\mathfrak{m})$. Then E (viewed as multiplications) is a subring of $\overline{R}$ which contains R properly, and M has an E-module structure that extends the action of R on M.*

PROOF. The natural inclusion $\operatorname{Hom}_R(M, \mathfrak{m}) \longrightarrow \operatorname{Hom}_R(M, R)$ is bijective, since a surjective homomorphism $M \longrightarrow R$ would produce a non-trivial free summand of M. Now $\operatorname{Hom}_R(M, \mathfrak{m})$ is an E-module via the action of E on $\mathfrak{m}$ by endomorphisms, and hence so is $M^* = \operatorname{Hom}_R(M, R)$. Therefore M^{**} is also an E-module, and since the canonical map $M \longrightarrow M^{**}$ is bijective (as R is Gorenstein and M is MCM), M is an E-module. The other assertions regarding E are left to the reader. (See Exercise 4.29. Note that the existence of the module M prevents R from being a discrete valuation ring.) □

Now we are ready for the main theorem of this chapter. We will not give a self-contained proof that the Drozd-Roĭter conditions imply finite CM type. Instead, we will reduce to a few special situations where the matrix decompositions of Green and Reiner [**GR78**] and Çimen [**Çim94, Çim98**] apply.

Note that the final statement of Theorem 4.10 verifies the second Brauer-Thrall conjecture (see Conjecture 15.1) for analytically unramified one-dimensional rings.

4.10. THEOREM. *Let $(R, \mathfrak{m}, k)$ be an analytically unramified local ring of dimension one. These are equivalent:*

(i) R has finite CM type.
(ii) R satisfies both (DR1) and (DR2).

Let n be an arbitrary positive integer. If either (DR1) or (DR2) fails, there is an indecomposable MCM R-module of constant rank n; moreover, if $|k|$ is infinite, there are $|k|$ pairwise non-isomorphic indecomposable MCM R-modules of constant rank n.

PROOF. By Theorem 4.6, $\overline{R}$ is a finite birational extension of R. The last statement of the theorem and the fact that (i) $\implies$ (ii) now follow immediately from Theorem 4.2 with $S = \overline{R}$.

Assume now that (DR1) and (DR2) hold. Let $A = R/\mathfrak{c}$ and $B = \overline{R}/\mathfrak{c}$, so that $R_{\text{art}} = (A \hookrightarrow B)$. Then R_{art} satisfies (dr1) and (dr2). By Proposition 4.7 it will suffice to prove that R_{art} has finite representation type. If every residue field of B is separable over k, then R_{art} has finite representation type by Theorem 3.23.

Now suppose that B has a residue field $\ell = B/\mathfrak{n}$ that is not separable over k. By (dr1), ℓ/k has degree 2 or 3, and ℓ is the only residue field of B that is not equal to k.

4.11. CASE. *ℓ/k is purely inseparable of degree 3.*

If B is reduced (that is, R is *seminormal*), we can appeal to Theorem 3.23. Suppose now that B is *not* reduced. A careful computation of lengths (see Exercise 4.28) shows that R is Gorenstein, with exactly one ring S (the seminormalization of R) strictly between R and $\overline{R}$. By Lemma 4.9, $E := \text{End}_R(\mathfrak{m}) \supseteq S$, and every non-free indecomposable MCM R-module is naturally an S-module. The Drozd-Roĭter conditions clearly pass to the seminormal ring S, which therefore has finite CM type. It follows that R itself has finite CM type.

4.12. CASE. *ℓ/k is purely inseparable of degree 2.*

In this case, we appeal to Çimen's *tour de force* [**Çim94, Çim98**], where he shows, by explicit matrix decompositions, that R_{art} has finite representation type. □

Let's insert here a few historical remarks. The conditions (DR1) and (DR2) were introduced by Drozd and Roĭter in a remarkable 1967 paper [**DR67**], where they classified the module-finite $\mathbb{Z}$-algebras having only finitely many indecomposable finitely generated torsion-free modules. Jacobinski [**Jac67**] obtained similar results at about the same time. The theorems of Drozd-Roĭter and Jacobinski imply the equivalence of (i) and (ii) in Theorem 4.10 for rings essentially module-finite over $\mathbb{Z}$. In the same paper they asserted the equivalence of (i) and (ii) in general. In 1978 Green and Reiner [**GR78**] verified the classification theorem of Drozd and Roĭter, giving more explicit details of the matrix decompositions needed to verify finite CM type. Their proof, like that of Drozd and Roĭter, depended crucially on arithmetic properties of algebraic number fields and thus did not provide immediate insight into the general problem. An important point here is that the matrix reductions of Green and Reiner work in arbitrary characteristics, as long as the integral closure $\overline{R}$ has no residue field properly extending that of R.

In 1989 R. Wiegand [**Wie89**] proved necessity of the Drozd-Roĭter conditions (DR1) and (DR2) for a general one-dimensional local ring $(R, \mathfrak{m}, k)$ and, via the separable descent argument in the proof of Theorem 3.23, sufficiency under the assumption that every residue field of the integral closure $\overline{R}$ is separable over k. By (DR1), this left only the case where k is imperfect of characteristic two or three. In [**Wie94**], he used the seminormality argument above to handle the case of characteristic three. Finally, in his 1994 Ph.D. dissertation [**Çim94**], Çimen solved the remaining case—characteristic two—by difficult matrix reductions. It is worth noting that Çimen's matrix decompositions work in all characteristics and therefore confirm the computations done by Green and Reiner in 1978. The existence of $|k|$ indecomposables of constant rank k, when $|k|$ is infinite and (DR) fails, was proved by Karr and Wiegand [**KW09**] in 2009.

§3. ADE singularities

Of course we have not really proved sufficiency of the Drozd-Roĭter conditions, since we have not presented the difficult matrix calculations of Green and Reiner [**GR78**] and Çimen [**Çim94, Çim98**]. If R contains the field of rational numbers, there is an alternate approach that uses the classification, which we present in Chapter 6, of the two-dimensional hypersurface singularities of finite Cohen-Macaulay type. First we recall the 1985 classification, by Greuel and Knörrer [**GK85**], of the complete, equicharacteristic-zero curve singularities of finite Cohen-Macaulay type. Suppose k is an algebraically closed field of characteristic different from $2, 3$ or 5. The complete ADE (or *simple*) plane curve singularities over k are the rings $k[\![x,y]\!]/(f)$, where f is one of the following polynomials:

$$
\begin{array}{lll}
(A_n): & x^2 + y^{n+1}, & n \geqslant 1 \\
(D_n): & x^2 y + y^{n-1}, & n \geqslant 4 \\
(E_6): & x^3 + y^4 & \\
(E_7): & x^3 + xy^3 & \\
(E_8): & x^3 + y^5 &
\end{array}
$$

We will encounter these singularities again in Chapter 6. Here we will discuss briefly their role in the classification of one-dimensional rings of finite CM type. Greuel and Knörrer [**GK85**] proved that the ADE singularities are exactly the complete plane curve singularities of finite CM type in equicharacteristic zero. In fact, they showed much more, obtaining, essentially, the conclusion of Theorem 4.2 in this context:

4.13. THEOREM (Greuel-Knörrer). *Let $(R, \mathfrak{m}, k)$ be a reduced complete local ring of dimension one containing $\mathbb{Q}$. Assume that k is algebraically closed.*

- (i) *R satisfies the Drozd-Roĭter conditions if and only if R is a finite birational extension of an ADE hypersurface singularity.*
- (ii) *Suppose that R has infinite CM type.*
  - (a) *There are infinitely many rings between R and its integral closure.*
  - (b) *For each integer $n \geqslant 1$ there are infinitely many isomorphism classes of indecomposable MCM R-modules of constant rank n.* □

Greuel and Knörrer used Jacobinski's computations in [**Jac67**] to prove that ADE singularities have finite CM type. The fact that finite CM type passes to finite birational extensions (in dimension one!) is recorded in Proposition 4.14 below. We note that (iia) can fail for infinite fields that are not algebraically closed. Suppose, for example, that ℓ/k is a separable field extension of degree $d > 3$. Put $R = k + x\ell[\![x]\!]$. Then $\overline{R} = \ell[\![x]\!]$ is minimally generated, as an R-module, by $\{1, x, \ldots, x^{d-1}\}$. Theorem 4.10 implies that R has infinite CM type. There are, however, only finitely many rings between R and $\overline{R}$. Indeed, the conductor square (4.1) shows that the intermediate rings correspond bijectively to the intermediate fields between k and ℓ.

In Chapter 8 we will use the classification of the two-dimensional complete hypersurface rings of finite CM type to show that the one-dimensional ADE singularities have finite CM type (even in characteristic p, as long as $p \geqslant 7$). Then, in Chapter 10, we will deduce that the Drozd-Roĭter conditions imply finite CM type for any one-dimensional local ring CM ring $(R, \mathfrak{m}, k)$ containing a field, provided k is perfect and of characteristic $\neq 2, 3, 5$. Together with Greuel and Knörrer's result and the next proposition, this will give a different, slightly roundabout, proof that the Drozd-Roĭter conditions are sufficient for finite CM type in dimension one.

4.14. PROPOSITION. *Let R and S be one-dimensional local rings, and suppose S is a finite birational extension of R.*

  (i) If M and N are MCM S-modules, then we have $\operatorname{Hom}_R(M, N) = \operatorname{Hom}_S(M, N)$.
  (ii) Every MCM S-module is a MCM R-module.
  (iii) An MCM S-module M is indecomposable over S if and only if M is indecomposable over R.
  (iv) If R has finite CM type, so has S.

PROOF. We may assume that R is CM, else $R = S$ and everything is boring.

(i) We need only verify that any R-homomorphism is S-linear. Let $\varphi \colon M \longrightarrow N$ be an R-homomorphism. Given any $s \in S$, write $s = r/t$, where $r \in R$ and t is a non-zerodivisor of R. Then, for any $x \in M$, we have $t\varphi(sx) = \varphi(rx) = r\varphi(x) = ts\varphi(x)$. Since N is torsion-free, we have $\varphi(sx) = s\varphi(x)$. Thus φ is S-linear.

(ii) If M is a MCM S-module, then M is finitely generated and torsion-free, hence MCM, over R.

(iii) is clear from (i) and the fact that $_SM$ is indecomposable if and only if $\operatorname{Hom}_S(M, M)$ contains no idempotents. Finally, (iv) is clear from (iii), (ii) and the fact that by (i) non-isomorphic MCM S-modules are non-isomorphic over R. □

§4. The analytically ramified case

Let $(R, \mathfrak{m})$ be a local Noetherian ring of dimension one, let K be the total quotient ring $\{\text{non-zerodivisors}\}^{-1}R$, and let $\overline{R}$ be the integral closure of R in K. Suppose $\overline{R}$ is *not* finitely generated over R. Then, since algebra-finite integral extensions are module-finite, no finite subset of $\overline{R}$ generates $\overline{R}$ as an R-algebra, and we can build an infinite ascending chain of finitely generated R-subalgebras of $\overline{R}$. Each algebra in the chain is a maximal Cohen-Macaulay R-module, and it is easy to see (Exercise 4.31) that no two of the algebras are isomorphic as R-modules. Moreover, each of these algebras is isomorphic, as an R-module, to a faithful ideal of R. Therefore R has an infinite family of pairwise non-isomorphic faithful ideals. It follows (Exercise 4.32) that R has infinite CM type. Now Theorem 4.6 implies the following result:

4.15. PROPOSITION. *Let $(R, \mathfrak{m}, k)$ be a one-dimensional CM local ring with finite CM type. Then R is analytically unramified.* □

In particular, this proposition shows that R itself is reduced; equivalently, R is an isolated singularity: $R_\mathfrak{p}$ is a regular local ring (a field!) for every non-maximal prime ideal $\mathfrak{p}$. See Theorem 7.12.

What if R is *not* Cohen-Macaulay? The next theorem ([**Wie94**, Theorem 1]) and Theorem 4.10 provide the full classification of one-dimensional local rings of finite Cohen-Macaulay type. We will leave the proof as an exercise.

4.16. THEOREM. *Let $(R, \mathfrak{m})$ be a local ring of dimension one, and let N be the nilradical of R. Then R has finite Cohen-Macaulay type if and only if*

(i) R/N has finite Cohen-Macaulay type, and
(ii) $\mathfrak{m}^i \cap N = (0)$ for $i \gg 0$. □

For example, $k[\![x, y]\!]/(x^2, xy)$ has finite Cohen-Macaulay type, since (x) is the nilradical and $(x, y)^2 \cap (x) = (0)$. However $k[\![x, y]\!]/(x^3, x^2 y)$ has infinite CM type: For each $i \geqslant 1$, xy^{i-1} is a non-zero element of $(x, y)^i \cap (x)$.

4.17. COROLLARY. *Let $(R, \mathfrak{m})$ be a local ring of dimension one. Then R has finite CM type if and only if the $\mathfrak{m}$-adic completion $\widehat{R}$ has finite CM type.* □

PROOF. Suppose first that R is analytically unramified. Since the bottom lines of the conductor squares for R and for $\widehat{R}$ are identical (Remark 4.8), it follows from (iii) of Proposition 4.7 that R has finite CM type if and only if $\widehat{R}$ has finite CM type.

For the general case, let N be the nilradical of R. Suppose R has finite CM type. The CM ring R/N then has finite CM type by Theorem 4.16 and hence is analytically unramified by Proposition 4.15. It follows that $\widehat{N}$ is the nilradical of $\widehat{R}$. By the first paragraph, $\widehat{R}/\widehat{N}$ has finite CM type; moreover, $\widehat{\mathfrak{m}}^i \cap \widehat{N} = (0)$ for $i \gg 0$. Therefore $\widehat{R}$ has finite CM type. For the converse, assume that $\widehat{R}$ has finite CM type. Since every MCM $\widehat{R}/\widehat{N}$-module is also a MCM $\widehat{R}$-module, we see that $\widehat{R/N} = \widehat{R}/\widehat{N}$ has finite CM type. Since R/N is CM, so is $\widehat{R/N}$, and now Theorem 4.15 implies that $\widehat{R/N}$ is reduced. By the first paragraph, R/N has finite CM type. Now $\widehat{N}$ is contained in the nilradical of $\widehat{R}$, so Theorem 4.16 implies that $\widehat{\mathfrak{m}}^i \cap \widehat{N} = (0)$ for $i \gg 0$. It follows that $\mathfrak{m}^i \cap N = (0)$ for $i \gg 0$, and hence that R has finite CM type. □

We shall see in Chapter 10 that finite CM type *always* descends from the completion, even in higher dimensions, but that there are counterexamples to ascent of finite CM type. It is interesting to note that the proof of the corollary does not depend on the characterization (Theorem 4.10) of the one-dimensional analytically unramified local rings of finite CM type. We remark that in higher dimensions finite CM type does not always ascend to the completion (see Example 10.12).

§5. Multiplicity two

Suppose $(R, \mathfrak{m})$ is an analytically unramified one-dimensional local ring and that $\dim_k(\overline{R}/\mathfrak{m}\overline{R}) = 2$. One can show (cf. Exercise 4.30) that R automatically satisfies (DR2) and therefore has finite CM type. Here we will give a direct proof of finite CM type in multiplicity two, using some results in Bass's "ubiquity" paper [**Bas63**]. We don't assume that $\overline{R}$ is a finitely generated R-module.

We refer the reader to Appendix A, §2 for basic stuff on multiplicity, particularly for one-dimensional rings.

4.18. THEOREM. *Let $(R, \mathfrak{m}, k)$ be a Cohen-Macaulay local ring of dimension one, with $\mathrm{e}(R) = 2$.*

  (i) *Every ideal of R is generated by at most two elements.*
  (ii) *Every ring S with $R \subseteq S \subsetneq \overline{R}$ and finitely generated over R is local and Gorenstein. In particular R itself is Gorenstein.*
  (iii) *Every MCM R-module is isomorphic to a direct sum of ideals of R. In particular, every indecomposable MCM R-module has multiplicity at most 2 and is generated by at most 2 elements.*
  (iv) *The ring R has finite CM type if and only if R is analytically unramified.*

PROOF. For (i) we quote Theorem A.29 (ii).

(ii) Let S be a module-finite R-algebra properly contained in $\overline{R}$. Then S has a maximal ideal $\mathfrak{n}$ such that $S_\mathfrak{n}$ is not a DVR. We claim that S is local. For this, we may assume, by passing to the faithfully flat extension $(R[x])_{\mathfrak{m}R[x]}$, that k is infinite. Choose a principal reduction (t) of $\mathfrak{m}$ (see Theorem A.20). Suppose $\mathfrak{n}'$ is another maximal ideal of S, and note that $t \in \mathfrak{n} \cap \mathfrak{n}'$. Now $\ell_R(S/tS) = e_R(S) \leqslant 2$ by Theorem A.23 and the fact that S is isomorphic to an ideal of R (clear denominators). It follows that $\mathfrak{n} \cap \mathfrak{n}' = St$. Localizing at $\mathfrak{n}$, we see that $\mathfrak{n} S_\mathfrak{n}$ is principal, a contradiction. Now that we know S is local we return to the general situation, where k is allowed to be finite. Every ideal I of S is isomorphic to an ideal of R and hence is two-generated as an R-module; therefore I is generated by two elements as an ideal of S. Since the maximal ideal of S is two-generated, Exercise 4.34 guarantees that S is Gorenstein.

(iii) Let M first be a faithful MCM R-module. As M is torsion-free, the map $j \colon M \longrightarrow K \otimes_R M$ is injective. Let $H = \{t \in K \mid tj(M) \subseteq j(M)\}$; then M is naturally an H-module. Since M is faithful, $H \hookrightarrow \operatorname{Hom}_R(M, M)$, and thus H is a module-finite extension of R contained in $\overline{R}$. Suppose first that $H = \overline{R}$. Then R is reduced by Lemma 4.6, and hence $\overline{R}$ is a principal ideal ring. It follows from the structure theory for modules over a principal ideal ring that M has a copy of H as a direct summand, and of course H is isomorphic to an ideal of R. If H is properly contained in $\overline{R}$, then, since H/R has finite length, we can apply Lemma 4.9 repeatedly, eventually getting a copy of some subring of H as a direct summand of M. In either case, we see that M has a faithful ideal of R as a direct summand.

Suppose, now, that M is an arbitrary MCM R-module, and let $I = \operatorname{Ann}(M)$. Then R/I embeds in a direct product of copies of M (one copy for each generator); therefore R/I has depth 1 and hence is a one-dimensional CM ring. Of course $e(R/I) \leqslant 2$, and, since M is a faithful MCM R/I-module, M has a non-zero ideal of R/I as a direct summand. To complete the proof, it will suffice to show that R/I is isomorphic to an ideal of R. Dualizing over R, we have $(R/I)^* \cong \operatorname{Ann}(I)$, an ideal of height 0 (since $I \neq 0$). Therefore $R/\operatorname{Ann}(I)$ has positive multiplicity and hence, by Theorem A.29 (iii), $\operatorname{Ann}(I)$ is a principal ideal. Therefore $(R/I)^*$ is a cyclic R-module. Choosing a surjection $R \twoheadrightarrow (R/I)^*$, we get an injection $(R/I)^{**} \hookrightarrow R$. But R/I, being a MCM module over a Gorenstein ring, is reflexive (Theorem 11.5), and we have $R/I \hookrightarrow R$.

(iv) The "only if" implication is Proposition 4.15. For the converse, we assume that R is analytically unramified, so that $\overline{R}$ is a finitely generated R-module by Theorem 4.6. It will suffice, by item (iii), to show that R has only finitely many ideals up to isomorphism. We first observe that every submodule of $\overline{R}/R$ is cyclic. Indeed, if H is an R-submodule of $\overline{R}$ and $H \supseteq R$, then H is isomorphic to an ideal of R, whence is generated by two elements, one of which can be chosen to be 1_R. Since $\overline{R}/R$ in particular is cyclic, it follows that

$\overline{R}/R \cong \overline{R}/(R :_R \overline{R}) = R/\mathfrak{c}$. Thus every submodule of $R/\mathfrak{c}$ is cyclic; but then $R/\mathfrak{c}$ is an Artinian principal ideal ring and hence $R/\mathfrak{c}$ has only finitely many ideals. Since $\overline{R}/R \cong R/\mathfrak{c}$, we see that there are only finitely many R-modules between R and $\overline{R}$.

Given a faithful ideal I of R, put $E = (I :_R I)$, the endomorphism ring of I. Then I is a projective E-module by Remark 4.4. Since E is semilocal, I is isomorphic to E as an E-module and therefore as an R-module. In particular, R has only finitely many faithful ideals up to R-isomorphism.

Suppose now that J is a non-zero unfaithful ideal; then R is not a domain. Notice that if R had more than two minimal primes $\mathfrak{p}_i$, the direct product of the $R/\mathfrak{p}_i$ would be an R-submodule of $\overline{R}$ requiring more than two generators. Therefore R has exactly two minimal prime ideals $\mathfrak{p}$ and $\mathfrak{q}$. Exercise 4.35 implies that J is a faithful ideal of either $R/\mathfrak{p}$ or $R/\mathfrak{q}$. Now $R/\mathfrak{p}$ and $R/\mathfrak{q}$ are discrete valuation rings: if, say, $R/\mathfrak{p}$ were properly contained in $\overline{R/\mathfrak{p}}$, then $\overline{R/\mathfrak{p}} \times R/\mathfrak{q}$ would need at least three generators as an R-module. Therefore there are, up to isomorphism, only two possibilities for J. $\square$

§6. Ranks of indecomposable MCM modules

Suppose $(R, \mathfrak{m}, k)$ is a reduced local ring of dimension one, and let $\mathfrak{p}_1, \ldots, \mathfrak{p}_s$ be the minimal prime ideals of R. Recall that the *rank* of a finitely generated R-module M is the s-tuple $\operatorname{rank}_R(M) = (r_1, \ldots, r_s)$, where r_i is the dimension of $M_{\mathfrak{p}_i}$ as a vector space over the field $R_{\mathfrak{p}_i}$. If R has finite CM type, it follows from (DR1) and Theorem A.29 that $s \leqslant \mathrm{e}(R) \leqslant 3$. There are universal bounds on the ranks of the indecomposable MCM R-modules, as R varies over one-dimensional reduced local rings with finite CM type. The precise ranks that occur have recently been worked out by Baeth and Luckas [**BL10**].

4.19. THEOREM. *Let $(R, \mathfrak{m})$ be a one-dimensional, analytically unramified local ring with finite CM type. Let $s \leqslant 3$ be the number of minimal prime ideals of R.*

  (i) *If R is a domain, then every indecomposable finitely generated torsion-free R-module has rank 1, 2, or 3.*
  (ii) *If $s = 2$, then the rank of every indecomposable finitely generated torsion-free R-module is among the following possibilities: $(0,1)$, $(1,0)$, $(1,1)$, $(1,2)$, $(2,1)$, or $(2,2)$.*
  (iii) *If $s = 3$, then one can choose a fixed ordering of the minimal prime ideals so that the rank of every indecomposable finitely*

generated torsion-free R-module is among the following possibilities: $(0,0,1)$, $(0,1,0)$, $(1,0,0)$, $(0,1,1)$, $(1,0,1)$, $(1,1,0)$, $(1,1,1)$, *or* $(2,1,1)$.

Moreover, there are examples showing that each of these possibilities actually occurs. □

The lack of symmetry in the last possibility is significant: One cannot have, for example, both an indecomposable of rank $(2,1,1)$ and one of rank $(1,2,1)$. An interesting consequence of the theorem is a universal bound on modules of constant rank, even in the non-local case. First we note the following local-global theorem [**WW94**]:

4.20. THEOREM (Wiegand and Wiegand). *Let R be a reduced ring of dimension one with finitely generated integral closure, let M be a finitely generated torsion-free R-module, and let r be a positive integer. If, for each maximal ideal $\mathfrak{m}$ of R, the $R_\mathfrak{m}$-module $M_\mathfrak{m}$ has a direct summand of constant rank r, then M has a direct summand of constant rank r.* □

4.21. COROLLARY. *Let R be a one-dimensional reduced ring with finitely generated integral closure. Assume that $R_\mathfrak{m}$ has finite Cohen-Macaulay type for each maximal ideal $\mathfrak{m}$ of R. Then every indecomposable finitely generated torsion-free R-module of* constant *rank has rank* 1, 2, 3, 4, 5 *or* 6. □

Theorem 4.19 and Corollary 4.21 correct an error in a 1994 paper of R. and S. Wiegand [**WW94**] where it was claimed that the sharp universal bounds were 4 in the local case and 12 in general.

If one allows non-constant ranks, there is no universal bound, even if one assumes that all localizations have multiplicity two [**Wie88**]. An interesting phenomenon is that in order to achieve rank $(r_1, \ldots, r_s)$ with *all* of the r_i large, one must have the ranks sufficiently spread out. For example [**BL10**, Theorem 5.5], if R has finite CM type locally and $n \geqslant 8$, every finitely generated torsion-free R module with local ranks between n and $2n-8$ has a direct summand of constant rank 6.

§7. Why MCM modules?

Looking ahead, one might wonder why we study modules that are maximal Cohen-Macaulay rather than, e.g., merely torsion-free. The reason is that, in higher dimensions, there are *always* big indecomposable torsion-free modules, so the question of finite CM type becomes uninteresting. This follows from another result due to Bass [**Bas62**].

4.22. THEOREM (Bass). *Let $(R, \mathfrak{m})$ be a Noetherian local domain of dimension $d \geqslant 2$. Then for each $r \geqslant 1$ there is an indecomposable torsion-free module of rank r.*

PROOF. It follows from the theory of Hilbert functions (see Appendix A, §2) that the function $n \mapsto \mu_R(\mathfrak{m}^n)$ is eventually given by a polynomial in n of degree $d - 1$ and with positive leading coefficient. Since $d - 1 \geqslant 1$, this function is unbounded, whence R has an ideal I with $\mu_R(I) = r + 1$. Write $I = (a_0, \ldots, a_r)$.

Let $\alpha = (a_0, \ldots, a_r) \in F = R^{(r+1)}$, let

$$L = \{x \in F \mid tx \in R\alpha \text{ for some non-zero } t \in R\},$$

and set $M = F/L$. We claim M is indecomposable. Suppose $M = U \oplus V$. Then $\mu_R(U) + \mu_R(V) = \mu_R(M) \leqslant r + 1$, and $\text{rank}(U) + \text{rank}(V) = \text{rank}(M) = r$. Since for all Z we have $\text{rank}(Z) \leqslant \mu_R(Z)$ with equality if and only if Z is free, it follows that at least one of U, V, M is free.

If M is free, then L is a proper direct summand of F and therefore is free of rank at most r (as R is local). But I is equal to the order ideal $\{f(\alpha) \mid f \in \text{Hom}_R(F, R)\}$, and since $\alpha \in L$ it follows that $\mu_R(I) \leqslant r$, a contradiction. Therefore we may assume that U is free. We have a surjection $f \colon F \longrightarrow U$ with $\alpha \in \ker f$. If $U \neq 0$, then $\ker f$ would be a proper free summand of F, and this would lead to a contradiction as before. $\square$

§8. Exercises

4.23. EXERCISE. Let $R = \mathbb{C}[x, y]_{(x,y)}/(y^2 - x^3 - x^2)$. Prove that the integral closure $\overline{R}$ is $R\left[\frac{y}{x}\right]$ and that $\overline{R}$ has two maximal ideals. Prove that the completion $\widehat{R} = \mathbb{C}[\![x, y]\!]/(y^2 - x^2 - x^3)$ has two minimal prime ideals. Show that the conductor square for R is

$$\begin{array}{ccc} R & \hookrightarrow & k[t]_U \\ \downarrow & & \downarrow \pi \\ k & \hookrightarrow & k \times k \\ & \Delta & \end{array}$$

where Δ is the diagonal embedding, U is a certain multiplicatively closed set, and the right-hand vertical map sends t to $(1, -1)$.

4.24. EXERCISE. Let R be a one-dimensional CM local ring with integral closure $\overline{R}$, and let M be a torsion-free R-module. Show that $\overline{R} \otimes_R M$ is torsion-free over $\overline{R}$ if and only if M is free.

4.25. EXERCISE. Let $c_1, \ldots, c_n$ be distinct real numbers, and let S be the subring of $\mathbb{R}[t]$ consisting of real polynomial functions f satisfying
$$f(c_1) = \cdots = f(c_n) \quad \text{and} \quad f^{(k)}(c_i) = 0$$
for all $i = 1, \ldots, n$ and $k = 1, \ldots, 3$, where $f^{(k)}$ denotes the kth derivative. Let S' be the semilocalization of S at the union of prime ideals $(t - c_1) \cup \cdots \cup (t - c_n)$. Let $\mathfrak{m} = \{f \in S \mid f(c_1) = 0\}$, and set $R = S_\mathfrak{m}$. Show that $\mathfrak{m}$ is a maximal ideal of S and that

$$\begin{array}{ccc} R & \hookrightarrow & S' \\ \downarrow & & \downarrow \pi \\ k & \hookrightarrow & k[t_1]/(t_1^4) \times \cdots \times k[t_n]/(t_n^4) \end{array}$$

is the conductor square for R.

4.26. EXERCISE. Let Λ be a ring (not necessarily commutative), and let M_1 and M_2 be Noetherian left Λ-modules. Suppose there exist surjective Λ-homomorphisms $M_1 \twoheadrightarrow M_2$ and $M_2 \twoheadrightarrow M_1$. Prove that $M_1 \cong M_2$. (Cf. Exercise 1.27.)

4.27. EXERCISE. Prove the "only if" direction of (iii) in Proposition 4.7. (Hint: Use the fact that any indecomposable R_{art}-module is weakly extended from R, and use KRS (Theorem 3.6). See Proposition 10.4 if you get stuck.)

4.28. EXERCISE ([**Wie94**, Lemma 4]). Let $(R, \mathfrak{m}, k)$ be a reduced local ring of dimension one satisfying (DR1) and (DR2). Assume that $\overline{R}$ has a maximal ideal $\mathfrak{n}$ such that $\ell = \overline{R}/\mathfrak{n}$ has degree 3 over k. Further, assume that R is not seminormal (equivalently, $\overline{R}/\mathfrak{c}$ is not reduced). Prove the following:
  (i) $\overline{R}$ is local and $\mathfrak{m}\overline{R} = \mathfrak{n}$.
  (ii) There is exactly one ring strictly between R and $\overline{R}$, namely $S = R + \mathfrak{n}$.
  (iii) R is Gorenstein.
  (iv) S is seminormal.

4.29. EXERCISE. Let $(R, \mathfrak{m})$ be a one-dimensional CM local ring which is not a discrete valuation ring. Let $\overline{R}$ be the integral closure of R in its total quotient ring K. Identify $E = \{c \in K \mid c\mathfrak{m} \subseteq \mathfrak{m}\}$ with $\text{End}_R(\mathfrak{m})$ via the isomorphism taking c to multiplication by c. Prove that $E \subseteq \overline{R}$ and that E contains R properly.

4.30. EXERCISE. Let $(R, \mathfrak{m}, k)$ be a one-dimensional reduced local ring for which $\overline{R}$ is generated by two elements as an R-module. Prove that R satisfies the second Drozd-Roĭter condition (DR2). (Hint: Pass to $R/\mathfrak{c}$ and count lengths carefully.)

4.31. EXERCISE. Let R be a commutative ring with total quotient ring $K = \{\text{non-zerodivisors}\}^{-1}R$.
- (i) Let M be an R-submodule of K. Assume that M contains a non-zerodivisor of R. Prove that $\mathrm{End}_R(M)$ is naturally identified with $\{\alpha \in K \mid \alpha M \subseteq M\}$, so that every endomorphism of M is given by multiplication by an element of K.
- (ii) ([**Wie94**, Lemma 1]) Suppose A and B are subrings of K with $R \subseteq A \cap B$. Prove that if A and B are isomorphic as R-modules then $A = B$.

4.32. EXERCISE. Let R be a reduced one-dimensional local ring. Suppose R has an infinite family of ideals $\{I_i\}$ that are pairwise non-isomorphic as R-modules. Prove that R has infinite CM type. (Hint: the *Goldie dimension* of R is the least integer s such that every ideal of R is a direct sum of at most s indecomposable ideals. Prove that $s < \infty$.)

4.33. EXERCISE. Prove Theorem 4.16.

4.34. EXERCISE ([**Bas63**, Theorem 6.4]). Let $(R, \mathfrak{m})$ be a CM local ring of dimension one, and suppose $\mathfrak{m}$ can be generated by two elements. Prove that R is Gorenstein.

4.35. EXERCISE. Let $(R, \mathfrak{m})$ be a reduced local ring of dimension one, and let M be a MCM R-module. Prove that $(0 :_R M)$ is the intersection of the minimal prime ideals $\mathfrak{p}$ for which $M_\mathfrak{p} \neq 0$.

CHAPTER 5

Invariant Theory

In this chapter we describe an abundant source of MCM modules coming from invariant theory. We consider finite subgroups G of the general linear group $\mathrm{GL}(n,k)$ with $|G|$ invertible in the field k, acting by linear changes of variable on the power series ring $S = k[\![x_1,\ldots,x_n]\!]$. The invariant subring $R = S^G$ is a complete local CM normal domain of dimension n, and comes equipped with a natural MCM module, namely the ring S considered as an R-module. The main goal of this chapter is a collection of one-one correspondences between:

(i) the indecomposable R-direct summands of S;
(ii) the indecomposable projective modules over the endomorphism ring $\mathrm{End}_R(S)$;
(iii) the indecomposable projective modules over the skew group ring $S\#G$; and
(iv) the irreducible k-representations of G.

We also introduce two directed graphs (quivers), the McKay quiver and the Gabriel quiver, associated with these data, and show that they are isomorphic.

In the next chapter we will specialize to the case $n = 2$, and show that in fact every indecomposable MCM R-module is a direct summand of S, so that the correspondences above classify all the MCM R-modules.

§1. The skew group ring

We begin with a little general invariant theory of arbitrary commutative rings, focusing on a central object: the skew group ring.

5.1. NOTATION. Fix the following notation for this section. Let S be an arbitrary commutative ring and $G \subseteq \mathrm{Aut}(S)$ a finite group of automorphisms of S. We always assume that $|G|$ is a unit in S. Let $R = S^G$ be the ring of invariants, so $s \in R$ if and only if $\sigma(s) = s$ for every $\sigma \in G$.

5.2. EXAMPLE. Two central examples are given by linear actions on polynomial and power series rings. Let k be a field and V a k-vector space of dimension n, with basis $x_1, \ldots, x_n$. Let G be a finite subgroup of $\mathrm{GL}(V) \cong \mathrm{GL}(n, k)$, acting naturally by linear changes of coordinates on V. We extend this action to monomials $x_1^{a_1} \cdots x_n^{a_n}$ multiplicatively, and then to all polynomials in $x_1, \ldots, x_n$ by linearity. This defines an action of G on the polynomial ring $k[x_1, \ldots, x_n]$. Extending the action of G to infinite sums in the obvious way, we obtain also an action on the power series ring $k[\![x_1, \ldots, x_n]\!]$. In either case we say that G acts on S via linear changes of variables.

It is an old result of Cartan [**Car57**] that when S is either the polynomial or the power series ring, we may assume that the action of an arbitrary subgroup $G \subseteq \mathrm{Aut}_k(S)$ is in fact linear. This is the first instance where the assumption that $|G|$ be invertible in S will be used; it will be essential throughout.

5.3. LEMMA (Cartan). *Let k be a field and let S be either the polynomial ring $k[x_1, \ldots, x_n]$ or the power series ring $k[\![x_1, \ldots, x_n]\!]$. Let $G \subseteq \mathrm{Aut}_k(S)$ a finite group of k-algebra automorphisms of S with $|G|$ invertible in k. Then there exists a finite group $G' \subseteq \mathrm{GL}(n, k)$, acting on S via linear changes of variables, such that $S^{G'} \cong S^G$.*

PROOF. Let $V = (x_1, \ldots, x_n)/(x_1, \ldots, x_n)^2$ be the vector space of linear forms of S. Then G acts on V, giving a group homomorphism $\varphi \colon G \longrightarrow \mathrm{GL}(V)$. Set $G' = \varphi(G)$, and extend the action of G' linearly to all of S by linear changes of variables as in Example 5.2.

Define a ring homomorphism $\theta \colon S \longrightarrow S$ by the rule

$$\theta(s) = \frac{1}{|G|} \sum_{\sigma \in G} \varphi(\sigma)^{-1} \sigma(s).$$

Since θ restricts to the identity on V, it is an automorphism of S. For an element $\tau \in G$, we have $\varphi(\tau) \circ \theta = \theta \circ \tau$ as automorphisms of S. Hence the actions of G and G' are conjugate, and it follows that $S^{G'} \cong S^G$. $\square$

Let S, G, and R be as in 5.1. The fact that $|G|$ is invertible allows us to define the *Reynolds operator* $\rho \colon S \longrightarrow R$ by sending $s \in S$ to the average of its orbit:

$$\rho(s) = \frac{1}{|G|} \sum_{\sigma \in G} \sigma(s).$$

Then ρ is R-linear, and it splits the inclusion $R \subseteq S$, thereby making R an R-direct summand of S. It follows (Exercise 5.27) that $IS \cap R = I$ for every ideal I of R, whence R is Noetherian, respectively local, respectively complete, if S is.

The extension $R \longrightarrow S$ is integral. Indeed, every element $s \in S$ is a root of the monic polynomial $\prod_{\sigma \in G}(x - \sigma(s))$, whose coefficients are elementary symmetric polynomials in the conjugates $\{\sigma(s)\}$, so are in R. In particular it follows that R and S have the same Krull dimension.

Suppose that S is a domain with quotient field K, and let F be the quotient field of R. Then G acts naturally on K, and by Exercise 5.28 the fixed field is F, so that K/F is a Galois extension with Galois group G.

If S is a Noetherian domain, then S is a finitely generated R-module. Since this fact seems not to be well-known in this generality, we give a proof here. We learned this argument from [**BD08**]. For the classical result that finite generation holds if S is a finitely generated algebra over a field, see Exercise 5.29.

5.4. PROPOSITION. *Let S be a Noetherian integral domain and let $G \subseteq \mathrm{Aut}(S)$ be a finite group with $|G|$ invertible in S. Set $R = S^G$. Then S is a finitely generated R-module of rank equal to $|G|$.*

PROOF. Let F and K be the quotient fields of R and S, respectively. The Reynolds operator $\rho \colon S \longrightarrow R$ extends to an operator $\rho \colon K \longrightarrow F$ defined by the same rule. (In fact ρ is nothing but a constant multiple of the usual trace from K to F.)

Fix a basis $\alpha_1, \ldots, \alpha_n$ for K over F. We may assume that $\alpha_i \in S$ for each i. Indeed, if $s/t \in K$ with s and t in S, we may multiply numerator and denominator by product of the distinct images of t under G to assume $t \in R$, then replace s/t by s without affecting the F-span.

By [**Lan02**, Corollary VI.5.3], there is a dual basis $\alpha'_1, \ldots, \alpha'_n$ such that $\rho(\alpha_i \alpha'_j) = \delta_{ij}$. Let M denote the R-module span of $\{\alpha'_1, \ldots, \alpha'_n\}$ in K. We claim that $S \subseteq M$, so that S is a submodule of a finitely generated R-module, hence is finitely generated.

Let $s \in S$, and write $s = \sum_i f_i \alpha'_i$ with $f_1, \ldots, f_n \in F$. It suffices to prove that $f_j \in R$ for each j. Note that since $\alpha_j \in S$ for each j, we have $\rho(s\alpha_j) \in R$ for $j = 1, \ldots, n$. But

$$\rho(s\alpha_j) = \sum_i f_i \alpha_i \alpha'_j = f_j$$

so that $S \subseteq M$, as claimed. The statement about the rank of S over R is immediate. $\square$

If in addition S is a normal domain then the same is true of R. Indeed, any element $\alpha \in F$ which is integral over R is also integral over S. Since S is integrally closed in K, we have $\alpha \in S \cap F = R$.

Finally, if S and R are local rings, then we have by Exercise 5.30 that depth $R \geqslant$ depth S. In particular R is CM if S is, and in this case S is a MCM R-module. (This statement, for example, is quite false if $|G|$ is divisible by char(k) [**Fog81**].)

We now introduce the skew, or twisted, group ring.

5.5. DEFINITION. Let S be a ring and $G \subseteq \mathrm{Aut}(S)$ a finite group of automorphisms with order invertible in S. Let $S\#G$ denote the *skew group ring* of S and G. As an S-module, $S\#G = \bigoplus_{\sigma \in G} S \cdot \sigma$ is free on the elements of G; the product of two elements $s \cdot \sigma$ and $t \cdot \tau$ is

$$(s \cdot \sigma)(t \cdot \tau) = s\sigma(t) \cdot \sigma\tau.$$

Thus moving σ past t "twists" the ring element.

5.6. REMARKS. In the notation of Definition 5.5, a left $S\#G$-module M is nothing but an S-module with a compatible action of G, in the sense that $\sigma(sm) = \sigma(s)\sigma(m)$ for all $\sigma \in G$, $s \in S$, $m \in M$. We have $\sigma(st) = \sigma(s)\sigma(t)$ for all s and t in S, and so S itself is a left $S\#G$-module. Of course $S\#G$ is also a left module over itself.

Similarly, an $S\#G$-linear map between left $S\#G$-modules is an S-module homomorphism $f \colon M \longrightarrow N$ respecting the action of G, so that $f(\sigma(m)) = \sigma(f(m))$. This allows us to define a left $S\#G$-module structure on $\mathrm{Hom}_S(M, N)$, when M and N are $S\#G$-modules, by $\sigma(f)(m) = \sigma(f(\sigma^{-1}(m)))$. It follows that an S-linear map $f \colon M \longrightarrow N$ between $S\#G$-modules is $S\#G$-linear if and only if it is invariant under the G-action. Indeed, if $\sigma(f) = f$ for all $\sigma \in G$, then $f(m) = \sigma(f(\sigma^{-1}(m)))$, so that $\sigma^{-1}(f(m)) = f(\sigma^{-1}(m))$ for all $\sigma \in G$. Concisely,

$$(5.1) \qquad \mathrm{Hom}_{S\#G}(M, N) = \mathrm{Hom}_S(M, N)^G.$$

Since the order of G is invertible in S, taking G-invariants of an $S\#G$-modules is an exact functor (Exercise 5.32). In particular, $-^G$ commutes with taking cohomology, so (5.1) extends to higher Exts:

$$(5.2) \qquad \mathrm{Ext}^i_{S\#G}(M, N) = \mathrm{Ext}^i_S(M, N)^G$$

for all $S\#G$-modules M and N and all $i \geqslant 0$. This has the following wonderful consequence, the easy proof of which we leave as an exercise.

5.7. PROPOSITION. *An $S\#G$-module M is projective if and only if it is projective as an S-module.* □

5.8. COROLLARY. *If S is a polynomial or power series ring in n variables, then the skew group ring $S\#G$ has global dimension equal to n.* □

We leave the proof of the corollary to the reader as well; the next example will no doubt be useful.

5.9. EXAMPLE. Set S be either the polynomial ring $k[x_1, \ldots, x_n]$ or the power series ring $k[\![x_1, \ldots, x_n]\!]$, with $G \subseteq \mathrm{GL}(n, k)$ acting by linear changes of variables as in Example 5.2. The Koszul complex $K_\bullet$ on the variables $\mathbf{x} = x_1, \ldots, x_n$ is a minimal $S\#G$-linear resolution of the residue field k of S (with trivial action of G). In detail, let $V = (x_1, \ldots, x_n)/(x_1, \ldots, x_n)^2$ be the k-vector space with basis $x_1, \ldots, x_n$, and
$$K_p = K_p(\mathbf{x}, S) = S \otimes_k \bigwedge^p V$$
for $p \geqslant 0$. The differential $\partial_p \colon K_p \longrightarrow K_{p-1}$ is given by
$$\partial_p(x_{i_1} \wedge \cdots \wedge x_{i_p}) = \sum_{j=1}^{p} (-1)^{j+1} x_{i_j} (x_{i_1} \wedge \cdots \wedge \widehat{x_{i_j}} \wedge \cdots \wedge x_{i_p}),$$
where $\{x_{i_1} \wedge \cdots \wedge x_{i_p}\}$, $1 \leqslant i_1 < i_2 < \cdots < i_p \leqslant n$, are the natural basis vectors for $\bigwedge^p V$. Since the x_i form an S-regular sequence, $K_\bullet$ is acyclic, minimally resolving k.

The exterior powers $\bigwedge^p V$ carry a natural action of G, by $\sigma(x_{i_1} \wedge \cdots \wedge x_{i_p}) = \sigma(x_{i_1}) \wedge \cdots \wedge \sigma(x_{i_p})$, and it's easy to see that the differentials ∂_p are $S\#G$-linear for this action. Since the modules appearing in $K_\bullet$ are free S-modules, they are projective over $S\#G$, so we see that $K_\bullet$ resolves the trivial module k over $S\#G$. Since every projective over $S\#G$ is free over S, the Depth Lemma then shows that $\mathrm{pd}_{S\#G} k$ cannot be any smaller than n.

5.10. REMARK. Let S and G be as in 5.1. The ring S sits inside $S\#G$ naturally via $S = S \cdot 1_G$. However, it also sits in a more symmetric fashion via a modified version of the Reynolds operator. Define $\widehat{\rho} \colon S \longrightarrow S\#G$ by
$$\widehat{\rho}(s) = \frac{1}{|G|} \sum_{\sigma \in G} \sigma(s) \cdot \sigma.$$
One checks easily that $\widehat{\rho}$ is an injective ring homomorphism, and that the image of $\widehat{\rho}$ is equal to $(S\#G)^G$, the fixed points of $S\#G$ under the left action of G. In particular, $\widehat{\rho}(1)$ is an idempotent of $S\#G$.

§2. The endomorphism algebra

The "twisted" multiplication on the skew group ring $S\#G$ is cooked up precisely so that the homomorphism
$$\gamma \colon S\#G \longrightarrow \mathrm{End}_R(S), \qquad \gamma(s \cdot \sigma)(t) = s\sigma(t),$$

is a ring homomorphism extending the group homomorphism $G \longrightarrow \operatorname{End}_R(S)$ that defines the action of G on S. In words, γ simply considers an element of $S\#G$ as an endomorphism of S.

In general, γ is neither injective nor surjective, even when S is a polynomial or power series ring. Under an additional assumption on the extension $R \longrightarrow S$, however, it is both, by a theorem due to Auslander [**Aus62**, Prop. 3.4]. We turn now to this additional assumption, explaining which will necessitate a brief detour through ramification theory. See Appendix B for the details.

Recall (Definition B.1) that a local homomorphism of local rings $(A, \mathfrak{m}, k) \longrightarrow (B, \mathfrak{n}, \ell)$ which is essentially of finite type is said to be *unramified* provided $\mathfrak{m}B = \mathfrak{n}$ and the induced homomorphism $A/\mathfrak{m} \longrightarrow B/\mathfrak{m}B$ is a finite separable field extension. Equivalently, the exact sequence

$$(5.3) \qquad 0 \longrightarrow \mathcal{J} \longrightarrow B \otimes_A B \xrightarrow{\mu} B \longrightarrow 0,$$

where $\mu \colon B \otimes_A B \longrightarrow B$ is the *diagonal map* defined by $\mu(b \otimes b') = bb'$ and $\mathcal{J}$ is the ideal of $B \otimes_A B$ generated by all elements of the form $b \otimes 1 - 1 \otimes b$, splits as $B \otimes_A B$-modules (this is Proposition B.9). We say that a ring homomorphism $A \longrightarrow B$ which is essentially of finite type is *unramified in codimension one* if the induced local homomorphism $A_{\mathfrak{q} \cap A} \longrightarrow B_{\mathfrak{q}}$ is unramified for every prime ideal $\mathfrak{q}$ of height one in B. If $A \longrightarrow B$ is module-finite, then it is equivalent to quantify over height-one primes in A.

In order to leverage codimension-one information to give a global conclusion, we will use a general lemma due to Auslander and Buchsbaum [**AB59**], which will reappear repeatedly in other contexts.

5.11. LEMMA. *Let A be a commutative Noetherian ring, and let $f \colon M \longrightarrow N$ be a homomorphism of finitely generated A-modules such that M satisfies the condition (S_2) and N satisfies (S_1). If $f_\mathfrak{p}$ is injective for every minimal prime ideal $\mathfrak{p}$ and surjective for every prime ideal $\mathfrak{p}$ of height 1, then f is an isomorphism.*

PROOF. Set $K = \ker f$ and $C = \operatorname{cok} f$. If $K \neq 0$, choose $\mathfrak{p} \in \operatorname{Ass} K$. Then $\mathfrak{p}$ has height at least 1, so depth $M_\mathfrak{p} \geqslant 1$. Since depth $K_\mathfrak{p} = 0$, the inclusion $K_\mathfrak{p} \hookrightarrow M_\mathfrak{p}$ gives a contradiction. Therefore $K = 0$, and the sequence

$$(5.4) \qquad 0 \longrightarrow M \xrightarrow{f} N \longrightarrow C \longrightarrow 0$$

is exact. If $C \neq 0$, choose $\mathfrak{p} \in \operatorname{Ass} C$, and localize (5.4) at $\mathfrak{p}$. Since $\mathfrak{p}$ has height at least two, we know that depth $N_\mathfrak{p} \geqslant 1$ and depth $M_\mathfrak{p} \geqslant 2$. But depth $C_\mathfrak{p} = 0$, and this contradicts the Depth Lemma. $\square$

§2. THE ENDOMORPHISM ALGEBRA

5.12. THEOREM (Auslander). *Let $(S, \mathfrak{n})$ be a normal domain and let G be a finite subgroup of $\mathrm{Aut}(S)$ with order invertible in S. Set $R = S^G$. If $R \longrightarrow S$ is unramified in codimension one, then the ring homomorphism $\gamma \colon S\#G \longrightarrow \mathrm{End}_R(S)$ defined by $\gamma(s \cdot \sigma)(t) = s\sigma(t)$ is an isomorphism.*

PROOF. Since $S\#G$ is isomorphic to a direct sum of copies of S as an S-module, it satisfies (S_2) over R. The endomorphism ring $\mathrm{End}_R(S)$ has depth at least $\min\{2, \mathrm{depth}\, S\}$ over each localization of R by Exercise 5.37, so satisfies (S_1). Thus by Lemma 5.11 it suffices to prove that γ is an isomorphism in height one. At height one primes, the extension is unramified, so we may assume for the proof that $R \longrightarrow S$ is unramified.

The strategy of the proof is to define a right splitting $\mathrm{End}_R(S) \longrightarrow S\#G$ for $\gamma \colon S\#G \longrightarrow \mathrm{End}_R(S)$ based on the diagram below.

(5.5)
$$\begin{array}{ccc} S\#G & \xrightarrow{\gamma} & \mathrm{End}_R(S) \\ \widetilde{\mu} \uparrow & & \downarrow {f \mapsto f \otimes \widehat{\rho}} \\ S \otimes_R (S\#G) & \xleftarrow{\mathrm{ev}_\epsilon} & \mathrm{Hom}_S(S \otimes_R S, S \otimes_R (S\#G)) \end{array}$$

We now define each of the arrows in (5.5) in turn. Recall from Remark 5.10 that the homomorphism

$$\widehat{\rho} \colon S \longrightarrow S\#G, \qquad \widehat{\rho}(s) = \frac{1}{|G|} \sum_{\sigma \in G} \sigma(s) \cdot \sigma$$

embeds S as the fixed points $(S\#G)^G$ of $S\#G$. Thus $- \otimes \widehat{\rho}$ defines the right-hand vertical arrow in (5.5).

Since we assume $R \longrightarrow S$ is unramified, the short exact sequence

(5.6) $$0 \longrightarrow \mathcal{J} \longrightarrow S \otimes_R S \xrightarrow{\mu} S \longrightarrow 0$$

splits as $S \otimes_R S$-modules, where again $\mu \colon S \otimes_R S \longrightarrow S$ is the diagonal map and $\mathcal{J}$ is generated by all elements of the form $s \otimes 1 - 1 \otimes s$ for $s \in S$. Tensoring (5.6) on the right with $S\#G$ thus gives another split exact sequence

(5.7) $$0 \longrightarrow \mathcal{J} \otimes_S (S\#G) \longrightarrow S \otimes_R (S\#G) \xrightarrow{\widetilde{\mu}} S\#G \longrightarrow 0$$

with $\widetilde{\mu}(t \otimes s \cdot \sigma) = ts \cdot \sigma \in S\#G$ defining the left-hand vertical arrow in (5.5).

Let $j \colon S \longrightarrow S \otimes_R S$ be a splitting for (5.6), and set $\epsilon = j(1)$. Then $\mu(\epsilon) = 1$ and

(5.8) $$(1 \otimes s - s \otimes 1)\epsilon = 0$$

for all $s \in S$. Evaluation at $\epsilon \in S \otimes_R S$ defines
$$\mathrm{ev}_\epsilon \colon \mathrm{Hom}_S(S \otimes_R S, S \otimes_R (S\#G)) \longrightarrow S \otimes_R (S\#G),$$
the bottom row of the diagram. Now we show that for an arbitrary $f \in \mathrm{End}_R(S)$, we have
$$\gamma\left(\widetilde{\mu}\left(\mathrm{ev}_\epsilon\left(f \otimes \widehat{\rho}\right)\right)\right) = \frac{1}{|G|} f.$$

Write $\epsilon = \sum_i x_i \otimes y_i$ for some elements $x_i, y_i \in S$. We claim first that
$$\sum_i x_i \sigma(y_i) = \begin{cases} 1 & \text{if } \sigma = 1_G, \text{ and} \\ 0 & \text{otherwise.} \end{cases}$$

To see this, note that
$$(s \otimes 1)\left(\sum_i x_i \otimes y_i\right) = (1 \otimes s)\left(\sum_i x_i \otimes y_i\right)$$
for every $s \in S$ by (5.8). Apply the endomorphism $1 \otimes \sigma$ to both sides, obtaining
$$\sum_i s x_i \otimes \sigma(y_i) = \sum_i x_i \otimes \sigma(s)\sigma(y_i).$$

Collapse the tensor products with $\mu \colon S \otimes_R S \longrightarrow S$, and factor each side, getting
$$s\left(\sum_i x_i \sigma(y_i)\right) = \sigma(s)\left(\sum_i x_i \sigma(y_i)\right).$$

This holds for every $s \in S$, so that either $\sigma = 1_G$ or $\sum_i x_i \sigma(y_i) = 0$, proving the claim.

Now fix $f \in \mathrm{End}_R(S)$ and $s \in S$. Unravelling all the definitions, we find
$$\gamma\left[\widetilde{\mu}\left[(f \otimes \widehat{\rho})(\epsilon)\right]\right](s) = \gamma\left[\widetilde{\mu}\left[(f \otimes \widehat{\rho})\left(\sum_i x_i \otimes y_i\right)\right]\right](s)$$
$$= \gamma\left[\widetilde{\mu}\left(\sum_i f(x_i) \otimes \widehat{\rho}(y_i)\right)\right](s)$$
$$= \gamma\left[\left(\sum_i f(x_i)\widehat{\rho}(y_i)\right)\right](s)$$
$$= \gamma\left[\left(\sum_i f(x_i)\left(\frac{1}{|G|}\sum_\sigma \sigma(y_i) \cdot \sigma\right)\right)\right](s)$$
$$= \frac{1}{|G|}\sum_i f(x_i)\left(\sum_\sigma \sigma(y_i)\sigma(s)\right).$$

Now, since the sum over σ is fixed by G, it lives in R, so

$$= \frac{1}{|G|} f\left(\sum_i x_i \left(\sum_\sigma \sigma(y_i)\sigma(s)\right)\right)$$

$$= \frac{1}{|G|} f\left(\sum_\sigma \left(\sum_i x_i \sigma(y_i)\right) \sigma(s)\right)$$

$$= \frac{1}{|G|} f\left(\sum_i x_i y_i s\right)$$

by the claim. By the definition of $\epsilon = \sum x_i \otimes y_i$, this last expression is equal to $\frac{1}{|G|} f(s)$, as desired. Therefore $\gamma \colon S\#G \longrightarrow \operatorname{End}_R(S)$ is a split surjection. Since both source and target of γ are torsion-free R-modules of rank equal to $|G|^2$, this forces γ to be an isomorphism. □

When S is a polynomial or power series ring and $G \subseteq \operatorname{GL}(n,k)$ acts linearly, the ramification of $R \longrightarrow S$ is explained by the presence of *pseudo-reflections* in G.

5.13. DEFINITION. Let k be a field. An element $\sigma \in \operatorname{GL}(n,k)$ of finite order is called a *pseudo-reflection* provided the fixed subspace $V^\sigma = \{v \in k^{(n)} \mid \sigma(v) = v\}$ has codimension one in $k^{(n)}$. Equivalently, $\sigma - I_n$ has rank 1. We say a subgroup $G \subseteq \operatorname{GL}(n,k)$ is *small* if it contains no pseudo-reflections.

If a pseudo-reflection σ is diagonalizable, then σ is similar to a diagonal matrix with diagonal entries $1, \ldots, 1, \lambda$ with $\lambda \neq 1$ a root of unity. In fact, one can show (Exercise 5.38) that a pseudo-reflection with order prime to $\operatorname{char}(k)$ is necessarily diagonalizable.

The importance of pseudo-reflections in invariant theory is highlighted by the foundational theorem of Chevalley-Shephard-Todd (Theorem B.27), which says that, in the case $S = k[\![x_1, \ldots, x_n]\!]$, the invariant ring $R = S^G$ is a regular local ring if and only if G is generated by pseudo-reflections. More relevant for our purposes, pseudo-reflections control the "large ramification" of invariant rings. We banish the proof of this fact to the Appendix (Theorem B.29).

5.14. THEOREM. *Let k be a field and let $G \subseteq \operatorname{GL}(n,k)$ be a finite group with order invertible in k. Let S be either the polynomial ring $k[x_1, \ldots, x_n]$ or the power series ring $k[\![x_1, \ldots, x_n]\!]$, with G acting by linear changes of variables. Set $R = S^G$. Then the extension $R \longrightarrow S$ is unramified in codimension one if and only if G is small.* □

In fact, by a theorem of Prill, we could always assume that G is small. Specifically, we may replace S and G by another power series ring S' (possibly with fewer variables) and finite group G', respectively, so that G' is small and $S'^{G'} \cong S^G$. See Proposition B.30 for this, which we will not use in this chapter.

In view of Theorem 5.14, we can restate Theorem 5.12 as follows for linear actions.

5.15. THEOREM. *Let k be a field and let $G \subseteq \mathrm{GL}(n,k)$ be a finite group with order invertible in k. Let S be either the polynomial ring $k[x_1, \ldots, x_n]$ or the power series ring $k[\![x_1, \ldots, x_n]\!]$, with G acting by linear changes of variables. Set $R = S^G$. If G contains no pseudo-reflections, then the natural homomorphism $\gamma \colon S\#G \longrightarrow \mathrm{End}_R(S)$ is an isomorphism.*

5.16. COROLLARY. *With notation as in Theorem 5.15, assume that G contains no pseudo-reflections. Then we have ring isomorphisms*

$$S\#G \xrightarrow{\iota} (S\#G)^{op} \xrightarrow{\nu} \mathrm{End}_{S\#G}(S\#G) \xrightarrow{\mathrm{res}} \mathrm{End}_R(S)$$

where $\iota(s \cdot \sigma) = \sigma^{-1}(s) \cdot \sigma^{-1}$, $\nu(s \cdot \sigma)(t \cdot \tau) = (t \cdot \tau)(s \cdot \sigma)$, and res is restriction to the subring $\widehat{\rho}(S) = (S\#G)^G$. The composition of these maps is the isomorphism γ. These isomorphisms induce one-one correspondences between

(i) the indecomposable direct summands of S as an R-module;
(ii) the indecomposable direct summands of $\mathrm{End}_R(S)$ as a left $\mathrm{End}_R(S)$-module; and
(iii) the indecomposable direct summands of $S\#G$ as a left $S\#G$-module.

Explicitly, if $P_0, \ldots, P_d$ are the indecomposable summands of $S\#G$, then P_j^G, for $j = 0, \ldots, d$, are the direct summands of S as an R-module. They are in particular MCM R-modules.

PROOF. It's easy to check that ι and ν are isomorphisms, and that the composition $\mathrm{res} \circ \nu \circ \iota$ is equal to γ. The primitive idempotents of $\mathrm{End}_R(S)$ correspond both to the indecomposable R-direct summands of S and to the indecomposable $\mathrm{End}_R(S)$-projectives, while those of $\mathrm{End}_{S\#G}(S\#G)$ correspond to the indecomposable projective $S\#G$-modules. The fact that $(S\#G)^G = S$ implies the penultimate statement, and the fact that S is MCM over R was observed already. □

We have not yet shown that the indecomposable direct summands of $S\#G$ as an $S\#G$-module are all the indecomposable projective $S\#G$-modules. This will follow from the first result of the next section, where we prove that projective modules over $S\#G$ (and hence over $\mathrm{End}_R(S)$) satisfy KRS when S is complete.

§3. Group representations and the McKay-Gabriel quiver

The module theory of the skew group ring $S\#G$, where the subgroup G of the general linear group $\mathrm{GL}(n,k)$ acts linearly on the power series ring S, faithfully reflects the representation theory of G. In this section we make this assertion precise.

Throughout the section, we consider linear group actions on power series rings, so that $G \subseteq \mathrm{GL}(n,k)$ is a finite group of order relatively prime to the characteristic of k, acting on $S = k[\![x_1, \ldots, x_n]\!]$ by linear changes of variables, with invariant ring $R = S^G$. Let $S\#G$ be the skew group ring.

5.17. DEFINITION. Let M be an $S\#G$-module and W a k-representation of G, that is, a module over the group algebra kG. Define an $S\#G$-module structure on $M \otimes_k W$ by the diagonal action
$$s\sigma(m \otimes w) = s\sigma(m) \otimes \sigma(w).$$

Define a functor $\mathcal{F}$ from the category of finite-dimensional k-representations W of G to that of finitely generated $S\#G$-modules by
$$\mathcal{F}(W) = S \otimes_k W$$
and similarly for homomorphisms. For any W, $\mathcal{F}(W)$ is obviously a free S-module and thus a projective $S\#G$-module.

In the opposite direction, let P be a finitely generated projective $S\#G$-module. Then $P/\mathfrak{n}P$ is a finite-dimensional k-vector space with an action of G, that is, a k-representation of G. Define a functor $\mathcal{G}$ from projective $S\#G$-modules to k-representations of G by
$$\mathcal{G}(P) = P/\mathfrak{n}P$$
and correspondingly on homomorphisms.

5.18. PROPOSITION. *The functors $\mathcal{F}$ and $\mathcal{G}$ form an adjoint pair, that is,*
$$\mathrm{Hom}_{kG}(\mathcal{G}(P), W) = \mathrm{Hom}_{S\#G}(P, \mathcal{F}(W)),$$
and are inverses of each other on objects. Concretely, for a projective $S\#G$-module P and a k-representation W of G, we have
$$S \otimes_k P/\mathfrak{n}P \cong P$$
and
$$(S \otimes_k W)/\mathfrak{n}(S \otimes_k W) \cong W.$$
In particular, there is a one-one correspondence between isomorphism classes of indecomposable projective $S\#G$-modules and irreducible k-representations of G.

PROOF. It is clear that $\mathcal{G}(\mathcal{F}(W)) \cong W$, since
$$(S \otimes_k W)/\mathfrak{n}(S \otimes_k W) \cong S/\mathfrak{n} \otimes_k W \cong W.$$
To show that the other composition is also the identity, let P be a projective $S\#G$-module. Then $\mathcal{F}(\mathcal{G}(P)) = S \otimes_k P/\mathfrak{n}P$ is a projective $S\#G$-module, with a natural projection onto $P/\mathfrak{n}P$. Of course, the original projective P also maps onto $P/\mathfrak{n}P$. This latter is in fact a projective cover of $P/\mathfrak{n}P$ (since idempotents in kG lift to $S\#G$ via the retraction $kG \longrightarrow S\#G \longrightarrow kG$). There is thus a lifting $S \otimes_k P/\mathfrak{n}P \longrightarrow P$, which is surjective modulo $\mathfrak{n}P$. NAK then implies that the lifting is surjective, so split, as P is projective. Comparing ranks over S, we must have $S \otimes_k P/\mathfrak{n}P \cong P$. $\square$

5.19. COROLLARY. *Let $V_0, \ldots, V_d$ be a complete set of pairwise non-isomorphic irreducible kG-modules. Then*
$$S \otimes_k V_0, \ldots, S \otimes_k V_d$$
is a complete set of non-isomorphic indecomposable finitely generated projective $S\#G$-modules. Furthermore, the category of finitely generated projective $S\#G$-modules satisfies the KRS property, i.e. each finitely generated projective P is isomorphic to a unique direct sum $\bigoplus_{i=0}^{d}(S \otimes_k V_i)^{(n_i)}$. $\square$

Putting together the one-one correspondences obtained so far, we have

5.20. COROLLARY. *Let k be a field, let $S = k[\![x_1, \ldots, x_n]\!]$, and let $G \subseteq \mathrm{GL}(n, k)$ be a finite group acting linearly on S without pseudo-reflections and such that $|G|$ is invertible in k. Then there are one-one correspondences between*

(i) *the indecomposable direct summands of S as an R-module;*
(ii) *the indecomposable finitely generated projective left $\mathrm{End}_R(S)$-modules;*
(iii) *the indecomposable finitely generated projective left $S\#G$-modules; and*
(iv) *the irreducible kG-modules.*

The correspondence between the first and last items is induced by $W \mapsto (S \otimes_k W)^G$ for a k-representation W of G.

Explicitly, if $V_0, \ldots, V_d$ are the non-isomorphic irreducible representations of G over k, then the modules of covariants
$$M_j = (S \otimes_k V_j)^G, \qquad j = 0, \ldots, d$$
are the indecomposable R-direct summands of S. They are in particular MCM R-modules. Furthermore, we have $\mathrm{rank}_R M_j = \dim_k V_j$. $\square$

§3. GROUP REPRESENTATIONS AND THE MCKAY-GABRIEL QUIVER

The one-one correspondence between projectives, representations, and certain MCM modules obtained so far extends to an isomorphism of two graphs naturally associated to these data, as we now explain. We will meet a third incarnation of these graphs in Chapter 13.

We keep all the notation established so far in this section, and additionally let $V_0, \ldots, V_d$ be a complete set of non-isomorphic irreducible k-representations of G, with V_0 the trivial representation k. The given linear action of G on S is induced from an n-dimensional representation of G on the space $V = \mathfrak{n}/\mathfrak{n}^2$ of linear forms.

5.21. DEFINITION. The *McKay quiver* of $G \subseteq \mathrm{GL}(V)$ has
- vertices $V_0, \ldots, V_d$, and
- m_{ij} arrows $V_i \longrightarrow V_j$ if the multiplicity of V_i in an irreducible decomposition of $V \otimes_k V_j$ is equal to m_{ij}.

In case k is algebraically closed, the multiplicities m_{ij} in the McKay quiver can also be computed from the characters $\chi, \chi_0, \ldots, \chi_d$ for the representations $V, V_0, \ldots, V_d$; see [**FH91**, 2.10]:

$$m_{ij} = \langle \chi_i, \chi\chi_j \rangle = \frac{1}{|G|} \sum_{\sigma \in G} \chi_i(\sigma)\chi(\sigma^{-1})\chi_j(\sigma^{-1}).$$

For each $i = 0, \ldots, d$, we set $P_i = S \otimes_k V_i$, the corresponding indecomposable projective $S\#G$-module. Then in particular $P_0 = S \otimes_k V_0 = S$, and $\{P_0, \ldots, P_d\}$ is a complete set of non-isomorphic indecomposable projective $S\#G$-modules by Proposition 5.18. The V_j are simple $S\#G$-modules via the surjection $S\#G \longrightarrow kG$, with minimal projective cover P_j. Since $\mathrm{pd}_{S\#G} V_j \leqslant n$ by Proposition 5.7, the minimal projective resolution of V_j over $S\#G$ thus has the form

$$0 \longrightarrow Q_{j,n} \longrightarrow Q_{j,n-1} \longrightarrow \cdots \longrightarrow Q_{j,1} \longrightarrow P_j \longrightarrow V_j \longrightarrow 0$$

with projective $S\#G$-modules $Q_{j,i}$ for $i = 1, \ldots, n$ and $j = 0, \ldots, d$.

5.22. DEFINITION. The *Gabriel quiver* of $G \subseteq \mathrm{GL}(V)$ has
- vertices $P_0, \ldots, P_d$, and
- m_{ij} arrows $P_i \longrightarrow P_j$ if the multiplicity of P_i in $Q_{j,1}$ is equal to m_{ij}.

5.23. THEOREM (Auslander). *The McKay quiver and the Gabriel quiver of R are isomorphic directed graphs.*

PROOF. First consider the trivial module $V_0 = k$. The minimal $S\#G$-resolution of k was computed in Example 5.9; it is the Koszul complex

$$K_\bullet: \quad 0 \longrightarrow S \otimes_k \bigwedge^n V \longrightarrow \cdots \longrightarrow S \otimes_k V \longrightarrow S \longrightarrow 0.$$

To obtain the minimal $S\#G$-resolution of V_j, we simply tensor the Koszul complex with V_j over k, obtaining the following:

$$0 \longrightarrow S\otimes_k \left(\bigwedge^n V \otimes_k V_j\right) \longrightarrow \cdots \longrightarrow S\otimes_k(V\otimes_k V_j) \longrightarrow S\otimes_k V_j \longrightarrow 0$$

This displays $Q_{j,1} = S\otimes_k (V \otimes_k V_j)$, so that the multiplicity of P_i in $Q_{j,1}$ is equal to that of V_i in $V \otimes_k V_j$. □

5.24. EXAMPLE. Take $n = 3$, and write $S = k[\![x,y,z]\!]$. Let $G = \mathbb{Z}/2\mathbb{Z}$, with the generator acting on $V = kx \oplus ky \oplus kz$ by negating each variable. Then $R = S^G = k[\![x^2, xy, xz, y^2, yz, z^2]\!]$. There are only two irreducible representations of G, namely the trivial representation k and its negative, which is isomorphic to the inverse determinant representation $V_1 = \det(V)^{-1} = \bigwedge^3 V^*$. The Koszul complex

$$0 \longrightarrow S \otimes \bigwedge^3 V \longrightarrow S \otimes_k \bigwedge^2 V \longrightarrow S \otimes_k V \longrightarrow S \longrightarrow 0$$

resolves k, while the tensor product

$$0 \longrightarrow S \otimes_k \left(\bigwedge^3 V \otimes_k \bigwedge^3 V^*\right) \longrightarrow S \otimes_k \left(\bigwedge^2 V \otimes_k \bigwedge^3 V^*\right)$$
$$\longrightarrow S \otimes_k \left(V \otimes_k \bigwedge^3 V^*\right) \longrightarrow S \otimes_k \bigwedge^3 V^* \longrightarrow 0$$

is canonically isomorphic to

$$0 \longrightarrow S \longrightarrow S \otimes_k V^* \longrightarrow S \otimes_k \bigwedge^2 V^* \longrightarrow S \otimes_k \bigwedge^3 V^* \longrightarrow 0.$$

Since the given representation $V = \left(\bigwedge^3 V^*\right)^{(3)}$ is just 3 copies of V_1, we obtain the McKay quiver

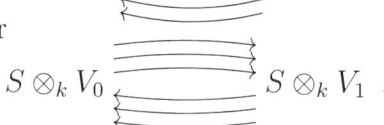

or the Gabriel quiver

Taking fixed points as specified in Corollary 5.20, we find MCM modules
$$M_0 \cong R \quad \text{and} \quad M_1 = (S \otimes_k V_1)^G.$$
Since V_1 is the negative of the trivial representation, the fixed points of $S \otimes_k V_1$, with the diagonal action, are generated over R by those elements $f \otimes \alpha$ such that $\sigma(f) = -f$. These are generated by the linear forms of S, so that M_1 is the submodule of S generated by

§3. GROUP REPRESENTATIONS AND THE MCKAY-GABRIEL QUIVER

(x, y, z). This is isomorphic to the ideal (x^2, xy, xz) of R. In particular we recover the obvious R-direct sum decomposition $S = R \oplus R(x, y, z)$ of S.

Observe that M_0 and M_1 are not the only indecomposable MCM R-modules, even though it turns out that R does have finite CM type; see Example 16.4.

From now on, we draw the McKay quiver for a group G, and refer to it as the McKay-Gabriel quiver.

5.25. EXAMPLE. Let $n = 2$ now, and write $S = k[\![u, v]\!]$. Let $r \geqslant 2$ be an integer not divisible by $\mathrm{char}(k)$, and choose $0 < q < r$ with $(q, r) = 1$. Take $G = \langle g \rangle \cong \mathbb{Z}/r\mathbb{Z}$ to be the cyclic group of order r generated by

$$g = \begin{pmatrix} \zeta_r & 0 \\ 0 & \zeta_r^q \end{pmatrix} \in \mathrm{GL}(2, k),$$

where ζ_r is a primitive r^{th} root of unity. Let $R = k[\![u, v]\!]^G$ be the corresponding ring of invariants, so that R is generated by the monomials $u^a v^b$ satisfying $a + bq \equiv 0 \bmod r$.

As G is Abelian, it has exactly r irreducible representations, each of which is one-dimensional. We label them $V_0, \ldots, V_{r-1}$, where the generator g is sent to ζ_r^i in V_i. The given representation V of G is isomorphic to $V_1 \oplus V_q$, so that for any j we have

$$V \otimes_k V_j \cong V_{j+1} \oplus V_{j+q},$$

where the indices are of course to be taken modulo r. The corresponding MCM R-modules are $M_j = (S \otimes_k V_j)^G$, each of which is an R-submodule of S:

$$M_j = R\left(u^a v^b \mid a + qb \equiv -j \mod r\right).$$

The general picture is a bit chaotic, so here are a few particular examples.

Take $r = 5$ and $q = 3$. Then $R = k[\![u^5, u^2v, uv^3, v^5]\!]$. The McKay-Gabriel quiver takes the following shape.

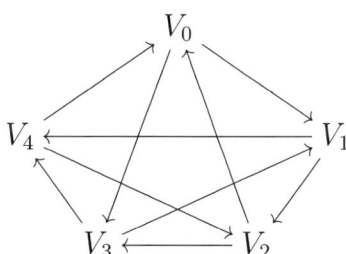

The corresponding indecomposable MCM R-modules appearing as R-direct summands of S are the ideals

$$M_0 = R$$
$$M_1 = R(u^4, uv, v^3) \cong (u^5, u^2v, uv^3)$$
$$M_2 = R(u^3, v) \cong (u^5, u^2v)$$
$$M_3 = R(u^2, uv^2, v^4) \cong (u^5, u^4v^2, u^3v^4)$$
$$M_4 = R(u, v^2) \cong (u^5, u^4v^2).$$

For another example, take $r = 8$ and $q = 5$, so that we obtain $R = k[\![u^8, u^3v, uv^3, v^8]\!]$. The McKay-Gabriel quiver looks like

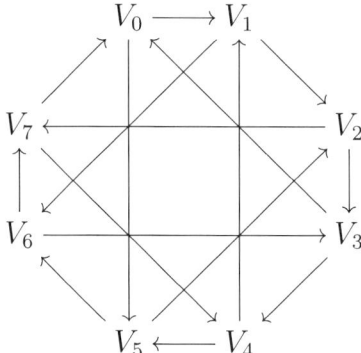

and the indecomposable MCM R-modules arising as direct summands of S are

$$M_0 = R$$
$$M_1 = R(u^7, u^2v, v^3) \cong (u^8, u^3v, uv^3)$$
$$M_2 = R(u^6, uv, v^6) \cong (u^8, u^3v, u^2v^6)$$
$$M_3 = R(u^5, v) \cong (u^8, u^3v)$$
$$M_4 = R(u^4, u^2v^2, v^4) \cong (u^8, u^6v^2, u^4v^4)$$
$$M_5 = R(u^3, uv^2, v^7) \cong (u^8, u^6v^2, u^5v^7)$$
$$M_6 = R(u^2, u^5v, v^2) \cong (u^2v^6, u^5v^7, v^8)$$
$$M_7 = R(u, v^5) \cong (uv^3, v^8).$$

Finally, take $r = n+1$ arbitrary, and $q = n$. Then

$$R = k[\![u^{n+1}, uv, v^{n+1}]\!] \cong k[\![x, y, z]\!]/(xz - y^{n+1})$$

is isomorphic to an (A_n) hypersurface singularity (see the next chapter). There are $n+1$ irreducible representations $V_0, \ldots, V_n$, and the McKay-Gabriel quiver looks like the one below.

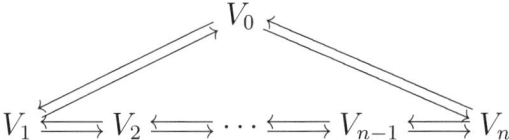

The non-free indecomposable MCM R-modules take the form
$$M_j = R\left(u^a v^b \mid b - a \equiv j \mod n+1\right)$$
for $j = 1, \ldots, n$. They have presentation matrices over $k[\![u^{n+1}uv, v^{n+1}]\!]$
$$\varphi_j = \begin{pmatrix} (uv)^{n+1-j} & -u^{n+1} \\ -v^{n+1} & (uv)^j \end{pmatrix}$$
or over $k[\![x, y, z]\!]/(xz - y^{n+1})$
$$\varphi_j = \begin{pmatrix} y^{n+1-j} & -x \\ -z & y^j \end{pmatrix}.$$

§4. Exercises

5.26. EXERCISE. Let k be the field with two elements, and define $\sigma\colon k[\![x]\!] \longrightarrow k[\![x]\!]$ by $x \mapsto \frac{x}{x+1} = x + x^2 + x^3 + \cdots$. What is the fixed ring of σ?

5.27. EXERCISE. Let $R \subseteq S$ be an extension of rings with an *algebra retraction*, that is, a ring homomorphism $S \longrightarrow R$ that restricts to the identity on R. Prove that $IS \cap R = I$ for every ideal I of R. Conclude that if S is Noetherian, or local, or complete, then the same holds for R. (Hint for completeness: If $\{x_i\}$ is a Cauchy sequence in R converging to $x \in S$, apply the Krull Intersection Theorem to $\sigma(x) - x_i$.)

5.28. EXERCISE. Let S be an integral domain with an action of a group $G \subseteq \mathrm{Aut}(S)$, and set $R = S^G$. Let F and K be the quotient fields of R and S, respectively. Prove that any element of K can be written as a fraction with denominator in R, and conclude that $K^G = F$.

5.29. EXERCISE. Suppose that S is a finitely generated algebra over a field k, let $G \subseteq \mathrm{Aut}_k(S)$ be a finite group, and set $R = S^G$. Prove that S is finitely generated as an R-module, and is finitely generated over k. (Hint: Let $A \subseteq S$ be the k-subalgebra generated by the coefficients of the monic polynomials satisfied by the generators of S, and prove that S is finitely generated over A. This argument goes back to Noether [**Noe15**].)

5.30. EXERCISE. Let S be a local ring, $G \subseteq \operatorname{Aut}(S)$ a finite group with order invertible in S, and $R = S^G$.
  (i) If $I \subseteq S$ is a G-stable ideal, prove that $(S/I)^G = R/(I \cap R)$.
  (ii) For $s \in S$ define the *norm* of s by $N(s) = \prod_{\sigma \in G} \sigma(s)$. Notice that $N(s) \in R$. If s is a non-zerodivisor in S, show that $N(s)S \cap R = N(s)R$.
  (iii) Use part (ii) to prove by induction on depth S that depth $R \geqslant$ depth S.

5.31. EXERCISE. Find an example of a non-CM local ring S and finite group acting such that the fixed ring R is CM. (There is an example with S one-dimensional complete local and R regular.)

5.32. EXERCISE. Let S and G be as in 5.1. Show that the fixed-point functor $-^G$ on $S\#G$-modules is exact. (Hint: left-exactness is easy. For right-exactness, take the average of the orbit of any preimage.)

5.33. EXERCISE. Let S be as in 5.1 and let M be an S-module. For each $\sigma \in G$, let $^\sigma M$ be the S-module with the same underlying Abelian group as M, and structure given by $s \cdot m = \sigma(s)m$. Prove that $S\#G \otimes_S M \cong \bigoplus_{\sigma \in G} {}^\sigma M$.

5.34. EXERCISE. Prove that in the situation of Proposition 5.7, a finitely generated $S\#G$-module M is projective if and only if it is projective as an S-module. Conclude that if S is regular of dimension d, then $S\#G$ has global dimension d.

5.35. EXERCISE. Set $R = k[\![x, y, z]\!]/(xy)$, the two-dimensional (A_∞) complete hypersurface singularity and let $\mathbb{Z}/r\mathbb{Z}$ act on R by letting the generator take (x, y, z) to $(x, \zeta_r y, \zeta_r z)$, where ζ_r is a primitive r^{th} root of unity. Give a presentation for the ring of invariants R^G. (Cf. Example 14.25.)

5.36. EXERCISE. Set $R = k[\![x, y, z]\!]/(x^2 y - z^2)$, the two-dimensional (D_∞) complete hypersurface singularity. Let $r = 2m + 1$ be an odd integer and let $\mathbb{Z}/r\mathbb{Z}$ act on R by $(x, y, z) \mapsto (\zeta_r^2 x, \zeta_r^{-1} y, \zeta_r^{m+2} z)$, where ζ_r is a primitive r^{th} root of unity. Find presentations for the ring of invariants R^G in the cases $m = 1$ and $m = 2$. Try to do $m = 4$. (Cf. Example 14.26.)

5.37. EXERCISE. Let A be a local ring and M, N two finitely generated A-modules. Then depth $\operatorname{Hom}_A(M, N) \geqslant \min\{2, \operatorname{depth} N\}$.

5.38. EXERCISE. Let $\sigma \in \mathrm{GL}(n,k)$ be a pseudo-reflection on $V = k^{(n)}$.
(1) Suppose $v \in V$ spans the image of $\sigma - 1_V$. Prove that σ is diagonalizable if and only if v is not fixed by σ.
(2) Use Maschke's Theorem to prove that if $|\sigma|$ is relatively prime to $\mathrm{char}(k)$, then V has a decomposition as kG-modules $V^\sigma \oplus kv$ and so σ is diagonalizable.

5.39. EXERCISE. If $k = \mathbb{R}$, show that any pseudo-reflection has order 2 (so is a *reflection*).

CHAPTER 6

Kleinian Singularities and Finite CM Type

In the previous chapter we saw that when $S = k[\![x_1,\ldots,x_n]\!]$ is a power series ring endowed with a linear action of a finite group G whose order is invertible in k, and $R = S^G$ is the invariant subring, then the R-direct summands of S are MCM R-modules and are closely linked to the representation theory of G. In dimension two, we shall see in this chapter that *every* indecomposable MCM R-modules is a direct summand of S. This is due to Herzog [**Her78b**]. Thus in particular two-dimensional rings of invariants under finite non-modular group actions have finite CM type. In the next chapter we shall prove that in fact every two-dimensional complete normal domain containing $\mathbb{C}$ and having finite CM type arises in this way.

In the present chapter, we first recall some basic facts on reflexive modules over normal domains, then prove the theorem of Herzog mentioned above. Next we discuss the two-dimensional invariant rings $k[\![u,v]\!]^G$ that are Gorenstein; by a result of Watanabe [**Wat74**] these are the ones for which $G \subseteq \mathrm{SL}(2,k)$. The finite subgroups of $\mathrm{SL}(2,\mathbb{C})$ are well-known, their classification going back to Klein, so here we call the resulting invariant rings *Kleinian singularities*, and we derive their defining equations following [**Kle93**]. It turns out that the resulting equations are precisely the three-variable versions of the ADE hypersurface rings from Chapter 4 §3. This section owes many debts to previous expositions, particularly [**Slo83**].

In the last two sections, we describe two incarnations of the *McKay correspondence*: first, the identification of the McKay-Gabriel quiver of $G \subseteq \mathrm{SL}(2,\mathbb{C})$ with the corresponding ADE Coxeter-Dynkin diagram, and then the original observation of McKay that both these are the same as the desingularization graph of $\mathrm{Spec}\, k[\![u,v]\!]^G$.

§1. Invariant rings in dimension two

In the last chapter we considered invariant rings of the form $R = k[\![x_1,\ldots,x_n]\!]^G$, where G is a finite group with order invertible in k acting linearly on the power series ring $S = k[\![x_1,\ldots,x_n]\!]$. In general, the direct summands of S as an R-module are MCM modules. Here we prove that in dimension two, every indecomposable MCM module is among the R-direct summands of S.

First we recall some background on reflexive modules over normal domains. See Chapter 14 for some extensions to the non-normal case.

6.1. REMARKS. Recall (from, for example, Appendix A) that for a normal domain R, if a finitely generated R-module M is MCM then it is *reflexive*, that is the natural map

$$\sigma_M \colon M \longrightarrow M^{**} = \mathrm{Hom}_R(\mathrm{Hom}_R(M, R), R),$$

defined by $\sigma_M(m)(f) = f(m)$, is an isomorphism. If in addition $\dim(R) = 2$, then the converse holds as well, so that M is MCM if and only if it is reflexive.

The first assertion of the next proposition is due to Herzog, and will imply that two-dimensional rings of invariants have finite CM type.

6.2. PROPOSITION. *Let $R \longrightarrow S$ be a module-finite extension of two-dimensional complete local rings which satisfy (S_2) and are Gorenstein in codimension one. Assume that R is a direct summand of S as an R-module. Then every finitely generated reflexive R-module is a direct summand of a finitely generated reflexive S-module. If in particular R is complete and S has finite CM type, then R has finite CM type as well.*

PROOF. Let M be a finitely generated reflexive R-module and set $M^* = \mathrm{Hom}_R(M, R)$. Then the split monomorphism $R \longrightarrow S$ induces a split monomorphism $M = \mathrm{Hom}_R(M^*, R) \longrightarrow \mathrm{Hom}_R(M^*, S)$. Now $\mathrm{Hom}_R(M^*, S)$ is naturally an S-module via the action on the codomain, and Exercise 5.37 shows that it satisfies (S_2) as an R-module, hence as an S-module, so is reflexive over S by Corollary A.13.

Let $N_1, \ldots, N_n$ be representatives for the isomorphism classes of indecomposable MCM S-modules. Then each N_i is a MCM R-module as well, so we write $N_i = M_{i,1} \oplus \cdots \oplus M_{i,m_i}$ for indecomposable MCM R-modules $M_{i,j}$. By the first statement of the Proposition, every indecomposable MCM R-module is a direct summand of a direct sum of copies of the N_i, so is among the $M_{i,j}$ by KRS over complete local rings (Theorem 1.9). □

6.3. THEOREM (Herzog). *Let $S = k[\![u, v]\!]$ be a power series ring in two variables over a field, G a finite subgroup of $\mathrm{GL}(2, k)$ acting linearly on S, and $R = S^G$. Assume that R is a direct summand of S as an R-module. Then every indecomposable finitely generated reflexive R-module is a direct summand of S as an R-module. In particular, R has finite CM type.*

PROOF. Note that S is module-finite over R, by Proposition 5.4. Let M be an indecomposable finitely generated reflexive R-module. By Proposition 6.2 M is an R-summand of a reflexive S-module N. But S is regular, so in fact N is free over S. Since R is complete, KRS implies that M is a direct summand of S. □

The one-one correspondences listed in Corollary 5.20 can thus be extended in dimension two.

6.4. COROLLARY. *Let k be a field, $S = k[\![x,y]\!]$, and $G \subseteq \mathrm{GL}(2,k)$ a finite group, with $|G|$ invertible in k, acting linearly on S without pseudo-reflections. Put $R = S^G$. Then there are one-one correspondences between*

- *the indecomposable reflexive (MCM) R-modules;*
- *the indecomposable direct summands of S as an R-module;*
- *the indecomposable projective $\mathrm{End}_R(S)$-modules;*
- *the indecomposable projective $S\#G$-modules; and*
- *the irreducible kG-modules.* □

Observe that while we need the assumption that $|G|$ be invertible in k for Corollary 6.4, Proposition 6.2 requires only the weaker assumption that R be a direct summand of S as an R-module. We will make use of this in Remark 6.22 below.

§2. Kleinian singularities

Having seen the privileged position that dimension two holds in the story so far, we are ready to define and study the two-dimensional hypersurface rings of finite CM type. These turn out to coincide with a class of rings ubiquitous throughout algebra and geometry, variously called *Kleinian singularities, Du Val singularities,* two-dimensional *rational double points,* and other names. Even more, they are the two-dimensional analogs of the ADE hypersurfaces seen in the previous chapter.

For historical reasons, we introduce the Kleinian singularities in a slightly opaque fashion. The rest of the section will clarify matters. For the first part of this chapter, we work over $\mathbb{C}$ for ease of exposition. We will in the end define the complete Kleinian singularities over any algebraically closed field of characteristic not 2, 3, or 5 (see Definition 6.21).

6.5. DEFINITION. A *complete complex Kleinian singularity* is a ring of the form $\mathbb{C}[\![u,v]\!]^G$, where G is a finite subgroup of $\mathrm{SL}(2,\mathbb{C})$.

The reason behind the restriction to $\mathrm{SL}(2,\mathbb{C})$ rather than $\mathrm{GL}(2,\mathbb{C})$ as in the previous chapter is the fact, due to Watanabe [**Wat74**], that $R = S^G$ is Gorenstein when $G \subseteq \mathrm{SL}(n,k)$, and the converse holds if G is small. Thus the complete Kleinian singularities are the two-dimensional complete Gorenstein rings of invariants of finite group actions.

In order to make sense of this definition, we recall the fact that the finite subgroups of $\mathrm{SL}(2,\mathbb{C})$ are the "binary polyhedral" groups, which are double covers of the rotational symmetry groups of the Platonic solids, together with two degenerate cases.

The classification of the Platonic solids goes back to Theaetetus around 400 BCE, and is at the center of Plato's *Timaeus*; the final book of Euclid's *Elements* is devoted to their properties. According to Bourbaki [**Bou02**], the determination of the finite groups of rotations in $\mathbb{R}^3$ goes back to Hessel, Bravais, and Möbius in the early 19th century, though they did not yet have the language of group theory. Jordan [**Jor77**] was the first to explicitly classify the finite groups of rotations of $\mathbb{R}^3$. Recall that $\mathrm{SO}(3)$ denotes the *special orthogonal group*, that is, the group of rotations of $\mathbb{R}^3$.

6.6. THEOREM. *The finite subgroups of $\mathrm{SO}(3)$ are up to conjugacy the following rotational symmetry groups.*

C_{n+1}: *The cyclic group of order $n+1$ for $n \geqslant 0$, the symmetry group of a pyramid with $(n+1)$-gonal base.*

D_{n-2}: *The dihedral group of order $2(n-2)$ for $n \geqslant 4$, the symmetry group of a beach ball ("hosohedron").*

T: *The symmetry group of a tetrahedron, which is isomorphic to the alternating group A_4 of order 12.*

O: *the symmetry group of the octahedron, which is isomorphic to the symmetric group S_4 of order 24.*

I: *The symmetry group of the icosahedron, which is isomorphic to the alternating group A_5 of order 60.* □

In order to leverage this classification into a description of the finite subgroups of $\mathrm{SL}(2,\mathbb{C})$, we recall some basics of classical group theory. Recall first that the *unitary group* $\mathrm{U}(n)$ is the subgroup of $\mathrm{GL}(n,\mathbb{C})$ consisting of unitary transformations, i.e. those preserving the standard Hermitian inner product on $\mathbb{C}^n$. The *special unitary group* $\mathrm{SU}(n)$ is $\mathrm{SL}(n,\mathbb{C}) \cap \mathrm{U}(n)$. We first observe that to classify the finite subgroups of $\mathrm{SL}(n,\mathbb{C})$, it suffices to classify those of $\mathrm{SU}(n)$.

6.7. LEMMA. *Every finite subgroup of $\mathrm{GL}(n,\mathbb{C})$, resp. $\mathrm{SL}(n,\mathbb{C})$, is conjugate to a subgroup of $\mathrm{U}(n)$, resp. $\mathrm{SU}(n)$.*

PROOF. Let G be a finite subgroup of $\operatorname{GL}(n, \mathbb{C})$. Denote the usual Hermitian inner product on $\mathbb{C}^n$ by $\langle\,,\,\rangle$. It suffices to define a new inner product $\{\,,\,\}$ on $\mathbb{C}^n$ such that $\{\sigma u,\, \sigma v\} = \{u,\, v\}$ for every $\sigma \in G$ and $u, v \in \mathbb{C}^n$. Indeed, if we find such an inner product, let $\mathcal{B}$ be an orthonormal basis for $\{\,,\,\}$, and let $\rho\colon \mathbb{C}^n \longrightarrow \mathbb{C}^n$ be the change-of-basis operator taking $\mathcal{B}$ to the standard basis. Then $\rho G \rho^{-1} \subseteq \operatorname{U}(n)$, as

$$\begin{aligned}\langle \rho\sigma\rho^{-1}u,\, \rho\sigma\rho^{-1}v \rangle &= \{\sigma\rho^{-1}u,\, \sigma\rho^{-1}v\} \\ &= \{\rho^{-1}u,\, \rho^{-1}v\} \\ &= \langle u,\, v \rangle \end{aligned}$$

for every $\sigma \in G$ and $u, v \in \mathbb{C}^n$. Define the desired new product by

$$\{u, v\} = \frac{1}{|G|} \sum_{\sigma \in G} \langle \sigma(u),\, \sigma(v) \rangle\,.$$

Then it is easy to check that $\{\,,\,\}$ is again an inner product on $\mathbb{C}^n$, and that $\{\sigma u,\, \sigma v\} = \{u,\, v\}$ for every σ, u, v. $\square$

The special unitary group $\operatorname{SU}(2)$ acts on the complex projective line $\mathbb{P}^1_{\mathbb{C}}$ by fractional linear transformations (Möbius transformations):

$$\begin{pmatrix} \alpha & -\beta \\ \beta & \overline{\alpha} \end{pmatrix} [z : w] = [\alpha z - \beta w : \overline{\beta} z + \overline{\alpha} w]\,.$$

Since the matrices $\pm I$ act trivially, the action factors through $\operatorname{PSU}(2) = \operatorname{SU}(2)/\{\pm I\}$. We claim now that $\operatorname{PSU}(2) \cong \operatorname{SO}(3)$, the group of symmetries of the 2-sphere S^2. Position S^2 with its south pole at the origin, and consider the stereographic projection onto the equatorial plane, which we identify with $\mathbb{C}$. Extend this to an isomorphism $S^2 \longrightarrow \mathbb{P}^1_{\mathbb{C}}$ by sending the north pole to the point at infinity. This isomorphism identifies the conformal transformations of $\mathbb{P}^1_{\mathbb{C}}$ with the rotations of the sphere, and gives a double cover of $\operatorname{SO}(3)$.

6.8. PROPOSITION. *There exists a surjective group homomorphism* $\pi\colon \operatorname{SU}(2) \longrightarrow \operatorname{SO}(3)$ *with kernel* $\{\pm I\}$. $\square$

6.9. LEMMA. *The only element of order 2 in* $\operatorname{SU}(2)$ *is* $-I$.

PROOF. This is a direct calculation using the general form $\begin{pmatrix} \alpha & -\beta \\ \beta & \overline{\alpha} \end{pmatrix}$ of an arbitrary element of $\operatorname{SU}(2)$. $\square$

6.10. LEMMA. *Let Γ be a finite subgroup of* $\operatorname{SU}(2)$. *Then either Γ is cyclic of odd order, or $|\Gamma|$ is even and $\Gamma = \pi^{-1}(\pi(\Gamma))$ is the preimage of a finite subgroup G of* $\operatorname{SO}(3)$.

PROOF. If Γ has odd order, then $-I \notin \Gamma$, so $\Gamma \cap \ker \pi = \{I\}$, and the restriction of π to Γ is an isomorphism of Γ onto its image. By the classification of finite subgroups of $\mathrm{SO}(3)$, we see that the only ones of odd order are the cyclic groups C_{n+1} with $n+1$ odd. If $|\Gamma|$ is even, then by Cauchy's Theorem there is an element of order 2 in Γ, which must be $-I$. Thus $\ker \pi \subseteq \Gamma$ and $\Gamma = \pi^{-1}(\pi(\Gamma))$. $\square$

6.11. THEOREM. *The finite non-trivial subgroups of* $\mathrm{SL}(2, \mathbb{C})$, *up to conjugacy, are the following groups, called* binary polyhedral groups. *Let ζ_r denote a primitive r^{th} root of unity in $\mathbb{C}$.*

$\mathcal{C}_m$: *The cyclic group of order m for $m \geqslant 2$, generated by*
$$\begin{pmatrix} \zeta_m & \\ & \zeta_m^{-1} \end{pmatrix}.$$

$\mathcal{D}_m$: *The binary dihedral group of order $4m$ for $m \geqslant 1$, generated by $\mathcal{C}_{2m}$ and*
$$\begin{pmatrix} & i \\ i & \end{pmatrix}.$$

$\mathcal{T}$: *The binary tetrahedral group of order 24, generated by $\mathcal{D}_2$ and*
$$\frac{1}{\sqrt{2}} \begin{pmatrix} \zeta_8 & \zeta_8^3 \\ \zeta_8 & \zeta_8^7 \end{pmatrix}.$$

$\mathcal{O}$: *The binary octahedral group of order 48, generated by $\mathcal{T}$ and*
$$\begin{pmatrix} \zeta_8^3 & \\ & \zeta_8^5 \end{pmatrix}.$$

$\mathcal{I}$: *The binary icosahedral group of order 120, generated by*
$$\frac{1}{\sqrt{5}} \begin{pmatrix} \zeta_5^4 - \zeta_5 & \zeta_5^2 - \zeta_5^3 \\ \zeta_5^2 - \zeta_5^3 & \zeta_5 - \zeta_5^4 \end{pmatrix} \quad \text{and} \quad \frac{1}{\sqrt{5}} \begin{pmatrix} \zeta_5^2 - \zeta_5^4 & \zeta_5^4 - 1 \\ 1 - \zeta_5 & \zeta_5^3 - \zeta_5 \end{pmatrix}. \quad \square$$

As abstract groups, the binary polyhedral groups can be presented in a uniform way: they are all generated by three elements a, b, and c subject to the relation $a^p = b^q = c^r = abc$, where $p \leqslant q \leqslant r$ constitute an integer solution to $\frac{1}{p} + \frac{1}{q} + \frac{1}{r} > 1$, namely one of $(1, q, r)$, $(2, 2, r)$, $(2, 3, 3)$, $(2, 3, 4)$, and $(2, 3, 5)$. (The integers p, q, r are not mysterious; they are just the orders of the stabilizers of a face, an edge, and a vertex of the corresponding Platonic solid.) The concrete presentations above are more useful for our purposes.

6.12. THEOREM. *The complete complex Kleinian singularities are the rings of invariants of the groups above acting linearly on the power series ring $S = \mathbb{C}[\![u, v]\!]$. We name them as follows:*

Singularity Name	Group Name
A_n	$\mathcal{C}_{n+1}$, cyclic ($n \geq 1$)
D_n	$\mathcal{D}_{n-2}$, binary dihedral ($n \geq 4$)
E_6	$\mathcal{T}$, binary tetrahedral
E_7	$\mathcal{O}$, binary octahedral
E_8	$\mathcal{I}$, binary icosahedral

□

At this point the naming system is utterly mysterious, but we continue anyway.

It is a classical fact from invariant theory that the Kleinian singularities "embed in codimension one," that is, are defined by a single equation.[1] We can make this explicit by writing down a set of generating invariants for each of the binary polyhedral groups. These calculations go back to Klein [**Kle93**], and are also found in Du Val's book [**DV64**]; for a more modern treatment see [**Lam86**]. We like the concreteness of having actual invariants in hand, so we present them here. The details of the derivations are quite involved, so we only sketch them.

6.13 (A_n). In this case, the only monomials fixed by the generator $(u,v) \mapsto (\zeta_{n+1}u, \zeta_{n+1}^{-1}v)$ are uv, u^{n+1}, and v^{n+1}. Thus we set

$$X_\mathcal{C}(u,v) = u^{n+1} + v^{n+1}, \qquad Y_\mathcal{C}(u,v) = uv,$$

and

$$Z_\mathcal{C}(u,v) = u^{n+1} - v^{n+1}.$$

These generate all the invariants, and satisfy the relation

$$Z_\mathcal{C}^2 = X_\mathcal{C}^2 - 4Y_\mathcal{C}^{n+1}.$$

6.14 (D_n). The cyclic subgroup $\mathcal{C}_{2(n-2)}$ of $\mathcal{D}_{n-2}$ has invariants $a = u^{2(n-2)} + v^{2(n-2)}$, $b = uv$, and $c = u^{2(n-2)} - v^{2(n-2)}$ as in the case above. The additional generator $(u,v) \mapsto (iv, iu)$ changes the sign of b, multiplies a by $(-1)^n$, and sends c to $-(-1)^n c$. Now we have two cases to consider depending on the parity of n. If n is even, then c, a^2, ab, and b^2 are all fixed, but we can throw out b^2 since $b^2 = c^2 - 4(a^2)^{n-2}$.

---

[1]Abstractly, we can see this from the connection with Platonic solids as follows [**McK01, Dic59**]: drawing a sphere around the platonic solid, we project from the north pole to the equatorial plane, which we interpret as $\mathbb{C}$. Thus the projection of each vertex v gives a complex number z_v, and we form the homogeneous polynomial $V(x,y) = \prod_v (x - z_v y)$. Similarly, the center of each edge e gives a complex number z_e, and the center of each face f a corresponding z_f, which we compile into the polynomials $E(x,y) = \prod_e (x - z_e y)$ and $F(x,y) = \prod_f (x - z_f y)$. These are three functions in two variables, and so there must be a relation $f(V, E, F) = 0$.

In the other case, when n is odd, similar considerations imply that the invariants are generated by b, a^2, and ac. Thus in this case we set
$$X_{\mathcal{D}}(u,v) = u^{2(n-2)} + (-1)^n v^{2(n-2)}, \qquad Y_{\mathcal{D}}(u,v) = u^2 v^2$$
$$Z_{\mathcal{D}}(u,v) = uv\left(u^{2(n-2)} - (-1)^n v^{2(n-2)}\right).$$
For these generating invariants we have the relation
$$Z_{\mathcal{D}}^2 = Y_{\mathcal{D}} X_{\mathcal{D}}^2 + 4(-Y_{\mathcal{D}})^{n-1}.$$

6.15 (E_6). The invariants (D_4) of the subgroup $\mathcal{D}_2$ are
$$u^4 + v^4, \qquad u^2 v^2, \qquad \text{and} \qquad uv\left(u^4 - v^4\right).$$
The third of these is invariant under the whole group $\mathcal{T}$, so we set
$$Y_{\mathcal{T}}(u,v) = uv\left(u^4 - v^4\right).$$
Searching for an invariant (or coinvariant) of the form $P(u,v) = X_{\mathcal{D}} + tY_{\mathcal{D}} = u^4 + tu^2 v^2 + v^4$, we find that if $t = \sqrt{-12}$, and we set
$$P(u,v) = u^4 + \sqrt{-12}\, u^2 v^2 + v^4 \quad \text{and} \quad \overline{P}(u,v) = u^4 - \sqrt{-12}\, u^2 v^2 + v^4,$$
then
$$X_{\mathcal{T}}(u,v) = P(u,v)\,\overline{P}(u,v) = u^8 + 14u^4 v^4 + v^8$$
is invariant.

Furthermore, $\left[\frac{1}{4}(t-2)\right]^3 = 1$, so that every linear combination of P^3 and $\overline{P}^3$, such as
$$Z_{\mathcal{T}}(u,v) = \frac{1}{2}\left[P^3 + \overline{P}^3\right]$$
$$= u^{12} - 33u^8 v^4 - 33u^4 v^8 + v^{12},$$
is invariant. These three invariants generate all others, and satisfy the relation
$$Z_{\mathcal{T}}^2 = X_{\mathcal{T}}^3 + 108 Y_{\mathcal{T}}^4.$$

6.16 (E_7). Begin with the above invariants for $\mathcal{T}$. The additional generator for $\mathcal{O}$ leaves $X_{\mathcal{T}}$ fixed but changes the signs of $Y_{\mathcal{T}}$ and $Z_{\mathcal{T}}$. We therefore obtain generating invariants
$$X_{\mathcal{O}}(u,v) = Y_{\mathcal{T}}(u,v)^2 = \left(u^5 v - uv^5\right)^2$$
$$Y_{\mathcal{O}}(u,v) = X_{\mathcal{T}}(u,v) = u^8 + 14u^4 v^4 + v^8$$
$$Z_{\mathcal{O}}(u,v) = Y_{\mathcal{T}}(u,v) Z_{\mathcal{T}}(u,v)$$
$$= uv\left(u^4 - v^4\right)(u^{12} - 33u^8 v^4 - 33u^4 v^8 + v^{12})$$
(of degrees 8, 12, and 18, respectively). These satisfy
$$Z_{\mathcal{O}}^2 = -X_{\mathcal{O}}\left(108 X_{\mathcal{O}}^2 - Y_{\mathcal{O}}^3\right).$$

6.17 (E_8). From the geometry of the 12 vertices of the icosahedron, Klein derives an invariant of degree 12:

$$Y_\mathcal{I}(u,v) = uv(u^5 + \varphi^5 v^5)(u^5 - \varphi^{-5} v^5)$$
$$= uv(u^{10} + 11u^5 v^5 + v^{10}),$$

where $\varphi = (1+\sqrt{5})/2$ is the golden ratio. The Hessian of this form is also invariant, and takes the form $-121 X_\mathcal{I}(u,v)$, where

$$X_\mathcal{I}(u,v) = \begin{vmatrix} \partial^2/\partial u^2 & \partial^2/\partial v \partial u \\ \partial^2/\partial u \partial v & \partial^2/\partial v^2 \end{vmatrix}$$
$$= (u^{20} + v^{20}) - 228\left(u^{15} v^5 - u^5 v^{15}\right) + 494 u^{10} v^{10}.$$

The Jacobian of these two forms (i.e. the determinant of the 2×2 matrix of partial derivatives) is invariant as well:

$$Z_\mathcal{I}(u,v) = \left(u^{30} + v^{30}\right) + 522\left(u^{25} v^5 - u^5 v^{25}\right) - 10005\left(u^{20} v^{10} + u^{10} v^{20}\right).$$

Now one checks that[2]

$$Z_\mathcal{I}^2 = X_\mathcal{I}^3 + 1728 Y_\mathcal{I}^5.$$

It's interesting to note that in each case above, we have $\deg X \cdot \deg Y = 2|G|$, namely $2(n+1)$, $8(n-2)$, 48, 96, 240. Since the defining equation in each case is obtained as a relation among homogeneous polynomials, we see that each equation is *quasi-homogeneous*, that is, there exist weights for the variables making the relation homogeneous. Specifically, the weights are the degrees of the generating invariants.

Adjusting the polynomials by certain n^th roots ($n \leq 5$), one obtains the following normal forms for the Kleinian singularities.

6.18. THEOREM. *The complete complex Kleinian singularities are the hypersurface rings defined by the following polynomials.*

(A_n): $\quad x^2 + y^{n+1} + z^2, \quad n \geq 1$
(D_n): $\quad x^2 y + y^{n-1} + z^2, \quad n \geq 4$
(E_6): $\quad x^3 + y^4 + z^2$
(E_7): $\quad x^3 + xy^3 + z^2$
(E_8): $\quad x^3 + y^5 + z^2$ □

We summarize the information we have on the Kleinian singularities so far in Table 6.19.

---

[2] tempting one to call E_8 the *great gross singularity* ($1728 = 12 \times 144$, a dozen gross, aka a great gross).

TABLE 6.19. Complete Kleinian Singularities

Name	$f(x,y,z)$	G	$\lvert G\rvert$	(p,q,r)
(A_n), $n \geqslant 1$	$x^2 + y^{n+1} + z^2$	$\mathcal{C}_{n+1}$, cyclic	$n+1$	$(1,1,n)$
(D_n), $n \geqslant 4$	$x^2 y + y^{n-1} + z^2$	$\mathcal{D}_{n-2}$, b. dihedral	$4n-8$	$(2,2,n-2)$
(E_6)	$x^3 + y^4 + z^2$	$\mathcal{T}$, b. tetrahedral	24	$(2,3,3)$
(E_7)	$x^3 + xy^3 + z^2$	$\mathcal{O}$, b. octahedral	48	$(2,3,4)$
(E_8)	$x^3 + y^5 + z^2$	$\mathcal{I}$, b. icosahedral	120	$(2,3,5)$

6.20. REMARK. Now we relax our requirement that we work over $\mathbb{C}$. Assume from now on only that k is an algebraically closed field of characteristic different from 2, 3, and 5.

With this restriction on the characteristic, the groups defined by generators in Theorem 6.11 exist equally well in $\mathrm{SL}(2,k)$, with two exceptions: $\mathcal{C}_n$ and $\mathcal{D}_n$ are not defined if $\operatorname{char} k$ divides n. Therefore, in positive characteristics ($\neq$ 2, 3, 5), we simply *define* the (A_{n-1}) and (D_{n+2}) singularities using the elements X, Y, and Z listed in 6.13 and 6.14 and derive the normal forms listed in Theorem 6.18.

6.21. DEFINITION. Let k be an algebraically closed field of characteristic not equal to 2, 3, or 5. The *complete Kleinian singularities* over k are the hypersurface rings $k[\![x,y,z]\!]/(f)$, where f is one of the polynomials listed in Theorem 6.18.

6.22. REMARK. There is one further technicality to address. In the cases $\mathcal{C}_n$ and $\mathcal{D}_n$ where n is divisible by the characteristic of k, we lose the ability to define the Reynolds operator. However, in each case we can verify that the Kleinian singularity is a direct summand of the regular ring $k[\![u,v]\!]$ by using the generating invariants X, Y, and Z.

The case (A_{n-1}) was mentioned in passing already in Example 5.25. Set $R = k[\![u^n, uv, v^n]\!]$. Then $k[\![u,v]\!]$ is isomorphic as an R-module to $\bigoplus_{j=0}^{n-1} M_j$, where M_j is the R-span of the monomials $u^a v^b$ such that $b - a \equiv j \bmod n$. In particular, R is a direct summand of $k[\![u,v]\!]$ in any characteristic.

For the case (D_{n+2}), we have $R = k[\![u^{2n} + v^{2n}, u^2 v^2, uv(u^{2n} - v^{2n})]\!]$. Then R is a direct summand of $A = k[\![u^{2n}, uv, v^{2n}]\!]$: observe that $A = R \oplus R(uv, u^{2n} - v^{2n})$ and that the second summand is generated by elements negated by $\tau\colon (u,v) \mapsto (v,-u)$. As A is an (A_{2n-1}) singularity, it is a direct summand of $k[\![u,v]\!]$ by the previous case.

Combined with Herzog's Theorem 6.3, these observations prove the following theorem.

6.23. THEOREM. *Let k be an algebraically closed field of characteristic not equal to 2, 3, or 5, and let R be a complete Kleinian singularity over k. Then R has finite CM type.* □

§3. McKay-Gabriel quivers of the Kleinian singularities

In this section we compute the McKay-Gabriel quivers (defined in Chapter 5) for the complete complex Kleinian singularities. We will recover McKay's observation that the underlying graphs of the quivers are exactly the *extended* (also *affine*, or *Euclidean*) *Coxeter-Dynkin diagrams* $\widetilde{A}_n$, $\widetilde{D}_n$, $\widetilde{E}_6$, $\widetilde{E}_7$, $\widetilde{E}_8$, corresponding to the name of the singularity from Table 6.19.

For background on the Coxeter-Dynkin diagrams A_n, D_n, E_6, E_7, E_8, and their extended counterparts $\widetilde{A}_n$, $\widetilde{D}_n$, $\widetilde{E}_6$, $\widetilde{E}_7$, $\widetilde{E}_8$, we recommend Reiten's survey article in the Notices [**Rei97**]. They have deep connections with more areas of mathematics than we can enumerate. Beyond the connections we will make explicitly in this and the next section, we will content ourselves with the following brief description. The extended ADE diagrams are the finite connected graphs with no loops (a loop is a single edge with both ends at the same vertex) bearing an additive function, i.e. a function f from the vertices $\{1, \ldots, n\}$ to $\mathbb{N}$ satisfying $2f(i) = \sum_j f(j)$ for every i, where the sum is taken over all neighbors j of i. Similarly, the (non-extended) ADE diagrams are the graphs bearing a sub-additive but not additive function, that is, one satisfying $2f(i) \geqslant \sum_j f(j)$ for each i, with strict inequality for at least one i. The non-extended diagrams are obtained by removing a single distinguished vertex and its incident edges from the extended ADE diagrams.

They're all listed in Table 6.24, with their (sub-)additive functions labeling the vertices. The distinguished vertex to be removed in obtaining the ordinary diagrams from the extended ones is circled. We shall see that, furthermore, the ranks of the irreducible representations (that is, indecomposable MCM modules) attached to each vertex of the quiver gives the (sub-)additive function on the diagram.

Recall from Definition 5.21 that the McKay-Gabriel quiver of a two-dimensional representation $G \hookrightarrow \mathrm{GL}(V)$ has for vertices the irreducible representations $V_0, \ldots, V_d$ of the group G, with an arrow $V_i \longrightarrow V_j$ for each copy of V_i in the direct-sum decomposition of $V \otimes_k V_j$. The number of arrows $V_i \longrightarrow V_j$ will (temporarily) be denoted m_{ij}. Recall

TABLE 6.24. ADE and Extended ADE Diagrams

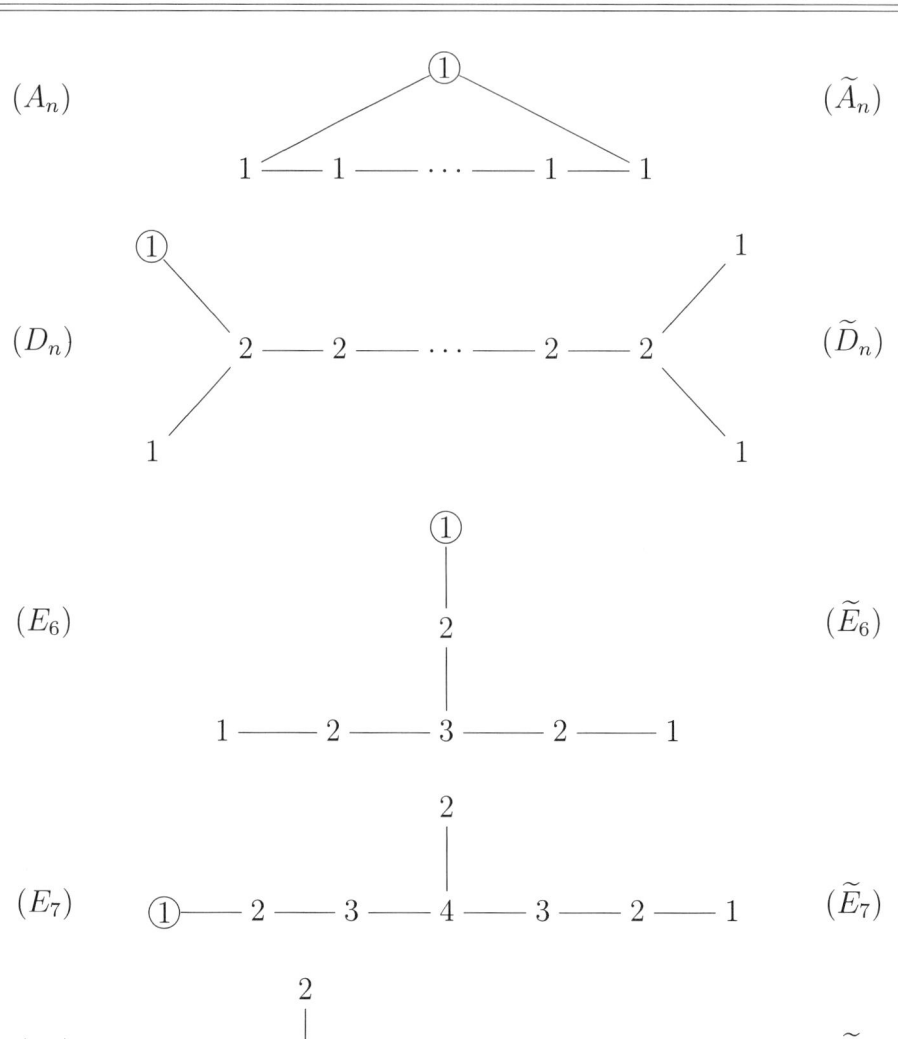

that when k is algebraically closed

$$m_{ij} = \langle \chi_i, \chi\chi_j \rangle = \frac{1}{|G|} \sum_{\sigma \in G} \chi_i(\sigma)\chi(\sigma^{-1})\chi_j(\sigma^{-1}),$$

where $\chi, \chi_0, \ldots, \chi_d$ are the characters of $V, V_0, \ldots, V_d$.

6.25. LEMMA. *Let G be a finite subgroup of $\mathrm{SL}(2,\mathbb{C})$ other than the two-element cyclic group. Then $m_{ij} \in \{0,1\}$ and $m_{ij} = m_{ji}$ for all*

$i, j = 1, \ldots, d$. *In other words, the arrows in the McKay-Gabriel quiver appear in opposed pairs.*

PROOF. Let G be one of the subgroups of $\mathrm{SL}(2, \mathbb{C})$ listed in Theorem 6.11; in particular, the given two-dimensional representation V is defined by the matrices listed there. By Schur's Lemma and Hom-tensor adjointness, we have
$$m_{ij} = \dim_\mathbb{C} \mathrm{Hom}_{\mathbb{C}G}(V \otimes_{\mathbb{C}G} V_j, V_i)$$
$$= \dim_\mathbb{C} \mathrm{Hom}_{\mathbb{C}G}(V_j, \mathrm{Hom}_{\mathbb{C}G}(V, V_j)).$$
The inner Hom has dimension equal to the number of copies of V_i appearing in the irreducible decomposition of V. These irreducible decompositions are easily read off from the listed matrices; the only one consisting of two copies of a single irreducible is (A_1), which corresponds to the two-element cyclic subgroup $\mathcal{C}_2$. Thus $\mathrm{Hom}_{\mathbb{C}G}(V_i, V)$ has dimension at most 1 for all i, and so $m_{ij} \leq 1$ for all i, j.

Since the trace of a matrix in $\mathrm{SL}(2, \mathbb{C})$ is the same as that of its inverse, the given representation V satisfies $\chi(\sigma^{-1}) = \chi(\sigma)$ for every σ. Thus
$$m_{ij} = \langle \chi_i, \chi \chi_j \rangle = \langle \chi_i \chi, \chi_j \rangle = m_{ji}$$
for every i and j. □

In displaying the McKay-Gabriel quivers for the Kleinian singularities, we replace each opposed pair of arrows by a simple edge. This has the effect, thanks to Lemma 6.25, of reducing the quiver to a simple graph with no multiple edges.

6.26 (A_n). We have already calculated the McKay-Gabriel quiver for the (A_n) singularities $xz - y^{n+1}$, for $n \geq 1$, in Example 5.25. Replacing the pairs of arrows there by single edges, we obtain

$$\begin{array}{c} V_0 \\ \diagup \quad \diagdown \\ V_1 \text{—} V_2 \text{—} \cdots \text{—} V_{n-1} \text{—} V_n \,. \end{array}$$

6.27 (D_n). The binary dihedral group $\mathcal{D}_{n-2}$ is generated by two elements
$$\alpha = \begin{pmatrix} \zeta_{2(n-2)} & \\ & \zeta_{2(n-2)}^{-1} \end{pmatrix} \quad \text{and} \quad \beta = \begin{pmatrix} & i \\ i & \end{pmatrix}$$
satisfying the relations
$$\alpha^{n-2} = \beta^2 = (\alpha\beta)^2, \quad \text{and} \quad \beta^4 = 1.$$
There are four natural one-dimensional representations as follows:

$$\begin{aligned} V_0 &: \alpha \mapsto 1, & \beta \mapsto 1;\\ V_1 &: \alpha \mapsto 1, & \beta \mapsto -1;\\ V_{n-1} &: \alpha \mapsto -1, & \beta \mapsto i;\\ V_n &: \alpha \mapsto -1, & \beta \mapsto -i\,. \end{aligned}$$

Furthermore, there is for each $j = 2, \ldots, n-2$ an irreducible two-dimensional representation V_j given by

$$a \mapsto \begin{pmatrix} \zeta_{2(n-2)}^{j-1} & \\ & \zeta_{2(n-2)}^{-j+1} \end{pmatrix} \quad \text{and} \quad b \mapsto \begin{pmatrix} & i^{j-1} \\ i^{j-1} & \end{pmatrix}.$$

In particular, the given representation V is isomorphic to V_2. It's easy to compute now that

$$V \otimes_k V_j \cong V_{j+1} \oplus V_{j-1}$$

for $2 \leqslant j \leqslant n-2$, leading to the McKay-Gabriel quiver for the (D_n) singularity.

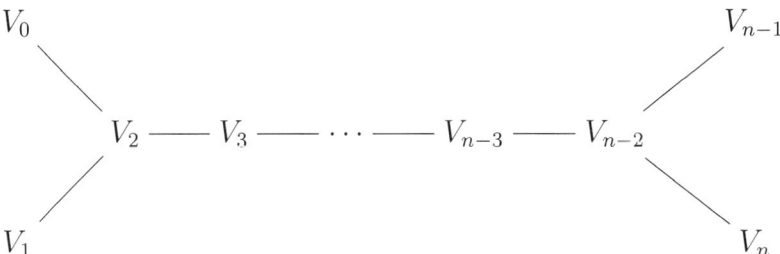

For the remaining examples, we will take the character table of G as given (see, for example, [**Hum94**], [**IN99**], or [**GAP08**]). From these data, we will be able to calculate the McKay-Gabriel quiver, since the character of a tensor product is the product of the characters and the irreducible representations are uniquely determined up to equivalence by their characters.

6.28 (E_6). The given presentation of $\mathcal{T}$ is defined by the generators

$$\alpha = \begin{pmatrix} i & \\ & -i \end{pmatrix}, \quad \beta = \begin{pmatrix} & i \\ i & \end{pmatrix}, \quad \text{and} \quad \gamma = \frac{1}{\sqrt{2}} \begin{pmatrix} \zeta_8 & \zeta_8^3 \\ \zeta_8 & \zeta_8^7 \end{pmatrix}.$$

§3. MCKAY-GABRIEL QUIVERS OF THE KLEINIAN SINGULARITIES   95

The character table has the following form.

representative	I	$-I$	β	γ	γ^2	γ^4	γ^5
|class|	1	1	6	4	4	4	4
order	1	2	4	6	3	3	6
V_0	1	1	1	1	1	1	1
V_1	2	-2	0	1	-1	-1	1
V_2	3	3	-1	0	0	0	0
V_3	2	-2	0	ζ_3	$-\zeta_3$	$-\zeta_3^2$	ζ_3^2
$V_3^\vee$	2	-2	0	ζ_3^2	$-\zeta_3^2$	$-\zeta_3$	ζ_3
V_4	1	1	1	ζ_3	ζ_3	ζ_3^2	ζ_3^2
$V_4^\vee$	1	1	1	ζ_3^2	ζ_3^2	ζ_3	ζ_3

Here $V = V_1$ is the given two-dimensional representation. Now one verifies for example that the character of $V_1 \otimes_k V_4$, that is the elementwise product of the second and sixth rows of the table, is equal to the character of V_3. Hence $V_1 \otimes_k V_4 \cong V_3$ and the McKay-Gabriel quiver contains an edge connecting V_3 and V_4. Similarly, $V_1 \otimes_k V_2 \cong V_1 \oplus V_3 \oplus V_3^\vee$, so V_2 is a vertex of degree three. Continuing in this way gives the following McKay-Gabriel quiver.

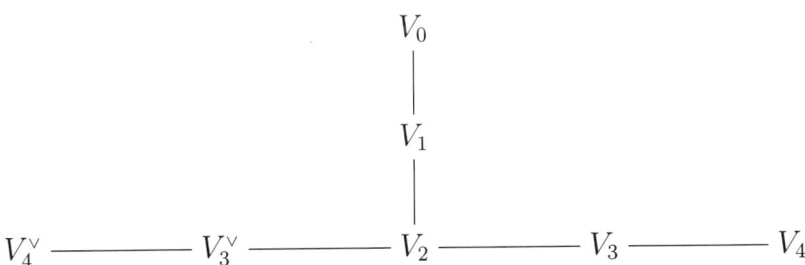

6.29. (E_7) The binary octahedral group $\mathcal{O}$ is generated by α, β, and γ from the previous case together with

$$\delta = \begin{pmatrix} \zeta_8^3 & \\ & \zeta_8^5 \end{pmatrix}.$$

This time the character table is as follows.

representative	I	$-I$	β	γ	γ^2	δ	$\beta\delta$	δ^3
|class|	1	1	6	8	8	6	12	6
order	1	2	4	6	3	8	4	8
V_0	1	1	1	1	1	1	1	1
V_1	2	-2	0	1	-1	$-\sqrt{2}$	0	$\sqrt{2}$
V_2	3	3	-1	0	0	1	-1	1
V_3	4	-4	0	-1	1	0	0	0
V_4	3	3	-1	0	0	-1	1	-1
V_5	2	-2	0	1	-1	$\sqrt{2}$	0	$-\sqrt{2}$
V_6	1	1	1	1	1	-1	-1	-1
V_7	2	2	2	-1	-1	0	0	0

Again $V = V_1$ is the given two-dimensional representation. Now we compute the McKay-Gabriel quiver to be the following.

$$\begin{array}{c} V_7 \\ | \\ V_0 \text{---} V_1 \text{---} V_2 \text{---} V_3 \text{---} V_4 \text{---} V_5 \text{---} V_6 \end{array}$$

6.30. (E_8) Finally, we consider the binary icosahedral group $\mathcal{I}$, generated by

$$\sigma = \frac{1}{\sqrt{5}} \begin{pmatrix} \zeta_5^4 - \zeta_5 & \zeta_5^2 - \zeta_5^3 \\ \zeta_5^2 - \zeta_5^3 & \zeta_5 - \zeta_5^4 \end{pmatrix} \quad \text{and} \quad \tau = \frac{1}{\sqrt{5}} \begin{pmatrix} \zeta_5^2 - \zeta_5^4 & \zeta_5^4 - 1 \\ 1 - \zeta_5 & \zeta_5^3 - \zeta_5 \end{pmatrix}.$$

Set $\varphi^+ = (1+\sqrt{5})/2$, the golden ratio, and $\varphi^- = (1-\sqrt{5})/2$. The character table for $\mathcal{I}$ is below.

representative	I	$-I$	σ	τ	τ^2	$\sigma\tau$	$(\sigma\tau)^2$	$(\sigma\tau)^3$	$(\sigma\tau)^4$
|class|	1	1	30	20	20	12	12	12	12
order	1	2	4	6	3	10	5	10	5
V_0	1	1	1	1	1	1	1	1	1
V_1	2	-2	0	1	-1	φ^+	$-\varphi^-$	φ^-	$-\varphi^+$
V_2	3	3	-1	0	0	φ^+	φ^-	φ^-	φ^+
V_3	4	-4	0	-1	1	1	-1	1	-1
V_4	5	5	1	-1	-1	0	0	0	0
V_5	6	-6	0	0	0	-1	1	-1	1
V_6	4	4	0	1	1	-1	-1	-1	-1
V_7	2	-2	0	1	-1	φ^-	$-\varphi^+$	φ^+	$-\varphi^-$
V_8	3	3	-1	0	0	φ^-	φ^+	φ^+	φ^-

We find that the McKay-Gabriel quiver of $\mathcal{I}$ is the extended Coxeter-Dynkin diagram $\widetilde{E}_8$.

$$\begin{array}{c}V_8\\|\\V_7 \text{—} V_6 \text{—} V_5 \text{—} V_4 \text{—} V_3 \text{—} V_2 \text{—} V_1 \text{—} V_0\end{array}$$

We have verified the first sentence of the following result, and the rest is straightforward to check from the definitions.

6.31. PROPOSITION. *The McKay-Gabriel quivers of the finite subgroups of* $\mathrm{SL}(2,\mathbb{C})$ *are the extended Coxeter-Dynkin diagrams. The dimensions of the irreducible representations appearing in the McKay-Gabriel quiver define an additive function on the quiver: Twice the dimension at a given vertex is equal to the sum of the dimensions at the neighboring vertices. In accordance with Corollary 5.20, these dimensions coincide with the ranks of the indecomposable MCM modules over the Kleinian singularity.* □

§4. Geometric McKay correspondence

The one-one correspondences derived in Chapter 5 in general, and in this chapter in dimension two, connect the representation theories of a finite subgroup of $\mathrm{SL}(2,k)$ and of its ring of invariants to the (extended) ADE Coxeter-Dynkin diagrams. These diagrams were known to be related to the geometry of the Kleinian singularities much earlier. Du Val's three-part 1934 paper [**DV34**] showed that the desingularization graphs of surfaces "not affecting the conditions of adjunction" are of ADE type; these are exactly the Kleinian singularities [**Art66**].

The first direct link between the representation theory of the complex Kleinian singularities and geometric data is due to Gonzalez-Sprinberg and Verdier [**GSV81**]. They constructed, on a case-by-case basis, a one-one correspondence between the irreducible representations of a binary polyhedral group and the irreducible components of the exceptional fiber in a minimal resolution of singularities of the invariant ring. (See below for definitions.) At the end of this section we describe Artin and Verdier's direct argument linking MCM modules and exceptional components.

This section is significantly more geometric than other parts of the book; in particular, we omit many of the proofs which would take us too far afield to justify. Most unexplained terminology can be found in [**Har77**].

Throughout the section, $(R, \mathfrak{m}, k)$ will be a two-dimensional normal local domain with algebraically closed residue field k. We do not assume char $k = 0$. Let $X = \operatorname{Spec} R$, a two-dimensional affine scheme, that is, a surface. In particular, since R is normal, X is regular in codimension one, so $\mathfrak{m}$ is the unique singular point of X.

A *resolution of singularities* of X is a non-singular surface Y with a proper birational map $\pi \colon Y \longrightarrow X$ such that the restriction of π to $Y \setminus \pi^{-1}(\mathfrak{m})$ is an isomorphism. Since $\dim(X) = 2$, resolutions of X exist as long as R is excellent [**Lip78**]. The *geometric genus* $g(X)$ of X is the k-dimension of the first cohomology group $\mathrm{H}^1(Y, \mathcal{O}_Y)$. This number is finite, and is independent of the choice of a resolution Y. Again since $\dim(X) = 2$, there is among all resolutions of X a *minimal resolution* $\pi \colon \widetilde{X} \longrightarrow X$ such that any other resolution factors through π.

6.32. DEFINITION. We say that X and R have (or are) *rational singularities* if $g(X) = 0$, that is, $\mathrm{H}^1(\widetilde{X}, \mathcal{O}_{\widetilde{X}}) = 0$.

We can rephrase this definition in a number of ways. Since $X = \operatorname{Spec} R$ is affine, the cohomology $\mathrm{H}^i(\widetilde{X}, \mathcal{O}_{\widetilde{X}})$ is isomorphic to the higher direct image $\mathrm{R}^i\pi_*(\mathcal{O}_{\widetilde{X}})$, so R has a rational singularity if and only if $\mathrm{R}^1\pi_*(\mathcal{O}_{\widetilde{X}}) = 0$. This is equivalent to the condition that $\mathrm{R}^i\pi_*(\mathcal{O}_{\widetilde{X}}) = 0$ for all $i \geqslant 1$, since the fibers of a resolution π are at most one-dimensional [**Har77**, III.11.2]. The direct image $\pi_*\mathcal{O}_{\widetilde{X}}$ itself is easy to compute: it is a coherent sheaf of R-algebras, so $S = \Gamma(X, \pi_*\mathcal{O}_{\widetilde{X}})$ is a module-finite R-algebra. But since π is birational, S has the same quotient field as R. Thus S is an integral extension, whence equal to R by normality, and so $\pi_*\mathcal{O}_{\widetilde{X}} = \mathcal{O}_X$.

Alternatively, recall that the *arithmetic genus* of a scheme Y is defined by $p_a(Y) = \chi(\mathcal{O}_Y) - 1$, where χ is the Euler characteristic, defined by the alternating sum of the k-dimensions of the $\mathrm{H}^i(Y, \mathcal{O}_Y)$. It follows from the Leray spectral sequence, for example, that if $\pi \colon \widetilde{X} \longrightarrow X$ is a resolution of singularities, then

$$p_a(X) - p_a(\widetilde{X}) = \dim_k \mathrm{H}^1(\widetilde{X}, \mathcal{O}_{\widetilde{X}}),$$

so that X is a rational singularity if and only if the arithmetic genus of X is not changed by resolving the singularity.

For a more algebraic criterion, assume momentarily that R is a non-negatively graded ring over a field $R_0 = k$ of characteristic zero. Flenner [**Fle81**] and Watanabe [**Wat83**] independently proved that R has a rational singularity if and only if the *a-invariant* $a(R)$ is negative. In general, $a(R)$ is the largest n such that the nth graded piece of the local cohomology module $\mathrm{H}_{\mathfrak{m}}^{\dim(R)}(R)$ is non-zero. For a two-dimensional quasi-homogeneous hypersurface singularity such as the Kleinian singularities in Theorem 6.18, the definition is particularly easy to apply:

$$a(k[x,y,z]/(f)) = \deg f - \deg x - \deg y - \deg z\,.$$

In particular, we check from Table 6.19 that the Kleinian singularities have rational singularities in characteristic zero.

More generally, any two-dimensional quotient singularity $k[u,v]^G$ or $k[\![u,v]\!]^G$, where G is a finite group with $|G|$ invertible in k, has rational singularities [**Bur74**, **Vie77**]. In fact, the restriction on $|G|$ is unnecessary for the Kleinian singularities: if S has rational singularities and R is a subring of S which is a direct summand as R-module, then R has rational singularities [**Bou87**]. Thus the Kleinian singularities have rational singularities in any characteristic in which they are defined.

As a final bit of motivation for the study of rational surface singularities, we point out that a normal surface $X = \operatorname{Spec} R$ is a rational singularity if and only if the divisor class group $\operatorname{Cl}(R)$ is finite, if and only if R has only finitely many rank-one MCM modules up to isomorphism [**Mum61**, **Lip69**].

Return now to our two-dimensional normal domain R, its spectrum X, and $\pi\colon \widetilde{X} \longrightarrow X$ the minimal resolution of singularities. With $0 \in X$ the unique singular point of X, set $E = \pi^{-1}(0)$, the *exceptional fiber* of π. Then E is connected by Zariski's Main Theorem [**Har77**, III.5.2], and is one-dimensional since π is birational. In other words, E is a union of irreducible curves on $\widetilde{X}$, so we write $E = \bigcup_{i=1}^n E_i$. The next result is [**Bri68**, Lemma 1.3].

6.33. LEMMA. *Let $\pi\colon \widetilde{X} \longrightarrow X$ be the minimal resolution of a rational singularity X, and let $E = \bigcup_{i=1}^n E_i$ be the exceptional fiber.*

(i) *Each E_i is non-singular, in particular reduced, and furthermore is a rational curve, i.e. $E_i \cong \mathbb{P}^1$.*
(ii) *$E_i \cap E_j \cap E_k = \emptyset$ for pairwise distinct i,j,k.*
(iii) *$E_i \cap E_j$ is either empty or a single reduced point for $i \neq j$, that is, the E_i meet transversely if at all.*
(iv) *E is cycle-free.* □

To describe the intersection properties of the exceptional curves more precisely, recall a bit of the intersection theory of curves on non-singular surfaces. Let C and D be curves on $\widetilde{X}$ with no common component. The *intersection multiplicity of C and D at a closed point $x \in \widetilde{X}$* is the length of the quotient $\mathcal{O}_{\widetilde{X},x}/(f,g)$, where $f = 0$ and $g = 0$

are local equations of C and D at x. The *intersection number* $C \cdot D$ of C and D is the sum of intersection multiplicities at all common points x. The *self-intersection* C^2, a special case, is defined to be the degree of the normal bundle to C in $\widetilde{X}$. Somewhat counter-intuitively, this can be negative; see [**Har77**, V.1.9.2] for an example.

The first part of the next theorem is due to Du Val [**DV34**] and Mumford [**Mum61**, **Hir95a**]; it immediately implies the second and third parts [**Art66**, Prop. 2 and Thm. 4].

6.34. THEOREM. *Let* $\pi \colon \widetilde{X} \longrightarrow X$ *be the minimal resolution of a surface singularity (not necessarily rational) with exceptional fiber* $E = \bigcup_{i=1}^{n} E_i$. *Define the* intersection matrix *of* X *to be the symmetric matrix* $E(X)_{ij} = (E_i \cdot E_j)$.

(i) *The matrix $E(X)$ is negative definite with off-diagonal entries either 0 or 1.*

(ii) *There exist positive divisors supported on E (that is, divisors of the form $Z = \sum_{i=1}^{n} m_i E_i$ with $m_i \geqslant 1$ for all i) such that $Z \cdot E_i \leqslant 0$ for all i.*

(iii) *Among all such Z as in (ii), there is a unique smallest one, which is called the* fundamental divisor *of X and denoted Z_f.*

□

To find the fundamental divisor there is a straightforward combinatorial algorithm: begin with $m_i = 1$ for all i, so that $Z_1 = \sum_i E_i$. If $Z_1 \cdot E_i \leqslant 0$ for each i, we set $Z_f = Z_1$ and stop; otherwise $Z_1 \cdot E_j > 0$ for some j. In that case, we put $Z_2 = Z_1 + E_j$ and continue. The process terminates by the negative definiteness of the matrix $E(X)$. See below for two examples.

For a rational singularity, we can identify Z_f more precisely, and this will allow us to identify the Gorenstein rational singularities.

6.35. PROPOSITION (Artin). *The fundamental divisor Z_f of a normal surface X with a rational singularity satisfies*

$$\left(\mathcal{O}_{\widetilde{X}} \otimes_{\mathcal{O}_X} \mathfrak{m}\right) / torsion = \mathcal{O}_{\widetilde{X}}(-Z_f).$$

In particular, we have formulas for the multiplicity and the embedding dimension $\mu_R(\mathfrak{m})$ of R:

$$\mathrm{e}(R) = -Z_f^2$$
$$\mathrm{embdim}(R) = -Z_f^2 + 1$$

□

6.36. COROLLARY. *A two-dimensional normal local domain R with a rational singularity has minimal multiplicity in the sense of Abhyankar:*
$$e(R) = \mu_R(\mathfrak{m}) - \dim(R) + 1.$$
□

6.37. COROLLARY. *Let $(R, \mathfrak{m})$ be a two-dimensional normal local domain, and assume that R is Gorenstein. If R is a rational singularity, then R is a hypersurface ring of multiplicity two.* □

Isolated singularities of multiplicity two are often called "double points."

PROOF OF COROLLARY 6.37. By the Proposition, we have $e(R) = -Z_f^2$ and $\mu_R(\mathfrak{m}) = -Z_f^2 + 1$. Since k is algebraically closed, by Theorem A.20 there exists a minimal reduction, that is, a regular sequence of length two in $\mathfrak{m} \setminus \mathfrak{m}^2$ such that the quotient $\overline{R}$ satisfies $e(R) = \ell(\overline{R})$ and $\mu_{\overline{R}}(\overline{\mathfrak{m}}) = \mu_R(\mathfrak{m}) - 2$. These together imply that $\mu_{\overline{R}}(\overline{\mathfrak{m}}) = \ell(\overline{R}) - 1$, so the Hilbert function of $\overline{R}$ is $(1, -Z_f^2 - 1, 0, \ldots)$. However, $\overline{R}$ is Gorenstein, so has socle dimension equal to 1. This forces $Z_f^2 = -2$, which gives $e(R) = 2$ and $\mu_R(\mathfrak{m}) = 3$. In particular R is a hypersurface ring. □

6.38. COROLLARY. *Let R be a Gorenstein rational surface singularity. The self-intersection number E_i^2 of each exceptional component is -2. Equivalently the normal bundle $\mathcal{N}_{E_i/\widetilde{X}}$ is $\mathcal{O}_{E_i}(-2)$.*

PROOF. This is a straightforward calculation using the adjunction formula and Riemann-Roch Theorem, see [**Dur79**, A3], together with $Z_f^2 = -2$. □

6.39. REMARK. At this point, we can describe the connection between Gorenstein rational surface singularities and the ADE Coxeter-Dynkin diagrams. To do this, we define the *desingularization graph* of a surface X to be the dual graph of the exceptional fiber in a minimal resolution of singularities. Precisely, let $\pi \colon \widetilde{X} \longrightarrow X$ be the minimal resolution of singularities, and let $E_1, \ldots, E_n$ be the irreducible components of the exceptional fiber. Then the desingularization graph has vertices $E_1, \ldots, E_n$, with an edge joining E_i to E_j for $i \neq j$ if and only if $E_i \cap E_j \neq \emptyset$.

Let $Z_f = \sum_i m_i E_i$ be the fundamental divisor of X, and define a function f from the vertices $\{E_1, \ldots, E_n\}$ to $\mathbb{N}$ by $f(E_i) = m_i$. Then for $i = 1, \ldots, n$ we have
$$0 \geqslant Z \cdot E_i = -2m_i + \sum_j m_j(E_i \cdot E_j) = -2m_i + \sum_j m_j,$$

where the sum is over all $j \neq i$ such that $E_i \cap E_j \neq \emptyset$. This gives $2f(E_i) \geqslant \sum_j f(E_j)$, and the negative definiteness of the intersection matrix (Theorem 6.34) implies that f is a sub-additive, non-additive function on the graph. Thus the graph is ADE.

We illustrate the general facts described so far with two examples of resolutions of rational double points: the (A_1) and (D_4) hypersurfaces. We will also draw the desingularization graphs for these two examples.

6.40. EXAMPLE. Let X be the hypersurface in $\mathbb{A}^3$ defined by the (A_1) polynomial $x^2 + y^2 + z^2$. To resolve the singularity of X at the origin, we blow up the origin in $\mathbb{A}^3$. Precisely, we set
$$\widetilde{\mathbb{A}}^3 = \left\{ ((x,y,z),(a:b:c)) \in \mathbb{A}^3 \times \mathbb{P}^2 \mid xb = ya,\ xc = za,\ yc = zb \right\}.$$
(See [**Har77**] for basics on blowups.) The projection $\varphi \colon \widetilde{\mathbb{A}}^3 \longrightarrow \mathbb{A}^3$ is an isomorphism away from the origin in $\mathbb{A}^3$, while $\varphi^{-1}(0,0,0)$ is the projective plane $\mathbb{P}^2 \subseteq \widetilde{\mathbb{A}}^3$.

Let $\widetilde{X}$ be the blowup of X at the origin. That is, $\widetilde{X}$ is the Zariski closure of $\varphi^{-1}(X \setminus (0,0,0))$ in $\widetilde{\mathbb{A}}^3$. Then $\widetilde{X}$ is defined in $\widetilde{\mathbb{A}}^2$ by the vanishing of $a^2 + b^2 + c^2$. The restriction of φ gives $\pi \colon \widetilde{X} \longrightarrow X$, and the exceptional fiber E is the preimage of $(0,0,0)$ in $\widetilde{X}$. We claim that $\widetilde{X}$ is smooth, and that E is a single projective line $\mathbb{P}^1$.

The blowup $\widetilde{X}$ is covered by three affine charts U_a, U_b, U_c, defined by $a \neq 0$, $b \neq 0$, $c \neq 0$ respectively, or equivalently by $a = 1$, $b = 1$, $c = 1$. In the chart U_a, we have $y = xb$ and $z = xc$, so that the defining equation of X becomes
$$x^2 + x^2 b^2 + x^2 z^2 = x^2 \left(1 + b^2 + c^2 \right)$$
Above $X \setminus (0,0,0)$, we have $x \neq 0$, so the preimage of $X \setminus (0,0,0)$ is defined by $x \neq 0$ and $1 + b^2 + c^2 = 0$. The Zariski closure of $\varphi^{-1}(X \setminus (0,0,0))$ is thus in this chart the cylinder $1 + b^2 + c^2 = 0$ in $U_a \cong \mathbb{A}^2$. Applying the same reasoning to the other charts, we conclude that $\widetilde{X}$ is smooth.

Remaining in the chart U_a, we see that the exceptional fiber E is defined in $\widetilde{X}$ by $x = 0$, so is defined in U_a by $1 + b^2 + c^2 = x = 0$, with similar equations in U_b and U_c. We conclude that E is smooth, and even rational, so $E \cong \mathbb{P}^1$.

Drawing the desingularization graph of X is thus quite trivial: it has a single node and no edges.

$$\bullet \atop E$$

Observe that this is the (A_1) Coxeter-Dynkin diagram. Since $E^2 = -2$ by Corollary 6.38, we find that $Z_f = E$ is the fundamental divisor.

§4. GEOMETRIC MCKAY CORRESPONDENCE

6.41. EXAMPLE. For a slightly more sophisticated example, consider the (D_4) hypersurface $X \subseteq \mathbb{A}^3$ defined by the vanishing of $x^2y + y^3 + z^2$. Again blowing up the origin in $\mathbb{A}^3$, we obtain as before

$$\widetilde{\mathbb{A}}^3 = \left\{ ((x,y,z),(a:b:c)) \in \mathbb{A}^3 \times \mathbb{P}^2 \mid xb = ya,\ xc = za,\ yc = zb \right\},$$

with projection $\varphi \colon \widetilde{\mathbb{A}}^3 \longrightarrow \mathbb{A}^3$. This time let X_1 be the Zariski closure of $\varphi^{-1}(X \setminus (0,0,0))$. In the affine chart U_a where $a = 1$, we again have $y = xb$ and $z = xc$, so the defining polynomial becomes

$$x^3 b + x^3 b^3 + x^2 c^2 = x^2 \left(x\left(b + b^3\right) + c^2 \right).$$

Thus X_1 is defined by $x(b+b^3)+c^2$ in U_a, so is a *singular* surface. In fact, an easy change of variables reveals that in this chart X_1 is isomorphic to an (A_1) hypersurface singularity (in the variables $\frac{1}{2}(x + (b + b^3))$, $\frac{i}{2}(x - (b + b^3))$, and c). In particular, X_1 has three singular points, with coordinates $x = c = 0$ and $b + b^3 = 0$. In the coordinates of $\widetilde{\mathbb{A}}^3$, they are at $((0,0,0),(1:b:0))$, where $b^3 = -b$. The exceptional fiber, which we denote E_1, corresponds in this chart to $x = 0$, whence $c = 0$, so is just the b-axis.

In the other charts, we find no further singularities. On U_b, the defining polynomial is

$$y^3 a + y^3 + y^2 c^2 = y^2 \left(ya + y + c^2 \right)$$

so that X_1 is defined in U_b by $ya + y + c^2 = 0$. This is also an (A_1) singularity, this time with a single singular point at $y = c = 0$. However, this point has $\widetilde{\mathbb{A}}^3$ coordinates $((0,0,0),(-1:1:0))$, so we've already seen it; it lies in U_a. The exceptional fiber here is the a-axis. Finally, in the chart U_c, we find

$$z^3 a^2 b + z^3 b^3 + z^2 = z^2 \left(za^2 b + zb^3 + 1 \right)$$

so that X_1 is smooth in this chart and E_1 is not visible. In particular we find that $E_1 \cong \mathbb{P}^1$.

Since the first blowup X_1 is not smooth, we continue, resolving the singularities of the surface $x(b + b^3) + c^2 = 0$ by blowing up its three singular points. Since each singular point is locally isomorphic to an (A_1) hypersurface, we appeal to the previous example to see that the resulting surface $\widetilde{X}$ is smooth, and that each of the three new exceptional fibers E_2, E_3, E_4 intersects the original one E_1 transversely.

The desingularization graph thus has the shape of the (D_4) Coxeter-Dynkin diagram:

$$
\begin{array}{c}
E_2 \\
| \\
E_3 \text{———} E_1 \text{———} E_4
\end{array}
$$

To compute the fundamental divisor Z_f, we begin with $Z_1 = E_1 + E_2 + E_3 + E_4$. Since $E_i^2 = -2$ and $E_j \cdot E_1 = 1$ for each $j = 2, 3, 4$, we find

$$Z_1 \cdot E_1 = -2 + 1 + 1 + 1 = 1 > 0.$$

Thus we replace Z_1 by $Z_2 = 2E_1 + E_2 + E_3 + E_4$. Now

$$Z_2 \cdot E_1 = -4 + 1 + 1 + 1 \leqslant 0,$$

and for $j = 2, 3, 4$ we have $Z_2 \cdot E_j = 2 - 2 + 0 + 0 \leqslant 0$, so that $Z_f = Z_2 = 2E_1 + E_2 + E_3 + E_4$ is the fundamental divisor.

The calculations in the examples can be carried out for each of the Kleinian singularities in Table 6.19, and one verifies the next result, which was McKay's original observation.

6.42. THEOREM (McKay). *Let G be a finite subgroup of $\mathrm{SL}(2,\mathbb{C})$ and $R = \mathbb{C}[\![u,v]\!]^G$ the corresponding ring of invariants. Then the desingularization graph of $X = \operatorname{Spec} R$ is an ADE Coxeter-Dynkin diagram. In particular, it is equal to the McKay-Gabriel quiver of G with the vertex corresponding to the trivial representation removed. Furthermore, the coefficients of the fundamental divisor Z_f coincide with the dimensions of the corresponding irreducible representations of G, and with the ranks of the corresponding indecomposable MCM R-modules.* □

We can now state the theorem of Artin and Verdier on the geometric McKay correspondence. Here is the notation in effect through the end of the section:

6.43. NOTATION. Let $(R, \mathfrak{m}, k)$ be a complete local normal domain of dimension two, which is a rational singularity. Let $\pi \colon \widetilde{X} \longrightarrow X = \operatorname{Spec} R$ be its minimal resolution of singularities, and $E = \pi^{-1}(\mathfrak{m})$ the exceptional fiber, with irreducible components $E_1, \ldots, E_n$. Let $Z_f = \sum_i m_i E_i$ be the fundamental divisor of X. We identify a reflexive R-module M with the associated coherent sheaf of $\mathcal{O}_X$-modules, and define the *strict transform* of M by

$$\widetilde{M} = (M \otimes_{\mathcal{O}_X} \mathcal{O}_{\widetilde{X}})/\text{torsion},$$

a sheaf on $\widetilde{X}$.

§4. GEOMETRIC MCKAY CORRESPONDENCE

6.44. THEOREM (Artin-Verdier). *With notation as above, assume in addition that R is Gorenstein. Then there is a one-one correspondence, induced by the first Chern class $c_1(-)$, between indecomposable non-free MCM R-modules and irreducible components E_i of the exceptional fiber. Precisely: Let M be an indecomposable non-free MCM R-module, and let $[C] = c_1(\widetilde{M}) \in \mathrm{Pic}(\widetilde{X})$. Then there is a unique index i such that $C \cdot E_i = 1$ and $C \cdot E_j = 0$ for $i \neq j$. Furthermore, we have $\mathrm{rank}_R(M) = C \cdot Z_f = m_i$.* □

The first Chern class mentioned in the Theorem is a mechanism for turning a locally free sheaf $\mathcal{E}$ into a divisor $c_1(\mathcal{E})$ in the Picard group $\mathrm{Pic}(\widetilde{X})$. In particular, $c_1(-)$ is additive on short exact sequences over $\widetilde{X}$.

The main ingredients of the proof of Theorem 6.44 are compiled in the next propositions. We refer to [**AV85**] for the proofs.

6.45. PROPOSITION. *With notation as in 6.43, $\widetilde{M}$ enjoys the following properties.*

(i) $\widetilde{M}$ is a locally free $\mathcal{O}_{\widetilde{X}}$-module, generated by its global sections.

(ii) $\Gamma(\widetilde{X}, \widetilde{M}) = M$ and $\mathrm{H}^1(\widetilde{X}, \widetilde{M}^*) = 0$.

(iii) There is a short exact sequence of sheaves on $\widetilde{X}$

(6.1) $$0 \longrightarrow \mathcal{O}_{\widetilde{X}}^{(r)} \longrightarrow \widetilde{M} \longrightarrow \mathcal{O}_C \longrightarrow 0,$$

where $r = \mathrm{rank}_R(M)$, and C is a closed one-dimensional subscheme of $\widetilde{X}$ which meets the exceptional fiber E transversely. Furthermore, the global sections of (6.1) give an exact sequence of R-modules:

(6.2) $$0 \longrightarrow R^{(r)} \longrightarrow M \longrightarrow \Gamma(\widetilde{X}, \mathcal{O}_C) \longrightarrow 0$$

□

Observe that the class $[C]$ of the curve C in the Picard group $\mathrm{Pic}(\widetilde{X})$ is equal to the first Chern class $c_1(\widetilde{M})$ of $\widetilde{M}$, since $c_1(-)$ is additive on short exact sequences and $c_1(\mathcal{L}) = [\mathcal{L}] \in \mathrm{Pic}(\widetilde{X})$ for any line bundle $\mathcal{L}$.

6.46. PROPOSITION. *Keep all the notation of 6.43, and assume in addition that R is Gorenstein. Fix a reflexive R-module M, and let C be the curve guaranteed by Proposition 6.45. Then*

(i) $C \cdot Z_f \leqslant r$, with equality if M has no non-trivial free direct summands.

(ii) If $C = C_1 \cup \cdots \cup C_s$ is the decomposition of C into irreducible components, then M decomposes accordingly: $M \cong M_1 \oplus \cdots \oplus$

M_s, with each M_i indecomposable and $c_1(\widetilde{M_i}) = [C_i]$ for each i. □

§5. Exercises

6.47. EXERCISE. Let $R \longrightarrow S$ be a module-finite extension of complete local rings, with S regular. Prove that if M is a reflexive R-module such that $\operatorname{Ext}_R^i(M^*, S) = 0$ for $i = 1, \ldots, n-2$, then $M \in \operatorname{add}_R(S)$.

6.48. EXERCISE. Let R be a reduced Noetherian ring and M, N, P finitely generated reflexive R-modules. Define the *reflexive product* of M and N by
$$M \cdot N = (M \otimes_R N)^{**}$$
Prove the following isomorphisms.
  (i) $M \cdot N \cong N \cdot M$.
  (ii) $\operatorname{Hom}_R(M \cdot N, P) \cong \operatorname{Hom}_R(M, \operatorname{Hom}_R(N, P))$.
  (iii) $M \cdot (N \cdot P) \cong (M \cdot N) \cdot P$.

6.49. EXERCISE. Let $(R, \mathfrak{m})$ be a reduced CM local ring of dimension two, $X = \operatorname{Spec} R$, $U = X \setminus \{\mathfrak{m}\}$, and $i \colon U \longrightarrow X$ the open embedding. Let $M \mapsto \widetilde{M}$ and $\mathcal{F} \mapsto \Gamma(\mathcal{F})$ be the usual sheafification and global section functors between R-modules and coherent sheaves on X.
  (i) If M is MCM, then the natural map $M \longrightarrow \Gamma(i_*i^*\widetilde{M})$ is an isomorphism. (Use the exact sequence $0 \longrightarrow \operatorname{H}^0_{\mathfrak{m}}(M) \longrightarrow M \longrightarrow \Gamma(i_*i^*\widetilde{M}) \longrightarrow \operatorname{H}^1_{\mathfrak{m}}(M) \longrightarrow 0$.)
  (ii) If M is torsion-free, $M^{**} \longrightarrow \Gamma(i_*i^*\widetilde{M})$ is an isomorphism. (Use the case above and $\lambda(\operatorname{H}^1_{\mathfrak{m}}(M)) < \infty$. Notice i^* is exact since i is an open embedding, and $i^* \operatorname{H}^1_{\mathfrak{m}}(M) = 0$, so get a square relating M to M^{**} and $\widetilde{M^{**}}$.)
  (iii) Assume R is normal, and let $\operatorname{VB}(U)$ be the category of locally free $\mathcal{O}_U$-modules. Then $i^* \colon \operatorname{CM}(R) \longrightarrow \operatorname{VB}(U)$ is an equivalence.

6.50. EXERCISE (Abhyankar). Let $(R, \mathfrak{m}, k)$ be a CM local ring of multiplicity $\operatorname{e}(R)$. Verify the inequality
$$\operatorname{e}(R) \geqslant \mu_R(\mathfrak{m}) - \dim(R) + 1\,.$$

6.51. EXERCISE. Generalize Corollary 6.37 by showing that any Gorenstein local ring $(R, \mathfrak{m}, k)$ satisfying $\operatorname{e}(R) = \mu_R(\mathfrak{m}) - \dim(R) + 1$ is a hypersurface of multiplicity two.

6.52. EXERCISE. Classify the finite subgroups of $\mathrm{GL}(2,\mathbb{C})$ by using the surjection $\mathbb{C}^* \times \mathrm{SL}(2,\mathbb{C}) \longrightarrow \mathrm{GL}(2,\mathbb{C})$ sending (d,σ) to $d\sigma$.

6.53. EXERCISE. Let G be a finite cyclic subgroup of $\mathrm{GL}(2,\mathbb{C})$, say $G = \langle \sigma \rangle$. Show that the ring of invariants $\mathbb{C}[\![u,v]\!]^G$ is generated by two invariants if and only if σ has an eigenvalue equal to 1.

CHAPTER 7

Isolated Singularities and Dimension Two

In this chapter we present a pair of celebrated theorems due originally to Auslander. The first, Theorem 7.12, states that a CM local ring of finite CM type has at most an isolated singularity. We give the simplified proof due to Huneke and Leuschke, which requires some easy general preliminaries on elements of Ext^1. The second, Theorem 7.19, gives a strong converse to Herzog's Theorem 6.3, namely that in dimension two over a field of characteristic zero, every CM complete local algebra having finite CM type is a ring of invariants.

§1. Miyata's theorem

The classical Yoneda correspondence (see for example [**ML95**]) allows us to identify elements of an Ext-module $\mathrm{Ext}^i_R(M, N)$ as equivalence classes of i-fold extensions of N by M. In the case $i = 1$, this is particularly simple: an element $\alpha \in \mathrm{Ext}^1_R(M, N)$ is an equivalence class of short exact sequences $0 \longrightarrow N \longrightarrow X \longrightarrow M \longrightarrow 0$, where we declare two such sequences, with middle terms X, X', to be equivalent if they fit into a commutative diagram

(7.1)
$$\begin{array}{ccccccccc} 0 & \longrightarrow & N & \longrightarrow & X & \longrightarrow & M & \longrightarrow & 0 \\ & & \parallel & & \downarrow & & \parallel & & \\ 0 & \longrightarrow & N & \longrightarrow & X' & \longrightarrow & M & \longrightarrow & 0 \,. \end{array}$$

It follows from the Snake Lemma that in this situation $X \cong X'$, so the middle term X_α is determined by the element α. The converse is false (cf. Exercise 7.22), but Miyata's Theorem [**Miy67**] gives a partial converse: if a short exact sequence is "apparently" split—the middle term is isomorphic to the direct sum of the other two—then it is split.

7.1. THEOREM (Miyata). *Let R be a commutative Noetherian ring and let*
$$\alpha: \quad N \xrightarrow{p} X_\alpha \xrightarrow{q} M \longrightarrow 0$$
be an exact sequence of finitely generated R-modules. If $X_\alpha \cong M \oplus N$, then α is a split short exact sequence.

PROOF. It suffices to show that $p\colon N \longrightarrow X_\alpha$ is a *pure* homomorphism, that is, $Z \otimes_R p\colon Z \otimes_R N \longrightarrow Z \otimes_R X_\alpha$ is injective for every finitely generated R-module Z. Indeed, taking $Z = R$ will show that p is injective, and by Exercise 7.23 (or Exercise 13.37), pure submodules with finitely-presented quotients are direct summands.

Fix a finitely generated R-module Z. To show that $Z \otimes_R p$ is injective, we may localize at a maximal ideal and assume that $(R, \mathfrak{m})$ is local. Suppose $c \in Z \otimes N$ is a non-zero element of the kernel of $Z \otimes_R p$. Take n so large that $c \notin \mathfrak{m}^n(Z \otimes_R N) = \mathfrak{m}^n Z \otimes_R N$. Tensoring further with $R/\mathfrak{m}^n$ gives the right-exact sequence

$$(Z/\mathfrak{m}^n Z) \otimes_R N \xrightarrow{\overline{p}} (Z/\mathfrak{m}^n Z) \otimes_R (M \oplus N) \longrightarrow (Z/\mathfrak{m}^n Z) \otimes_R M \longrightarrow 0$$

of finite length R-modules. Counting lengths shows that $\overline{p}$ is injective, contradicting the presence of the nonzero element $\overline{c}$ in the kernel. □

Let

$$\alpha\colon \quad 0 \longrightarrow N \longrightarrow X_\alpha \longrightarrow M \longrightarrow 0$$

$$\beta\colon \quad 0 \longrightarrow N \longrightarrow X_\beta \longrightarrow M \longrightarrow 0$$

be two extensions of N by M, with $X_\alpha \cong X_\beta$. As mentioned above, α and β need not represent the same element of $\operatorname{Ext}_R^1(M, N)$. In the rest of this section we describe a result of Striuli [**Str05**] giving a partial result in that direction.

7.2. REMARK. We recall briefly a few more details of the Yoneda correspondence for Ext^1. First, recall that if $\alpha \in \operatorname{Ext}_R^1(M, N)$ is represented by the short exact sequence

$$\alpha\colon \quad 0 \longrightarrow N \longrightarrow X_\alpha \longrightarrow M \longrightarrow 0,$$

then for $r \in R$, the product $r\alpha$ can be computed via either a pullback or a pushout. Precisely, $r\alpha$ is represented either by the top row of the diagram

$$\begin{array}{ccccccccc} r\alpha\colon & 0 & \longrightarrow & N & \longrightarrow & P & \longrightarrow & M & \longrightarrow 0 \\ & & & \| & & \downarrow & & \downarrow r & \\ \alpha\colon & 0 & \longrightarrow & N & \xrightarrow{p} & X & \xrightarrow{q} & M & \longrightarrow 0 \end{array}$$

or the bottom row of the diagram

$$\begin{array}{ccccccccc} \alpha\colon & 0 & \longrightarrow & N & \xrightarrow{p} & X & \xrightarrow{q} & M & \longrightarrow 0 \\ & & & \downarrow r & & \downarrow & & \| & \\ r\alpha\colon & 0 & \longrightarrow & N & \longrightarrow & Q & \longrightarrow & M & \longrightarrow 0 \end{array}$$

where
$$P = \{(x,m) \in X \oplus M \mid q(x) = rm\}$$
and
$$Q = X \oplus N / \langle (p(n), -rn) \mid n \in N \rangle \, .$$

More generally, the same sorts of diagrams define actions of $\operatorname{End}_R(M)$ and $\operatorname{End}_R(N)$ on $\operatorname{Ext}^1_R(M,N)$, on the right and left respectively, replacing r by an endomorphism of the appropriate module.

Pullbacks and pushouts also define the connecting homomorphisms δ in the long exact sequences of Ext. If $\alpha \in \operatorname{Ext}^1_R(M,N)$ is as above, then for an R-module Z the long exact sequence looks like

$$\cdots \longrightarrow \operatorname{Hom}_R(Z, X) \xrightarrow{q_*} \operatorname{Hom}_R(Z, M) \xrightarrow{\delta} \operatorname{Ext}^1_R(Z, N) \longrightarrow \cdots \, .$$

The image of a homomorphism $g \colon Z \longrightarrow M$ in $\operatorname{Ext}^1_R(M,N)$ is the top row of the pullback diagram below.

$$\begin{array}{ccccccccc}
0 & \longrightarrow & N & \longrightarrow & U & \longrightarrow & Z & \longrightarrow & 0 \\
& & \| & & \downarrow & & \downarrow g & & \\
0 & \longrightarrow & N & \xrightarrow{p} & X & \xrightarrow{q} & M & \longrightarrow & 0
\end{array}$$

In particular, when $Z = M$ we find that $\delta(1_M) = \alpha$. Similar considerations apply for the long exact sequence attached to $\operatorname{Hom}_R(-, Z)$.

Here is the result that will occupy the rest of the section. In fact this result holds for arbitrary Noetherian rings; we leave the straightforward extension to the interested reader.

7.3. THEOREM (Striuli). *Let R be a local ring. Let*

$$\alpha: \quad 0 \longrightarrow N \longrightarrow X_\alpha \longrightarrow M \longrightarrow 0$$

$$\beta: \quad 0 \longrightarrow N \longrightarrow X_\beta \longrightarrow M \longrightarrow 0$$

be two short exact sequences of finitely generated R-modules. Suppose that $X_\alpha \cong X_\beta$ and that $\beta \in I \operatorname{Ext}^1_R(M,N)$ for some ideal I of R. Then the complex $\alpha \otimes_R R/I$ is a split exact sequence.

We need one preliminary result.

7.4. PROPOSITION. *Let $(R, \mathfrak{m})$ be a local ring and I an ideal of R. Let*

$$\alpha: \quad 0 \longrightarrow N \xrightarrow{p} X_\alpha \xrightarrow{q} M \longrightarrow 0$$

be a short exact sequence of finitely generated R-modules, and denote by $\overline{\alpha} = \alpha \otimes_R R/I$ the complex

$$\overline{\alpha}: \quad 0 \longrightarrow N/IN \xrightarrow{\overline{p}} X_\alpha/IX_\alpha \xrightarrow{\overline{q}} M/IM \longrightarrow 0.$$

If $\alpha \in I \operatorname{Ext}^1_R(M, N)$, then $\overline{\alpha}$ is a split exact sequence.

PROOF. By Miyata's Theorem 7.1 it suffices to show that X_α/IX_α is isomorphic to $M/IM \oplus N/IN$. Let

$$\xi: \quad 0 \longrightarrow Z \xrightarrow{i} F_0 \xrightarrow{d_0} M \longrightarrow 0$$

be the beginning of a minimal resolution of M over R, so that $Z = \operatorname{syz}^R_1(M)$ is the first syzygy of M. Then applying $\operatorname{Hom}_R(-, N)$ gives a surjection $\operatorname{Hom}_R(Z, N) \longrightarrow \operatorname{Ext}^1_R(M, N)$. In particular $I \operatorname{Hom}_R(Z, N)$ maps onto $I \operatorname{Ext}^1_R(M, N)$, so there exists $\varphi \in I \operatorname{Hom}_R(Z, N)$ such that α is obtained from the pushout diagram below.

$$\begin{array}{ccccccccc}
\xi: & 0 & \longrightarrow & Z & \xrightarrow{i} & F_0 & \xrightarrow{d_0} & M & \longrightarrow 0 \\
& & & \varphi\downarrow & & \psi\downarrow & & \| & \\
\alpha: & 0 & \longrightarrow & N & \xrightarrow{p} & X_\alpha & \xrightarrow{q} & M & \longrightarrow 0
\end{array}$$

In particular, we have $\varphi(Z) \subseteq IN$. The pushout diagram also induces an exact sequence

$$\nu: \quad 0 \longrightarrow Z \xrightarrow{\begin{bmatrix} i \\ -\varphi \end{bmatrix}} F_0 \oplus N \xrightarrow{[\psi\ p]} X_\alpha \longrightarrow 0.$$

Let L be an arbitrary R/I-module of finite length, and tensor both ξ and ν with L:

$$Z \otimes_R L \xrightarrow{i \otimes 1_L} F_0 \otimes_R L \xrightarrow{d_0 \otimes 1_L} M \otimes_R L \longrightarrow 0$$

$$Z \otimes_R L \xrightarrow{\begin{bmatrix} i \otimes 1_L \\ -\varphi \otimes 1_L \end{bmatrix}} (F_0 \otimes_R L) \oplus (N \otimes_R L) \xrightarrow{\begin{bmatrix} \psi \otimes 1_L \\ p \otimes 1_L \end{bmatrix}^T} X_\alpha \otimes_R L \longrightarrow 0.$$

Since $\varphi(Z) \subset IN$ and $IL = 0$, the image of $-\varphi \otimes 1_L$ is zero in $N \otimes_R L$. Denoting the image of $i \otimes 1$ by K, we get exact sequences

$$0 \longrightarrow K \longrightarrow F_0 \otimes_R L \longrightarrow M \otimes_R L \longrightarrow 0$$

$$0 \longrightarrow K \longrightarrow (F_0 \otimes_R L) \oplus (N \otimes_R L) \longrightarrow X_\alpha \otimes_R L \longrightarrow 0.$$

Counting lengths (over either R or R/I, equally) now gives

$$\ell(X_\alpha \otimes_R L) = \ell(M \otimes_R L) + \ell(N \otimes_R L).$$

In particular, since L is an R/I-module, we have

$$\ell(X_\alpha/IX_\alpha \otimes_{R/I} L) = \ell(M/IM \otimes_{R/I} L) + \ell(N/IN \otimes_{R/I} L).$$

Exercise 7.25 now applies, since L was arbitrary, to give $X_\alpha/IX_\alpha \cong M/IM \oplus N/IN$. $\square$

PROOF OF THEOREM 7.3. Since $\beta \in I \operatorname{Ext}_R^1(M, N)$, we have, by Proposition 7.4, $X_\beta/IX_\beta \cong M/IM \oplus N/IN$ and hence $X_\alpha/IX_\alpha \cong M/IM \oplus N/IN$. Now Miyata's Theorem 7.1 implies that $\alpha \otimes_R R/I$ is split exact. $\square$

Here is an amusing consequence.

7.5. COROLLARY. *Let $(R, \mathfrak{m})$ be a local ring and M a non-free finitely generated module. Let α be the short exact sequence*

$$\alpha: \quad 0 \longrightarrow M_1 \longrightarrow F \longrightarrow M \longrightarrow 0,$$

where F is a finitely generated free module and $M_1 \subseteq \mathfrak{m}F$. Then α is a part of a minimal generating set of $\operatorname{Ext}_R^1(M, M_1)$.

PROOF. If $\alpha \in \mathfrak{m} \operatorname{Ext}_R^1(M, M_1)$, then $\overline{\alpha} = \alpha \otimes R/\mathfrak{m}$ is split exact. But since $M_1 \subseteq \mathfrak{m}F$, the image of $M_1 \otimes R/\mathfrak{m}$ is zero, a contradiction. $\square$

7.6. EXAMPLE. The converse of Proposition 7.4 fails. Consider the one-dimensional (A_2) singularity $R = k[\![t^2, t^3]\!]$. Since R is Gorenstein, $\operatorname{Ext}_R^1(k, R) \cong k$, and so every nonzero element of $\operatorname{Ext}_R^1(k, R)$ is part of a basis. Define α to be the bottom row of the pushout diagram

$$\begin{array}{ccccccccc}
0 & \longrightarrow & \mathfrak{m} & \longrightarrow & R & \longrightarrow & k & \longrightarrow & 0 \\
& & \varphi \downarrow & & \downarrow & & \| & & \\
0 & \longrightarrow & R & \longrightarrow & X & \longrightarrow & k & \longrightarrow & 0
\end{array}$$

where φ is defined by $\varphi(t^2) = t^3$ and $\varphi(t^3) = t^4$. Then α is non-split, since there is no map $R \longrightarrow R$ extending φ, whence $\alpha \notin \mathfrak{m} \operatorname{Ext}_R^1(k, R)$. On the other hand, $\mu(X) = 2$ and hence $X/\mathfrak{m}X \cong k \oplus k$. It follows that $\overline{\alpha}$ is split exact.

These results raise the following question, which will be particularly relevant in Chapter 15.

7.7. QUESTION. *Let $(R, \mathfrak{m})$ be a CM local ring and let M and N be MCM R-modules. Take a maximal regular sequence $\mathbf{x}$ on R, M, and N, and take $\alpha \in \operatorname{Ext}_R^1(M, N)$. Is it true that $\alpha \in \mathbf{x} \operatorname{Ext}_R^1(M, N)$ if and only if $\alpha \otimes R/(\mathbf{x})$ is split exact?*

§2. Isolated singularities

Now we come to the first major theorem in the general theory of CM local rings of finite CM type: that they have at most isolated singularities. The result is due originally to Auslander [**Aus86a**] for complete local rings, though as Yoshino observed, the original proof relies only on the KRS property, hence works equally well for Henselian rings by Theorem 1.8. Auslander's argument is a tour de force of functorial imagination, and an early vindication of the use of almost split sequences in commutative algebra (cf. Chapter 13). Here we give a simple argument due to Huneke and Leuschke [**HL02**], valid for all CM local rings, using the results of the previous section.

7.8. DEFINITION. Let $(R, \mathfrak{m})$ be a local ring. We say that R is, or has, an *isolated singularity* provided $R_\mathfrak{p}$ is a regular local ring for all non-maximal prime ideals $\mathfrak{p}$.

Note that we include the case where R is regular under the definition above. We also say R has "at most" an isolated singularity to explicitly allow this possibility.

The next lemma is standard, and we leave its proof as an exercise (Exercise 7.27).

7.9. LEMMA. *Let $(R, \mathfrak{m})$ be a CM local ring. Then the following conditions are equivalent.*

(i) *The ring R has at most an isolated singularity.*
(ii) *Each MCM R-module is locally free on the punctured spectrum.*
(iii) *For all MCM R-modules M and N, $\mathrm{Ext}^1_R(M, N)$ has finite length.* □

7.10. LEMMA. *Let $(R, \mathfrak{m})$ be a local ring, $r \in \mathfrak{m}$, and*

$$
\begin{array}{ccccccccc}
\alpha: & 0 & \longrightarrow & N & \xrightarrow{i} & X_\alpha & \longrightarrow & M & \longrightarrow 0 \\
& & & {\scriptstyle r}\downarrow & & {\scriptstyle f}\downarrow & & \parallel & \\
r\alpha: & 0 & \longrightarrow & N & \xrightarrow{j} & X_{r\alpha} & \longrightarrow & M & \longrightarrow 0
\end{array}
$$

a commutative diagram of short exact sequences of finitely generated R-modules. Assume that $X_\alpha \cong X_{r\alpha}$ (not necessarily via the map f). Then $\alpha \in r\,\mathrm{Ext}^1_R(M, N)$.

Note that the case $r = 0$ is Miyata's Theorem 7.1.

PROOF. The pushout diagram gives an exact sequence

$$ 0 \longrightarrow N \xrightarrow{\left[\begin{smallmatrix} r \\ -i \end{smallmatrix}\right]} N \oplus X_\alpha \xrightarrow{[j\ f]} X_{r\alpha} \longrightarrow 0. $$

Since $N \oplus X_\alpha \cong N \oplus X_{r\alpha}$, Miyata's Theorem 7.1 implies that the sequence splits. In particular, the induced map on Ext,

$$ \left[\begin{smallmatrix} r \\ -i_* \end{smallmatrix}\right] : \mathrm{Ext}^1_R(M, N) \longrightarrow \mathrm{Ext}^1_R(M, N) \oplus \mathrm{Ext}^1_R(M, X_\alpha), $$

is a split injection. Let h be a left inverse for $\begin{bmatrix} r \\ -i_* \end{bmatrix}$.

Now apply $\operatorname{Hom}_R(M, -)$ to α, getting an exact sequence

$$\cdots \longrightarrow \operatorname{Hom}_R(M, M) \xrightarrow{\delta} \operatorname{Ext}^1_R(M, N) \xrightarrow{i_*} \operatorname{Ext}^1_R(M, X_\alpha) \longrightarrow \cdots.$$

The connecting homomorphism δ takes 1_M to α, so $i_*(\alpha) = 0$. Thus

$$\alpha = h(r\alpha, 0) = rh(\alpha, 0) \quad \in r\operatorname{Ext}^1_R(M, N).\qquad\square$$

7.11. THEOREM. *Let $(R, \mathfrak{m})$ be local and M, N finitely generated R-modules. Suppose there are only finitely many isomorphism classes of modules X for which there exists a short exact sequence*

$$0 \longrightarrow N \longrightarrow X \longrightarrow M \longrightarrow 0.$$

Then $\operatorname{Ext}^1_R(M, N)$ has finite length.

PROOF. Let $\alpha \in \operatorname{Ext}^1_R(M, N)$, and let $r \in \mathfrak{m}$. By Exercise 7.26, it will suffice to prove that $r^n \alpha = 0$ for $n \gg 0$. For any integer $n \geqslant 0$, we consider a representative for $r^n \alpha$, namely

$$r^n\alpha : \quad 0 \longrightarrow N \longrightarrow X_n \longrightarrow M \longrightarrow 0.$$

Since there are only finitely many isomorphism classes of such X_n, there exists an infinite sequence $n_1 < n_2 < \cdots$ such that $X_{n_i} \cong X_{n_j}$ for every i, j. Set $\beta = r^{n_1}\alpha$, and let $i > 1$. Note that $r^{n_i}\alpha = r^{n_i - n_1}\beta$. Hence we get the commutative diagram

$$\begin{array}{ccccccccc}
\beta : & & 0 & \longrightarrow & N & \longrightarrow & X_{n_1} & \longrightarrow & M & \longrightarrow & 0 \\
& & & & \Big\downarrow r^{n_i - n_1} & & \Big\downarrow & & \Big\| & & \\
r^{n_i - n_1}\beta : & & 0 & \longrightarrow & N & \longrightarrow & X_{n_i} & \longrightarrow & M & \longrightarrow & 0
\end{array}$$

for each i. By Lemma 7.10, $X_{n_1} \cong X_{n_i}$ implies $\beta \in r^{n_i - n_1}\operatorname{Ext}^1_R(M, N)$ for every i. This implies $\beta \in \mathfrak{m}^t \operatorname{Ext}^1_R(M, N)$ for every $t \geqslant 1$, whence, by the Krull Intersection Theorem, $\beta = 0$. $\square$

If R has finite CM type, then for all MCM modules M and N, there exist only finitely many MCM modules X generated by at most $\mu_R(M) + \mu_R(N)$ elements, thus finitely many potential middle terms for short exact sequences. Thus we obtain Auslander's theorem:

7.12. THEOREM (Auslander). *Let $(R, \mathfrak{m})$ be a CM ring with finite CM type. Then R has at most an isolated singularity.* $\square$

7.13. REMARK. A non-commutative version of Theorem 7.12 is easy to state, and the same proof applies. This was Auslander's original context [**Aus86a**]. Specifically, Auslander considers the following situation: Let T be a complete regular local ring and let Λ be a (possibly non-commutative) T-algebra which is a finitely generated free

T-module. Say that Λ is *non-singular* if $\operatorname{gldim}\Lambda = \dim(T)$, and that Λ has *finite representation type* if there are only finitely many isomorphism classes of indecomposable finitely generated (left) Λ-modules that are free as T-modules. If Λ has finite representation type, then $\Lambda_{\mathfrak{p}}$ is non-singular for all non-maximal primes $\mathfrak{p}$ of T.

We mention here a few further applications of Theorem 7.11, all based on the same elementary observation. Suppose that R is a CM local ring and M is a MCM R-module such that there are only finitely many non-isomorphic MCM modules of multiplicity less than or equal to $\mu_R(M) \cdot \operatorname{e}(R)$; then M is locally free on the punctured spectrum. This follows immediately from Theorem 7.12 upon taking N to be the first syzygy of M in a minimal free resolution. If in addition R is a domain, then the criterion simplifies to the existence of only finitely many MCM modules of *rank* at most $\mu_R(M)$.

Obvious candidates for M are the canonical module ω, the conormal module I/I^2 of a regular presentation $R = A/I$, and the module of Kähler differentials $\Omega^1_{R/k}$ if R is essentially of finite type over a field k. Since the freeness of these modules implies that R is Gorenstein, respectively complete intersection [**Mat89**, 19.9], respectively regular [**Kun86**, Theorem 7.2], we obtain the following corollaries.

7.14. COROLLARY. *Let $(R, \mathfrak{m})$ be a CM local ring with canonical module ω. If R has only finitely many non-isomorphic MCM modules of multiplicity up to $r(R)\operatorname{e}(R)$, where $r(R) = \dim_k \operatorname{Ext}_R^{\dim(R)}(k, R)$ denotes the Cohen-Macaulay type of R, then R is Gorenstein on the punctured spectrum.* □

7.15. COROLLARY. *Let $(A, \mathfrak{n})$ be a regular local ring, and suppose $I \subseteq \mathfrak{n}^2$ is an ideal such that $R = A/I$ is CM. Assume that I/I^2 is a MCM R-module. If R has only finitely many non-isomorphic MCM modules of multiplicity at most $\mu_A(I) \cdot \operatorname{e}(R)$, then R is complete intersection on the punctured spectrum.* □

7.16. COROLLARY. *Let k be a field of characteristic zero, and let R be a k-algebra essentially of finite type. Let $\Omega^1_{R/k}$ be the module of Kähler differentials of R over k. Assume that Ω is a MCM R-module. If R has only finitely many non-isomorphic MCM modules of multiplicity up to $\operatorname{embdim}(R) \cdot \operatorname{e}(R)$, then R has at most an isolated singularity.* □

The second corollary raises the question of when the conormal module I/I^2 is MCM over A/I for an ideal I in a regular local ring A. Herzog [**Her78a**] showed that this is the case if A/I is Gorenstein and

I has height three; see [**HU89**] and [**Buc81**, 6.2.10] for some further results in this direction.

§3. Two-dimensional CM rings of finite CM type

Our aim in this section is to prove a converse to Herzog's Theorem 6.3, which states that rings of invariants of two-dimensional regular local rings have finite CM type. The result, due to Auslander [**Aus86b**] and Esnault [**Esn85**], is that if a complete local ring R of dimension two, with a coefficient field k of characteristic zero, has finite CM type, then $R \cong k[\![u,v]\!]^G$ for some finite group $G \subseteq \mathrm{GL}(n,k)$.

Auslander's proof relies on a deep result of Mumford in topology (see [**Mum61**] and [**Hir95b**]). We give Mumford's theorem below, followed by the interpretation and more general statement in commutative algebra due to Flenner [**Fle75**] (see also [**CS93**]).

7.17. THEOREM (Mumford). *Let V be a normal complex space of dimension 2 and $x \in V$ a point. Then the following properties hold.*
  (i) *The local fundamental group $\pi(V,x)$ is finitely generated.*
  (ii) *If the local homology group $H_1(V,x)$ vanishes, then $\pi(V,x)$ is isomorphic to the fundamental group of a valued tree having negative definite intersection matrix.*
  (iii) *If $\pi(V,x) = \{1\}$ is trivial, then x is a regular point.* □

To translate Mumford's result into commutative algebra, we recall the definition of the *étale fundamental group*, also called the algebraic fundamental group. See [**Mil08**] for more details. (We will not attempt maximal generality in this brief sketch; in particular, we will ignore the need to choose a base point.) For a connected normal scheme X, the étale fundamental group $\pi_1^{et}(X)$ classifies the finite étale coverings of X in a manner analogous to the usual fundamental group classifying the covering spaces of a topological space.

The construction of π_1^{et} is clearest when $X = \mathrm{Spec}\, A$ for a normal domain A. Let K be the quotient field of A, and fix an algebraic closure Ω of K. Then $\pi_1^{et}(X) \cong \mathrm{Gal}(L/K)$, where L is the union of all the finite separable field extensions K' of K contained in Ω, and such that the integral closure of A in K' is étale over A. There is a Galois correspondence between subgroups $H \subseteq \pi_1^{et}(X)$ of finite index and finite étale covers $A \longrightarrow B$ of A. In particular, $\pi_1^{et}(X) = 0$ if and only if A has no non-trivial finite étale covers.

With some extra work, the étale fundamental group can be defined for arbitrary schemes X. In particular, one may take X to be the punctured spectrum $\mathrm{Spec}^\circ A = \mathrm{Spec}\, A \setminus \{\mathfrak{m}\}$ of a local ring $(A, \mathfrak{m})$. We say that the local ring $(A, \mathfrak{m})$ is *pure* if the induced morphism of étale fundamental groups $\pi_1^{et}(\mathrm{Spec}^\circ A) \longrightarrow \pi_1^{et}(\mathrm{Spec}\, A)$ is an isomorphism. (Unfortunately this usage of the word "pure" has nothing to do with the usage of the same word earlier in this chapter.) The point is the surjectivity: A is pure if and only if every étale cover of the punctured spectrum extends to an étale cover of the whole spectrum.

7.18. THEOREM (Flenner). *Let $(A, \mathfrak{m}, k)$ be an excellent Henselian local normal domain of dimension two. Assume that $\mathrm{char}\, k = 0$. Consider the following conditions.*
   (i) $\pi_1^{et}(\mathrm{Spec}^\circ A) = 0$;
   (ii) A is pure;
   (iii) A is a regular local ring.
Then (i) $\implies$ (ii) $\iff$ (iii), and the three conditions are equivalent if k is algebraically closed. □

The implication "A regular $\implies$ A pure" is a restatement of the theorem on the purity of the branch locus (Theorem B.12). The content of the theorem of Mumford and Flenner is in the other implications, in particular, a converse to purity of the branch locus.

Now we come to Auslander and Esnault's characterization of the equicharacteristic zero, two-dimensional, complete local rings having finite CM type.

7.19. THEOREM (Auslander, Esnault). *Let R be a complete CM local ring of dimension two with coefficient field k. Assume that k has characteristic zero. If R has finite CM type, then there exists a power series ring $S = k[\![u, v]\!]$ and a finite group G acting on S by linear changes of variables such that $R \cong S^G$.*

PROOF. First, notice that by Theorem 7.12 R is regular in codimension one, whence a normal domain.

Let K be the quotient field of the normal domain R, and fix an algebraic closure Ω. Consider the family of all finite field extensions K' of K, contained in Ω, and such that the integral closure of R in K' is unramified in codimension one over R. Let L be the field generated by all these K', and let S be the integral closure of R in L.

We will show that L is a finite Galois extension of K, so that in particular S is a module-finite R-algebra [**Mat89**, p. 262, Lemma 1]. Furthermore, S is a local ring since R is Henselian. Observe that if we show that S is a local ring module-finite over R, then by construction S has no module-finite ring extensions which are unramified in codimension one; indeed, any such ring extension would also be module-finite and unramified in codimension one over R. (See Appendix B.) In other words, we will have $\pi_1^{et}(\operatorname{Spec} S \setminus \{\mathfrak{m}_S\}) = 0$ and it will follow that S is a regular local ring, hence $S \cong k[\![u,v]\!]$.

To show that L/K is a finite Galois extension, assume that there is an infinite ascending chain
$$K \subsetneq L_1 \subsetneq L_2 \subsetneq \cdots \subsetneq L$$
of finite Galois extensions of K inside L. Let S_i be the integral closure of R in L_i. Then we have a corresponding infinite ascending chain
$$R \subsetneq S_1 \subsetneq S_2 \subsetneq \cdots \subsetneq S$$
of module-finite ring extensions. Each S_i is a normal domain, so in particular a reflexive R-module. By Exercise 4.31, the S_i are pairwise non-isomorphic as R-modules, contradicting the assumption that R has finite CM type. Thus L/K is finite, and it's easy to see it is a Galois extension. Let G be the Galois group of L over K. Then G acts on S with fixed ring R, and the argument of Lemma 5.3 allows us to assume the action is linear. $\square$

Theorem 7.19 is false in positive characteristic. Artin [**Art77**] has given counterexamples to Mumford's characterization of smoothness in characteristic $p > 0$; the simplest is the (A_{p-1}) singularity $x^2 + y^p + z^2 = 0$, which has trivial étale fundamental group. Even though $k[\![x,y,z]\!]/(x^2 + y^p + z^2)$ has finite CM type by Theorem 6.23, it is not a ring of invariants when k has characteristic p.

Among other things, Auslander's Theorem 7.19 implies that the two-dimensional CM local rings of finite CM type with residue field $\mathbb{C}$ have rational singularities (see Definition 6.32). This suggests the following conjecture.

7.20. CONJECTURE. *Let $(R, \mathfrak{m})$ be a CM local ring of dimension at least two. Assume that R has finite CM type. Then R has rational singularities.*

The assumption $\dim(R) \geqslant 2$ is necessary to allow for the existence of non-normal, that is, non-regular, one-dimensional rings of finite CM type.

To add some evidence for this conjecture, we recall that by results of Mumford [**Mum61**] (in characteristic zero) and Lipman [**Lip69**] (in characteristic $p > 0$), a normal surface singularity $X = \operatorname{Spec} R$ has a rational singularity if and only if there are only finitely many rank one MCM R-modules up to isomorphism.

Here is a weaker version of Conjecture 7.20. This problem was first raised in print by Eisenbud and Herzog [**EH88**].

7.21. CONJECTURE. *Let $(R, \mathfrak{m})$ be a CM local ring of dimension at least two. If R has finite CM type, then R has minimal multiplicity, that is,*
$$\mathrm{e}(R) = \mu_R(\mathfrak{m}) - \dim(R) + 1\,.$$

Recall that rational singularity implies minimal multiplicity, Corollary 6.36. We will prove Conjecture 7.21 for hypersurfaces in Chapter 9, §3, and in fact Conjecture 7.20 for the hypersurface case will follow from the classification in Chapter 9.

§4. Exercises

7.22. EXERCISE. Prove that the $p-1$ inequivalent non-zero extensions in $\mathrm{Ext}^1_{\mathbb{Z}}(\mathbb{Z}/p\mathbb{Z}, \mathbb{Z}/p\mathbb{Z})$ all have isomorphic middle terms. Find an example of abelian groups A and B and two elements of $\mathrm{Ext}^1_{\mathbb{Z}}(A, B)$ with isomorphic middle terms but different annihilators. (See [**Str05**] for one example, due to Caviglia.)

7.23. EXERCISE. Let $N \subset M$ be modules over a commutative ring R. Prove that N is a pure submodule of M if and only if the following condition is satisfied: Whenever $x_1, \ldots, x_t$ is a sequence of elements in N, and $x_i = \sum_{j=1}^{s} r_{ij} m_j$ for some $r_{ij} \in R$ and $m_j \in M$, there exist $y_1, \ldots, y_s \in N$ such that $x_i = \sum_{j=1}^{s} r_{ij} y_j$ for $i = 1, \ldots, t$. Conclude that if M/N is finitely presented and $N \subset M$ is pure, then the inclusion of N into M splits. (See also Exercise 13.37.)

7.24. EXERCISE. Let R be a commutative Artinian ring and let M, N be two finitely generated R-modules. Prove that $M \cong N$ if and only if $\ell(\mathrm{Hom}_R(M, X)) = \ell(\mathrm{Hom}_R(N, X))$ for every finitely generated R-module X. (Hint: It suffices by induction on $\ell(\mathrm{Hom}_R(N, N))$ to show that M and N have a non-zero direct summand in common. To show this, take generators $f_1, \ldots, f_r$ for $\mathrm{Hom}_R(M, N)$ to define a homomorphism $F \colon M^{(r)} \longrightarrow N$, and show that F splits.) See [**Bon89**].

7.25. EXERCISE. Prove that the following conditions are equivalent for finitely generated modules M and N over a local ring $(R, \mathfrak{m})$.
   (i) $M \cong N$;
   (ii) $\ell(\mathrm{Hom}_R(M, L)) = \ell(\mathrm{Hom}_R(N, L))$ for every R-module L of finite length;
   (iii) $\ell(M \otimes_R L) = \ell(N \otimes_R L)$ for every R-module L of finite length.

(Hint: Use Matlis duality for (ii) $\implies$ (iii). Assuming (ii), reduce modulo $\mathfrak{m}^n$ and conclude from Exercise 7.24 that $M/\mathfrak{m}^n M \cong N/\mathfrak{m}^n N$ for every n, then use Corollary 1.14.)

7.26. EXERCISE. Let $(R, \mathfrak{m})$ be local, and let M be a finitely generated R-module. Show that M has finite length if and only if for all $r \in \mathfrak{m}$ and for all $x \in M$, there exists an integer n such that $r^n x = 0$.

7.27. EXERCISE. Consider these conditions on a local ring R.
 (i) The ring R has at most an isolated singularity.
 (ii) Every MCM R-module is locally free on the punctured spectrum.
 (iii) For all MCM R-modules M and N, $\ell(\operatorname{Ext}^1_R(M, N)) < \infty$.

Prove a slightly more general version of Lemma 7.9: We have (i) $\implies$ (ii) $\implies$ (iii), and (iii) $\implies$ (i) if R is CM.

CHAPTER 8

The Double Branched Cover

In this chapter we introduce two key tools in the representation theory of hypersurface rings: matrix factorizations and the double branched cover. We fix the following notation for the entire chapter.

8.1. CONVENTIONS. Let $(S, \mathfrak{n}, k)$ be a regular local ring and let f be a non-zero element of $\mathfrak{n}^2$. Put $R = S/(f)$ and $\mathfrak{m} = \mathfrak{n}/(f)$. We let $d = \dim(R) = \dim(S) - 1$.

§1. Matrix factorizations

With the notation of 8.1, suppose M is a MCM R-module. Then M has depth d when viewed as an R-module or as an S-module. By the Auslander-Buchsbaum formula, M has projective dimension 1 over S. Therefore the minimal free resolution of M as an S-module is of the form

$$(8.1) \qquad 0 \longrightarrow G \xrightarrow{\varphi} F \longrightarrow M \longrightarrow 0,$$

where G and F are finitely generated free S-modules. Since $f \cdot M = 0$, M is a torsion S-module, so $\operatorname{rank}_S G = \operatorname{rank}_S F$.

For any $x \in F$, the image of fx in M vanishes, so there is a unique element $y \in G$ such that $\varphi(y) = fx$. Since the element y is linearly determined by x, we get a homomorphism $\psi \colon F \longrightarrow G$ satisfying $\varphi\psi = f\, 1_F$. It follows from the injectivity of the map φ that $\psi\varphi = f\, 1_G$ too. This construction motivates the following definition [**Eis80**].

8.2. DEFINITION. Let $(S, \mathfrak{n}, k)$ be a regular local ring, and let f be a non-zero element of $\mathfrak{n}^2$. A *matrix factorization* of f is a pair (φ, ψ) of homomorphisms between free S-modules of the same rank, $\varphi \colon G \longrightarrow F$ and $\psi \colon F \longrightarrow G$, such that

$$\psi\varphi = 1_G \qquad \text{and} \qquad \varphi\psi = 1_F \,.$$

Equivalently (after choosing bases), φ and ψ are square matrices of the same size over S, say $n \times n$, such that

$$\psi\varphi = \varphi\psi = I_n \,.$$

Let (φ, ψ) be a matrix factorization of f as in Definition 8.2. Since f is a non-zerodivisor, it follows that φ and ψ are injective, and we have short exact sequences

(8.2)
$$0 \longrightarrow G \xrightarrow{\varphi} F \longrightarrow \operatorname{cok}\varphi \longrightarrow 0$$
$$0 \longrightarrow F \xrightarrow{\psi} G \longrightarrow \operatorname{cok}\psi \longrightarrow 0$$

of S-modules. As $fF = \varphi\psi(F)$ is contained in the image of φ, the cokernel of φ is annihilated by f. Similarly, $f \cdot \operatorname{cok}\psi = 0$. Thus $\operatorname{cok}\varphi$ and $\operatorname{cok}\psi$ are naturally finitely generated modules over $R = S/(f)$.

8.3. PROPOSITION. *Let $(S, \mathfrak{n})$ be a regular local ring and let f be a non-zero element of $\mathfrak{n}^2$.*
  (i) *For every MCM R-module, there is a matrix factorization (φ, ψ) of f with $\operatorname{cok}\varphi \cong M$.*
  (ii) *If (φ, ψ) is a matrix factorization of f, then $\operatorname{cok}\varphi$ and $\operatorname{cok}\psi$ are MCM R-modules.*

PROOF. Only the second statement needs verification. The exact sequences (8.2) and the fact that $f \cdot \operatorname{cok}\varphi = 0 = f \cdot \operatorname{cok}\psi$ imply that the cokernels have projective dimension one over S. By the Auslander-Buchsbaum formula, they have depth equal to $\dim(S) - 1 = \dim(R)$ and therefore are MCM R-modules. $\square$

8.4. NOTATION. When we wish to emphasize the provenance of a presentation matrix φ as half of a matrix factorization (φ, ψ), we write $\operatorname{cok}(\varphi, \psi)$ in place of $\operatorname{cok}\varphi$. We also write $(\varphi \colon G \longrightarrow F, \psi \colon F \longrightarrow G)$ to include the free S-module G and F in the notation.

There are two distinguished *trivial* matrix factorizations of any element $f \in S$, namely $(f, 1)$ and $(1, f)$. Note that $\operatorname{cok}(1, f) = 0$, while $\operatorname{cok}(f, 1) \cong R$.

8.5. DEFINITION. Consider a pair of matrix factorizations of $f \in S$, say $(\varphi \colon G \longrightarrow F, \psi \colon F \longrightarrow G)$ and $(\varphi' \colon G' \longrightarrow F', \psi' \colon F' \longrightarrow G')$. A *homomorphism of matrix factorizations* between (φ, ψ) and (φ', ψ') is a pair of S-module homomorphisms $\alpha \colon F \longrightarrow F'$ and $\beta \colon G \longrightarrow G'$ rendering the diagram

(8.3)
$$\begin{array}{ccccc} F & \xrightarrow{\psi} & G & \xrightarrow{\varphi} & F \\ {\scriptstyle\alpha}\downarrow & & {\scriptstyle\beta}\downarrow & & \downarrow{\scriptstyle\alpha} \\ F' & \xrightarrow{\psi'} & G' & \xrightarrow{\varphi'} & F' \end{array}$$

commutative. (In fact, commutativity of just one of the squares is sufficient; see Exercise 8.34.)

A homomorphism $(\alpha, \beta)\colon (\varphi, \psi) \longrightarrow (\varphi', \psi')$ of matrix factorizations induces a homomorphism of R-modules $\operatorname{cok}(\varphi, \psi) \longrightarrow \operatorname{cok}(\varphi', \psi')$, which we denote $\operatorname{cok}(\alpha, \beta)$:

$$\begin{array}{ccccccccc}
0 & \longrightarrow & G & \xrightarrow{\varphi} & F & \longrightarrow & \operatorname{cok}(\varphi, \psi) & \longrightarrow & 0 \\
 & & \downarrow \beta & & \downarrow \alpha & & \downarrow \operatorname{cok}(\alpha, \beta) & & \\
0 & \longrightarrow & G' & \xrightarrow{\varphi'} & F' & \longrightarrow & \operatorname{cok}(\varphi', \psi') & \longrightarrow & 0
\end{array}$$

Conversely, every S-linear map from $\operatorname{cok}(\varphi, \psi)$ to $\operatorname{cok}(\varphi', \psi')$ lifts to give a homomorphism of matrix factorizations.

Two matrix factorizations (φ, ψ) and (φ', ψ') are *equivalent* if there is a homomorphism of matrix factorizations $(\alpha, \beta)\colon (\varphi, \psi) \longrightarrow (\varphi', \psi')$ in which both α and β are isomorphisms.

Direct sums of matrix factorizations are defined in the natural way:

$$(\varphi, \psi) \oplus (\varphi', \psi') = \left(\begin{pmatrix} \varphi & \\ & \varphi' \end{pmatrix}, \begin{pmatrix} \psi & \\ & \psi' \end{pmatrix} \right).$$

We say that a matrix factorization is *reduced* provided it is not equivalent to a matrix factorizations having a trivial direct summand $(f, 1)$ or $(1, f)$. It's straightforward to see that (φ, ψ) is reduced if and only if all the entries of φ and ψ are in the maximal ideal of S. See Exercise 8.35. In particular, φ has no unit entries if and only if $\operatorname{cok}(\varphi, \psi)$ has no non-zero R-free direct summands.

With overlines denoting reduction modulo f, a matrix factorization $(\varphi\colon G \longrightarrow F,\ \psi\colon F \longrightarrow G)$ induces a complex

$$(8.4) \qquad \cdots \longrightarrow \overline{G} \xrightarrow{\overline{\varphi}} \overline{F} \xrightarrow{\overline{\psi}} \overline{G} \xrightarrow{\overline{\varphi}} \overline{F} \longrightarrow \operatorname{cok}(\varphi, \psi) \longrightarrow 0$$

in which $\overline{G}$ and $\overline{F}$ are finitely generated free modules over $R = S/(f)$. In fact (Exercise 8.36), this complex is exact, hence is a free resolution of $\operatorname{cok}(\varphi, \psi)$. If (φ, ψ) is a reduced matrix factorization, then (8.4) is a minimal R-free resolution of $\operatorname{cok}(\varphi, \psi)$.

The reversed pair (ψ, φ) is also a matrix factorization of f, and the resolution (8.4) exhibits $\operatorname{cok}(\psi, \varphi)$ as a first syzygy of $\operatorname{cok}(\varphi, \psi)$ and vice versa:

$$(8.5) \qquad \begin{array}{c} 0 \longrightarrow \operatorname{cok}(\psi, \varphi) \longrightarrow \overline{F} \longrightarrow \operatorname{cok}(\varphi, \psi) \longrightarrow 0 \\ 0 \longrightarrow \operatorname{cok}(\varphi, \psi) \longrightarrow \overline{G} \longrightarrow \operatorname{cok}(\psi, \varphi) \longrightarrow 0 \end{array}$$

are exact sequences of R-modules. This gives the first assertion of the next result; we leave the rest, and the proof of the theorem following, as exercises. Recall that an R-module M is *stable* provided it does not have a direct summand isomorphic to R. We remark that a direct sum of two stable modules is again stable, by Lemma 1.2 (i) (or directly, Exercise 8.37).

8.6. PROPOSITION. *Keep the notation of 8.1.*

(i) *Let M be a MCM R-module. Then M has a free resolution which is periodic of period at most two.*
(ii) *Let M be a stable MCM R-module. Then the minimal free resolution of M is periodic of period at most two.*
(iii) *Let M be a MCM R-module. Then $\operatorname{syz}_1^R M$ is a stable MCM R-module. If M is indecomposable and not free, so is $\operatorname{syz}_1^R M$.*
(iv) *Let M be a finitely generated R-module. Then the minimal free resolution of M is eventually periodic of period at most two. In particular the minimal free resolution of M is bounded.*
(v) *Let M and N be R-modules with M finitely generated. For each $i \geqslant \dim(R) - \operatorname{depth} M$, we have $\operatorname{Ext}_R^i(M,N) \cong \operatorname{Ext}_R^{i+2}(M,N)$ and $\operatorname{Tor}_i^R(M,N) \cong \operatorname{Tor}_{i+2}^R(M,N)$.* □

In the next chapter we will see a converse to (iv): If every minimal free resolution over a local ring R is bounded, then (the completion of) R is a hypersurface ring.

8.7. THEOREM (Eisenbud). *Keep the notation of 8.1. The association*

$$(\varphi, \psi) \longleftrightarrow \operatorname{cok}(\varphi, \psi)$$

induces an equivalence of categories between reduced matrix factorizations of f up to equivalence and of stable MCM R-modules up to isomorphism. In particular, it gives a bijection between equivalence classes of reduced matrix factorizations and isomorphism classes of stable MCM modules. □

8.8. REMARK. If in addition f is a prime/irreducible element of S, so that R is an integral domain, then from $\varphi\psi = f \cdot I_n$ it follows that both $\det\varphi$ and $\det\psi$ are, up to units, powers of f. Specifically, we must have $\det\varphi = uf^k$ and $\det\psi = u^{-1}f^{n-k}$ for some unit $u \in S$ and $k \leqslant n$. In this case the R-module $\operatorname{cok}(\varphi,\psi)$ has rank k, while $\operatorname{cok}(\psi,\varphi)$ has rank $n - k$. To see this, localize at the prime ideal (f). Then over the discrete valuation ring $S_{(f)}$, φ is equivalent to $f \cdot 1_k \oplus 1_{n-k}$ and so $\operatorname{cok}\varphi$ has rank k over the field $R_{(f)}$.

Similar remarks hold when f is merely reduced, provided we consider rank M as the tuple $(\operatorname{rank}_{R_{\mathfrak{p}}} M_{\mathfrak{p}})$ as $\mathfrak{p}$ runs over the minimal primes in R.

8.9. REMARK. Let
$$(\varphi\colon G \longrightarrow F,\ \psi\colon F \longrightarrow G) \quad \text{and} \quad (\varphi'\colon G' \longrightarrow F',\ \psi'\colon F' \longrightarrow G')$$
be two matrix factorizations of f. Put $M = \operatorname{cok}(\varphi,\psi)$, $N = \operatorname{cok}(\psi,\varphi)$, $M' = \operatorname{cok}(\varphi',\psi')$, and $N' = \operatorname{cok}(\psi',\varphi')$. Then any homomorphism of matrix factorizations $(\alpha,\beta)\colon (\psi,\varphi) \longrightarrow (\varphi',\psi')$ (note the order!) defines a pushout diagram

(8.6)
$$\begin{array}{ccccccccc} 0 & \longrightarrow & N & \longrightarrow & \overline{F} & \longrightarrow & M & \longrightarrow & 0 \\ & & \downarrow & & \downarrow & & \| & & \\ 0 & \longrightarrow & M' & \longrightarrow & Q & \longrightarrow & M & \longrightarrow & 0 \end{array}$$

of R-modules, the bottom row of which is the image of $\operatorname{cok}(\alpha,\beta)$ under the surjective connecting homomorphism
$$\operatorname{Hom}_R(N, M') \longrightarrow \operatorname{Ext}^1_R(M, M')\,.$$
In particular, every extension of M' by M arises in this way.

The middle module Q is of course MCM as well. Splicing (8.6) together with the R-free resolutions of N and M', we obtain a morphism of exact sequences

(8.7)
$$\begin{array}{ccccccccccc} \cdots & \xrightarrow{\overline{\varphi}} & \overline{F} & \xrightarrow{\overline{\psi}} & \overline{G} & \xrightarrow{\overline{\varphi}} & \overline{F} & \longrightarrow & M & \longrightarrow & 0 \\ & & \beta\downarrow & & \alpha\downarrow & & \downarrow & & \| & & \\ \cdots & \xrightarrow[\overline{\psi'}]{} & \overline{G'} & \xrightarrow[\overline{\varphi'}]{} & \overline{F'} & \longrightarrow & Q & \longrightarrow & M & \longrightarrow & 0 \end{array}$$

defined, after the first step, by α and β. The *mapping cone* of (8.7) is thus the exact complex
$$\cdots \longrightarrow \overline{F'}\oplus\overline{F} \xrightarrow{\begin{bmatrix}\overline{\psi'} & \beta \\ & -\overline{\varphi}\end{bmatrix}} \overline{G'}\oplus\overline{G} \xrightarrow{\begin{bmatrix}\overline{\varphi'} & \alpha \\ & -\overline{\psi}\end{bmatrix}} \overline{F'}\oplus\overline{F} \longrightarrow Q\oplus M \longrightarrow M \longrightarrow 0.$$

We may cancel the two occurrences of M (since the map between them is the identity) and find that
$$Q \cong \operatorname{cok}\left(\begin{pmatrix}\overline{\varphi'} & \alpha \\ & -\overline{\psi}\end{pmatrix}, \begin{pmatrix}\overline{\psi'} & \beta \\ & -\overline{\varphi}\end{pmatrix}\right).$$

§2. The double branched cover

We continue with the notation and conventions established in 8.1 and assume, in addition, that S is complete. Thus $(S, \mathfrak{n}, k)$ is a complete regular local ring of dimension $d+1$, $0 \neq f \in \mathfrak{n}^2$, and $R = S/(f)$. We will refer to a ring R of this form as a *complete hypersurface singularity*.

8.10. DEFINITION. The *double branched cover* of R is
$$R^\sharp = S[\![z]\!]/(f + z^2)\,,$$
a complete hypersurface singularity of dimension $d+1$.

8.11. WARNING. It is important to have a particular defining equation in mind, since different equations defining the same hypersurface R can lead to non-isomorphic rings $R^\sharp$. For example, we have $\mathbb{R}[\![x]\!]/(x^2) = \mathbb{R}[\![x]\!]/(-x^2)$, yet $\mathbb{R}[\![x, z]\!]/(x^2 + z^2) \not\cong \mathbb{R}[\![x, z]\!]/(-x^2 + z^2)$. (One is a domain; the other is not.) Exercise 8.39 shows that such oddities cannot occur if k is algebraically closed and of characteristic different from two.

We want to compare the MCM modules over $R^\sharp$ with those over R. Observe that we have a surjection $R^\sharp \longrightarrow R$, killing the class of z. There is no homomorphism the other way in general. However, $R^\sharp$ is a finitely generated free S-module, generated by the images of 1 and z; cf. Exercise 8.41.

8.12. DEFINITION. Let N be a MCM $R^\sharp$-module. Set
$$N^\flat = N/zN\,,$$
a MCM R-module. Contrariwise, let M be a MCM R-module. View M as an $R^\sharp$-module via the surjection $R^\sharp \longrightarrow R$, and set
$$M^\sharp = \operatorname{syz}_1^{R^\sharp} M\,.$$

Notice that there is no conflict of notation if we view R as an R-module and sharp it: Since z is a non-zerodivisor of $R^\sharp$ (cf. Exercise 8.40), we have a short exact sequence
$$0 \longrightarrow R^\sharp \xrightarrow{z} R^\sharp \longrightarrow R \longrightarrow 0\,.$$
Thus $R^\sharp$ is indeed the first syzygy of R as an $R^\sharp$-module.

8.13. NOTATION. Let $\varphi \colon G \longrightarrow F$ be a homomorphism of finitely generated free S-modules, or equivalently a matrix with entries in S. We use the same symbol φ for the induced homomorphism $S[\![z]\!] \otimes_S G \longrightarrow S[\![z]\!] \otimes_S F$; as matrices, they are identical. In particular we abuse the notation 1_F, using it also for the identity map $S[\![z]\!] \otimes_S 1_F$.

Furthermore let $\widetilde{\varphi}\colon \widetilde{G} \longrightarrow \widetilde{F}$ denote the corresponding homomorphism over $R^\sharp$, obtained via the composition of the natural homomorphisms $S \longrightarrow S[\![z]\!] \longrightarrow S[\![z]\!]/(f+z^2) = R^\sharp$. Finally, as in §1, we let $\overline{\varphi}\colon \overline{G} \longrightarrow \overline{F}$ denote the matrix over $R = S/(f)$ obtained via the natural map $S \longrightarrow R$. Thus $\overline{F} = \widetilde{F}/z\widetilde{F}$.

8.14. LEMMA. *Let* $(\varphi\colon G \longrightarrow F, \psi\colon F \longrightarrow G)$ *be a matrix factorization of* f, *let* $M = \mathrm{cok}(\varphi, \psi)$, *and let* $\pi\colon \widetilde{F} \twoheadrightarrow M$ *be the composition* $\widetilde{F} \twoheadrightarrow \overline{F} \twoheadrightarrow M$.

(i) *There is an exact sequence*

$$\widetilde{F}\oplus\widetilde{G} \xrightarrow{\begin{bmatrix} \widetilde{\psi} & -z1_{\widetilde{G}} \\ z1_{\widetilde{F}} & \widetilde{\varphi} \end{bmatrix}} \widetilde{G}\oplus\widetilde{F} \xrightarrow{[\widetilde{\varphi}\ z1_{\widetilde{F}}]} \widetilde{F} \xrightarrow{\pi} M \longrightarrow 0$$

of $R^\sharp$-*modules*.
(ii) *The matrices over* $S[\![z]\!]$

$$\begin{pmatrix} \psi & -z1_G \\ z1_F & \varphi \end{pmatrix} \quad \text{and} \quad \begin{pmatrix} \varphi & z1_F \\ -z1_G & \psi \end{pmatrix}$$

form a matrix factorization of $f+z^2$ *over* $S[\![z]\!]$.
(iii) *We have*

$$M^\sharp \cong \mathrm{cok}\left(\begin{pmatrix} \psi & -z1_G \\ z1_F & \varphi \end{pmatrix}, \begin{pmatrix} \varphi & z1_F \\ -z1_G & \psi \end{pmatrix}\right).$$

(iv) *The* $R^\sharp$-*module* $M^\sharp$ *is stable if and only if* $_R M$ *is stable, in which case*

$$\mathrm{syz}_1^{R^\sharp}(M^\sharp) \cong M^\sharp.$$

PROOF. The proof of (ii) amounts to matrix multiplication, and the isomorphism in (iii) is an immediate consequence of (i), (ii), and the matrix calculation

$$\begin{pmatrix} & 1 \\ 1 & \end{pmatrix} \begin{pmatrix} \varphi & z1_F \\ -z1_G & \psi \end{pmatrix} \begin{pmatrix} & 1 \\ 1 & \end{pmatrix} = \begin{pmatrix} \psi & -z1_G \\ z1_F & \varphi \end{pmatrix}$$

over $S[\![z]\!]$. Item (iv) follows from (iii) and Exercise 8.35, since the entries of the matrix factorization for $M^\sharp$ are those in the matrix factorization for M, together with z. It thus suffices to prove (i). First we note that z is a non-zerodivisor of $R^\sharp$ (Exercise 8.40). Therefore the columns of the following commutative diagram are exact.

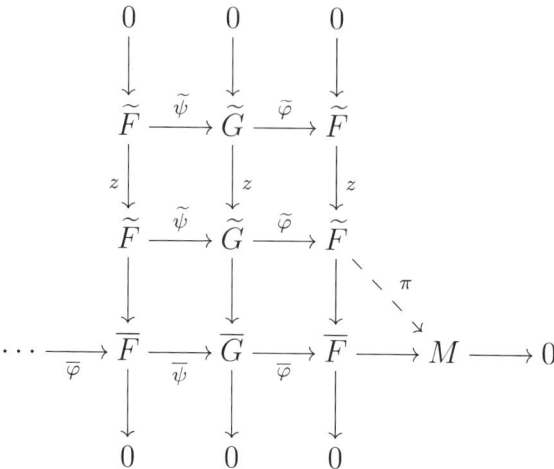

The bottom row is also exact by (8.4), but the first two rows aren't even complexes. In fact,

(8.8) $$\widetilde{\varphi}\widetilde{\psi} = -z^2 \, 1_{\widetilde{F}} \, .$$

An easy diagram chase shows that $\ker \pi = \operatorname{im} \widetilde{\varphi} + z\widetilde{F} = \operatorname{im} [\widetilde{\varphi} \ z 1_{\widetilde{F}}]$. Also,

$$\ker [\widetilde{\varphi} \ z 1_{\widetilde{F}}] \supseteq \operatorname{im} \begin{bmatrix} \widetilde{\psi} & -z 1_{\widetilde{G}} \\ z 1_{\widetilde{F}} & \widetilde{\varphi} \end{bmatrix}$$

by (8.8). For the opposite inclusion, let $\begin{bmatrix} x \\ y \end{bmatrix} \in \ker [\widetilde{\varphi} \ z 1_{\widetilde{F}}]$, so that $\widetilde{\varphi}(x) = -zy$. A diagram chase yields elements $a \in \widetilde{F}$ and $b \in \widetilde{G}$ such that $[\widetilde{\psi} \ -z 1_{\widetilde{G}}] \begin{bmatrix} a \\ b \end{bmatrix} = x$. We need to show that $[z 1_{\widetilde{F}} \ \widetilde{\varphi}] \begin{bmatrix} a \\ b \end{bmatrix} = y$. Using (8.8), we obtain the equations

$$z(za + \widetilde{\varphi}(b)) = -\widetilde{\varphi}\widetilde{\psi}(a) + z\widetilde{\varphi}(b) = -\widetilde{\varphi}(\widetilde{\psi}(a) - zb) = -\widetilde{\varphi}(x) = zy \, .$$

Cancelling the non-zerodivisor z, we get the desired result. □

This allows us already to prove one "natural" relation between sharping and flatting.

8.15. PROPOSITION. *Let M be a MCM R-module. Then*
$$M^{\sharp\flat} \cong M \oplus \operatorname{syz}_1^R M \, .$$

PROOF. The R-module $M^{\sharp\flat}$ is presented by the matrix factorization $(\Phi \otimes_{R^\sharp} R, \Psi \otimes_{R^\sharp} R)$, where (Φ, Ψ) is the matrix factorization for $M^\sharp$ given in Lemma 8.14. Killing z in that matrix factorization gives

$$M^{\sharp\flat} \cong \operatorname{cok}\left(\begin{pmatrix} \varphi & \\ & \psi \end{pmatrix}, \begin{pmatrix} \psi & \\ & \varphi \end{pmatrix}\right),$$

as desired. □

8.16. COROLLARY. *Let M be an indecomposable stable MCM R-module.*

(i) $M^\sharp$ is a direct sum of either one or two indecomposable $R^\sharp$-modules.

(ii) M is a direct summand of $N^\flat$ for some indecomposable non-free R-module N.

PROOF. If $M^\sharp$ were a direct sum of three or more non-trivial R-modules, then $M^{\sharp\flat}$ would be too. But by Proposition 8.15 and Proposition 8.6 (iii), $M^{\sharp\flat}$ is a direct sum of exactly two indecomposable modules.

For the second statement, we note that $M \oplus \mathrm{syz}_1^R(M)$ is stable by (iii) of Proposition 8.6 and Exercise 8.37. Write $M^\sharp \cong N_1 \oplus \cdots \oplus N_t$, where the N_i are indecomposable MCM $R^\sharp$-modules, none of the free by Propostion 8.15. Then $M \oplus \mathrm{syz}_1^R(M) \cong M^{\sharp\flat} \cong N_1^\flat \oplus \cdots \oplus N_t^\flat$. By KRS, M is isomorphic to a direct summand of some $N_i^\flat$. □

The question of whether $M^\sharp$ is indecomposable or a direct sum of two indecomposable modules will be answered in Proposition 8.30. Now we turn to the other "natural" relation. Recall that $R^\sharp$ is a free S-module of rank 2; in particular any MCM $R^\sharp$-module is a finitely generated free S-module.

8.17. LEMMA. *Let N be a MCM $R^\sharp$-module. Let $\varphi \colon N \longrightarrow N$ be an S-linear homomorphism representing multiplication by z on N.*

(i) The pair $(\varphi, -\varphi)$ is a matrix factorization of f with
$$\mathrm{cok}(\varphi, -\varphi) \cong N^\flat.$$

(ii) If N is stable, then
$$N^\flat \cong \mathrm{syz}_1^R(N^\flat),$$
and hence $N^\flat$ is stable too.

(iii) Consider $z\, 1_N \pm \varphi$ as an endomorphism of $S[\![z]\!] \otimes_S N$, a finitely generated free $S[\![z]\!]$-module. Then
$$(z\, 1_N - \varphi,\ z\, 1_N + \varphi)$$
is a matrix factorization of $f + z^2$ with $\mathrm{cok}(z\, 1_N - \varphi, z\, 1_N + \varphi) \cong N$. If N is stable, then $(z\, 1_N - \varphi,\ z\, 1_N + \varphi)$ is a reduced matrix factorization.

PROOF. On the S-module N, $-\varphi^2$ corresponds to multiplication by $-z^2$. But since N is an $R^\sharp$-module, the action of $-z^2$ on N agrees with that of f. In other words, $-\varphi^2 = f\, 1_N$. Now φ and $-\varphi$ obviously have isomorphic cokernels, each isomorphic to $N/zN = N^\flat$. Items (i) and (ii) follow. We leave the first assertion of (iii) as Exercise 8.42. For the final sentence, note that if $z\,1_N - \varphi$ contains a unit of $S[\![z]\!]$, then φ contains a unit of S as an entry. But then $z\,1_N + \varphi$ has a unit entry, so that the trivial matrix factorization $(f + z^2, 1)$ is a direct summand of $(z\,1_N - \varphi, z\,1_N + \varphi)$ up to equivalence. This exhibits $R^\sharp$ as a direct summand of N, contradicting the stability of N. □

8.18. PROPOSITION. *Let N be a stable MCM $R^\sharp$-module. Assume that* char $k \neq 2$. *Then*
$$N^{\flat\sharp} \cong N \oplus \operatorname{syz}_1^{R^\sharp} N \,.$$

PROOF. Let $\varphi\colon N \longrightarrow N$ be the homomorphism of free S-modules representing multiplication by z as in Lemma 8.17. Then $(\varphi, -\varphi)$ is a matrix factorization of f with $\operatorname{cok}(\varphi, -\varphi) \cong N^\flat$ by the Lemma, so that
$$N^{\flat\sharp} = \operatorname{syz}_1^{R^\sharp}(N^\flat)$$
$$\cong \operatorname{cok}\left(\begin{pmatrix} -\varphi & -z\,1_N \\ z\,1_N & \varphi \end{pmatrix}, \begin{pmatrix} \varphi & z\,1_N \\ -z\,1_N & -\varphi \end{pmatrix}\right)$$
by (iii) of Lemma 8.14. Noting that $\frac{1}{2} \in R$ and hence that the matrix $\begin{bmatrix} 1 & 1 \\ -1 & 1 \end{bmatrix}$ is invertible over R, we pass to an equivalent matrix
$$\begin{bmatrix} z\,1_N - \varphi & 0 \\ 0 & z\,1_N + \varphi \end{bmatrix} = \frac{1}{2}\begin{bmatrix} 1 & 1 \\ -1 & 1 \end{bmatrix}\begin{bmatrix} -\varphi & -z\,1_N \\ z\,1_N & \varphi \end{bmatrix}\begin{bmatrix} 1 & 1 \\ -1 & 1 \end{bmatrix}.$$
Then
$$N^{\flat\sharp} \cong \operatorname{cok}(z\,1_N - \varphi,\ z\,1_N + \varphi) \oplus \operatorname{cok}(z\,1_N + \varphi,\ z\,1_N - \varphi) \cong N \oplus \operatorname{syz}_1^{R^\sharp} N$$
by (iii) of Lemma 8.17. □

The proof of the next result is essentially identical to that of Corollary 8.16:

8.19. COROLLARY. *Assume that the characteristic of k is different from two. Let N be an indecomposable non-free MCM $R^\sharp$-module.*

(i) *$N^\flat$ is a direct sum of either one or two indecomposable R-modules.*

(ii) *N is a direct summand of $M^\sharp$ for some indecomposable non-free R-module M.* □

We now have all the machinery we need to verify that R has finite CM type if and only if $R^\#$ does. The same arguments take care of the case of *countable* CM type (see Definition 14.1), so in anticipation of Chapter 14 we prove both statements simultaneously.

8.20. THEOREM (Knörrer). *Let $(S, \mathfrak{n}, k)$ be a complete regular local ring, f a non-zero element of $\mathfrak{n}^2$, and $R = S/(f)$.*

(i) If $R^\sharp$ has finite (respectively countable) CM type, then so has R.

(ii) If R has finite (respectively countable) CM type and char $k \neq 2$, then $R^\sharp$ has finite CM type.

PROOF. We will prove (i), leaving the almost identical proof of (ii) to the reader. Let $\{N_i\}_{i \in I}$ be a representative list of the indecomposable non-free MCM $R^\sharp$-modules, where I is a finite (respectively countable) index set. Write $N_i^\flat = M_{i1} \oplus \cdots \oplus M_{ir_i}$, where each M_{ij} is an indecomposable R-module. (We are ignoring the first assertion of Corollary 8.19 here, since we're allowing the residue field to have characteristic two.) By Corollary 8.16, every indecomposable non-free MCM R-module is a direct summand of some N_i and therefore, by KRS, must occur on the finite (respectively countable) list $\{M_{ij}\}$. □

8.21. EXAMPLE. One cannot remove the assumption on the characteristic in (ii). For example, let $R = k[\![x]\!]/(x^2)$, and notice that $R^\sharp$ has infinite CM type in characteristic two by Proposition 4.15.

8.22. COROLLARY (ADE Redux). *Let $(R, \mathfrak{m}, k)$ be an ADE (or simple) plane curve singularity (cf. Chapter 4, §3) over an algebraically closed field k of characteristic different from 2, 3 or 5. Then R has finite CM type.*

PROOF. The hypersurface $R^\sharp$ is a complete Kleinian singularity and therefore has finite CM type by Theorem 6.23. By Theorem 8.20, R has finite CM type. □

8.23. EXAMPLE. Assume k is a field with char $k \neq 2$, and let n and d be integers with $n \geqslant 1$ and $d \geqslant 0$. Put $R_{n,d} = k[\![x, z_1, \ldots, z_d]\!]/(x^{n+1} + z_1^2 + \cdots + z_d^2)$. The ring $R_{n,0} = k[\![x]\!]/(x^{n+1})$ obviously has finite CM type (see Theorem 3.3). By applying Theorem 8.20 repeatedly, we see that the d-dimensional (A_n)-singularity $R_{n,d}$ has finite CM type for every d. If k contains a square root of -1, the ring $R = k[\![x_1, \ldots, x_t, y_1, \ldots, y_t]\!]/(x_1 y_1 + \cdots + x_t y_t)$ also has finite CM type: The change of variables $x_i = u_i + \sqrt{-1}\, v_i$, $y_i = u_i - \sqrt{-1}\, v_i$ shows that $R \cong R_{1, 2d+2}$.

§3. Knörrer's periodicity

The results of the previous section on the double branched cover imply that if M and N are indecomposable non-free MCM modules over R and $R^\sharp$, respectively, then $M^{\sharp\flat}$ and $N^{\flat\sharp}$ both decompose into precisely two indecomposable MCM modules. However, we do not yet know whether this splitting occurs on the way up or the way down. In this section we clarify this point, and use the result to prove Knörrer's theorem that the MCM modules over R are in bijection with those over the *double* double branched cover $R^{\sharp\sharp}$.

8.24. NOTATION. We keep all the notations of the last section, so that $(S, \mathfrak{n}, k)$ is a complete regular local ring, f is a non-zero element of $\mathfrak{n}^2$, and $R = S/(f)$ is the corresponding complete hypersurface singularity. In addition, we assume throughout this section that k is an *algebraically closed* field of characteristic different from 2.

We first prove a sort of converse to Lemma 8.17.

8.25. LEMMA. *Let M be a MCM R-module such that $M \cong \operatorname{syz}_1^R M$. Then $M \cong \operatorname{cok}(\varphi_0, \varphi_0)$ for a square matrix φ_0 satisfying $\varphi_0^2 = f I_n$.*

PROOF. We may assume that M is indecomposable, and write $M = \operatorname{cok}(\varphi \colon G \longrightarrow F, \psi \colon F \longrightarrow G)$ by Theorem 8.7. By assumption there is an equivalence of matrix factorizations $(\alpha, \beta) \colon (\varphi, \psi) \longrightarrow (\psi, \varphi)$, i.e. a commutative diagram of free S-modules

$$\begin{array}{ccccc} F & \xrightarrow{\psi} & G & \xrightarrow{\varphi} & F \\ \downarrow{\alpha} & & \downarrow{\beta} & & \downarrow{\alpha} \\ G & \xrightarrow{\varphi} & F & \xrightarrow{\psi} & G \end{array}$$

with α and β isomorphisms. Thus $\operatorname{cok}(\beta\alpha, \alpha\beta)$ is an automorphism of M. Since M is indecomposable and R is complete, $\operatorname{End}_R(M)$ is an nc-local ring. Furthermore, $\operatorname{End}_R(M)/\mathcal{J}(\operatorname{End}_R(M)) \cong k$ since k is algebraically closed. Hence we may write

$$\operatorname{cok}(\beta\alpha, \alpha\beta) = c 1_M + \rho,$$

where $c \in k^\times$ and $\rho \in \mathcal{J}(\operatorname{End}_R(M))$. By replacing α by $c^{-1}\alpha$, we may assume that $c = 1$. Now, putting $\rho_1 = \beta\alpha - 1_F$ and $\rho_2 = \alpha\beta - 1_G$, we have

$$(\beta\alpha, \alpha\beta) = (1_F, 1_G) + (\rho_1, \rho_2)$$

with $\operatorname{cok}(\rho_1, \rho_2) \in \mathcal{J}(\operatorname{End}_R(M))$. Then $\alpha\rho_1 = \rho_2\alpha$ and $\beta\rho_2 = \rho_1\beta$.

Represent the function $(1 + x)^{-1/2}$ by its Maclaurin series and set

$$\alpha' = \alpha(1_F + \rho_1)^{-1/2} = (1_G + \rho_2)^{-1/2}\alpha \colon F \longrightarrow G$$
$$\beta' = \beta(1_G + \rho_2)^{-1/2} = (1_F + \rho_1)^{-1/2}\beta \colon G \longrightarrow F.$$

Then the homomorphism of matrix factorizations $(\alpha', \beta') \colon (\varphi, \psi) \longrightarrow (\psi, \varphi)$ satisfies $\beta'\alpha'\beta = \beta$ and $\alpha'\beta'\alpha = \alpha$. Therefore $\beta'\alpha' = 1_F$ and

$\alpha'\beta' = 1_G$. Finally, set $\varphi_0 = \varphi\alpha'\colon F \longrightarrow F$. Then $\varphi_0 = \beta'\psi$, and we have
$$\varphi_0^2 = \varphi\alpha'\beta'\psi = \varphi\psi = fI_n\,,$$
and $\mathrm{cok}(\varphi_0, \varphi_0) \cong M$. $\square$

Let $R^\sharp = S[\![z]\!]/(f+z^2)$ be the double branched cover of the previous section. Then $R^\sharp$ carries an involution σ, which fixes S and sends z to $-z$. Denote by $R^\sharp[\sigma]$ the *skew group ring* of the two-element group generated by σ (cf. Chapter 5), i.e. $R^\sharp[\sigma] = R^\sharp \oplus (R^\sharp \cdot \sigma)$ as $R^\sharp$-modules, with multiplication
$$(r + s\sigma)(r' + s'\sigma) = (rr' + s\sigma(s')) + (rs' + s\sigma(r'))\sigma\,.$$
The modules over $R^\sharp[\sigma]$ are precisely the $R^\sharp$-modules N carrying a compatible action of the involution σ:
$$\sigma(rx) = \sigma(r)\sigma(x)$$
for $r \in R^\sharp$ and $x \in N$. (Note that $R^\sharp$ itself is naturally an $R^\sharp[\sigma]$-module.) We will call an $R^\sharp[\sigma]$-module N maximal Cohen-Macaulay (MCM, as usual) if it is MCM as an $R^\sharp$-module.

Let N be a finitely generated $R^\sharp[\sigma]$-module, and set
$$N^+ = \{x \in M \mid \sigma(x) = x\}$$
$$N^- = \{x \in M \mid \sigma(x) = -x\}\,.$$
Then $N = N^+ \oplus N^-$ as $R^\sharp$-modules. If N is a MCM $R^\sharp[\sigma]$-module, then it follows that N^+ and N^- are MCM modules over $(R^\sharp)^+ = S$, i.e. free S-modules of finite rank.

8.26. DEFINITION. Let R, $R^\sharp$, and $R^\sharp[\sigma]$ be as above.

(i) Let N be a MCM $R^\sharp[\sigma]$-module. Define a MCM R-module $\mathcal{A}(N)$ as follows. Multiplication by z, respectively $-z$, defines an S-linear map between finitely generated free S-modules
$$\varphi\colon N^+ \longrightarrow N^-, \quad \text{resp.} \quad \psi\colon N^- \longrightarrow N^+$$
which together constitute a matrix factorization of f. Set
$$\mathcal{A}(N) = \mathrm{cok}(\varphi, \psi)\,.$$

(ii) Let M be a MCM R-module, and define a MCM $R^\sharp$-module $\mathcal{B}(M)$ with compatible σ-action by the following recipe. Write $M = \mathrm{cok}(\varphi\colon G \longrightarrow F,\ \psi\colon F \longrightarrow G)$ with F and G finitely generated free S-modules. Set
$$\mathcal{B}(M) = G \oplus F\,,$$

with multiplication by z defined via
$$z(x,y) = (-\psi(y), \varphi(x))$$
and σ-action
$$\sigma(x,y) = (x,-y).$$

8.27. PROPOSITION. *The mappings $\mathcal{A}(-)$ and $\mathcal{B}(-)$ induce mutually inverse bijections between the isomorphism classes of stable MCM R-modules and the isomorphism classes of MCM $R^\sharp[\sigma]$-modules having no direct summand $R^\sharp[\sigma]$-isomorphic to $R^\sharp$.*

PROOF. It is easy to verify that $\mathcal{A}(R^\sharp) = \operatorname{cok}(1,f) = 0$, that $\mathcal{B}(R) = R^\sharp$, and that $\mathcal{AB}$ and $\mathcal{BA}$ are naturally the identities otherwise. □

In fact $\mathcal{A}$ and $\mathcal{B}$ can be used to define equivalences of categories between the MCM $R^\sharp[\sigma]$-modules and the matrix factorizations of f, though we will not need this fact.

8.28. LEMMA. *Let M be a MCM R-module. Then*
$$M^\sharp \cong \mathcal{B}(M)$$
as $R^\sharp$-modules (we ignore the action of σ on the right-hand side). Thus $M^\sharp$ acquires, via this isomorphism, the structure of an $R^\sharp[\sigma]$-module.

PROOF. Write $M = \operatorname{cok}(\varphi \colon G \longrightarrow F, \psi \colon F \longrightarrow G)$, so that $\mathcal{B}(M) = G \oplus F$ as S-modules, with $z(x,y) = (-\psi(y), \varphi(x))$. Since $M^\sharp = \operatorname{syz}_1^{R^\sharp}(M)$, we want to build an $R^\sharp$-isomorphism between $\mathcal{B}(M)$ and the kernel of the map $\pi \colon \widetilde{F} \longrightarrow M$ in Lemma 8.14 (i). First we compute the kernel: Given an element $u \in \widetilde{F}$, write $u = a + bz$, with $a, b \in F$. In the notation of 8.13 π is the composition $\widetilde{F} \longrightarrow \overline{F} \longrightarrow M$, where $\widetilde{F} \longrightarrow \overline{F}$ kills z and $\overline{F} \longrightarrow M$ is the cokernel of $\overline{\varphi} \colon \overline{G} \longrightarrow \overline{F}$. Thus $u \in \ker \pi \iff a \in \varphi(G) + fF$.

Define $\xi \colon \mathcal{B}(M) \longrightarrow \widetilde{F}$ by $\xi(x,y) = \varphi(x) + yz$ for $x \in G$ and $y \in F$. One checks that $\xi(z(x,y)) = z\xi(x,y)$, so ξ is $R^\sharp$-linear. Clearly $\operatorname{im} \xi \subseteq \ker \pi$. For the reverse inclusion, let $u = a + bz \in \ker \pi$, write $a = \varphi(x') + fy'$, with $x' \in G$ and $y' \in F$, and check that $\xi(x', b - y'z) = u$. □

8.29. PROPOSITION. *Let N be a stable MCM $R^\sharp$-module. Then N is in the image of $(-)^\sharp$, that is, $N \cong M^\sharp$ for some MCM R-module M, if and only if $N \cong \operatorname{syz}_1^{R^\sharp} N$.*

PROOF. If $N \cong M^\sharp$, then $N \cong \operatorname{syz}_1^{R^\sharp} N$ by Lemma 8.14(iv).

For the converse, it suffices to show that if N is an indecomposable MCM $R^\sharp$-module such that $N \cong \operatorname{syz}_1^{R^\sharp} N$, then N has the structure of an $R^\sharp[\sigma]$-module. Indeed, in that case $N \cong \mathcal{B}(\mathcal{A}(N)) \cong \mathcal{A}(N)^\sharp$ by Proposition 8.27 and Lemma 8.28, so that N is in the image of $(-)^\sharp$.

By assumption, there is an isomorphism of $R^\sharp$-modules $\xi\colon N \longrightarrow \operatorname{syz}_1^{R^\sharp} N$. Now Lemma 8.17 (iii) implies that $\operatorname{syz}_1^{R^\sharp}(N)$ has the same underlying Abelian group as N, with $R^\sharp$-structure defined via σ, i.e. $r \cdot x = \sigma(r)x$. Therefore we may consider ξ as a function $N \longrightarrow N$ satisfying $\xi(rn) = \sigma(r)\xi(n)$. As N is indecomposable, $R^\sharp$ is complete, and k is algebraically closed we may assume, by shenanigans similar to those in Lemma 8.25, that ξ satisfies $\xi^2 = 1_N$. Therefore ξ defines an action of σ on N, whence N has a structure of $R^\sharp[\sigma]$-module. $\square$

Now we can say exactly which modules decompose upon sharping or flatting.

8.30. PROPOSITION. *Keep all the notation of 8.24. In particular, k is an algebraically closed field of characteristic not equal to 2.*

(i) Let M be an indecomposable non-free MCM R-module. Then $M^\sharp$ is decomposable if, and only if, $M \cong \operatorname{syz}_1^R M$. In this case $M^\sharp \cong N \oplus \operatorname{syz}_1^{R^\sharp} N$ for an indecomposable $R^\sharp$-module N such that $N \not\cong \operatorname{syz}_1^{R^\sharp} N$.

(ii) Let N be a non-free indecomposable MCM $R^\sharp$-module. Then $N^\flat$ is decomposable if, and only if, $N \cong \operatorname{syz}_1^{R^\sharp} N$. In this case $N^\flat \cong M \oplus \operatorname{syz}_1^R M$ for an indecomposable R-module M such that $M \not\cong \operatorname{syz}_1^R M$.

PROOF. First let $_R M$ be indecomposable, MCM, and non-free. If $M \cong \operatorname{syz}_1^R M$, then $M \cong \operatorname{cok}(\varphi, \varphi)$ for some φ by Lemma 8.25, so that by Lemma 8.14

$$M^\sharp \cong \operatorname{cok}\left(\begin{pmatrix} \varphi & -z1_F \\ z1_F & \varphi \end{pmatrix}, \begin{pmatrix} \varphi & z1_F \\ -z1_F & \varphi \end{pmatrix}\right)$$
$$\cong \operatorname{cok}(\varphi + iz1_F, \varphi - iz1_F) \oplus \operatorname{cok}(\varphi - iz1_F, \varphi + iz1_F),$$

via a diagonalization similar to that in the proof of Proposition 8.18. (Here i is a square root of -1 in k.) Conversely, suppose $M^\sharp \cong N_1 \oplus N_2$ for non-zero MCM $R^\sharp$-modules N_1 and N_2. Then

$$N_1^\flat \oplus N_2^\flat \cong M^{\sharp\flat} \cong M \oplus \operatorname{syz}_1^R M$$

by Proposition 8.15. Since M is indecomposable and R is complete, by KRS we may interchange N_1 and N_2 if necessary to assume that $N_1^\flat \cong M$ and $N_2^\flat \cong \operatorname{syz}_1^R M$. Note that N_1 is stable since M is not free. Then $\operatorname{syz}_1^R(N_1^\flat) \cong N_1^\flat$ by Lemma 8.17(ii), so

$$M \cong N_1^\flat \cong \operatorname{syz}_1^R(N_1^\flat) \cong \operatorname{syz}_1^R M,$$

as desired.

Next let N be a non-free indecomposable MCM $R^\sharp$-module. By Proposition 8.29, if $N \cong \mathrm{syz}_1^{R^\sharp} N$ then $N \cong M^\sharp$ for some $_R M$, whence
$$N^\flat \cong M^{\sharp\flat} \cong M \oplus \mathrm{syz}_1^R M$$
is decomposable by Proposition 8.15. The converse is shown as above.

To complete the proof of (i), suppose $M \cong \mathrm{syz}_1^R M$, so that $M^\sharp \cong N \oplus \mathrm{syz}_1^{R^\sharp} N$ for some $_{R^\sharp} N$. Then $M^{\sharp\flat} \cong M \oplus \mathrm{syz}_1^R M$ is a direct sum of exactly two indecomposable modules, so $N^\flat$ must be indecomposable. Hence $N \not\cong \mathrm{syz}_1^{R^\sharp} N$ by the part of (ii) we have already proved. The last sentence of (ii) follows similarly. $\square$

8.31. DEFINITION. In the notation of 8.24, set
$$R^{\sharp\sharp} = S[\![u,v]\!]/(f+uv).$$
(Since we assume k is algebraically closed of characteristic not 2, this is isomorphic to $(R^\sharp)^\sharp$.) For a MCM R-module M given by the matrix factorization $M = \mathrm{cok}(\varphi\colon G \longrightarrow F,\ \psi\colon F \longrightarrow G)$, we define a MCM $R^{\sharp\sharp}$-module $M^{\boldsymbol{\times}}$ by
$$M^{\boldsymbol{\times}} = \mathrm{cok}\left(\begin{pmatrix} \varphi & -v\,1_F \\ u\,1_G & \psi \end{pmatrix},\ \begin{pmatrix} \psi & v\,1_G \\ -u\,1_F & \varphi \end{pmatrix}\right).$$
Here we continue our convention (cf. 8.13) of using 1_F and 1_G for the identity maps on the free $S[\![u,v]\!]$-modules induced from F and G.

We leave verification of the next lemma as an exercise.

8.32. LEMMA. *Keep the notation of the Definition.*
  (i) $(M^\sharp)^\sharp \cong M^{\boldsymbol{\times}} \oplus \mathrm{syz}_1^{R^{\sharp\sharp}}(M^{\boldsymbol{\times}})$.
  (ii) $(M^{\boldsymbol{\times}})^{\flat\flat} \cong M \oplus \mathrm{syz}_1^R M$.
  (iii) $(\mathrm{syz}_1^R M)^{\boldsymbol{\times}} \cong \mathrm{syz}_1^{R^{\sharp\sharp}}(M^{\boldsymbol{\times}})$.

Now we can prove a more precise version of Theorem 8.20.

8.33. THEOREM (Knörrer). *The association $M \mapsto M^{\boldsymbol{\times}}$ defines a bijection between the isomorphism classes of indecomposable non-free MCM modules over R and over $R^{\sharp\sharp}$.*

PROOF. Let M be an indecomposable MCM R-module which is not free. Then $M^{\sharp\sharp}$ splits into precisely two indecomposable summands by Proposition 8.30(i), so that $M^{\boldsymbol{\times}}$ is indecomposable by Lemma 8.32(i).

If M' is another indecomposable MCM R-module with $(M')^{\boldsymbol{\times}} \cong M^{\boldsymbol{\times}}$, then by Lemma 8.32(ii) we have either $M' \cong M$ or $M' \cong \mathrm{syz}_1^R M$. Assume $M \not\cong M' \cong \mathrm{syz}_1^R M$. Then by Proposition 8.30 $M^\sharp$ is indecomposable. Therefore the two indecomposable direct summands of $M^{\sharp\sharp}$ are non-isomorphic by Proposition 8.30, and it follows from Lemma 8.32, parts (i) and (iii) that

$$M^{\maltese} \not\cong \mathrm{syz}_1^{R^{\sharp\sharp}}(M^{\maltese}) \cong (\mathrm{syz}_1^R)^{\maltese} \cong (M')^{\maltese},$$

a contradiction.

Finally let N be an indecomposable non-free MCM $R^{\sharp\sharp}$-module. We must show that N is a direct summand of $M^{\maltese}$ for some ${}_RM$. From Lemma 8.32(i) we find

$$(N^{\flat\flat})^{\sharp\sharp} \cong (N^{\flat\flat})^{\maltese} \oplus \mathrm{syz}_1^{R^{\sharp\sharp}}((N^{\flat\flat})^{\maltese})$$
$$\cong (N^{\flat\flat} \oplus \mathrm{syz}_1^{R^{\sharp\sharp}}(N^{\flat\flat}))^{\maltese}.$$

On the other hand,

$$(N^{\flat\flat})^{\sharp\sharp} \cong \left((N^{\flat})^{\flat\sharp}\right)^{\sharp}$$
$$\cong \left(N^{\flat} \oplus \mathrm{syz}_1^{R^{\sharp}}(N^{\flat})\right)^{\sharp}$$
$$\cong N^{\flat\sharp} \oplus \mathrm{syz}_1^{R^{\sharp}}(N^{\flat\sharp})$$
$$\cong N^{(2)} \oplus (\mathrm{syz}_1^{R^{\sharp}} N)^{(2)}.$$

Hence N is in the image of $(-)^{\maltese}$. □

We will not prove Knörrer's stronger result than in fact $M \longleftrightarrow M^{\maltese}$ induces an equivalence between the stable categories of MCM modules; see [**Knö87**] for details.

§4. Exercises

8.34. EXERCISE. Prove that commutativity of one of the squares in the diagram (8.3) implies commutativity of the other.

8.35. EXERCISE. Prove that a matrix factorization (φ, ψ) is reduced if and only if all entries of φ and ψ are in the maximal ideal $\mathfrak{n}$ of S.

8.36. EXERCISE. Verify exactness of the sequence (8.4).

8.37. EXERCISE. Let Λ be a ring, possibly non-commutative, having exactly one maximal left ideal. Let M and N be left Λ-modules. If $M \oplus N$ has a direct summand isomorphic to ${}_\Lambda\Lambda$, then either M or N has a direct summand isomorphic to ${}_\Lambda\Lambda$. Is this still true if, instead, Λ has exactly one maximal two-sided ideal?

8.38. EXERCISE. Fill in the details of the proofs of Proposition 8.6 and Theorem 8.7.

8.39. EXERCISE. Let $(S, \mathfrak{n}, k)$ be a complete local ring, let $f \in \mathfrak{n}^2 \setminus \{0\}$, and put $g = uf$, where u is a unit of R. If k is closed under square roots and has characteristic different from 2, show that $S[\![z]\!]/(f+z^2) \cong S[\![z]\!]/(g+z^2)$.

8.40. EXERCISE. Prove that z is a non-zerodivisor of $R^\sharp = S[\![z]\!]/(f+z^2)$.

8.41. EXERCISE. Prove that the natural inclusion $S[z]/(f+z^2) \longrightarrow S[\![z]\!]/(f+z^2)$ is an isomorphism. In particular, $R^\sharp$ is a free S-module with basis $\{1, z\}$. Show by example that if S is not assumed to be complete then $S[\![z]\!]/(f+z^2)$ need not be finitely generated as an S-module.

8.42. EXERCISE. With notation as in the proof of Lemma 8.17 (iii), show that the sequence
$$S[\![z]\!]^{(n)} \xrightarrow{zI_n - \varphi} S[\![z]\!]^{(n)} \longrightarrow N \longrightarrow 0$$
is exact. (Hint: Use Exercise 8.41 and choose bases.)

8.43. EXERCISE. In Exercise 8.21 we gave an example of a zero-dimensional ring R with finite CM type such that $R^\sharp$ has infinite CM type. Find higher-dimensional examples of this behavior.

8.44. EXERCISE. Prove Lemma 8.32.

CHAPTER 9

Hypersurfaces with Finite CM Type

In this chapter we will show that the complete, equicharacteristic hypersurface singularities with finite CM type are exactly the ADE singularities. In any characteristic but two, Theorem 9.7 shows that such a hypersurface of dimension $d \geqslant 2$ is the double branched cover of one of dimension $d - 1$. In Theorem 9.8, proved in 1987 by Buchweitz-Greuel-Schreyer and Knörrer [**Knö87, BGS87**], we restrict to rings having an algebraically closed coefficient field of characteristic different from 2, 3, and 5, and show that finite CM type is equivalent to *simplicity* (Definition 9.1), and to being an ADE singularity. We'll also prove Herzog's Theorem 9.15: Gorenstein rings of finite CM type are abstract hypersurfaces [**Her78b**]. In §4 we derive matrix factorizations for the Kleinian singularities (two-dimensional ADE hypersurface singularities). At the end of the chapter we will discuss the situation in characteristics 2, 3 and 5. Later, in Chapter 10, we will see how to eliminate the assumption that R be complete, and also we'll weaken "algebraically closed" to "perfect".

We will do a few things in slightly greater generality than strictly needed in this chapter, so that they will apply also to the study of *countable* CM type in Chapter 14.

§1. Simple singularities

9.1. DEFINITION. Let $(S, \mathfrak{n})$ be a regular local ring, and let $R = S/(f)$, where $0 \neq f \in \mathfrak{n}^2$. We call R a *simple* (respectively *countably simple*) singularity (relative to the presentation $R = S/(f)$) provided there are only finitely (respectively countably) many ideals L of S such that $f \in L^2$.

9.2. THEOREM (Buchweitz-Greuel-Schreyer). *Let $R = S/(f)$, where $(S, \mathfrak{n})$ is a regular local ring and $0 \neq f \in \mathfrak{n}^2$. If R has finite (respectively countable) CM type, then R is a simple (respectively countably simple) singularity.*

PROOF. Let $\mathcal{M}$ be a complete set of representatives for the equivalence classes of reduced matrix factorizations of f. By Theorem 8.7,

$\mathcal{M}$ is finite (respectively countable). For each $(\varphi, \psi) \in \mathcal{M}$, let $L(\varphi, \psi)$ be the ideal of S generated by the entries of $[\varphi \,|\, \psi]$. Let $\mathcal{S}$ be the set of ideals that are ideal sums of finite subsets of $\{L(\varphi, \psi) \mid (\varphi, \psi) \in \mathcal{M}\}$. Then $\mathcal{S}$ is finite (respectively countable), and we claim that every proper ideal L for which $f \in L^2$ belongs to $\mathcal{S}$. To see this, let $a_0, \ldots, a_r$ generate L, and write $f = a_0 b_0 + \cdots + a_r b_r$, with $b_i \in L$. For $0 \leqslant s \leqslant r$, let $f_s = a_0 b_0 + \cdots + a_s b_s$. Put $\sigma_0 = a_0$, $\tau_0 = b_0$, and for $1 \leqslant s \leqslant r$ define, inductively, a $2^s \times 2^s$ matrix factorization of f_s by

$$(9.1) \quad \sigma_s = \begin{bmatrix} a_s I_{2^{s-1}} & \sigma_{s-1} \\ \tau_{s-1} & -b_s I_{2^{s-1}} \end{bmatrix} \quad \text{and} \quad \tau_s = \begin{bmatrix} b_s I_{2^{s-1}} & \sigma_{s-1} \\ \tau_{s-1} & -a_s I_{2^{s-1}} \end{bmatrix}.$$

Letting $\sigma = \sigma_r$ and $\tau = \tau_r$, we see that (σ, τ) is a reduced matrix factorization of f with $L(\sigma, \tau) = L$. Then (σ, τ) is equivalent to $(\varphi_1, \psi_1)^{(n_1)} \oplus \cdots \oplus (\varphi_t, \psi_t)^{(n_t)}$, where $(\varphi_j, \psi_j) \in \mathcal{M}$ and $n_j > 0$ for each j. Finally, we have $L = L(\sigma, \tau) = \sum_{j=1}^{t} L(\varphi_j, \psi_j) \in \mathcal{S}$. □

The following lemma, together with the Weierstrass Preparation Theorem, will show that every simple singularity of dimension $d \geqslant 2$ is a double branched cover of a $(d-1)$-dimensional simple singularity:

9.3. LEMMA. *Let $(S, \mathfrak{n}, k)$ be a regular local ring and $R = S/(f)$ a singularity with $d = \dim(R) \geqslant 1$.*

  (i) *Suppose R is a simple singularity and k is an infinite field. Then:*
    (a) *R is reduced, i.e. for each $g \in \mathfrak{n}$ we have $g^2 \nmid f$.*
    (b) *$\mathrm{e}(R) \leqslant 3$.*
    (c) *If k is algebraically closed and $d \geqslant 2$, then $\mathrm{e}(R) = 2$.*
  (ii) *Suppose R is a countably simple singularity and that k is an uncountable field. Then:*
    (a) *For each $g \in \mathfrak{n}$ we have $g^3 \nmid f$.*
    (b) *$\mathrm{e}(R) \leqslant 3$.*
    (c) *If k is algebraically closed and $d \geqslant 2$, then $\mathrm{e}(R) = 2$.*

PROOF. (ia) Suppose f has a repeated factor, so that we can write $f = g^2 h$, where $g \in \mathfrak{n}$ and $h \in S$. Now $\dim(S/(g)) = d \geqslant 1$, so $S/(g)$ has infinitely many ideals. Therefore S has infinitely many ideals that contain g, and f is in the square of each, a contradiction.

(iia) Suppose f is divisible by the cube of some $g \in \mathfrak{n}$. Let $\Lambda \subset R$ be a complete set of coset representatives for $k^\times$. If $S/(g)$ is not a DVR, let $\{\xi, \eta\}$ be part of a minimal generating set for $\mathfrak{n}/(g)$, and lift ξ, η to $x, y \in \mathfrak{n}$. For $\lambda \in \Lambda$, put $L_\lambda = (x + \lambda y, g)$. We have $L_\lambda \neq L_\mu$ if $\lambda \neq \mu$, for if $L = L_\lambda = L_\mu$ and $\lambda \not\equiv \mu \bmod \mathfrak{m}$, then L contains $(\lambda - \mu)y$, hence y, and hence x; this means that $L/(g) = (\xi, \eta) = (\xi + \lambda \eta)$, contradicting the choice of ξ and η. For each $\lambda \in k$, we have $f \in L_\lambda^3 \subseteq L_\lambda^2$, a contradiction.

Now assume that $S/(g)$ *is* a DVR. Then $\dim(S) = 2$ and $g \notin \mathfrak{n}^2$. Write $\mathfrak{n} = (g, h)$, and note that g and h are non-associate irreducible elements of S. For $\lambda \in \Lambda$, put $I_\lambda = (g + \lambda h^2, gh)$. Suppose $I := I_\lambda = I_\mu$ with $\lambda \neq \mu$. Then $I = (g, h^2)$. Writing

$$(9.2) \qquad g = (g + \lambda h^2)p + ghq,$$

with $p, q \in S$, we see that $g \mid \lambda h^2 p$, whence $g \mid p$. Write $p = gs$, with $s \in S$, plug this into (9.2), and cancel g, getting the equation $1 = (g + \lambda h^2)s + hq \in \mathfrak{n}$, a contradiction. Thus $L_\lambda \neq L_\mu$ when $\lambda \neq \mu$. Moreover, we have

$$g^3 = g(g + \lambda h^2)^2 - \lambda h(g + \lambda h^2)(gh) - \lambda(gh)^2 \in I_\lambda^2$$

for each λ. Thus $f \in I_\lambda^2$ for each λ, again contradicting countable simplicity.

(ib) and (iib) Suppose $e(R) \geqslant 4$. Then $f \in \mathfrak{n}^4$ (cf. Exercise 9.29). If L is any ideal such that $\mathfrak{n}^2 \subsetneq L \subsetneq \mathfrak{n}$, then $f \in L^2$. These ideals correspond to non-zero proper subspaces of the k-vector space $\mathfrak{n}/\mathfrak{n}^2$, so there are infinitely (respectively uncountably) many of them, a contradiction.

(ic) and (iic) We know that $e(R)$ is either 2 or 3, so we suppose $e(R) = 3$, that is, $f \in \mathfrak{n}^3 \backslash \mathfrak{n}^4$. Let f^* be the coset of f in $\mathfrak{n}^3/\mathfrak{n}^4$. Then f^* is a cubic form in the associated graded ring $G = k \oplus \mathfrak{n}/\mathfrak{n}^2 \oplus \mathfrak{n}^2/\mathfrak{n}^3 \oplus \ldots = k[x_0, \ldots, x_d]$, where $(x_0, \ldots, x_d) = \mathfrak{n}$. The zero set Z of f^* is an infinite (respectively uncountable) subset of $\mathbb{P}_k^d$. Fix a point $\lambda = (\lambda_0 : \lambda_1 : \cdots : \lambda_d) \in Z$, and let $\{L_1, \ldots, L_d\}$ be a basis for the k-vector space of linear forms vanishing at λ. These forms generate the ideal of G consisting of polynomials vanishing at λ, and it follows that

$$(9.3) \qquad f^* \in (L_1, \ldots, L_d)(x_0, \ldots, x_d)^2 G.$$

Now we lift each $L_i \in \mathfrak{n}/\mathfrak{n}^2$ to a element $\widetilde{L}_i \in \mathfrak{n}\backslash\mathfrak{n}^2$, and put $I_\lambda = (\widetilde{L}_1, \ldots, \widetilde{L}_d)S + \mathfrak{n}^2$. Pulling (9.3) back to S and using the fact that $f^* = f + \mathfrak{n}^4$, we get $f \in (\widetilde{L}_1, \ldots, \widetilde{L}_d)\mathfrak{n}^2 + \mathfrak{n}^4$, whence $f \in I_\lambda^2$. Since $I_\lambda \neq I_\mu$ if λ and μ are distinct points of Z, we have contradicted simplicity (respectively countable simplicity). $\square$

The next lemma will be used to control the order of the higher-degree terms in the defining equations of (countably) simple singularities:

9.4. LEMMA. *Let $R = S/(f)$ be a hypersurface singularity of positive dimension, where $(S, \mathfrak{n}, k)$ is a regular local ring and $0 \neq f \in \mathfrak{n}^2$.*

Assume either that k is infinite and R is a simple singularity, or that k is uncountable and R is a countably simple singularity. Let $\alpha, \beta \in \mathfrak{n}$. Then $f \notin (\alpha, \beta^2)^3$.

PROOF. Suppose $f \in (\alpha, \beta^2)^3$. Let $\Lambda \subseteq S$ be a complete set of representatives for the residue field k, and for each $\lambda \in \Lambda$ put $I_\lambda = (\alpha + \lambda\beta^2, \beta^3)$. One checks easily that $(\alpha, \beta^2)^3 \subseteq I_\lambda^2$ (Exercise 9.31). Therefore it will suffice to show that $I_\lambda \neq I_\mu$ whenever λ and μ are distinct elements of Λ.

Suppose $\lambda, \mu \in \Lambda$, $\lambda \neq \mu$, and $I_\lambda = I_\mu$. Since $\lambda - \mu$ is a unit of S, we see that $\beta^2 \in I_\lambda$. Writing $\beta^2 = s(\alpha + \lambda\beta^2) + t\beta^3$, with $s, t \in S$, we see that $\beta^2(1 - t\beta) \in (\alpha + \lambda\beta^2)$. Therefore $\beta^2 \in (\alpha + \lambda\beta^2)$, and it follows that $I_\lambda = (\alpha + \lambda\beta^2)$. Now $f \in (\alpha, \beta^2)^3 = I_\lambda^3 = (\alpha + \lambda\beta^2)^3$, and this contradicts (ia) or (iia) of Lemma 9.3. □

§2. Hypersurfaces in good characteristics

For this section k is an algebraically closed field and d is a positive integer. Put $S = k[\![x_0, \ldots, x_d]\!]$ and $\mathfrak{n} = (x_0, \ldots, x_d)$. We will consider d-dimensional hypersurface singularities: rings of the form $S/(f)$, where $0 \neq f \in \mathfrak{n}^2$.

We refer to [**Lan02**, Chapter IV, Theorem 9.2] for the following version of the Weierstrass Preparation Theorem:

9.5. THEOREM (WPT). *Let $(D, \mathfrak{m})$ be a complete local ring, and let $g \in D[\![x]\!]$. Suppose $g = a_0 + a_1 x + \cdots + a_e x^e +$ higher degree terms, with $a_0, a_1, \ldots, a_{e-1} \in \mathfrak{m}$ and $a_e \in D \setminus \mathfrak{m}$. Then there exist $b_1, \ldots, b_e \in \mathfrak{m}$ and a unit $u \in D[\![x]\!]$ such that $g = (x^e + b_1 x^{e-1} + \cdots + b_e)u$.* □

The conclusion is equivalent to asserting that $D[\![x]\!]/(g)$ is a free D-module of rank e, with basis $1, x, \ldots, x^{e-1}$.

9.6. COROLLARY. *Let ℓ be an infinite field, and let g be a non-zero power series in $S = \ell[\![x_0, \ldots, x_n]\!]$, $n \geq 1$. Assume that the order e of g is at least 2 and is not a multiple of $\operatorname{char}(\ell)$. Then, after a change of coordinates, we have $g = (x_n^e + b_2 x_n^{e-2} + b_3 x_n^{e-3} + \cdots + b_{e-1} x_n + b_e)u$, where $b_2, \ldots, b_e$ are non-units of $D := \ell[\![x_0, \ldots, x_{n-1}]\!]$ and u is a unit of S.*

PROOF. We make a linear change of variables, following Zariski and Samuel [**ZS75**, p. 147], so that Theorem 9.5 applies with respect to the new variables. Write $g = g_e + g_{e+1} + \cdots$, where each g_j is a homogeneous polynomial of degree j and $g_e \neq 0$. Then $x_n g_e \neq 0$, and, since ℓ is infinite, there is a point $(c_0, c_1, \ldots, c_n) \in \ell^{n+1}$ such that $x_n g_e$ does not vanish when evaluated at $(c_0, \ldots, c_n)$. Then $c_n \neq 0$, and since

$x_n g_e$ is homogeneous we can scale and assume that $c_n = 1$. We change variables as follows:
$$\varphi \colon x_i \mapsto \begin{cases} x_i + c_i x_n & \text{if } i < n \\ x_n & \text{if } i = n. \end{cases}$$

Now, $\varphi(g) = \varphi(g_e) +$ higher-order terms, and $\varphi(g_e)$ contains the term $g_e(c_0, c_1, \ldots, c_{n-1}, 1)x_n^e = cx_n^e$, where $c \in \ell^\times$. It follows that $\varphi(g)$ has the form required in Theorem 9.5, with $x = x_n$. Replacing g by $\varphi(g)$, we now have $g = (x_n^e + b_1 x_n^{e-1} + \cdots + b_e)u$, where the b_i are non-units of D and u is a unit of S. Finally, we make the substitution $x_n \mapsto x_n - \frac{b_1}{e} x_n^{e-1}$ to eliminate the coefficient b_1. □

Here is the main theorem of this chapter, due to Buchweitz-Greuel-Schreyer and Knörrer [**Knö87, BGS87**].

9.7. THEOREM. *Let k be an algebraically closed field of characteristic different from 2, and put $S = k[\![x_0, \ldots, x_d]\!]$, where $d \geqslant 2$. Let $R = S/(f)$, where $0 \neq f \in (x_0, \ldots, x_d)^2$. Then R has finite CM type if and only if there is a non-zero element $g \in (x_0, x_1)^2 k[\![x_0, x_1]\!]$ such that $k[\![x_0, x_1]\!]/(g)$ has finite CM type and $R \cong S/(g + x_2^2 + \cdots + x_d^2)$.*

PROOF. The "if" direction follows from Theorem 8.20 and induction on d. For the converse, we assume that R has finite CM type. Then R is a simple singularity (Theorem 9.2), and (ic) of Lemma 9.3 implies that $e(R) = 2$. Since $\text{char}(k) \neq 2$, we may assume, by Corollary 9.6, that $f = x_d^2 + b$, with $b \in (x_0, \ldots, x_{d-1})^2 k[\![x_0, x_1, \ldots, x_{d-1}]\!]$. (The unit u in the conclusion of the corollary does not affect the isomorphism class of R.) Note that $b \neq 0$, by (ia) of Lemma 9.3. Then $R = A^\#$, where $A = k[\![x_0, x_1, \ldots, x_{d-1}]\!]/(b)$. Now Theorem 8.20 implies that A has finite CM type. If $d = 2$ we set $g = b$, and we're done. Otherwise, after a change of coordinates we have $b = (x_{d-1}^2 + c)u$, where $c \in (x_0, \ldots, x_{d-2})^2 k[\![x_0, \ldots, x_{d-2}]\!] \setminus \{0\}$ and u is a unit of $k[\![x_0, x_1, \ldots, x_{d-1}]\!]$. Now $f = (x_d^2 u^{-1} + x_{d-1}^2 + c)u$. Since S is complete, it is in particular Henselian by Hensel's Lemma (Corollary 1.9), and since $\text{char}(k) \neq 2$ the classical definition of Henselianness (Corollary A.31) provides a unit v such that $v^2 = u^{-1}$. After replacing x_d by $x_d v$ and discarding the unit u, we now have $R \cong S/(c + x_{d-1}^2 + x_d^2)$. Repeat! □

If the characteristic of k is different from 2, 3 and 5, we get a more explicit version of the theorem.

9.8. THEOREM. *Let k be an algebraically closed field of characteristic different from 2, 3, and 5, let d be a positive integer, and let $R = k[\![x, y, x_2, \ldots, x_d]\!]/(f)$, where $0 \neq f \in (x, y, x_2, \ldots, x_d)^2$. These are equivalent:*

(i) R has finite CM type.
(ii) R is a simple singularity.
(iii) $R \cong k[\![x, y, x_2, \ldots, x_d]\!]/(g + x_2^2 + \cdots + x_d^2)$, where $g \in k[\![x, y]\!]$ defines a one-dimensional ADE singularity (cf. Chapter 4, §3).
$\square$

A consequence of this theorem is that, in this context, simplicity of R depends only on the isomorphism class of R, not on the presentation $R = S/(f)$. We know of no proof of this fact in general. The proof of the theorem will occupy the rest of the section.

PROOF. (i) $\implies$ (ii) by Theorem 9.2.

(ii) $\implies$ (iii): Suppose first that $d \geqslant 2$; then $e(R) = 2$ by (ic) of Lemma 9.3. By Corollary 9.6, we may assume that $f = x_d^2 + b$, where b is a non-zero non-unit of $k[\![x_0, x_1, \ldots, x_{d-1}]\!]$. Then $R = A^\#$, where $A = k[\![x_0, x_1, \ldots, x_{d-1}]\!]/(b)$. Simplicity passes from R to A: If there were an infinite number of ideals L_i of $k[\![x_0, x_1, \ldots, x_{d-1}]\!]$ with $b \in L_i^2$ for each i, we would have $x_d^2 + b \in (L_i S + x_d S)^2$ for each i, where $S = k[\![x_0, \ldots, x_d]\!]$. Since $(L_i S + x_d S) \cap k[\![x_0, x_1, \ldots, x_{d-1}]\!] = L_i$, the extended ideals would be distinct, contradicting simplicity of R. Thus we can continue the process, dropping dimensions till we reach dimension one. It suffices, therefore, to prove that (ii) $\implies$ (iii) when $d = 1$.

Changing notation, we set $S = k[\![y, x]\!]$ and $\mathfrak{n} = (y, x)S$. (The silly ordering of the variables stems from the choice of the normal forms for the ADE singularities in Chapter 4, §3.) We have a power series $f \in \mathfrak{n}^2 \setminus \{0\}$ which is contained in the squares of only finitely many ideals, and we want to show that $R = S/(f)$ is an ADE singularity. We will follow Yoshino's proof of [**Yos90**, Proposition 8.5] closely, adding a few details and making a few necessary modifications (some of them to accommodate non-zero characteristic $p > 5$).

Suppose first that $e(R) = 2$. By Corollary 9.6, we may assume that $f = x^2 + g$, where $g \in yk[\![y]\!]$. Then $g \neq 0$ by (ia) of Lemma 9.3, and we write $g = y^t u$, where $u \in k[\![y]\!]^\times$. Then $t \geqslant 2$, else R would be a discrete valuation ring. Replacing f by $u^{-1}f$, we now have $f = u^{-1}x^2 + y^t$. Now we let $v \in k[\![y]\!]^\times$ be a square root of u^{-1} using Hensel's Lemma (Corollaries 1.9 and A.31) and make the change of variables $x \mapsto vx$. Then $f = x^2 + y^t$, so R is an (A_{t-1})-singularity.

Before taking on the more challenging case $e(R) = 3$, we pause for a primer on tangent directions of the branches of an analytic curve. Given any non-zero, non-unit power series $g \in K[\![x, y]\!]$, where K is any algebraically closed field, let g_e be the initial form of g. Thus

g_e is a non-zero homogeneous polynomial of degree $e \geqslant 1$ and $g = g_e +$ higher-degree forms. We can factor g_e as a product of powers of distinct linear forms:
$$g_e = \ell_1^{m_1} \cdots \ell_h^{m_h},$$
where each $m_i > 0$ and the linear forms ℓ_i are not associates in $K[x, y]$. (To do this, dehomogenize, then factor, then homogenize.) The *tangent lines* to the curve $g = 0$ are the lines $\ell_i = 0$, $1 \leqslant i \leqslant h$. We will need the "Tangent Lemma" (cf. [**Abh90**, p. 141]):

9.9. LEMMA. *Let g be a non-zero non-unit in $K[\![x,y]\!]$, where K is an algebraically closed field. If g is irreducible, then g has a unique tangent line.*

PROOF. Let e be the order of g. By the Weierstrass Preparation Theorem 9.5 we may assume that $g = y^e + b_1 y^{e-1} + \cdots + b_{e-1} y + b_e$, where the $b_i \in B := K[\![x]\!]$. Since $x^i \mid b_i$ (else the order of g would be smaller than e), we may write
$$g = y^e + c_1 x y^{e-1} + \cdots + c_{e-1} x^{e-1} y + c_e x^e,$$
with $c_i \in B$. Let us assume that the curve $g = 0$ has more than one tangent line. Then we can factor the leading form
$$g_e = y^e + c_1(0) x y^{e-1} + \cdots + c_{e-1}(0) x^{e-1} y + c_e(0) x^e$$
as a product of linear factors $y - a_i x$ with not all a_i equal. Dehomogenizing (setting $x = 1$), we have
$$y^e + c_1(0) y^{e-1} + \cdots + c_{e-1}(0) y + c_e(0) = \prod_{i=1}^{e} (y - a_i).$$
By grouping the factors intelligently, we can write
$$(9.4) \qquad y^e + c_1(0) y^{e-1} + \cdots + c_{e-1}(0) y + c_e(0) = pq,$$
where p and q are relatively prime monic polynomials of positive degree.

Put $z = \frac{y}{x}$. Then z is transcendental over B, and we have
$$g = x^e h, \text{ where } h = z^e + c_1 z^{e-1} + \cdots + c_{e-1} z + c_e \in B[z].$$
By (9.4), the reduction of h modulo xB factors as the product of two relatively prime monic polynomials of positive degree. Since B is Henselian (cf. Theorem A.30 and Corollary 1.9), we can write
$$h = (z^m + u_1 z^{m-1} + \cdots + u_0)(z^n + v_1 z^{n-1} + \cdots + v_0).$$
with $u_i, v_j \in B$ and with both m and n positive. Then
$$g = (y^m + u_1 x y^{m-1} + \cdots + u_0 x^m)(y^n + v_1 x y^{n-1} + \cdots + v_0 x^n)$$
is the desired factorization of g. $\square$

The lemma is exemplified by the nodal cubic $g = y^2 - x^2 - x^3 = y^2 - x^2(1+x)$, which, though irreducible in $K[x,y]$, factors in $K[\![x,y]\!]$ as long as $\mathrm{char}(K) \neq 2$. It has two distinct tangent lines, $x + y = 0$ and $x - y = 0$; and indeed it factors: If h is a square root of $1 + x$ (obtained from the Taylor expansion of $(1+x)^{\frac{1}{2}}$, or via Hensel's Lemma: Corollaries 1.9 and A.31), then $g = (y + xh)(y - xh)$.

Now assume $\mathrm{e}(R) = 3$, and write $f = x^3 + xa + b$, where $a, b \in yk[\![y]\!]$. Since f has order 3, we have $a \in y^2 k[\![y]\!]$ and $b \in y^3 k[\![y]\!]$.

9.10. CASE. *f is irreducible.*

Then $b \neq 0$. The initial form f_3 of f is a power of a single linear form by Lemma 9.9, and it follows that $f_3 = x^3$. Therefore the order of a is at least 3, and b has order $n \geqslant 4$. If $a = 0$ we have $f = x^3 + uy^n$, where $u \in k[\![y]\!]^\times$. By extracting a cube root of u^{-1} (using Corollary A.31), we may assume that $f = x^3 + y^n$. Now Lemma 9.4 implies that n must be 4 or 5, and R is an (E_6) or (E_8) singularity. If $a \neq 0$ we can assume that $f = x^3 + uxy^m + y^n$, where $m \geqslant 3$ and $u \in k[\![y]\!]^\times$. Suppose for a moment that $m = 3$ and $n \geqslant 5$. Then one can find a root $\xi \in k[\![y]\!]^\times$ of $T^3 + uT^2 + y^{2n-9} = 0$ by lifting the simple root $-\overline{u}$ of $T^3 + \overline{u}T^2 \in k[T]$. One checks that then $x = \xi^{-1} y^{m-3}$ is a root of f, contradicting irreducibility. Thus either $m \geqslant 4$ or $n = 4$.

Suppose $n = 4$, so $f = x^3 + uxy^m + y^4$. After the transformation $y \mapsto y - \frac{1}{4}uxy^{m-3}$, f takes the form

$$f = \begin{cases} x^3 + bx^2 y^2 + y^4 & (b \in k[\![x,y]\!]) & \text{if } m > 3 \\ vx^3 + cx^2 y^2 + y^4 & (c \in k[\![x,y]\!],\ v \in k[\![x,y]\!]^\times) & \text{if } m = 3. \end{cases}$$

If $m = 3$, we use the transformation $x \mapsto v^{-\frac{1}{3}} x$ to eliminate the unit v (modifying c along the way). Thus in either case we have $f = x^3 + bx^2 y^2 + y^4$, and now the transformation $x \mapsto x - \frac{1}{3} by^2$ puts f into the form $f = x^3 + wy^4$, where $w \in k[\![x,y]\!]^\times$. Replacing y by $w^{-\frac{1}{4}} y$, we obtain the (E_6)-singularity.

Now assume that $n \neq 4$ (and, consequently, $m \geqslant 4$). Lemma 9.4 implies that $n = 5$. The transformation $y \mapsto y - \frac{1}{5} uxy^{m-4}$ (with a unit adjustment to x if $m = 4$) puts f in the form $x^3 + bx^2 y^3 + y^5$. The change of variable $x \mapsto x - \frac{1}{3} by^3$ now transforms f to $x^3 + wy^5$, where $w \in k[\![x,y]\!]^\times$. On replacing y with $w^{-\frac{1}{5}} y$, we obtain the (E_8) singularity, finishing this case.

9.11. CASE. *f is reducible but has only one tangent line.*

Changing notation, we may assume that $f = x(x^2 + ax + b)$, where a and b are non-units of $k[\![y]\!]$. As before, x^3 must be the initial form of f, so $f = x(x^2 + cxy^2 + dy^3)$, where $c, d \in k[\![y]\!]$. By Lemma 9.4 d must be a unit. After replacing y by $d^{-\frac{1}{3}}y$, we can write $f = x(x^2 + exy^2 + y^3)$, where $e \in k[\![y]\!]$. Next do the change of variable $y \mapsto y - \frac{1}{3}ex$ to eliminate the y^2 term. Now $f = x(ux^2 + y^3)$, where $u \in k[\![x,y]\!]^\times$. Replacing x by $u^{-\frac{1}{2}}x$, we have, up to a unit multiple, an (E_7) singularity.

9.12. CASE. *f is reducible and has more than one tangent line.*

Write $f = \ell q$, where ℓ is linear in x and q is quadratic. If the tangent line of ℓ happens to be a tangent line of q, then, by Lemma 9.9, q factors as a product of two linear polynomials with distinct tangent lines. In any case, we can write $f = (x-r)(x^2 + sx + t)$, where $r, s, t \in yk[\![y]\!]$, and where the tangent line to $x - r$ is *not* a tangent line of $x^2 + sx + t$. After the usual changes of variables and multiplication by a unit, we may assume that $f = (x-r)(x^2 + y^n)$, where $n \geq 2$. If $n = 2$, then f is a product of three distinct lines, and we get (D_4). Assume now that $n \geq 3$. Then $x = 0$ is the tangent line to $x^2 + y^n$ and therefore cannot be the tangent line to $x - r$. Hence $r = uy$ for some unit $u \in k[\![y]\!]^\times$. We make the coordinate change $y \mapsto \frac{x-y}{u}$. Now $f = y(ax^2 + bxy^{n-1} + cy^n)$, where a and c are units of $k[\![x,y]\!]$. Better, up to the unit multiple c, we have $f = y(ac^{-1}x^2 + bc^{-1}xy^{n-1} + y^n)$. Replace x by $(ac^{-1})^{-\frac{1}{2}}x$; now $f = y(x^2 + dxy^{n-1} + y^n)$. After the change of coordinates $x \mapsto x - \frac{1}{2}dy^{n-1}$, we have $f = y(x^2 - \frac{1}{4}d^2y^{2n-2} + y^n)$. Since $2n - 2 > n$, we can rewrite this as $f = y(x^2 + ey^n)$, where $e \in k[\![x,y]\!]^\times$. Finally, we factor out e and replace x by $e^{\frac{1}{2}}x$, bringing f into the form $y(x^2 + y^n)$, the equation for the (D_{n+2}) singularity.

To finish the cycle and complete the proof of Theorem 9.8, we now show that (iii) $\implies$ (i). If $d = 1$ we invoke Corollary 8.22. Assuming inductively that $k[\![x_0, \ldots, x_r]\!]/(g + z_2^2 + \cdots + z_r^2)$ has finite CM type for some $r \geq 1$, we see, by (ii) of Theorem 8.20, that $k[\![x_0, \ldots, x_{r+1}]\!]/(g + z_2^2 + \cdots + z_{r+1}^2)$ has finite CM type as well. $\square$

9.13. REMARK. Inspecting the proof, we see that the demonstration of (ii) $\implies$ (iii) in the one-dimensional case of Theorem 9.8 uses only the following three properties of a simple singularity $R = S/(f)$, where $(S, \mathfrak{n})$ is a two-dimensional regular local ring:

(i) R is reduced;
(ii) $e(R) \leq 3$; and
(iii) $f \notin (\alpha, \beta^2)^3$ for every $\alpha, \beta \in \mathfrak{n}$.

Since the one-dimensional ADE hypersurfaces obviously satisfy these properties, it follows that f defines a simple singularity if and only if these three conditions are satisfied.

§3. Gorenstein rings of finite CM type

In this section we will prove Herzog's theorem [**Her78b**] stating that the rings of the title are abstract hypersurfaces, that is, the completion of such a ring is a hypersurface singularity. Before giving the proof, we establish the following result (also from [**Her78b**]) of independent interest. Recall that a MCM module M is *stable* provided it has no non-zero free summands.

9.14. LEMMA. *Let $(R, \mathfrak{m})$ be a CM local ring, let M be a stable MCM R-module, and let $N = \mathrm{syz}_1^R(M)$.*
  (i) *N is stable.*
  (ii) *Assume M is indecomposable, that $\mathrm{Ext}_R^1(M, R) = 0$, and that $R_\mathfrak{p}$ is Gorenstein for every prime ideal $\mathfrak{p}$ of R with height $\mathfrak{p} \leqslant 1$. Then N is indecomposable.*

PROOF. We have a short exact sequence

$$(9.5) \qquad 0 \longrightarrow N \longrightarrow F \longrightarrow M \longrightarrow 0 \,,$$

where F is free and $N \subseteq \mathfrak{m}F$. Let $(\underline{x}) = (x_1, \ldots, x_d)$ be a maximal R-regular sequence in $\mathfrak{m}$. Since M is MCM, $(\underline{x})$ is M-regular, and it follows that the map $N/\underline{x}N \longrightarrow F/\underline{x}F$ is injective. We therefore have an injection $N/\underline{x}N \hookrightarrow \mathfrak{m}F/\underline{x}F$. Since $(\underline{x})$ is a maximal N-regular sequence, $\mathfrak{m} \in \mathrm{Ass}(N/\underline{x}N)$, so $\mathfrak{m} \in \mathrm{Ass}(\mathfrak{m}F/\underline{x}F) = \mathrm{Ass}(\mathfrak{m}/(\underline{x}))$. It follows that $\mathfrak{m}/\underline{x}$ is an unfaithful $R/(\underline{x})$-module and hence that $N/\underline{x}N$ is unfaithful too. But then $N/\underline{x}N$ cannot have have $R/(\underline{x})$ as a direct summand, and item (i) follows.

For the second statement, we note at the outset that both M and N are reflexive R-modules, by Corollary A.14. We dualize (9.5), using the vanishing of $\mathrm{Ext}_R^1(M, R)$, to get an exact sequence

$$(9.6) \qquad 0 \longrightarrow M^* \longrightarrow F^* \longrightarrow N^* \longrightarrow 0 \,.$$

Suppose $N = N_1 \oplus N_2$, with both summands non-zero. By (i), neither summand is free. Since N is reflexive, neither N_1^* nor N_2^* is free, and it follows from (9.6) that M^* decomposes non-trivially. As M is reflexive, this contradicts indecomposability of M. □

9.15. THEOREM (Herzog). *Let $(R, \mathfrak{m}, k)$ be a Gorenstein local ring with a bound on the number of generators required for indecomposable MCM modules. Then $\widehat{R}$ is a hypersurface ring.*

PROOF. Let $M = \mathrm{syz}_d^R(k)$, and write $M = M_1 \oplus \cdots \oplus M_t$, where each M_i is indecomposable and the summands are indexed so that $M_i \cong R$ if and only if $i > s$. By Lemma 9.14, $\mathrm{syz}_j^R(M)$ is a direct sum of at most s indecomposable modules for $j > d$. (The requisite vanishing of Ext follows from the Gorenstein hypothesis.) It follows that the Betti numbers of k are bounded. By a theorem of Tate and Gulliksen [**Tat57, Gul80**] (see also [**Eis80**, Corollary 6.2] or [**Avr98**]) this forces $\widehat{R}$ to be a hypersurface ring. □

Combining Theorem 9.15 with Theorem 9.8, we have characterized finite CM type for complete Gorenstein algebras over an algebraically closed field. See Corollary 10.19 for the final improvement.

9.16. THEOREM. *Let k be an algebraically closed field of characteristic different from 2, 3, and 5. Let $(R, \mathfrak{m}, k)$ be a Gorenstein complete local ring containing k as a coefficient field. If R has finite CM type, then R is a complete ADE hypersurface singularity.* □

The ADE classification of Theorem 9.16 allows us to verify Conjectures 7.20 and 7.21 in this case.

9.17. COROLLARY. *Let R be as in Theorem 9.16. Then R has minimal multiplicity. If $\operatorname{char}(k) = 0$, then R has a rational singularity.* □

§4. Matrix factorizations for the Kleinian singularities

Theorem 6.23 is the statement that the complete Kleinian singularities $k[\![x, y, z]\!]/(f)$ have finite CM type, where f is one of the polynomials listed in Table 6.19 and k is an algebraically closed field of characteristic not 2, 3, or 5. This was the key step in the classification of Gorenstein rings of finite CM type in the previous section. Given their central importance, it is worthwhile to have a complete listing of the matrix factorizations for the indecomposable MCM modules over these rings.

To describe the matrix factorizations, we return to the setup of Definition 6.5: Let G be a finite subgroup of $\operatorname{SL}(2, \mathbb{C})$, that is, one of the binary polyhedral groups of Theorem 6.11. Let G act linearly on the power series ring $S = \mathbb{C}[\![u, v]\!]$, and set $R = S^G$. Then R is generated over $\mathbb{C}$ by three invariants $x(u, v)$, $y(u, v)$, and $z(u, v)$, which satisfy a relation $z^2 + g(x, y) = 0$ for some polynomial g depending on G, so that $R \cong \mathbb{C}[\![x, y, z]\!]/(z^2 + g(x, y))$.

Set $A = \mathbb{C}[\![x(u, v), y(u, v)]\!] \subset R$. Then A is a power series ring, in particular a regular local ring. Since $z^2 \in A$, we see that as in Chapter 8, R is a free A-module of rank 2. Moreover, any MCM R-module is A-free as well. It is known [**ST54, Coh76**] that A is also a ring of invariants of a finite group $G' \subset \operatorname{U}(2)$, generated by complex reflections of order 2 and containing G as a subgroup of index 2.

Let $V_0, \ldots, V_d$ be a full set of the non-isomorphic irreducible representations of G; then we know from Corollary 5.20 and Theorem 6.3 that $M_j = (S \otimes_{\mathbb{C}} V_j)^G$, for $j = 0, \ldots, d$, are precisely the direct summands of S as R-module and are also precisely the indecomposable MCM R-modules. To get a handle on the M_j, we can express them as $(S \otimes_{\mathbb{C}} \operatorname{Ind}_G^{G'} V_j)^{G'}$. Being free over A, each M_j will have a basis of G'-invariants. These, and the identities of the representations $\operatorname{Ind}_G^{G'} V_j$, are computed in [**GSV81**].

Now we show how to obtain the matrix factorization corresponding to each M_j, following [**GSV81**]. The proof of the next proposition is a straightforward verification, mimicking the proof (see Remark B.6(i)) that the kernel of the diagonal map $\mu \colon B \otimes_A B \longrightarrow B$ is generated by all elements of the form $b \otimes 1 - 1 \otimes b$. The essential observation is that $z^2 = -g(x, y) \in A$.

9.18. PROPOSITION. *Define an R-module endomorphism $\sigma \colon S \longrightarrow S$ by sending z to $-z$, and let $^\sigma S$ be the R-module with underlying abelian group S, but with R-module structure given by $r \cdot s = \sigma(r)s$. Then we have two exact sequences of R-modules:*

$$0 \longrightarrow {}^\sigma S \xrightarrow{i^-} R \otimes_A S \xrightarrow{p^+} S \longrightarrow 0$$

and

$$0 \longrightarrow S \xrightarrow{i^+} R \otimes_A S \xrightarrow{p^-} {}^\sigma S \longrightarrow 0,$$

where $i^-(s) = i^+(s) = z \otimes s - 1 \otimes zs$, $j^+(r \otimes s) = rs$, and $j^-(r \otimes s) = \sigma(r)s$.

From this proposition one deduces the following theorem. We omit the details.

9.19. THEOREM. *Let $S = \mathbb{C}[\![u, v]\!]$, G a finite subgroup of $\operatorname{SL}(2, \mathbb{C})$ acting linearly on S, and $R = S^G$. Let x, y, and z be generating invariants for R satisfying the relation $z^2 + g(x, y) = 0$, and let $A = \mathbb{C}[\![x, y]\!]$. Then the R-free resolution of S has the form*

$$\cdots \xrightarrow{T^-} R \otimes_A S \xrightarrow{T^+} R \otimes_A S \xrightarrow{T^-} R \otimes_A S \xrightarrow{p^+} S \longrightarrow 0,$$

where

$$T^{\pm}(r \otimes s) = zr \otimes s \pm r \otimes zs.$$

Moreover, the R-free resolution of each indecomposable R-direct summand M_j of S is the direct summand of the above resolution of the form

$$\cdots \xrightarrow{T_j^-} R \otimes_A M_j \xrightarrow{T_j^+} R \otimes_A M_j \xrightarrow{T_j^-} R \otimes_A M_j \xrightarrow{p_j^+} M_j \longrightarrow 0.$$

§4. MATRIX FACTORIZATIONS FOR THE KLEINIAN SINGULARITIES

In terms of matrices, the resolution and corresponding matrix factorization of the MCM R-module M_j can be deduced from the theorem as follows. Let $\Phi \colon S \longrightarrow S$ denote the R-linear homomorphism given by multiplication by z, and let $\Phi_j \colon M_j \longrightarrow M_j$ be the restriction to M_j. Then each Φ_j is an A-linear map of free A-modules. Choose a basis and represent Φ_j by an $n \times n$ matrix φ_j with entries in x and y. Then φ_j^2 is equal to multiplication by $z^2 = -g(x,y) \in A$ on M_j, so that
$$(zI_n - \varphi_j,\ zI_n + \varphi_j)$$
is a matrix factorization of $z^2 + g(x,y)$ with cokernel M_j.

Our task is thus reduced to computing the matrix representing multiplication by z on each M_j. As in Chapter 6, we treat each case separately.

9.20 (A_n). We have already computed the presentation matrices of the MCM modules over $\mathbb{C}[\![x,y,z]\!]/(xz - y^{n+1})$ in Example 5.25, but we illustrate Theorem 9.19 in this easy case before proceeding to the more involved ones below. The cyclic group $\mathcal{C}_{n+1}$, generated by
$$\epsilon_{n+1} = \begin{pmatrix} \zeta_{n+1} & 0 \\ 0 & \zeta_{n+1}^{-1} \end{pmatrix},$$
has invariants $x = u^{n+1} + v^{n+1}$, $y = uv$, and $z = u^{n+1} - v^{n+1}$, satisfying
$$z^2 - (x^2 - 4y^{n+1}) = 0.$$
Set $A = \mathbb{C}[\![x,y]\!] \subset R = k[\![x,y,z]\!]$. Then $A = \mathbb{C}[\![u^{n+1} + v^{n+1}, uv]\!]$ is an invariant ring of the group G' generated by ϵ_{n+1} and the additional reflection $s = \begin{pmatrix} & 1 \\ 1 & \end{pmatrix}$.

Let V_j, for $j = 0, \ldots, n$, be the irreducible representation of $\mathcal{C}_{n+1}$ with character $\chi_j(g) = \zeta_{n+1}^j$. Then the MCM R-modules $M_j = (S \otimes_\mathbb{C} V_j)^G$ are generated over R by the monomials $u^a v^b$ such that $b - a \equiv j \mod n+1$. Over A, each M_j is freely generated by u^j and v^{n+1-j}. Since
$$zu^j = (u^{n+1} - v^{n+1})u^j = (u^{n+1} + v^{n+1})u^j - 2(uv)^j v^{n+1-j}$$
and
$$zv^{n+1-j} = (u^{n+1} - v^{n+1})v^{n+1-j} = 2(uv)^{n+1-j}u^j - (u^{n+1} + v^{n+1})v^{n+1-j},$$
the matrix φ_j representing the action of z on M_j is
$$\varphi_j = \begin{pmatrix} x & 2y^{n+1-j} \\ -2y^j & -x \end{pmatrix}.$$

One checks that $\varphi_j^2 = (x^2 - 4y^{n+1})I_2$, so $(zI_2 - \varphi_j, zI_2 + \varphi_j)$ is the matrix factorization corresponding to M_j.

Making a linear change of variables, we find that the indecomposable matrix factorizations of the (A_n) singularity defined by $x^2 + y^{n+1} + z^2 = 0$ are $(zI_2 - \varphi_j, zI_2 + \varphi_j)$, where

$$\varphi_j = \begin{pmatrix} ix & y^{n+1-j} \\ -y^j & -ix \end{pmatrix},$$

for $j = 0, \ldots, n$, and where i denotes a square root of -1.

9.21 (D_n). The dihedral group $\mathcal{D}_{n-2}$ is generated by

$$\alpha = \begin{pmatrix} \zeta_{2(n-2)} & 0 \\ 0 & \zeta_{2(n-2)}^{-1} \end{pmatrix} \quad \text{and} \quad \beta = \begin{pmatrix} 0 & i \\ i & 0 \end{pmatrix},$$

where again i denotes a square root of -1. The invariants of α and β are $x = u^{2(n-2)} + (-1)^n v^{2(n-2)}$, $y = u^2 v^2$, and $z = uv(u^{2(n-2)} - (-1)^n v^{2(n-2)})$, which satisfy

$$z^2 - y(x^2 - 4(-1)^n y^{n-2}) = 0.$$

Again we set $A = \mathbb{C}[\![x,y]\!] = \mathbb{C}[\![u^{2(n-2)} + (-1)^n v^{2(n-2)}, u^2 v^2]\!]$ and again A is the ring of invariants of the group G' generated by α, β, and $s = \begin{pmatrix} & 1 \\ 1 & \end{pmatrix}$.

In the matrices below, we will implicitly make the linear changes of variable necessary to put the defining equation of R into the form

$$z^2 - \left(-y\left(x^2 + y^{n-2}\right)\right) = 0.$$

Consider first the one-dimensional representation V_1 given by $\alpha \mapsto 1$ and $\beta \mapsto -1$. The MCM R-module $M_1 = (S \otimes_{\mathbb{C}} V_1)^G$ has A-basis given by $(uv, u^{2(n-2)} - (-1)^n v^{2(n-2)})$, and after the change of variable the matrix φ_1 for multiplication by z is

$$\varphi_1 = \begin{pmatrix} 0 & -x^2 - y^{n-1} \\ y & 0 \end{pmatrix}.$$

Next consider the two-dimensional irreducible representations V_j, for $j = 2, \ldots, n-2$, given by

$$\alpha \mapsto \begin{pmatrix} \zeta_{2(n-2)}^{j-1} & 0 \\ 0 & \zeta_{2(n-2)}^{-j+1} \end{pmatrix} \quad \text{and} \quad \beta \mapsto \begin{pmatrix} 0 & i^{j-1} \\ i^{j-1} & 0 \end{pmatrix}.$$

For each j, the corresponding MCM R-module M_j has A-basis

$$(u^{j-1}, uv^{2n-j-2}, u^j v, v^{2n-j-3}).$$

The matrix φ_j depends on the parity of j; for j even, it is

$$\varphi_j = \begin{pmatrix} & & -xy & -y^{n-1-j/2} \\ & & -y^{j/2} & x \\ x & y^{n-1-j/2} & & \\ y^{j/2} & -xy & & \end{pmatrix}$$

while if j is odd we have

$$\varphi_j = \begin{pmatrix} & & -xy & -y^{n-1-(j-1)/2} \\ & & -y^{(j+1)/2} & xy \\ x & y^{n-2-(j-1)/2} & & \\ y^{(j-1)/2} & -x & & \end{pmatrix}.$$

Finally consider V_{n-1} and V_n, which are the irreducible components of the two-dimensional reducible representation

$$\alpha \mapsto \begin{pmatrix} -1 & 0 \\ 0 & -1 \end{pmatrix}, \quad \beta \mapsto \begin{pmatrix} 0 & i \\ i & 0 \end{pmatrix}.$$

The MCM R-modules M_{n-1} and M_n have bases over A given by

$$(uv(u^{n-2} + (-1)^{n+1}v^{n-2}), u^{n-2} + (-1)^n v^{n-2})$$

and

$$(uv(u^{n-2} + (-1)^n v^{n-2}), u^{n-2} + (-1)^{n+1} v^{n-2}),$$

respectively. Again the corresponding matrices φ_{n-1} and φ_n depend on parity: for n odd we have

$$\varphi_{n-1} = \begin{pmatrix} iy^{(n-1)/2} & -x \\ xy & -iy^{(n-1)/2} \end{pmatrix} \text{ and } \varphi_n = \begin{pmatrix} iy^{(n-1)/2} & -xy \\ x & -iy^{(n-1)/2} \end{pmatrix},$$

and for n even

$$\varphi_{n-1} = \begin{pmatrix} 0 & -x - iy^{(n-2)/2} \\ xy - iy^{n/2} & 0 \end{pmatrix}$$

and

$$\varphi_n = \begin{pmatrix} 0 & -x + iy^{(n-2)/2} \\ xy + iy^{n/2} & 0 \end{pmatrix}.$$

For the E-series examples, we suppress the details of the complex reflection group G' and the A-bases for the M_j. The interested reader should see [**ST54**] and [**GSV81**].

9.22 (E_6). The defining equation of the (E_6) hypersurface singularity is $z^2 - (-x^3 - y^4) = 0$. For each of the six non-trivial irreducible representations V_1, V_2, V_3, $V_3^\vee$, V_4, and $V_4^\vee$, one can choose A-bases for M_j so that multiplication by z is given by the following matrices. The matrix factorizations for the corresponding MCM R-modules are given by $(zI_n - \varphi, zI_n + \varphi)$.

$$\varphi_1 = \begin{pmatrix} & & -x^2 & -y^3 \\ & & -y & x \\ x & y^3 & & \\ y & -x^2 & & \end{pmatrix} \qquad \varphi_2 = \begin{pmatrix} & & & -x^2 & -y^3 & xy^2 \\ & & & xy & -x^2 & -y^3 \\ & & & -y^2 & xy & -x^2 \\ x & 0 & y^2 & & & \\ y & x & 0 & & & \\ 0 & y & x & & & \end{pmatrix}$$

$$\varphi_3 = \begin{pmatrix} iy^2 & 0 & -x^2 & 0 \\ 0 & iy^2 & -xy & -x^2 \\ x & 0 & -iy^2 & 0 \\ -y & x & 0 & -iy^2 \end{pmatrix} \qquad \varphi_3^{\vee} = \begin{pmatrix} -iy^2 & 0 & -x^2 & 0 \\ 0 & -iy^2 & -xy & -x^2 \\ x & 0 & iy^2 & 0 \\ -y & x & 0 & iy^2 \end{pmatrix}$$

$$\varphi_4 = \begin{pmatrix} iy^2 & -x^2 \\ x & -iy^2 \end{pmatrix} \qquad \varphi_4^{\vee} = \begin{pmatrix} -iy^2 & -x^2 \\ x & iy^2 \end{pmatrix}$$

9.23 (E_7). The (E_7) singularity is defined by $z^2 - (-x^3 - xy^3) = 0$. There are 7 non-trivial irreducible representations $V_1, \ldots, V_7$, and the matrices φ_j corresponding to multiplication by z are given below. The matrix factorizations for the corresponding MCM R-modules are given by $(zI_n - \varphi, zI_n + \varphi)$.

$$\varphi_1 = \begin{pmatrix} & & -x^2 & -xy^2 \\ & & -y & x \\ x & xy^2 & & \\ y & -x^2 & & \end{pmatrix}$$

$$\varphi_2 = \begin{pmatrix} & & & -x^2 & -xy^2 & x^2y \\ & & & xy & -x^2 & -xy^2 \\ & & & -y^2 & xy & -x^2 \\ x & 0 & xy & & & \\ y & x & 0 & & & \\ 0 & y & x & & & \end{pmatrix}$$

$$\varphi_3 = \begin{pmatrix} & & & & 0 & 0 & -x^2 & -xy^2 \\ & & & & 0 & 0 & -xy & x^2 \\ & & & & -x & -y^2 & 0 & xy \\ & & & & -y & x & -x & 0 \\ 0 & -xy & x^2 & xy^2 & & & & \\ x & 0 & xy & -x^2 & & & & \\ x & y^2 & 0 & 0 & & & & \\ y & -x & 0 & 0 & & & & \end{pmatrix}$$

$$\varphi_4 = \begin{pmatrix} & & & xy & -x^2 & -xy^2 \\ & & & -y^2 & xy & -x^2 \\ & & & -x & -y^2 & xy \\ 0 & xy & x^2 & & & \\ x & 0 & xy & & & \\ y & x & 0 & & & \end{pmatrix}$$

$$\varphi_5 = \begin{pmatrix} & & -xy & -x^2 \\ & & -x & y^2 \\ y^2 & x^2 & & \\ x & -xy & & \end{pmatrix}$$

$$\varphi_6 = \begin{pmatrix} 0 & y^3 + x^2 \\ -x & 0 \end{pmatrix} \qquad \varphi_7 = \begin{pmatrix} & & -x^2 & -xy^2 \\ & & -xy & x^2 \\ x & y^2 & & \\ y & -x & & \end{pmatrix}$$

9.24 (E_8). The defining equation of the hypersurface(E_8) singularity is $z^2 - (-x^3 - y^5) = 0$. Here are the matrices φ_j representing multiplication by z on the 8 non-trivial indecomposable MCM R-modules. The matrix factorizations are given by $(zI_n - \varphi, zI_n + \varphi)$.

$$\varphi_1 = \begin{pmatrix} & & -x^2 & -y^4 \\ & & -y & x \\ x & y^4 & & \\ y & -x^2 & & \end{pmatrix} \qquad \varphi_2 = \begin{pmatrix} & & & -x^2 & -y^4 & xy^3 \\ & & & xy & -x^2 & -y^4 \\ & & & -y^2 & xy & -x^2 \\ x & 0 & y^3 & & & \\ y & x & 0 & & & \\ 0 & y & x & & & \end{pmatrix}$$

$$\varphi_3 = \begin{pmatrix} & & & & xy & -y^2 & -x^2 & 0 \\ & & & & -y^3 & 0 & 0 & -x \\ & & & & x^2 & 0 & 0 & -y^2 \\ & & & & 0 & x & -y^3 & -y \\ 0 & y^2 & -x & 0 & & & & \\ y^3 & xy & 0 & -x^2 & & & & \\ x & 0 & -y & y^2 & & & & \\ 0 & x^2 & y^3 & 0 & & & & \end{pmatrix}$$

$$\varphi_4 = \begin{pmatrix} & & & & & -y^3 & x^2 & 0 & 0 & 0 \\ & & & & & 0 & y^3 & -x^2 & xy^2 & -y^4 \\ & & & & & 0 & -xy & -y^3 & -x^2 & xy^2 \\ & & & & & y^2 & 0 & xy & -y^3 & -x^2 \\ & & & & & -x & -y^2 & 0 & 0 & 0 \\ y^2 & 0 & 0 & 0 & x^2 & & & & & \\ -x & 0 & 0 & 0 & y^3 & & & & & \\ 0 & x & y^2 & 0 & 0 & & & & & \\ y & 0 & x & y^2 & 0 & & & & & \\ 0 & y & 0 & x & y^2 & & & & & \end{pmatrix}$$

$$\varphi_5 = \begin{pmatrix} & & & & & & 0 & 0 & 0 & -x^2 & xy^2 & -y^4 \\ & & & & & & 0 & 0 & 0 & -y^3 & -x^2 & xy^2 \\ & & & & & & 0 & 0 & 0 & xy & -y^3 & -x^2 \\ & & & & & & -x & -y^2 & 0 & 0 & 0 & y^3 \\ & & & & & & 0 & -x & -y^2 & y^2 & 0 & 0 \\ & & & & & & -y & 0 & -x & 0 & y^2 & 0 \\ 0 & 0 & y^3 & x^2 & -xy^2 & y^4 & & & & & & \\ y^2 & 0 & 0 & y^3 & x^2 & -xy^2 & & & & & & \\ 0 & y^2 & 0 & -xy & y^3 & x^2 & & & & & & \\ x & y^2 & 0 & 0 & 0 & 0 & & & & & & \\ 0 & x & y^2 & 0 & 0 & 0 & & & & & & \\ y & 0 & x & 0 & 0 & 0 & & & & & & \end{pmatrix}$$

$$\varphi_6 = \begin{pmatrix} & & & & 0 & -y^3 & -x^2 & 0 \\ & & & & -y^2 & 0 & xy & -x^2 \\ & & & & -x & -y^2 & 0 & y^3 \\ & & & & 0 & -x & y^2 & 0 \\ 0 & y^3 & x^2 & -xy^2 & & & & \\ y^2 & 0 & 0 & x^2 & & & & \\ x & 0 & 0 & -y^3 & & & & \\ y & x & -y^2 & 0 & & & & \end{pmatrix}$$

$$\varphi_7 = \begin{pmatrix} & & -y^3 & -x^2 \\ & & x & -y^2 \\ y^2 & -x^2 & & \\ x & y^3 & & \end{pmatrix} \quad \varphi_8 = \begin{pmatrix} & & & -x^2 & xy^2 & -y^4 \\ & & & -y^3 & -x^2 & xy^2 \\ & & & xy & -y^3 & -x^2 \\ x & y^2 & 0 & & & \\ 0 & x & y^2 & & & \\ y & 0 & x & & & \end{pmatrix}$$

9.25. REMARK. We observe that the forms above for the indecomposable matrix factorizations over the two-dimensional ADE singularities make it easy to find the indecomposable matrix factorizations in dimension one. When the matrix φ (involving only x and y) has the distinctive anti-diagonal block shape, the pair of non-zero blocks constitutes (up to a sign) an indecomposable matrix factorization for the one-dimensional ADE polynomial in x and y. When the matrix φ does not have block form, $(\varphi, -\varphi)$ is an indecomposable matrix factorization. See §3 of Chapter 13.

§5. Bad characteristics

Here we describe, without proofs, the classification of hypersurfaces of finite CM type in characteristics 2, 3 and 5. If the characteristic of k is different from 2, Theorem 9.7 reduces the classification to the case of dimension one. We quote the following two theorems due to Greuel and Kröning [**GK90**] (cf. also the paper [**KS85**] by Kiyek and Steinke):

9.26. THEOREM (Characteristic 3). *Let k be an algebraically closed field of characteristic 3, let $d \geqslant 1$, and let $R = k[\![x, y, x_2, \ldots, x_d]\!]/(f)$, where $0 \neq f \in (x, y, x_2, \ldots, x_d)^2$. Then R has finite CM type if and only if $R \cong k[\![x, y, x_2, \ldots, x_d]\!]/(g + x_2^2 + \cdots + x_d^2)$, where $g \in k[x, y]$ is one of the following:*

$$
\begin{array}{ll}
(A_n): & x^2 + y^{n+1}, \quad n \geqslant 1 \\
(D_n): & x^2 y + y^{n-1}, \quad n \geqslant 4 \\
(E_6^0): & x^3 + y^4 \\
(E_6^1): & x^3 + y^4 + x^2 y^2 \\
(E_7^0): & x^3 + xy^3 \\
(E_7^1): & x^3 + xy^3 + x^2 y^2 \\
(E_8^0): & x^3 + y^5 \\
(E_8^1): & x^3 + y^5 + x^2 y^3 \\
(E_8^2): & x^3 + y^5 + x^2 y^2
\end{array}
$$
□

9.27. THEOREM (Characteristic 5). *Let k be an algebraically closed field of characteristic 5, let $d \geqslant 1$, and let $R = k[\![x, y, x_2, \ldots, x_d]\!]/(f)$, where $0 \neq f \in (x, y, x_2, \ldots, x_d)^2$. Then R has finite CM type if and only if $R \cong k[\![x, y, x_2, \ldots, x_d]\!]/(g + x_2^2 + \cdots + x_d^2)$, where $g \in k[x, y]$ is one of the following:*

$$
\begin{array}{ll}
(A_n): & x^2 + y^{n+1}, \quad n \geqslant 1 \\
(D_n): & x^2 y + y^{n-1}, \quad n \geqslant 4 \\
(E_6): & x^3 + y^4 \\
(E_7): & x^3 + xy^3 \\
(E_8^0): & x^3 + y^5
\end{array}
$$

(E_8^1): $x^3 + y^5 + xy^4$ □

In characteristics different from two, notice that $S[\![u,v]\!]/(f + u^2 + v^2) \cong S[\![u,v]\!]/(f + uv)$, via the transformation $u \mapsto \frac{u+v}{2}$, $v \mapsto \frac{u-v}{2\sqrt{-1}}$. Thus, if one does not mind skipping a dimension, one can transfer finite CM type up and down along the iterated double branched cover $R^{\sharp\sharp} = S[\![u,v]\!]/(f+uv)$, where $R = S/(f)$. Remarkably, this works in characteristic two as well [**Sol89**, **GK90**].

9.28. THEOREM (Solberg, Greuel-Kröning). *Let k be an algebraically closed field of arbitrary characteristic, let $d \geqslant 3$, and define $R = k[\![x_0, \ldots, x_d]\!]/(f)$, where $0 \neq f \in (x_0, \ldots, x_d)^2$. Then R has finite CM type if and only if there exists a non-zero non-unit $g \in k[\![x_0, \ldots, x_{d-2}]\!]$ such that $R \cong k[\![x_0, \ldots, x_d]\!]/(g + x_{d-1}x_d)$ and $k[\![x_0, \ldots, x_{d-2}]\!]/(g)$ has finite CM type.*   □

Solberg proved the "if" direction in his 1987 dissertation [**Sol89**]. He showed, in fact, that, for any non-zero non-unit $g \in k[\![x_0, \ldots, x_{d-2}]\!]$, the hypersurface ring $k[\![x_0, \ldots, x_{d-2}]\!]/(g)$ has finite CM type if and only if $k[\![x_0, \ldots, x_d]\!]/(g + x_{d-1}x_d)$ has finite CM type. The proof, which uses the theory of AR sequences (cf. Chapter 13), is quite unlike the proof in characteristics different from two, in that there seems to be no nice correspondence between MCM R-modules and MCM $R^{\sharp\sharp}$-modules (such as in Theorem 8.33). In 1988 Greuel and Kröning [**GK90**] used deformation theory to show that if R as in the theorem has finite CM type, then $R \cong k[\![x_0, \ldots, x_d]\!]/(g + x_{d-1}x_d)$ for a suitable non-zero non-unit element $g \in k[\![x_0, \ldots, x_{d-2}]\!]$, thereby establishing the converse of the theorem.

In order to finish the classification of complete hypersurface singularities of finite CM type in characteristic two, one needs to classify those singularities in dimensions one and two. The normal forms are listed in Section 5 of [**Sol89**] and in [**GK90**] and depend on earlier work of Artin [**Art77**], Artin-Verdier [**AV85**], and Kiyek-Steinke [**KS85**].

§6. Exercises

9.29. EXERCISE. Let $(S, \mathfrak{n})$ be a regular local ring, and $f \in \mathfrak{n}^r \setminus \mathfrak{n}^{r+1}$. Show that the hypersurface ring $S/(f)$ has multiplicity r. (Hint: pass to the associated graded ring and compute the Hilbert function of $S/(f)$. See Appendix A for an alternative approach.)

9.30. EXERCISE. Let S be a regular local ring and $R = S/(f)$ a hypersurface singularity of dimension at least two. If R is simple, prove that R is an integral domain.

9.31. EXERCISE. In the notation of Lemma 9.4, prove that
$$(\alpha^3, \alpha^2\beta^2, \alpha\beta^4, \beta^6) \subseteq (\alpha + \lambda\beta^2, \beta^3)^2$$
for any λ. (Hint: start with β^6 and work backwards.)

CHAPTER 10

Ascent and Descent

We have seen in Chapter 9 that the hypersurface rings $(R, \mathfrak{m}, k)$ of finite Cohen-Macaulay type have a particularly nice description when R is complete, k is algebraically closed and R contains a field of characteristic different from 2, 3, and 5. In this section we will see to what extent finite CM type ascends to and descends from faithfully flat extensions such as the completion or Henselization, and how it behaves with respect to residue field extension. In 1987 Schreyer [**Sch87**] conjectured that a local ring $(R, \mathfrak{m}, k)$ has finite CM type if and only if the $\mathfrak{m}$-adic completion $\widehat{R}$ has finite CM type. We have already seen that Schreyer's conjecture is true in dimension one (Corollary 4.17). We shall see that the "if" direction holds in general, and the "only if" direction holds when R is excellent and Cohen-Macaulay. For rings that are neither excellent nor CM, there are counterexamples (see 10.12). Schreyer also conjectured ascent and descent of finite CM type along extensions of the residue field (see Theorem 10.14 below). We shall prove descent in general, and ascent in the separable case. Inseparable extensions, however, can cause problems (see Example 10.17). We will revisit some of these issues in Chapter 17, where we consider ascent and descent of *bounded* CM type.

§1. Descent

Recall from Chapter 2 that for a finitely generated R-module M, we denote by $\mathrm{add}_R(M)$ the full subcategory of R-mod containing modules that are isomorphic to direct summands of direct sums of copies of M. When $A \longrightarrow B$ is a faithfully flat ring extension and M and N are finitely generated R-modules, we have $M \in \mathrm{add}_R(N)$ if and only if $S \otimes_R M \in \mathrm{add}_S(S \otimes_R N)$ (Proposition 2.18). Furthermore, when R is local and M is finitely generated, $\mathrm{add}_R(M)$ contains only finitely many isomorphism classes of indecomposable modules (Theorem 2.2).

Here is the main result on descent ([**Wie98**, Theorem 1.5]).

10.1. THEOREM. *Let $(R, \mathfrak{m}) \longrightarrow (S, \mathfrak{n})$ be a flat local homomorphism such that $S/\mathfrak{m}S$ is Cohen-Macaulay. If S has finite CM type, then so has R.*

PROOF. The hypothesis that the closed fiber $S/\mathfrak{m}S$ is CM guarantees that $S\otimes_R M$ is a MCM S-module whenever M is a MCM R-module (see Exercise 10.20). Let $\mathcal{U}$ be the class of MCM S-modules that occur in direct-sum decompositions of extended MCM modules; thus $Z \in \mathcal{U}$ if and only if there is a MCM R-module X such that Z is isomorphic to an S-direct-summand of $S \otimes_R X$. Let $Z_1, \ldots, Z_t$ be a complete set of representatives for isomorphism classes of indecomposable modules in $\mathcal{U}$. Choose, for each i, a MCM R-module X_i such that $Z_i \mid S \otimes_R X_i$, and put $Y = X_1 \oplus \cdots \oplus X_t$.

Suppose now that L is an indecomposable MCM R-module. Then $S \otimes_R L \cong Z_1^{(a_1)} \oplus \cdots \oplus Z_t^{(a_t)}$ for suitable non-negative integers a_i, and it follows that $S \otimes_R L$ is isomorphic to a direct summand of $S \otimes_R Y^{(a)}$, where $a = \max\{a_1, \ldots, a_t\}$. By Proposition 2.18, L is a direct summand of a direct sum of copies of Y. Then, by Theorem 2.2, there are only finitely many possibilities for L, up to isomorphism. $\square$

By the way, the class $\mathcal{U}$ in the proof of Theorem 10.1 is *not* necessarily the class of all MCM S-modules. For example, consider the extension $R = k[\![t^2]\!] \longrightarrow k[\![t^2, t^3]\!] = S$; in this case, the only extended MCM modules are the free ones. (Cf. also Exercise 14.29). The first order of business in the next section will be to find situations where this unfortunate behavior cannot occur, that is, where *every* MCM S-module is a direct summand of an extended MCM module.

§2. Ascent to the completion

It's a long way to the completion of a local ring, so we will make a stop at the Henselization. In this section and the next, we will need to understand the behavior of finite CM type under direct limits of étale and, more generally, unramified extensions. We will recall the basic definitions here and refer to Appendix B for details, in particular, for reconciling our definitions with others in the literature.

10.2. DEFINITION. A local homomorphism $(R, \mathfrak{m}, k) \longrightarrow (S, \mathfrak{n}, \ell)$ of local rings is *unramified* provided S is essentially of finite type over R (that is, S is a localization of some finitely generated R-algebra) and the following properties hold.

(i) $\mathfrak{m}S = \mathfrak{n}$, and
(ii) $S/\mathfrak{m}S$ is a finite separable field extension of $R/\mathfrak{m}$.

If, in addition, φ is flat, then we say φ is *étale*. (We say also that S is an *unramified*, respectively, *étale* extension of R.) Finally, a *pointed étale neighborhood* is an étale extension $(R, \mathfrak{m}, k) \longrightarrow (S, \mathfrak{n}, \ell)$ inducing an isomorphism on residue fields.

By Proposition B.9, properties (i) and (ii) are equivalent to the single requirement that the *diagonal map* $\mu\colon S\otimes_R S \longrightarrow R$ (taking $s_1 \otimes s_2$ to $s_1 s_2$) splits as $S \otimes_R S$-modules (equivalently, $\ker(\mu)$ is generated by an idempotent).

It turns out (see [**Ive73**] for details) that the isomorphism classes of pointed étale neighborhoods of a local ring $(R, \mathfrak{m})$ form a direct system. The remarkable fact that makes this work is that if $R \longrightarrow S$ and $R \longrightarrow T$ are pointed étale neighborhoods then there is *at most one* homomorphism $S \longrightarrow T$ making the obvious diagram commute.

10.3. DEFINITION. The *Henselization* R^{h} of R is the direct limit of a set of representatives of the isomorphism classes of pointed étale neighborhoods of R.

The Henselization is, conveniently, a Henselian ring (Chapter 1, §2 and Appendix A, §3).

Suppose $R \hookrightarrow S$ is a flat local homomorphism. As in Chapter 4, §1, we say that a finitely generated S-module M is *weakly extended* (from R) provided there is a finitely generated R-module N such that M is isomorphic to a direct summand of the S-module $S \otimes_R N$. In this case we also say that M is weakly extended from N.

Our immediate goal is to show, in Theorem 10.8, that if R has finite CM type then R^{h} does too. We prove in Proposition 10.4 that it will suffice to show that every MCM R^{h}-module is weakly extended from a MCM R-module. In Proposition 10.5 we show, under certain conditions, that it is enough to show that every finitely generated R^{h}-module is weakly extended. Then, in Proposition 10.7 we verify that these conditions are satisfied and that, indeed, every finitely generated R^{h}-module is weakly extended from R [**HW09**, Theorem 5.2]. The proof depends on the fact (Theorem 7.12) that rings of finite CM type have isolated singularities. Some results here include details that will not be needed in this chapter but will be used in the study of bounded and countable CM type.

10.4. PROPOSITION. *Let $(R, \mathfrak{m}) \longrightarrow (S, \mathfrak{n})$ be a local homomorphism. Assume that every MCM S-module is weakly extended from a MCM R-module. If R has finite CM type, so has S.*

PROOF. Let $L_1, \ldots, L_t$ be a complete list of representatives for the isomorphism classes of indecomposable MCM R-modules. Let $L = L_1 \oplus \cdots \oplus L_t$, and put $V = S \otimes_R L$. Given a MCM S-module M, we choose a MCM R-module N such that $M \mid S \otimes_R N$. Writing $N = L_1^{(a_1)} \oplus \cdots \oplus L_t^{(a_t)}$, we see that $N \in \mathrm{add}_R(L)$ and hence that

$M \in \mathrm{add}_S(V)$. Thus every MCM S-module is contained in $\mathrm{add}_S(V)$; now Theorem 2.2 completes the proof. $\square$

10.5. PROPOSITION. *Let* $(R, \mathfrak{m}) \longrightarrow (S, \mathfrak{n})$ *be a flat local homomorphism of CM local rings. Assume that*

(i) *the closed fiber* $S/\mathfrak{m}S$ *is Artinian;*
(ii) $S_{\mathfrak{q}}$ *is Gorenstein for each prime ideal* $\mathfrak{q} \neq \mathfrak{n}$; *and*
(iii) *every finitely generated S-module is weakly extended from R.*

Then every MCM S-module is weakly extended from a MCM R-module. In particular, if R has finite CM type, so has S.

PROOF. Note that $\dim(R) = \dim(S)$ by [**BH93**, (A.11)]; let d be the common value. Let M be a MCM S-module. As S is Gorenstein on the punctured spectrum, Corollary A.15 implies that M is a dth syzygy of some finitely generated S-module U. We choose a finitely generated R-module V such that $U \mid S \otimes_R V$, say, $U \oplus X \cong S \otimes_R V$. Let W be a dth syzygy of V. Then W is MCM by the Depth Lemma. Since $R \longrightarrow S$ is flat, $S \otimes_R W$ is a dth syzygy of $S \otimes_R V$, as is $M \oplus L$, where L is a dth syzygy of X. By Schanuel's Lemma (A.10) there are finitely generated free S-modules G_1 and G_2 such that $(S \otimes_R W) \oplus G_1 \cong (L \oplus M) \oplus G_2$. Of course G_1 is extended from a free R-module F. Putting $N = W \oplus F$, we see that $M \mid S \otimes_R N$. This proves the first assertion, and the second follows from Proposition 10.4. $\square$

In dimension one we can get by with fewer hypotheses:

10.6. PROPOSITION. *Let* $(R, \mathfrak{m}) \longrightarrow (S, \mathfrak{n})$ *be a local homomorphism of one-dimensional CM rings, and let M be a MCM S-module. If M is weakly extended from R, then M is weakly extended from a MCM R-module.*

PROOF. Recall that over a one-dimensional CM ring a non-zero finitely generated module is MCM if and only if it is torsion-free. We have

(10.1) $$M \oplus X \cong S \otimes_R N$$

for some finitely generated R-module N. Let T be the torsion submodule of N. Then $S \otimes_R T$ is a torsion S-module, since it is killed by a regular element. Moreover, $(S \otimes_R N)/(S \otimes_R T) \cong S \otimes_R (N/T)$, which is a MCM S-module by Exercise 10.20. It follows that $S \otimes_R T$ is exactly the torsion submodule of $S \otimes_R N$. Killing the torsion in (10.1), we have $M \oplus X/\mathrm{tors}(X) \cong S \otimes_R (N/T)$. $\square$

Let's pause for a moment to recall a few definitions. First, a Noetherian ring A is *regular* provided $A_{\mathfrak{m}}$ is a regular local ring for each maximal ideal $\mathfrak{m}$ of A. A Noetherian ring A containing a field k is *geometrically regular over k* provided $\ell \otimes_k A$ is a regular ring for every finite algebraic extension ℓ of k. A homomorphism $\varphi \colon A \longrightarrow B$ of Noetherian rings is *regular* provided φ is flat, and for each $\mathfrak{p} \in \mathrm{Spec}(A)$ the fiber $B_\mathfrak{p}/\mathfrak{p}B_\mathfrak{p}$ is geometrically regular over the field $A_\mathfrak{p}/\mathfrak{p}A_\mathfrak{p}$. Finally, A is *excellent* provided

  (i) A is Noetherian,
  (ii) $A_\mathfrak{p} \longrightarrow (A_\mathfrak{p})\widehat{}$ is a regular homomorphism for each prime ideal $\mathfrak{p}$ of A,
  (iii) the non-singular locus of B is open in $\mathrm{Spec}(B)$ for every finitely generated A-algebra B, and
  (iv) A is universally catenary.

A local homomorphism $(R, \mathfrak{m}) \longrightarrow (S, \mathfrak{n})$ is *absolutely flat* [**Fer72**] provided both $R \longrightarrow S$ and the diagonal map $S \otimes_R S \longrightarrow S$ are flat homomorphisms. Equivalently [**Fer72**, Theorem 4.1], $R \longrightarrow S$ is flat, and for each $\mathfrak{p} \in \mathrm{Spec}(R)$ the fiber map $R_\mathfrak{p}/\mathfrak{p}R_\mathfrak{p} \longrightarrow S_\mathfrak{p}/\mathfrak{p}S_\mathfrak{p}$ is absolutely flat.

10.7. PROPOSITION. *Let $(R, \mathfrak{m}) \longrightarrow (S, \mathfrak{n})$ be a flat local homomorphism of CM local rings, and assume that S is the direct limit of a system $\{(S_\alpha, \mathfrak{n}_\alpha)\}_{\alpha \in \Lambda}$ of étale extensions of $(R, \mathfrak{m})$.*

  (i) Every finitely generated S-module is weakly extended from R.
  (ii) If $\mathfrak{q}$ is a prime ideal of S and $\mathfrak{p} = \mathfrak{q} \cap R$, then $\mathfrak{p}S_\mathfrak{q} = \mathfrak{q}S_\mathfrak{q}$. In particular, $\mathfrak{m}S = \mathfrak{n}$.
  (iii) If R is Gorenstein (respectively, regular) on the punctured spectrum, then so is S.
  (iv) If R is excellent and reduced, so is S.
  (v) $\dim R = \dim S$.

PROOF. Given an arbitrary finitely generated S-module, we choose a matrix A whose cokernel is M. Since all of the entries of A live in some étale extension T of R, we see that $M = S \otimes_T N$ for some finitely generated T-module N. Refreshing notation, we may assume that $\varphi \colon R \longrightarrow S$ is étale. We apply $- \otimes_S M$ to the diagonal map $\mu \colon S \otimes_R S \longrightarrow S$, getting a commutative diagram

(10.2)
$$\begin{array}{ccc} S \otimes_R S \otimes_S M & \xrightarrow{\mu \otimes 1_M} & S \otimes_S M \\ \cong \Big\downarrow & & \Big\downarrow \cong \\ S \otimes_R M & \longrightarrow & M \end{array}$$

in which the horizontal maps are split surjections of S-modules. The S-module structure on $S \otimes_R M$ comes from the S-action on S, not on

M. (The distinction is important; see Exercise 10.23.) Thus we have a split injection of S-modules $j\colon M \longrightarrow S \otimes_R M$. Now write $_RM$ as a directed union of finitely generated R-modules N_α. The flatness of φ implies that $S \otimes_R M$ is the directed union of the modules $S \otimes_R N_\alpha$. Since $j(M)$ is a finitely generated S-module, there is an index α_0 such that $j(M) \subseteq S \otimes_R N_{\alpha_0}$. We put $N = N_{\alpha_0}$. Since $j(M)$ is a direct summand of $S \otimes_R M$, it must be a direct summand of the smaller module $S \otimes_R N$. This proves (i).

To prove (ii), (iii), and (v), let $\mathfrak{q}$ be an arbitrary prime ideal of S. Put $\mathfrak{q}_\alpha = \mathfrak{q} \cap S_\alpha$ for each α and $\mathfrak{p} = \mathfrak{q} \cap R$. Each extension $R_\mathfrak{p} \longrightarrow (S_\alpha)_{\mathfrak{q}_\alpha}$ is étale by Exercise 10.22, and hence $\mathfrak{p}(S_\alpha)_{\mathfrak{q}_\alpha} = \mathfrak{q}_\alpha (S_\alpha)_{\mathfrak{q}_\alpha}$ for each α. Item (ii) follows, and now [**BH93**, (A.11)] gives (v).

Suppose R is Gorenstein on the punctured spectrum, and assume $\mathfrak{q} \neq \mathfrak{n}$. We see from (ii) that $\mathfrak{p} \neq \mathfrak{m}$, so $R_\mathfrak{p}$ is Gorenstein. Since the closed fiber $S_\mathfrak{q}/\mathfrak{p} S_\mathfrak{q}$ is a field, [**BH93**, (3.3.15)] implies that $S_\mathfrak{q}$ is Gorenstein. If, on the other hand, $R_\mathfrak{p}$ is regular, we see that the maximal ideal of $S_\mathfrak{q}$ can be generated by $\dim(R_\mathfrak{p})$ elements. Since by [**BH93**, (A.11)] $\dim(S_\mathfrak{q}) = \dim(R_\mathfrak{p})$, we conclude that $S_\mathfrak{q}$ is regular.

To prove (iv), we show first that S is reduced. Since S is CM, it is enough, by Proposition A.8, to show that $S_\mathfrak{q}$ is a field if $\mathfrak{q}$ is a minimal prime ideal of S. By the "going-down theorem" [**BH93**, Lemma A.9], $\mathfrak{p} := \mathfrak{q} \cap R$ is a minimal prime ideal of R. Therefore $R_\mathfrak{p}$ is a field, and now (ii) implies that $S_\mathfrak{q}$ is a field too. Next we observe that $S \otimes_R S \longrightarrow S$, being a direct limit of split maps, is flat, that is, $R \longrightarrow S$ is absolutely flat. (One could also use [**Fer72**, Theorem 4.1], since each fiber is a finite direct product of separable field extensions.) Finally, we apply [**Gre76**, Theorem 5.3] to conclude that S is excellent. $\square$

10.8. THEOREM. *Let $(R, \mathfrak{m}) \longrightarrow (S, \mathfrak{n})$ be a local homomorphism of CM local rings such that S is the direct limit of a system $\{(S_\alpha, \mathfrak{n}_\alpha)\}_{\alpha \in \Lambda}$ of étale extensions of $(R, \mathfrak{m})$. Assume R has finite CM type. Then every MCM S-module is weakly extended from a MCM R-module, so S too has finite CM type. In particular, the Henselization R^h has finite CM type.*

PROOF. By Theorem 7.12, R has at most an isolated singularity and therefore is Gorenstein on the punctured spectrum. By (iii) of Proposition 10.7, this property ascends to S. Now Propositions 10.7 and 10.5 guarantee that every MCM S-module is weakly extended from a MCM R-module. Proposition 10.4 completes the proof. $\square$

Finally, we prove ascent of finite CM type to the completion for excellent rings. Actually, we don't need the full strength of excellence; we just need $R \longrightarrow \widehat{R}$ to be a regular homomorphism.

We will need the following consequences of regularity of a ring homomorphism. The first assertion is clear from the definition, while the second follows from the first and from [**Mat89**, (32.2)].

10.9. PROPOSITION. *Let $A \longrightarrow B$ be a regular homomorphism, $\mathfrak{q} \in \mathrm{Spec}(B)$, and put $\mathfrak{p} = \mathfrak{q} \cap A$.*

(i) The homomorphism $A_\mathfrak{p} \longrightarrow B_\mathfrak{q}$ is regular.
(ii) If $A_\mathfrak{p}$ is a regular local ring, so is $B_\mathfrak{q}$. □

We'll also need the following remarkable theorem of Elkik [**Elk73**].

10.10. THEOREM (Elkik). *Let $(R, \mathfrak{m})$ be a local ring and M a finitely generated $\widehat{R}$-module. If $M_\mathfrak{p}$ is a free $R_\mathfrak{p}$-module for each non-maximal prime ideal $\mathfrak{p}$ of $\widehat{R}$, then M is extended from the Henselization R^h.* □

10.11. COROLLARY. *Let $(R, \mathfrak{m})$ be a CM local ring with $\mathfrak{m}$-adic completion $\widehat{R}$.*

(i) If $\widehat{R}$ has finite CM type, so has R.
(ii) Suppose R has finite CM type and $R \longrightarrow \widehat{R}$ is regular. Then every MCM $\widehat{R}$-module is weakly extended from a MCM R-module, and hence $\widehat{R}$ has finite CM type.

In particular, if R is excellent, then R has finite CM type if and only if $\widehat{R}$ has finite CM type.

PROOF. The first assertion is a special case of Theorem 10.1.

Suppose now that $R \longrightarrow \widehat{R}$ is regular and that R has finite CM type. Then R has at most an isolated singularity by Theorem 7.12, and it follows from Proposition 10.9 that $\widehat{R}$ too has at most an isolated singularity.

Now let M be an arbitrary MCM $\widehat{R}$-module. Then $M_\mathfrak{q}$ is a free $\widehat{R}_\mathfrak{q}$-module for each non-maximal prime ideal $\mathfrak{q}$ of $\widehat{R}$. By Theorem 10.10, M is extended from the Henselization, that is, there is an R^h-module N such that $M \cong N \otimes_{R^h} \widehat{R}$; moreover, N is a MCM R^h-module by Exercise 10.20. By Theorem 10.8, N is weakly extended from a MCM R-module, and therefore the same is true for M. Proposition 10.4 implies that $\widehat{R}$ has finite CM type. □

It is unknown whether or not Corollary 10.11 would be true without the hypothesis that R be CM, or without the hypothesis that $R \longrightarrow \widehat{R}$

be regular. The following example from [**LW00**] shows, however, that we can't omit *both* hypotheses:

10.12. EXAMPLE. Let $T = k[\![x,y,z]\!]/((x^3 - y^7) \cap (y,z))$, where k is any field. We claim that T has infinite CM type. To see this, set $R = k[\![x,y]\!]/(x^3 - y^7) \cong k[\![t^3, t^7]\!]$. Then R has infinite CM type by Theorem 4.10, since (DR2) fails for this ring. Further, $R[\![z]\!]$ has infinite CM type: the map $R \longrightarrow R[\![z]\!]$ is flat with CM closed fiber, and Theorem 10.1 applies. Now $R[\![z]\!] \cong T/(x^3 - y^7)$. As T and $T/(x^3 - y^7)$ have the same dimension, every MCM $T/(x^3 - y^7)$-module is MCM over T. Since $T/(x^3 - y^7)$ has infinite CM type, the claim follows.

It is easy to check that the image of x is a non-zerodivisor in T. By [**Lec86**, Theorem 1], T is the completion of some local integral domain A. Then A has finite CM type; in fact, it has no MCM modules at all! For if A had a MCM module, then A would be universally catenary [**Hoc73**, Section 1]. But this would imply, by [**Mat89**, Theorem 31.7], that A is formally equidimensional, that is, all minimal primes of $\widehat{A}\,(= T)$ have the same dimension. But the two minimal primes of T obviously have dimensions two and one, contradiction.

Another example of this behavior, using a very different construction, can be found in [**LW00**].

§3. Ascent along separable field extensions

Let $(R, \mathfrak{m}, k)$ be a local ring and ℓ/k a field extension. We want to lift the extension $k \hookrightarrow \ell$ to a flat local homomorphism $(R, \mathfrak{m}, k) \longrightarrow (S, \mathfrak{n}, \ell)$ with certain nice properties. The type of ring extension we seek is dubbed a *gonflement* by Bourbaki [**Bou06**, Appendice]. Translations of the term range from the innocuous "inflation" to the provocative "swelling" or "intumescence". To avoid choosing one, we stick with the French word.

10.13. DEFINITION. Let $(R, \mathfrak{m}, k)$ be a local ring.
(i) An *elementary gonflement* of R is either
   (a) a purely transcendental extension $R \longrightarrow (R[x])_{\mathfrak{m}R[x]}$ (where x is a single indeterminate), or
   (b) an extension $R \longrightarrow R[x]/(f)$, where f is a monic polynomial whose reduction modulo $\mathfrak{m}$ is irreducible in $k[x]$.
(ii) A *gonflement* is an extension $(R, \mathfrak{m}, k) \longrightarrow S$ with the following property: There is a well-ordered family $\{R_\alpha\}_{0 \leqslant \alpha \leqslant \lambda}$ of local extensions $(R, \mathfrak{m}, k) \hookrightarrow (R_\alpha, \mathfrak{m}_\alpha, k_\alpha)$ such that
   (a) $R_0 = R$ and $R_\lambda = S$,
   (b) $R_\beta = \bigcup_{\alpha < \beta} R_\alpha$ if β is a limit ordinal, and

(c) $R_{\alpha+1}$ is an elementary gonflement of R_α if $\alpha \neq \lambda$.

Elementary gonflements of type (ia) are often used to pass to a local ring with infinite residue field. (See Theorem A.20 for the reason why, and Proposition 4.3 for an application.) In this section we will need gonflements that are iterations of elementary gonflements of type (ib).

The following theorem ([**Bou06**, Appendice, §2]) summarizes the basic properties of gonflements.

10.14. THEOREM. *Let $(R, \mathfrak{m}, k)$ be a local ring.*
  (i) *Let $(R, \mathfrak{m}, k) \longrightarrow S$ be a gonflement.*
    (a) *S is local with maximal ideal $\mathfrak{n} := \mathfrak{m}S$. In particular, S is Noetherian.*
    (b) *$R \longrightarrow S$ is a flat local homomorphism.*
    (c) *$\dim R = \dim S$.*
    (d) *With the notation as in the definition, if $\alpha \leqslant \beta \leqslant \lambda$, then $R_\alpha \longrightarrow R_\beta$ is a gonflement.*
  (ii) *If $k \longrightarrow \ell$ is an arbitrary field extension, there is a gonflement $(R, \mathfrak{m}, k) \longrightarrow (S, \mathfrak{n}, \ell)$ lifting $k \longrightarrow \ell$.* □

We will prove ascent of finite CM type along gonflements with separable residue field growth. The next result is annoyingly similar to Proposition 10.7, but it's a bit different. (Notice, for example, that if $k \longrightarrow \ell$ is an uncountably generated algebraic extension, then one cannot represent ℓ as a well-ordered union of finite extensions.) The proof, which we omit, is pretty much the same as that of Proposition 10.7 except for the mechanical details of transfinite induction. The approach is standard: We want to prove (in the notation of Definition 10.13) that $S = R_\lambda$ has a certain property. We fix $\beta \leqslant \lambda$, assume that R_α has the property for all $\alpha < \beta$, then show, as in the proof of Proposition 10.7, that R_β has the property. Then we set $\beta = \lambda$ to complete the proof.

10.15. PROPOSITION. *Let $(R, \mathfrak{m}, k) \longrightarrow (S, \mathfrak{n}, \ell)$ be a gonflement, and assume that $k \longrightarrow \ell$ is a separable algebraic extension.*
  (i) *Every finitely generated S-module is weakly extended from R.*
  (ii) *If $\mathfrak{q}$ is a prime ideal of S and $\mathfrak{p} = \mathfrak{q} \cap R$, then $\mathfrak{p}S_\mathfrak{q} = \mathfrak{q}S_\mathfrak{q}$.*
  (iii) *If R is Gorenstein (respectively, regular) on the punctured spectrum, then so is S.*
  (iv) *If R is excellent and reduced, so is S.* □

10.16. THEOREM. *Let $(R, \mathfrak{m}, k) \longrightarrow (S, \mathfrak{n}, \ell)$ be a gonflement. Assume that R is CM and that $k \longrightarrow \ell$ is a separable algebraic extension.*
  (i) *If R is Gorenstein on the punctured spectrum, then every MCM S-module is weakly extended from a MCM R-module.*

(ii) R has finite CM type if and only if S has finite CM type.

PROOF. To prove (i), we appeal to Propositions 10.15 and 10.5. The "if" direction of item (ii) is a special case of Theorem 10.1. For the converse, suppose R has finite CM type. Then R is an isolated singularity by Theorem 7.12 and in particular is Gorenstein on the punctured spectrum. By item (i) and Proposition 10.4, S has finite CM type. □

The separability condition in 10.16 cannot be omitted. Indeed, here is an example of a local ring R with finite CM type and an elementary gonflement $R \longrightarrow S$ such that S has infinite CM type [**Wie98**, Example 3.4].

10.17. EXAMPLE. Let k be an imperfect field of characteristic 2, and let $\alpha \in k - k^2$. Put $R = k[\![x,y]\!]/(x^2 + \alpha y^2)$. Then R is a CM local domain of dimension one and multiplicity two, so by Theorem 4.18 R has finite CM type. However, by Proposition 4.15, $S = R \otimes_k k(\sqrt{\alpha}) = k(\sqrt{\alpha})[\![x,y]\!]/(x + \sqrt{\alpha}y)^2$ does not have finite Cohen-Macaulay type, since it is Cohen-Macaulay but not reduced.

Recall that we did not give a self-contained proof of Theorem 4.10. Here we describe a proof, independent of the matrix decompositions in [**GR78**], in an important special case.

10.18. THEOREM. *Let $(R, \mathfrak{m}, k)$ be an analytically unramified local ring of dimension one. Assume R contains a field and that k is perfect with* char $k \neq 2$, 3 *or* 5. *Then R has finite CM type if and only if R satisfies the Drozd-Roĭter conditions (DR1) and (DR2) of Chapter 4.*

PROOF. A complete proof of the "only if" direction is in Chapter 4. For the converse, we may assume, by Theorems 10.14 and 10.16, that k is algebraically closed. Corollary 4.17 (whose proof did not depend on Theorem 4.10!) allows us to assume that R is complete. Then $\overline{R} = k[\![t_1]\!] \times \cdots \times k[\![t_s]\!]$, where $s \leqslant 3$ and the t_i are analytic indeterminates. An elementary but tedious computation [**Yos90**, pages 72–73] now shows that R is a finite birational extension of an ADE singularity A. Since A has finite CM type (Corollary 8.22), Proposition 4.14 implies that R has finite CM type too. □

§4. Equicharacteristic Gorenstein singularities

We now assemble the pieces and obtain a nice characterization of the equicharacteristic Gorenstein singularities of finite CM type.

10.19. COROLLARY. *Let $(R, \mathfrak{m}, k)$ be an excellent, Gorenstein local ring containing a field. Let K be an algebraic closure of k. Assume $d := \dim(R) \geqslant 1$ and that k is perfect with $\mathrm{char}(k) \neq 2, 3$ or 5. Then R has finite CM type if and only if there is a non-zero non-unit $f \in k[\![x_0, \ldots x_d]\!]$ such that $\widehat{R} \cong k[\![x_0, \ldots, x_d]\!]/(f)$ and $K[\![x_0, \ldots, x_d]\!]/(f)$ is a complete ADE hypersurface singularity (see Chapter 9).*

PROOF. Using [**Mat89**, Theorem 22.5], we see that for any non-unit $f \in k[\![x_0, \ldots, x_d]\!]$, the hypersurface singularity $K[\![x_0, \ldots, x_d]\!]/(f)$ is flat over $k[\![x_0, \ldots, x_d]\!]/(f)$. The "if" direction now follows from Theorem 10.1 and the fact (Theorem 9.8) that simple singularities have finite CM type.

For the converse, suppose R has finite CM type. Since R is CM and excellent, the completion $\widehat{R}$ has finite CM type by Corollary 10.11. Moreover, Theorem 9.15 implies, since R is Gorenstein, that $\widehat{R}$ is a hypersurface, that is, $\widehat{R} \cong k[\![x_0, \ldots, x_d]\!]/(f)$ for a non-zero non-unit f.

The extension $\widehat{R} \longrightarrow K \otimes_k \widehat{R}$ is a gonflement lifting the field extension $k \longrightarrow K$ (cf. Exercise 10.25), and Theorem 10.16 implies that $A := K \otimes_k R$ has finite CM type. Moreover, A is excellent, by Proposition 10.15. Corollary 10.11 implies that $\widehat{A}$ has finite CM type. But $\widehat{A} \cong K[\![x_0, \ldots, x_d]\!]/(f)$, and by Theorem 9.8 $K[\![x_0, \ldots, x_d]\!]/(f)$ is a complete ADE hypersurface singularity. □

§5. Exercises

10.20. EXERCISE. Let $(R, \mathfrak{m}, k) \longrightarrow (S, \mathfrak{n}, \ell)$ be a flat local homomorphism, and let M be a finitely generated R-module. Prove that $S \otimes_R M$ is a MCM S-module if and only if M is MCM and the closed fiber $S/\mathfrak{m}S$ is a CM ring. (See [**BH93**, (1.2.16) and (A.11)].)

10.21. EXERCISE. Let $(R, \mathfrak{m}) \longrightarrow (S, \mathfrak{n})$ be a local homomorphism. Prove that the following two conditions are equivalent:

  (i) The induced map $R/\mathfrak{m} \longrightarrow S/\mathfrak{m}S$ is an isomorphism.
  (ii) The induced map $R/\mathfrak{m} \longrightarrow S/\mathfrak{n}$ is an isomorphism and $\mathfrak{m}S = \mathfrak{n}$.

10.22. EXERCISE. Let $\varphi \colon (R, \mathfrak{m}) \longrightarrow (S, \mathfrak{n})$ be a flat local homomorphism that is essentially of finite type (that is, S is a localization of a finitely generated R-algebra).

  (i) Prove that $S/\mathfrak{m}S$ is Artinian.
  (ii) Let $\mathfrak{q}$ be a prime ideal of S, and put $\mathfrak{p} = \varphi^{-1}(\mathfrak{q})$. If $R \longrightarrow S$ is étale, prove that $R_\mathfrak{p} \longrightarrow S_\mathfrak{q}$ is étale.

10.23. EXERCISE. Find an example of an étale homomorphism between local rings $R \longrightarrow S$ and a finitely generated S-module M such that the two S-actions on $S \otimes_R M$ (one via the action on S, the other via the action on M) give non-isomorphic S-modules.

10.24. EXERCISE. Suppose $(R, \mathfrak{m}, k) \longrightarrow (S, \mathfrak{n}, \ell)$ is a flat local homomorphism such that $\mathfrak{m}S = \mathfrak{n}$. Let M be a finitely generated R-module. Prove that $e_R(M) = e_S(S \otimes_R M)$. Show by example that this can fail without assumption that $\mathfrak{m}S = \mathfrak{n}$.

10.25. EXERCISE. Let $(R, \mathfrak{m})$ be a local ring with a coefficient field k, and let K/k be an algebraic field extension. Prove that $K \otimes_k R$ is a gonflement of R and that K is a coefficient field for R. (First do the case where $k \longrightarrow K$ is an elementary gonflement of type (ib) in Definition 10.13.)

10.26. EXERCISE. Let $(R, \mathfrak{m}, k)$ be a one-dimensional local ring satisfying the Drozd-Roĭter conditions (DR1) and (DR2) of Chapter 4, and let $R \longrightarrow (S, \mathfrak{n}, \ell)$ be a gonflement. Prove, without reference to finite CM type, that S satisfies (DR1) and (DR2).

CHAPTER 11

Auslander-Buchweitz Theory

We now turn to a celebrated tool in the study of CM representation types, and even more generally in the representation theory of local rings, namely MCM approximations. The slogan here is that any finitely generated module over a CM local ring with canonical module can be approximated by a MCM module, in a precise sense due originally to Auslander and Buchweitz [**AB89**]. The theory as originally constructed in [**AB89**] is quite abstract, and has since been further generalized. In keeping with our general strategy, we adopt a stubbornly concrete point of view. We deal exclusively with CM local rings, finitely generated modules, and approximations by MCM modules. We also use the more limited terminology of *MCM approximations* and *FID hulls*, rather than the general notions of (pre)covers and (pre)envelopes, though we touch on this technology in the next chapter.

In the first section we recall some basics on finitely generated modules of finite injective dimension, and particularly canonical modules, which occupy the central spot in the theory. We then detail the theory of MCM approximations and FID hulls, following Auslander and Buchweitz's original construction. Finally, we give some applications in terms of Auslander's δ-invariant. Other applications will appear in later chapters.

§1. Canonical modules

Here we give a quick primer on finitely generated modules of finite injective dimension over local rings and the most distinguished of such modules, the canonical module.

We point out first that over CM local rings, finitely generated modules of finite injective dimension exist.

11.1. PROPOSITION. *Let $(R, \mathfrak{m}, k)$ be a CM local ring. Then R admits a non-zero finitely generated module of finite injective dimension.*

PROOF. Let $\mathbf{x}$ be a system of parameters for R and $\overline{R}$ the quotient $R/(\mathbf{x})$. The injective hull $E = E_{\overline{R}}(k)$ of the residue field of $\overline{R}$ has finite length over $\overline{R}$ and hence over R. It follows that $M = \operatorname{Hom}_R(\overline{R}, E)$ is finitely generated over R, and dualizing the Koszul resolution of $\overline{R}$ into E displays $\operatorname{injdim} M < \infty$. □

As an aside, we take note here of the conjecture of Bass [**Bas63**] that the converse holds as well: "It seems conceivable that, say for A local, there exist finitely generated $M \neq 0$ with finite injective dimension only if A is a Cohen-Macaulay ring." This conjecture was established for local rings of prime characteristic or essentially of finite type over a field of characteristic zero by Peskine and Szpiro [**PS73**] using their Intersection Theorem. Since Roberts has proved the Intersection Theorem for all local rings [**Rob87**], Bass' Conjecture holds in general.

The first hint of a connection between modules of finite injective dimension and MCM modules comes courtesy of the next result, due to Ischebeck [**Isc69**], and its consequence below. We omit the proofs.

11.2. THEOREM. *Let $(R, \mathfrak{m}, k)$ be a local ring and M, N non-zero finitely generated R-modules with $\operatorname{injdim}_R N < \infty$. Then*

$$\operatorname{depth} R - \operatorname{depth} M = \sup \left\{ i \mid \operatorname{Ext}_R^i(M, N) \neq 0 \right\}. \quad \square$$

11.3. PROPOSITION. *Let $(R, \mathfrak{m}, k)$ be a CM local ring and M, N non-zero finitely generated R-modules. Then*

 (i) *M is MCM if and only if $\operatorname{Ext}_R^i(M, Y) = 0$ for all $i > 0$ and all finitely generated R-modules Y of finite injective dimension, and*
 (ii) *N has finite injective dimension if and only if $\operatorname{Ext}_R^i(X, N) = 0$ for all $i > 0$ and all MCM R-modules X.* □

Colloquially, we interpret Proposition 11.3 as the statement that MCM modules and finitely generated modules of finite injective dimension are "orthogonal." It will transpire that the intersection is "spanned" by a single module, namely the canonical module, to which we now turn. See [**BH93**, Chapter 3] for the details we omit.

11.4. DEFINITION. Let $(R, \mathfrak{m}, k)$ be a CM local ring of dimension d. A finitely generated R-module ω is a *canonical module* for R if ω is MCM, has finite injective dimension, and satisfies $\dim_k \operatorname{Ext}_R^d(k, \omega) = 1$.

The condition on $\operatorname{Ext}_R^d(k, \omega)$ is a sort of rank-one normalizing assumption: taking into account the calculation of both depth and injective dimension in terms of $\operatorname{Ext}_R^i(k, -)$, we can write Definition 11.4

compactly as

$$\operatorname{Ext}^i_R(k,\omega) \cong \begin{cases} k & \text{if } i = \dim R, \text{ and} \\ 0 & \text{otherwise.} \end{cases}$$

We need a laundry list of properties of canonical modules. Define the *codepth* of an R-module M to be $\operatorname{depth} R - \operatorname{depth} M$.

11.5. THEOREM. *Let $(R, \mathfrak{m}, k)$ be a CM local ring and ω a canonical module for R. Then*

(i) *ω is unique up to isomorphism, and R is Gorenstein if and only if $\omega \cong R$;*

(ii) *$\operatorname{End}_R(\omega) \cong R$.*

(iii) *For a CM R-module M of codepth t, set $M^\vee = \operatorname{Ext}^t_R(M, \omega)$. Then*

    (a) *$M^\vee$ is also CM of codepth t;*

    (b) *$\operatorname{Ext}^t_R(M, \omega) = 0$ for $i \neq t$; and*

    (c) *$M^{\vee\vee}$ is naturally isomorphic to M.*

(iv) *The canonical module behaves well with respect to factoring out a regular sequence, completion, and localization.* □

It is a result of Sharp, Foxby, and Reiten [**Sha71**, **Fox72**, **Rei72**] that a CM local ring R has a canonical module if and only if R is a homomorphic image of a Gorenstein local ring. In particular, by Cohen's Structure Theorems any complete local ring is a homomorphic image of a regular local ring, so admits a canonical module.

The stipulation that $\operatorname{Ext}^{\dim R}_R(k, \omega_R) \cong k$ is, as we observed, a kind of rank-one condition. Indeed, under a mild additional condition it forces ω_R to be isomorphic to an ideal of R. We say that R is *generically Gorenstein* if $R_\mathfrak{p}$ is Gorenstein for each minimal prime $\mathfrak{p}$ of R.

11.6. PROPOSITION. *Let R be a CM local ring and ω a canonical module for R. If R is generically Gorenstein, then ω is isomorphic to an ideal of R, and conversely. In this case, ω has constant rank 1, ω is an ideal of pure height one (that is, every associated prime of ω has height one), and R/ω is a Gorenstein ring of dimension $\dim R - 1$.*

PROOF. As $R_\mathfrak{p}$ is Gorenstein for every minimal $\mathfrak{p}$, we conclude that $\omega_\mathfrak{p}$ is free of rank one for those primes. In particular if we denote by K the total quotient ring, obtained by inverting the complement of the union of those minimal primes, then $\omega \otimes_R K$ is a rank-one projective module over the semilocal ring K. Thus $\omega \otimes_R K \cong K$. Fixing an isomorphism and composing with the natural map gives an R-homomorphism $\omega \longrightarrow K$, which is injective as ω is torsion-free.

Multiplying the image by a carefully chosen non-zerodivisor clears the denominators and knocks the image down into R, where it is an ideal. Being locally free at the minimal primes, it has height at least one.

Since ω is MCM, the Depth Lemma applied to the short exact sequence
$$0 \longrightarrow \omega \longrightarrow R \longrightarrow R/\omega \longrightarrow 0$$
yields $\operatorname{depth}(R/\omega) \geqslant \dim R - 1$, and since $\operatorname{height} \omega \geqslant 1$ we have that $\dim R/\omega \leqslant \dim R - 1$. Thus R/ω is a CM ring, in particular, unmixed, so ω has pure height one. Furthermore, R/ω is a CM R-module of codepth 1. Applying $\operatorname{Hom}_R(-, \omega)$ thus gives an exact sequence
$$\operatorname{Hom}_R(R/\omega, \omega) \longrightarrow \omega \longrightarrow R \longrightarrow \operatorname{Ext}^1_R(R/\omega, \omega) \longrightarrow 0$$
and $\operatorname{Ext}^1_R(R/\omega, \omega) = (R/\omega)^\vee$ is the canonical module for R/ω by the discussion after Theorem 11.5. Since R/ω is torsion and ω is torsion-free, the leftmost term in the exact sequence vanishes, whence $(R/\omega)^\vee$ is isomorphic to R/ω itself, so R/ω is Gorenstein.

For the converse, assume that ω is embedded into R as an ideal. Then as before we see that $\operatorname{height} \omega \geqslant 1$, so ω is not contained in any minimal prime and $R_\mathfrak{p}$ is Gorenstein for every minimal $\mathfrak{p}$. $\square$

We quickly observe, using this result, that there does indeed exist a CM local ring which is not a homomorphic image of a Gorenstein local ring, and hence does not admit a canonical module. This was first constructed by Ferrand and Raynaud [**FR70**]. Specifically, they construct a one-dimensional local domain $(R, \mathfrak{m})$ such that the completion $\widehat{R}$ is not generically Gorenstein. If R were to have a canonical module ω_R, it would be embeddable as an $\mathfrak{m}$-primary ideal of R. The completion $\widehat{\omega_R}$ is then a canonical module for $\widehat{R}$, and is an ideal of $\widehat{R}$. But this contradicts the criterion above.

We finish the section with the promised identification of the intersection of the class of MCM modules with that of modules of finite injective dimension.

11.7. PROPOSITION. *Let R be a CM local ring with canonical module ω and let M be a finitely generated R-module. If M is both MCM and of finite injective dimension, then M is isomorphic to a direct sum of copies of ω.*

PROOF. Let F be a free module mapping onto the canonical dual $M^\vee = \operatorname{Hom}_R(M, \omega)$ with kernel K. Dualizing gives a short exact sequence
$$0 \longrightarrow M \longrightarrow F^\vee \longrightarrow K^\vee \longrightarrow 0$$

where $K^\vee$ is MCM as K is. Proposition 11.3(ii) implies that the sequence splits as $\operatorname{injdim}_R M < \infty$, making M a direct summand of $F^\vee$. Dualizing again displays $M^\vee$ as a direct summand of the free module $F \cong F^{\vee\vee}$, whence $M^\vee$ is free and M is a direct sum of copies of ω. □

If R is not assumed to have a canonical module, the MCM modules of finite injective dimension are called *Gorenstein modules*. Should any exist, there is one of minimal rank and all others are direct sums of copies of the minimal one. See Corollary A.15 for an application of Gorenstein modules.

§2. MCM approximations and FID hulls

Throughout this section, $(R, \mathfrak{m}, k)$ denotes a CM local ring with canonical module ω.

We continue to think of the MCM modules and modules of finite injective dimension over R as orthogonal subspaces of the space of all finitely generated modules, with intersection spanned by the canonical module ω. Guided by memories of basic linear algebra, we hope to be able to project any R-module onto these subspaces.

11.8. DEFINITION. Let M be a finitely generated R-module. An exact sequence of finitely generated R-modules

$$0 \longrightarrow Y \longrightarrow X \longrightarrow M \longrightarrow 0$$

is a *MCM approximation* of M if X is MCM and $\operatorname{injdim}_R Y < \infty$. Dually, an exact sequence

$$0 \longrightarrow M \longrightarrow Y' \longrightarrow X' \longrightarrow 0$$

is a *hull of finite injective dimension* or *FID hull* if $\operatorname{injdim} Y' < \infty$ and either X' is MCM or $X' = 0$.

We sometimes abuse language and refer to the modules X and Y' as the MCM approximation and FID hull of M, rather than the whole extensions.

The orthogonality relations between MCM modules and modules of finite injective dimension translate into lifting properties for the MCM approximations and FID hulls.

11.9. PROPOSITION. *Let $0 \longrightarrow Y \longrightarrow X \longrightarrow M \longrightarrow 0$ be a MCM approximation of M and let $\varphi \colon Z \longrightarrow M$ be a homomorphism with Z MCM. Then φ factors through X. Any two liftings of φ are homotopic, i.e. their difference factors through Y.*

PROOF. Applying $\mathrm{Hom}_R(Z,-)$ to the approximation gives the exact sequence
$$\mathrm{Hom}_R(Z,Y) \longrightarrow \mathrm{Hom}_R(Z,X) \longrightarrow \mathrm{Hom}_R(Z,M) \longrightarrow \mathrm{Ext}^1_R(Z,Y)\,,$$
the rightmost term of which vanishes by Proposition 11.3(ii). Thus $\varphi \in \mathrm{Hom}_R(Z,M)$ lifts to an element of $\mathrm{Hom}_R(Z,X)$. The final assertion follows as well from exactness. □

We leave it as an exercise for the reader to state and prove the dual statement for FID hulls.

The lifting property of Proposition 11.9 allows a Schanuel-type result: if $0 \longrightarrow Y_1 \longrightarrow X_1 \longrightarrow M \longrightarrow 0$ and $0 \longrightarrow Y_2 \longrightarrow X_2 \longrightarrow M \longrightarrow 0$ are two MCM approximations of the same module M, then $X_1 \oplus Y_2 \cong X_2 \oplus Y_1$. We leave the details to the reader. (One can also proceed directly, via the orthogonality relation $\mathrm{Ext}^1_R(X_i, Y_j) = 0$; compare with Lemma A.10.) Just as for free resolutions, this motivates a notion of minimality for MCM approximations.

11.10. DEFINITION. Let $s\colon 0 \longrightarrow Y \xrightarrow{i} X \xrightarrow{p} M \longrightarrow 0$ be a MCM approximation of a non-zero finitely generated R-module M. We say that s is *minimal* provided Y and X have no non-zero direct summand in common via i. In other words, for any direct-sum decomposition $X = X_0 \oplus X_1$ with $X_0 \subseteq \mathrm{im}\, i$, we must have $X_0 = 0$.

Observe that any common direct summand of Y and X is both MCM and of finite injective dimension, so by Proposition 11.7 is a direct sum of copies of the canonical module ω.

While the definition of minimality above is quite natural, in practice a more technical notion is useful.

11.11. DEFINITION. Let A be a ring and $f\colon P \longrightarrow Q$ a homomorphism of A-modules. We say that f is *right minimal* if whenever $\varphi\colon P \longrightarrow P$ is an endomorphism such that $f\varphi = f$, in fact φ is an automorphism. If $s\colon 0 \longrightarrow N \longrightarrow P \xrightarrow{f} Q \longrightarrow 0$ is a short exact sequence, we say that s is right minimal if f is.

The equivalence of minimality and right minimality for an MCM approximation is "well-known to experts"; the proof we give is due to Hashimoto and Shida [**HS97**] (see also [**Yos93**]). It turns out that passing to the completion is essential to the argument.

11.12. LEMMA. *Let $s\colon 0 \longrightarrow Y \xrightarrow{i} X \xrightarrow{p} M \longrightarrow 0$ be a MCM approximation of a non-zero R-module M. Let $\widehat{s}\colon 0 \longrightarrow \widehat{Y} \xrightarrow{\widehat{i}} \widehat{X} \xrightarrow{\widehat{p}} \widehat{M} \longrightarrow 0$ be the completion of s. Then $\widehat{s}$ is a MCM approximation of $\widehat{M}$, and the following are equivalent.*

(i) $\widehat{s}$ *is right minimal;*
(ii) s *is right minimal;*
(iii) s *is minimal;*
(iv) $\widehat{s}$ *is minimal.*

PROOF. That $\widehat{s}$ is a MCM approximation of $\widehat{M}$ is trivial; the real matter is the equivalence.

(i) $\implies$ (ii) Assume that $\widehat{s}$ is right minimal, and $\varphi \in \operatorname{End}_R(X)$ satisfies $p\varphi = p$. Then $\widehat{p}\widehat{\varphi} = \widehat{p}$, so $\widehat{\varphi}$ is an automorphism by hypothesis, whence φ is an automorphism as well.

(ii) $\implies$ (iii) If $X = X_0 \oplus X_1$ is a direct-sum decomposition of X with $X_0 \subseteq \operatorname{im} i$, then the idempotent $\varphi \colon X \twoheadrightarrow X_1 \hookrightarrow X$ obtained from the projection onto X_1 satisfies $p\varphi = p$. Thus $X_0 \neq 0$ implies that s is not right minimal.

(iii) $\implies$ (iv) Assume that $\widehat{s}$ is not minimal, so that $\widehat{Y}$ and $\widehat{X}$ have a common non-zero direct summand via i. We have already observed that such a direct summand must be a direct sum of copies of the canonical module $\widehat{\omega}$, so there exist homomorphisms $\sigma \colon \widehat{X} \longrightarrow \widehat{\omega}$ and $\tau \colon \widehat{\omega} \longrightarrow \widehat{Y}$ such that

$$\sigma \widehat{i} \tau \colon \widehat{\omega} \longrightarrow \widehat{Y} \longrightarrow \widehat{X} \longrightarrow \widehat{\omega}$$

is the identity on $\widehat{\omega}$. Write $\sigma = \sum_j a_j \widehat{\sigma}_j$ and $\tau = \sum_k b_k \widehat{\tau}_k$, where $\sigma_j \in \operatorname{Hom}_R(X, \omega)$, $\tau_k \in \operatorname{Hom}_R(\omega, Y)$, and $a_j, b_k \in \widehat{R}$. Then

$$\sum_{j,k} a_j b_k \widehat{\sigma_j i \tau_k} = 1 \quad \in \quad \operatorname{End}_{\widehat{R}}(\widehat{\omega}) \cong \widehat{R}.$$

Since $\widehat{R}$ is local, at least one of the summands $a_j b_k \widehat{\sigma_j i \tau_k}$ is a unit of $\widehat{R}$. It follows that $\sigma_j i \tau_k$ is a unit of R, that is, $\sigma_k i \tau_k \colon \omega \longrightarrow \omega$ is an isomorphism. Thus s is not minimal.

(iv) $\implies$ (i) We assume that $R = \widehat{R}$ is complete. Let $\varphi \colon X \longrightarrow X$ be a non-isomorphism satisfying $p\varphi = p$. Let $\Lambda \subset \operatorname{End}_R(X)$ be the subring generated by R and φ, and observe that Λ is commutative and is a finitely generated R-module.

As φ carries the kernel of p into itself, s is naturally a short exact sequence of (finitely generated) Λ-modules. In particular, multiplication by $\varphi \in \Lambda$ is the identity on the non-zero module M, so by NAK φ is not contained in the radical of Λ. On the other hand, φ is not an isomorphism on X, so is not a unit of Λ. Thus Λ is not an nc-local ring. Since R is Henselian, it follows that Λ contains a non-trivial idempotent $e \neq 0, 1$.

Now $\varphi \in R + (1-\varphi)\Lambda$, so $R + (1-\varphi)\Lambda = \Lambda$. In particular, $\overline{\Lambda} := \Lambda/(1-\varphi)\Lambda$ is a quotient of R, so is a local ring. Replacing e

by $1 - e$ if necessary, we may assume that $\bar{e} = \bar{1}$ in $\bar{\Lambda}$. Since φ acts as the identity on M, we see that M is naturally a $\bar{\Lambda}$-module, and in particular e also acts as the identity on M.

Set $X_0 = \operatorname{im}(1 - e) = \ker(e) \subseteq X$. Then X_0 is a non-zero direct summand of X, and $p(X_0) = 0$ since e acts trivially on M. Thus s is not minimal. □

11.13. PROPOSITION. *If a finitely generated module M admits a MCM approximation, then there is a minimal one, which moreover is unique up to isomorphism of exact sequences inducing the identity on M.*

PROOF. Removing any direct summands common to Y and X via i in a given MCM approximation of M, we arrive at a minimal one. For uniqueness, suppose we have two minimal approximations $s\colon 0 \longrightarrow Y \xrightarrow{i} X \xrightarrow{p} M \longrightarrow 0$ and $s'\colon 0 \longrightarrow Y' \xrightarrow{i'} X' \xrightarrow{p'} M \longrightarrow 0$. The lifting property delivers a commutative diagram with exact rows

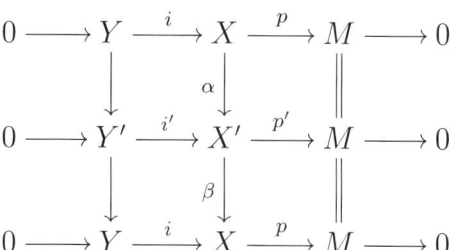

in which, in particular, $p\beta\alpha = p$. Since minimality implies right minimality, $\beta\alpha$ is an isomorphism. A similar diagram shows that $\alpha\beta$ is an isomorphism as well, so that s and s' are isomorphic exact sequences via an isomorphism which is the identity on M. □

Here is yet a third notion of minimality for MCM approximations, introduced by Hashimoto and Shida [**HS97**] and used to good effect by Simon and Strooker [**SS02**]. Set $d = \dim R$. It's immediate from the definition that a MCM approximation $0 \longrightarrow Y \longrightarrow X \longrightarrow M \longrightarrow 0$ induces isomorphisms

$$\operatorname{Ext}_R^i(k, M) \cong \begin{cases} \operatorname{Ext}_R^{i+1}(k, Y) & \text{for } 0 \leqslant i \leqslant d - 2 \text{ and} \\ \operatorname{Ext}_R^i(k, X) & \text{for } i \geqslant d + 1, \end{cases}$$

and a 4-term exact sequence

$$0 \longrightarrow \operatorname{Ext}_R^{d-1}(k, M) \longrightarrow \operatorname{Ext}_R^d(k, Y) \longrightarrow \operatorname{Ext}_R^d(k, X) \longrightarrow \operatorname{Ext}_R^d(k, M) \longrightarrow 0.$$

We will call the approximation Ext-*minimal* if the induced map of k-vector spaces $\operatorname{Ext}_R^d(k, Y) \longrightarrow \operatorname{Ext}_R^d(k, X)$ in the middle of this exact sequence is the zero map. Equivalently, one (and hence both) of the natural maps $\operatorname{Ext}_R^{d-1}(k, M) \longrightarrow \operatorname{Ext}_R^d(k, Y)$ and $\operatorname{Ext}_R^d(k, X) \longrightarrow \operatorname{Ext}_R^d(k, M)$ is an isomorphism. This means in particular that the Bass numbers of M are completely determined by X and Y.

If in a MCM approximation of M there is a non-zero indecomposable direct summand of Y carried isomorphically to a summand of X, then we've already seen that the summand must be isomorphic to ω, and so $\operatorname{Ext}_R^d(k, Y) \longrightarrow \operatorname{Ext}_R^d(k, X)$ has as a summand the identity map on $k = \operatorname{Ext}_R^d(k, \omega)$. Thus Ext-minimality implies minimality as defined above. In fact, all three notions of minimality are equivalent. As the proof of this fact uses some local cohomology, we relegate it to the Exercises.

11.14. PROPOSITION. *Let $(R, \mathfrak{m})$ be a CM local ring with canonical module, and let M be a non-zero finitely generated R-module. For a given MCM approximation of M, minimality, right minimality, and Ext-minimality are equivalent.* □

The considerations above are exactly paralleled on the FID hull side. A FID hull $0 \longrightarrow M \xrightarrow{j} Y \xrightarrow{q} X \longrightarrow 0$ is *minimal* if Y and X have no non-zero direct summand in common via q, is *left minimal* if every endomorphism $\psi \in \operatorname{End}_R(Y)$ such that $\psi j = j$ is in fact an automorphism, and is Ext-*minimal* if the induced linear map $\operatorname{Ext}_R^d(k, Y) \longrightarrow \operatorname{Ext}_R^d(k, X)$ is zero. The three notions are equivalent by arguments exactly similar to those above, and if a FID hull for M exists, then there is a minimal one, which is unique up to an isomorphism of exact sequences which is the identity on M.

We turn now to existence. The construction of MCM approximations is most transparent when the approximated module is CM, so we state that case separately. In particular, the construction below applies when M is an R-module of finite length, for example $M = R/\mathfrak{m}^n$ for some $n \geqslant 1$. We will return to this example in §4.

11.15. Proposition. *Let $(R, \mathfrak{m})$ be a CM local ring with canonical module ω, and let M be a CM R-module. Then M has a minimal MCM approximation.*

PROOF. Let $t = \operatorname{codepth} M$. By Theorem 11.5, $M^\vee = \operatorname{Ext}_R^t(M, \omega)$ is again CM of codepth t. In a truncated minimal free resolution of $M^\vee$

$$0 \longrightarrow \operatorname{syz}_t^R(M^\vee) \longrightarrow F_{t-1} \longrightarrow \cdots \longrightarrow F_1 \longrightarrow F_0 \longrightarrow M^\vee \longrightarrow 0$$

the tth syzygy $\operatorname{syz}_t^R(M^\vee)$ is MCM. Apply $\operatorname{Hom}_R(-, \omega)$ to get a complex

$$0 \longrightarrow F_0^\vee \longrightarrow F_1^\vee \longrightarrow \cdots \longrightarrow F_{t-1}^\vee \longrightarrow \operatorname{syz}_t^R(M^\vee)^\vee \longrightarrow 0$$

with homology $\operatorname{Ext}_R^i(M^\vee, \omega)$, which is $M^{\vee\vee} \cong M$ for $i = t$ and trivial otherwise. Inserting the homology at the rightmost end, and defining K to be the kernel, we get a short exact sequence

(11.1) $\qquad 0 \longrightarrow K \longrightarrow \operatorname{syz}_t^R(M^\vee)^\vee \longrightarrow M \longrightarrow 0,$

in which the middle term is MCM. Since K has a finite resolution by direct sums of copies of $R^\vee = \omega$, it has finite injective dimension, so that (11.1) is a MCM approximation of M.

It is easy to see that our initial choice of a minimal resolution forces the obtained approximation to be minimal as well. $\square$

As an aside, we mention here that in the setup of Proposition 11.15, if R is generically Gorenstein (so that ω_R has constant rank 1), then the MCM approximation $\operatorname{syz}_{\dim R}^R(k^\vee)^\vee$ of the residue field has constant rank as well, computable as

(11.2) $\qquad \operatorname{rank}\left(\operatorname{syz}_{\dim R}^R(k^\vee)^\vee\right) = \sum_{i=0}^{\dim R - 1} (-1)^{\dim R - i - 1} \beta_i^R(k),$

where $\beta_i^R(k)$ indicates the appropriate Betti number.

11.16. Question (Buchweitz). *Is the number defined in (11.2) the minimal possible rank for a non-free MCM module which occurs as the syzygy module of some R-module of finite length?*

For the general case, we give an independent construction of a MCM approximation of a finitely generated module, which simultaneously produces an FID hull as well. This argument is essentially that of [**AB89**], though in a more concrete setting. There are two other constructions: the *pitchfork construction*, originally due also to Auslander and Buchweitz (for which see Construction 12.11), and the *gluing construction* of Herzog and Martsinkovsky [**HM93**].

§2. MCM APPROXIMATIONS AND FID HULLS

11.17. THEOREM. *Let $(R, \mathfrak{m}, k)$ be a CM local ring with canonical module ω, and let M be a finitely generated R-module. Then M admits a MCM approximation and a FID hull.*

PROOF. We construct the approximation and hull by induction on codepth M. When M is MCM itself, the MCM approximation is trivial. For a FID hull, take a free module F mapping onto the dual $M^\vee = \mathrm{Hom}_R(M, \omega)$ as in the proof of Proposition 11.15. In the short exact sequence
$$0 \longrightarrow \mathrm{syz}_1^R(M^\vee) \longrightarrow F \longrightarrow M^\vee \longrightarrow 0,$$
the syzygy module $\mathrm{syz}_1^R(M^\vee)$ is again MCM, so applying $\mathrm{Hom}_R(-, \omega)$ gives another exact sequence
$$0 \longrightarrow M \longrightarrow F^\vee \longrightarrow \mathrm{syz}_1^R(M^\vee)^\vee \longrightarrow 0$$
in which $F^\vee \cong \omega^{(n)}$ has finite injective dimension and $\mathrm{syz}_1^R(M^\vee)^\vee$ is MCM.

Suppose now that codepth $M = t \geqslant 1$. Taking a syzygy of M in a minimal free resolution
$$0 \longrightarrow \mathrm{syz}_1^R(M) \longrightarrow F \longrightarrow M \longrightarrow 0$$
we have by induction a FID hull of $\mathrm{syz}_1^R(M)$
$$0 \longrightarrow \mathrm{syz}_1^R(M) \longrightarrow Y' \longrightarrow X' \longrightarrow 0.$$
Construct the pushout diagram from these two sequences.

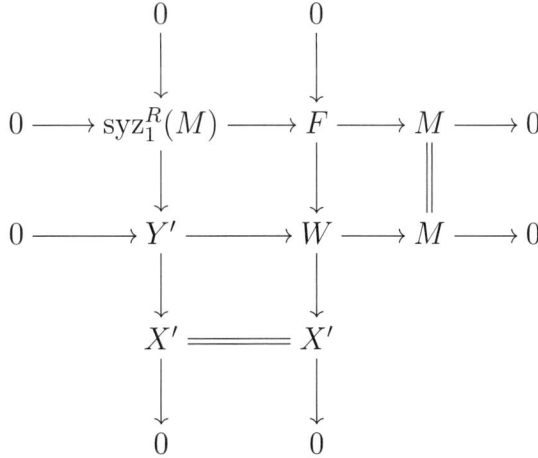

As X' is MCM and F is free, the exact middle column forces W to be MCM, so that the middle row is a MCM approximation of M.

A FID hull for W exists by the base case of the induction:
$$0 \longrightarrow W \longrightarrow Y'' \longrightarrow X'' \longrightarrow 0$$

and constructing another pushout

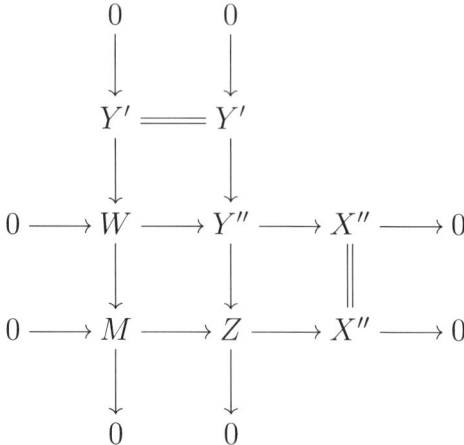

we see from the middle column that Z has finite injective dimension, so the bottom row is a FID hull for M. □

11.18. NOTATION. Having now established both the existence and the uniqueness of minimal MCM approximations and FID hulls, we introduce some notation for them. The minimal MCM approximation of M is denoted by
$$0 \longrightarrow Y_M \longrightarrow X_M \longrightarrow M \longrightarrow 0,$$
while the minimal FID hull of M is denoted
$$0 \longrightarrow M \longrightarrow Y^M \longrightarrow X^M \longrightarrow 0.$$

To show off the new notation, here is the final diagram of the proof of Theorem 11.17.

(11.3)
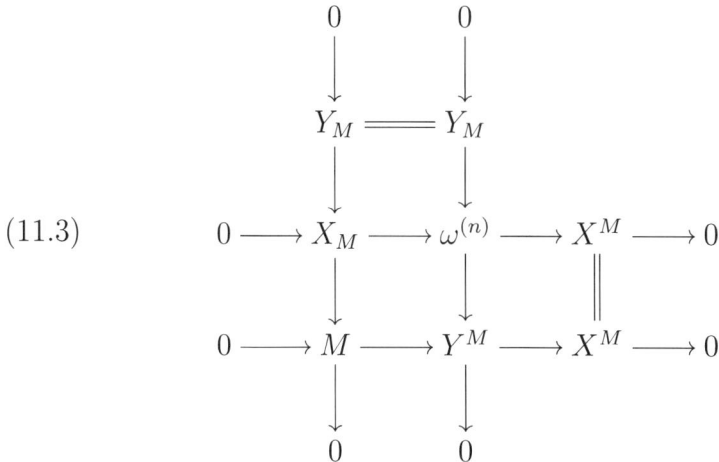

Here $n = \mu_R(X_M^\vee)$ as the middle row is an FID hull for X_M.

As in the last paragraph of the proof of Theorem 11.17, the central square in this diagram is a pushout. It therefore induces the exact sequence
$$0 \longrightarrow X_M \longrightarrow M \oplus \omega^{(n)} \longrightarrow Y^M \longrightarrow 0,$$
exhibiting the given module M, up to canonical summands, as an extension of a MCM module by a module of finite injective dimension. This is the "Approximation Theorem" of Auslander [**Aus67**, Chapter 3, Prop. 8], [**AB69**, 4.27], as observed by Buchweitz [**Buc86**, (5.3.2)].

We also record a few curiosities that arose in the proof of Theorem 11.17.

11.19. PROPOSITION. *Up to adding or deleting direct summands isomorphic to ω, we have*
  (i) $Y_M \cong Y^{\mathrm{syz}_1^R(M)}$;
  (ii) $X^M \cong X^{X_M}$; *and*
  (iii) X_M *is an extension of a free module by* $X^{\mathrm{syz}_1^R(M)}$, *that is, there is a short exact sequence* $0 \longrightarrow F \longrightarrow X_M \longrightarrow X^{\mathrm{syz}_1^R(M)} \longrightarrow 0$ *with F free.*

In particular, if R is Gorenstein then we have as well
  (iv) $X_M \cong X^{\mathrm{syz}_1^R(M)}$;
  (v) $X_M \cong \mathrm{syz}_1^R(X^M)$; *and*
  (vi) $Y_M \cong \mathrm{syz}_1^R(Y^M)$. □

We see already that the case of a Gorenstein local ring is special. In this case, finite injective dimension coincides with finite projective dimension, making the theory more tractable. In particular, the minimal MCM approximation of a module of finite projective dimension over a Gorenstein local ring is very simple to describe. We leave the proof of this fact as an exercise.

11.20. PROPOSITION. *Let R be a Gorenstein local ring and M a finitely generated R-module of finite projective dimension. The the minimal MCM approximation of M is* $0 \longrightarrow \mathrm{syz}_1^R(M) \longrightarrow F \longrightarrow M \longrightarrow 0$, *where F is a free module of minimal rank mapping onto M.* □

We will see more advantages of the Gorenstein condition in §4 and in Chapter 12; see also Exercise 11.48.

We also record here for later reference the case of codepth 1.

11.21. PROPOSITION. *Let R be a CM local ring with canonical module ω and let M be an R-module of codepth 1. Let $\xi_1, \ldots, \xi_t$ be a minimal set of generators for the (nonzero) module $\mathrm{Ext}_R^1(M, \omega)$, and let E be the extension of M by $\omega^{(t)}$ corresponding to the element*

$\xi = (\xi_1, \ldots, \xi_t) \in \operatorname{Ext}^1_R(M, \omega^{(t)}) \cong \operatorname{Ext}^1_R(M, \omega)^{(t)}$. Then E is a MCM module and
$$\xi\colon 0 \longrightarrow \omega^t \longrightarrow E \longrightarrow M \longrightarrow 0$$
is the minimal MCM approximation of M. In particular, this construction coincides with that of Proposition 11.15 if M is CM, i.e. if $\operatorname{Hom}_R(M, \omega) = 0$. □

To close out this section, we have a few more words to say about uniqueness. Since every MCM module is its own MCM approximation, the function $M \rightsquigarrow X_M$ is in general surjective, but not injective. However, we may restrict to CM modules of a fixed codepth and ask whether every MCM module X is a MCM approximation of a CM module of codepth r. For $r = 1$ and $r = 2$, these questions have essentially been answered by Yoshino-Isogawa [**YI00**] and Kato [**Kat07**]. Here is the criterion for $r = 1$.

11.22. PROPOSITION. *Let R be a CM local ring with canonical module ω, and assume that R is generically Gorenstein. Let X be a MCM R-module. Then X is a MCM approximation of some CM module M of codepth 1 if and only if X has constant rank.*

PROOF. First assume that X has constant rank s. Then there is a short exact sequence
$$0 \longrightarrow R^{(s)} \longrightarrow X \longrightarrow N \longrightarrow 0$$
in which N is a torsion module. In particular, N has dimension at most $\dim R - 1$. However, the Depth Lemma ensures that N has depth at least $\dim R - 1$, so N is CM of codepth 1. As R is generically Gorenstein, the canonical module ω embeds into R as an ideal of pure height one (Proposition 11.6). We therefore have embeddings $\omega^{(s)} \hookrightarrow R^{(s)}$ and $R^{(s)} \hookrightarrow X$ fitting into a commutative diagram

$$\begin{array}{ccccccccc}
0 & \longrightarrow & \omega^{(s)} & \longrightarrow & X & \longrightarrow & M & \longrightarrow & 0 \\
& & \downarrow & & \| & & \downarrow & & \\
0 & \longrightarrow & R^{(s)} & \longrightarrow & X & \longrightarrow & N & \longrightarrow & 0.
\end{array}$$

The Snake Lemma delivers an isomorphism from the kernel of $M \longrightarrow N$ onto $(R/\omega)^{(s)}$, and hence an exact sequence
$$0 \longrightarrow (R/\omega)^{(s)} \longrightarrow M \longrightarrow N \longrightarrow 0.$$
Therefore M is also CM of codepth 1, and the top row of the diagram is a MCM approximation of M.

For the converse, suppose that M is CM of codepth 1 and that X is a MCM approximation of M. Then $X \cong X_M \oplus \omega^{(t)}$ for some $t \geqslant 0$. In the minimal MCM approximation
$$0 \longrightarrow Y_M \longrightarrow X_M \longrightarrow M \longrightarrow 0,$$
we see that M is torsion, whence of rank zero, and Y_M is isomorphic to a direct sum of copies of ω. As R is generically Gorenstein, Y_M has constant rank, and so X_M and X do as well. $\square$

It's clear that a local ring R is a domain if and only if every finitely generated R-module has constant rank. If in addition R is CM, then it follows that R is a domain if and only if every MCM module has constant rank. (Take a high syzygy of an arbitrary finitely generated module M and compute the rank of M as an alternating sum.) These observations prove the following corollary.

11.23. COROLLARY. *Let R be a CM local ring with a canonical module and assume that R is generically Gorenstein. The following statements are equivalent.*

(i) *For every MCM R-module X, there exists a CM module M of codepth 1 such that X is MCM approximation of M.*
(ii) *R is a domain.* $\square$

The question of the injectivity of the function $M \rightsquigarrow X_M$ for modules M of a fixed codepth is, as far as we can tell, still open. The corresponding question for FID hulls, however, has a positive answer when R is Gorenstein, due to Kato [**Kat99**].

§3. Numerical invariants

Since the minimal MCM approximation and minimal FID hull of a module M are uniquely determined up to isomorphism by M, any numerical information we derive from X_M, Y_M, X^M, and Y^M are invariants of M. For example, if R is Henselian we might consider the number of indecomposable direct summands appearing in a direct-sum decomposition of X_M or Y^M as a kind of measure of the complexity of M, or if R is generically Gorenstein we might consider rank Y^M. All these possibilities were pointed out by Buchweitz [**Buc86**], but seem not to have gotten much attention. In this section we introduce two other numerical invariants of M, namely $\delta(M)$, first defined by Auslander; and $\gamma(M)$, defined by Herzog and Martsinkovsky.

Throughout, $(R, \mathfrak{m})$ is still a CM local ring with canonical module ω. For an arbitrary finitely generated R-module Z, we define the *free rank* of Z, denoted f-rank Z, to be the rank of a maximal free direct summand of Z. In other words, $Z \cong \underline{Z} \oplus R^{(\text{f-rank}\, Z)}$ with $\underline{Z}$ stable, i.e. having no non-trivial free direct summands. Dually, the *canonical rank* of Z, ω-rank Z, is the largest integer n such that $\omega^{(n)}$ is a direct summand of Z.

11.24. DEFINITION. Let M be a finitely generated R-module, and let $0 \longrightarrow Y_M \longrightarrow X_M \longrightarrow M \longrightarrow 0$ be the minimal MCM approximation for M. Then we define

$$\delta(M) = \text{f-rank}\, X_M \quad \text{and} \quad \gamma(M) = \omega\text{-rank}\, X_M\,.$$

For the rest of the section, we fix once and for all the minimal MCM approximation

$$0 \longrightarrow Y_M \xrightarrow{i} X_M \xrightarrow{p} M \longrightarrow 0$$

of a chosen finitely generated R-module M. Note first that since we chose our approximation to be (Ext-)minimal, we have

$$\text{Ext}_R^d(k, X_M) \cong \text{Ext}_R^d(k, M)\,,$$

where $d = \dim R$. This, together with the fact (see Exercise 11.51) that $\text{Ext}_R^d(k, Z) \neq 0$ for every non-zero finitely generated R-module Z, immediately gives the following crude bounds.

11.25. PROPOSITION. *Set $s = \dim_k \text{Ext}_R^d(k, M)$. Then*

(i) *$\delta(M) \cdot \dim_k \text{Ext}_R^d(k, R) \leqslant s$, with equality if and only if X_M is free. In particular, if $\dim_k \text{Ext}_R^d(k, M) < \dim_k \text{Ext}_R^d(k, R)$, then $\delta(M) = 0$.*
(ii) *$\gamma(M) \leqslant s$, with equality if and only if M has finite injective dimension.* $\square$

Note that the question of which modules M satisfy "X_M is free" is quite subtle. One situation in which it holds is when R is Gorenstein and M has finite projective dimension; see Proposition 11.20. However, it may hold in other cases as well, for example $M = R/\omega$, where ω is embedded as an ideal of height one as in Proposition 11.6.

To obtain sharper bounds, as well as a better understanding of what exactly each invariant measures, we consider them separately. Of the two, $\delta(M)$ has received more attention, so we begin there.

11.26. LEMMA. *Let M be a finitely generated R-module. Decompose $X_M = \underline{X} \oplus F$, where F is a free module of rank $\delta(M)$ and $\underline{X}$ is stable. Then*

$$\delta(M) = \mu_R\left(M/p\left(\underline{X}\right)\right)\,.$$

§3. NUMERICAL INVARIANTS

PROOF. The commutative diagram of short exact sequences

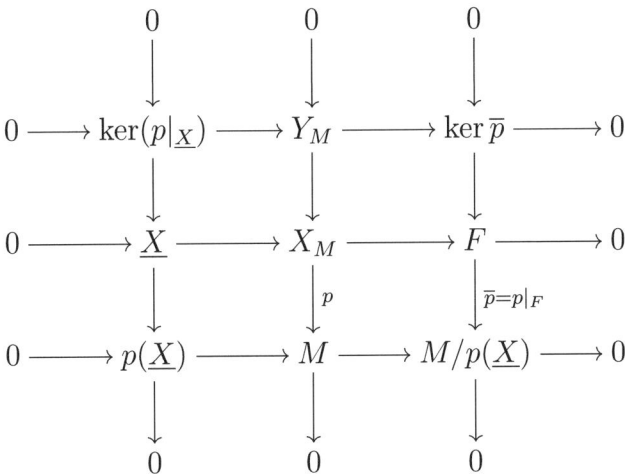

shows that $\delta(M) = \operatorname{rank} F \geqslant \mu_R(M/p(\underline{X}))$. If $\operatorname{rank} F > \mu_R(M/p(\underline{X}))$, then $\ker \overline{p}$ has a non-zero free direct summand. Since Y_M maps onto $\ker \overline{p}$, Y_M also has a free summand, which we easily see is a common direct summand of Y_M and X_M. As our approximation was chosen minimal, this is a contradiction. $\square$

The lemma allows us to characterize $\delta(M)$ without referring to the MCM approximation of M.

11.27. PROPOSITION. *Let M be a finitely generated R-module. The delta-invariant $\delta(M)$ is the minimum free rank of all MCM modules Z admitting a surjective homomorphism onto M.*

PROOF. Denote the minimum by $\delta' = \delta'(M)$, and set $\delta = \delta(M)$. Then evidently $\delta' \leqslant \delta$. For the other inequality, let $\varphi \colon Z \longrightarrow M$ be a surjection with Z MCM and f-rank $Z = \delta'$. Write $Z = \underline{Z} \oplus R^{(\delta')}$ and $X_M = \underline{X} \oplus R^{(\delta)}$. The lifting property applied to $\varphi|_{\underline{Z}}$ gives a homomorphism $\alpha \colon \underline{Z} \longrightarrow \underline{X} \oplus R^{(\delta)}$ fitting into a commutative diagram

$$\begin{array}{ccccccc}
0 & \longrightarrow & \ker \varphi|_{\underline{Z}} & \longrightarrow & \underline{Z} & \xrightarrow{\varphi|_{\underline{Z}}} & M \\
& & \downarrow & & \alpha \downarrow & & \parallel \\
0 & \longrightarrow & Y_M & \longrightarrow & \underline{X} \oplus R^{(\delta)} & \xrightarrow{p} & M & \longrightarrow 0
\end{array}$$

As $\underline{Z}$ has no free direct summands, the image of the composition $\underline{Z} \longrightarrow \underline{X} \oplus R^{(\delta)} \twoheadrightarrow R^{(\delta)}$ is contained in $\mathfrak{m}R^{(\delta)}$. Thus $\alpha(\underline{Z})$ contributes no minimal generators to $M/p(\underline{X})$, and therefore

$$\delta = \mu_R\left(M/p(\underline{X})\right) \leqslant \mu_R\left(M/p\alpha(\underline{Z})\right) \leqslant \delta'. \quad \square$$

In particular, Proposition 11.27 implies that for a MCM module X, we have $\delta(X) = $ f-rank X, and for M arbitrary, $\delta(M) = 0$ if and only if M is a homomorphic image of a stable MCM module. We also obtain some basic properties of δ.

11.28. COROLLARY. *Let M and N be finitely generated R-modules.*
(i) $\delta(M \oplus N) = \delta(M) + \delta(N)$.
(ii) $\delta(N) \leqslant \delta(M)$ if there is a surjection $M \twoheadrightarrow N$.
(iii) $\delta(M) \leqslant \mu_R(M)$.

PROOF. Since minimality is equivalent to Ext-minimality, the direct sum of minimal MCM approximations of M and N is again minimal. Thus $X_{M \oplus N} \cong X_M \oplus X_N$. The free rank of $X_M \oplus X_N$ is the sum of those of X_M and X_N, since a direct sum has a free summand if and only if one summand does. The second and third statements are clear from the Proposition. □

11.29. REMARK. Corollary 11.28 is central to the early history of the delta-invariant. Suppose that R is Gorenstein and that M is a finitely generated R-module admitting a surjection $M \twoheadrightarrow N$, where N has finite projective dimension. Since the minimal MCM approximation of N is simply a free cover (Proposition 11.20), we have $\delta(N) > 0$, and hence $\delta(M) > 0$. It was at first conjectured that $\delta(M) > 0$ if and only if M has a non-zero quotient module of finite projective dimension, but a counterexample was given by Ding [**Din94**]. Ding proved a formula for $\delta(R/I)$, where R is a one-dimensional Gorenstein local ring and I is an ideal of R containing a non-zerodivisor:

$$\delta(R/I) = 1 + \ell\left(\operatorname{soc}(R/I)\right) - \mu_R(I^*).$$

He then took $R = k[\![t^3, t^4]\!]$, where k is a field, and $I = (t^8 + t^9, t^{10})$. He showed that $\delta(R/I) = 1$ and that I is not contained in any proper principal ideal of R, so R/I cannot map onto a non-zero module of finite projective dimension.

We also mention here in passing a remarkable application of the δ-invariant, due to Martsinkovsky [**Mar90, Mar91**]. Let k be an algebraically closed field of characteristic zero and let $S = k[\![x_1, \ldots, x_n]\!]$ be a power series ring over k. Let $f \in S$ be a polynomial such that the hypersurface ring $R = S/(f)$ is an isolated singularity. The Jacobian ideal $j(f)$, generated by the partial derivatives of f, and its image $\overline{j(f)}$ in R, are thus primary to the respective maximal ideals. Then Martsinkovsky shows that $\delta\left(R/\overline{j(f)}\right) = 0$ if and only if $f \in j(f)$. In fact, these are equivalent to $f \in (x_1, \ldots, x_n)j(f)$, which by a foundational result of Saito [**Sai71**] occurs if and only if f is *quasi-homogeneous*, i.e.

there is an integral weighting of the variables $x_1, \ldots, x_n$ under which f is homogeneous.

Turning attention now to $\gamma(M) = \omega\text{-rank}\, X_M$, we have an analog of Lemma 11.26, the proof of which is similar enough that we skip it.

11.30. LEMMA. *Let M be a finitely generated R-module, and write $X_M = \overline{X} \oplus \omega^{(\gamma(M))}$, where $\overline{X}$ has no direct summand isomorphic to ω. Then*
$$\gamma(M) \cdot \mu_R(\omega) = \mu_R\left(M/p(\overline{X})\right).$$
□

As a consequence, we find an unexpected restriction on the R-modules of finite injective dimension.

11.31. PROPOSITION. *Let M be a finitely generated R-module of finite injective dimension. Then $\gamma(M) \cdot \mu_R(\omega) = \mu_R(M)$. In particular, $\mu_R(M)$ is an integer multiple of the Cohen-Macaulay type $\mu_R(\omega)$ of R.*
□

There is obviously no direct analog of Proposition 11.27 for $\gamma(M)$; as long as R is not Gorenstein, every M is a homomorphic image of a MCM module without ω-summands, namely, a free module. Still, we do retain additivity, and in certain cases the other assertions of Corollary 11.28.

11.32. PROPOSITION. *Let M and N be R-modules. Then $\gamma(M \oplus N) = \gamma(M) + \gamma(N)$.*
□

11.33. PROPOSITION. *Let $N \subseteq M$ be R-modules, both of finite injective dimension. Then $\gamma(M/N) \leqslant \gamma(M) - \gamma(N)$.*

PROOF. Since each of M, N, and M/N has finite injective dimension, Proposition 11.25 allows us to compute $\gamma(-)$ as $\dim_k \operatorname{Ext}_R^d(k, -)$. The long exact sequence of Ext ends with
$$\operatorname{Ext}_R^d(k, N) \longrightarrow \operatorname{Ext}_R^d(k, M) \longrightarrow \operatorname{Ext}_R^d(k, M/N) \longrightarrow 0,$$
and a k-dimension count gives the inequality.
□

This result fails without the assumption of finite injective dimension. For example, consider a non-Gorenstein ring R and a free module F mapping onto the canonical module ω. We have $\gamma(F) = 0$ and $\gamma(\omega) = 1$.

In case M has codepth 1, the explicit construction of MCM approximations in Proposition 11.21 allows us to compute $\gamma(M)$ directly. We leave the proof as yet another exercise.

11.34. PROPOSITION. *Let M be an R-module of codepth 1 (not necessarily Cohen-Macaulay). Then we have $\gamma(M) = \mu_R(\operatorname{Ext}^1_R(M,\omega))$.* □

For CM modules, the δ- and γ-invariants are dual. This follows easily from the construction of MCM approximations in this case.

11.35. PROPOSITION. *Let M be a CM R-module of codepth t, and write $M^\vee = \operatorname{Ext}^t_R(M,\omega)$ as usual. Then $\delta(M^\vee) = \gamma(\operatorname{syz}^R_t(M))$.* □

In fact, one can show, using the gluing construction of Herzog and Martsinkovsky [**HM93**], that $\delta(\operatorname{syz}_i(M^\vee)) = \gamma(\operatorname{syz}_{t-i}(M))$ for $i = 0, \ldots, t$.

When R is Gorenstein, δ and γ coincide, allowing us to combine all the above results, and enabling new ones. Here is an example.

11.36. PROPOSITION. *Assume that R is a Gorenstein ring, and let M be a finitely generated R-module. Then*

$$\delta(M) = \mu_R\left(Y^M\right) - \mu_R\left(X^M\right).$$

PROOF. Consider the diagram (11.3) following the construction of MCM approximations and FID hulls. In the Gorenstein situation, the $\omega^{(n)}$ in the center becomes a free module $R^{(n)}$. Thus

$$\delta(M) = \operatorname{f-rank} X_M = n - \mu_R(X^M).$$

The middle column implies $n \geqslant \mu_R(Y^M)$, but in fact we have equality: the image of the vertical arrow $Y_M \longrightarrow R^{(n)}$ is contained in $\mathfrak{m}R^{(n)}$ by the minimality of the left-hand column. Combining these gives the formula of the statement. □

§4. The index and applications to finite CM type

Once again, in this section $(R, \mathfrak{m}, k)$ is a CM local ring with canonical module ω. As a warm-up exercise, here is a straightforward result attributed to Auslander.

11.37. PROPOSITION. *The following conditions are equivalent.*
(i) R is a regular local ring.
(ii) $\delta(\operatorname{syz}^R_n(k)) > 0$ for all $n \geqslant 0$, i.e. no syzygy of k is a homomorphic image of a stable MCM module.
(iii) $\delta(k) = 1$.
(iv) $\gamma(\operatorname{syz}^R_{\dim(R)}(k)) > 0$.

PROOF. If R is a regular local ring, then every MCM module is free, so $\delta(M) > 0$ for every module M. In particular (ii) holds. Statement (ii) implies (iii) trivially. If R is non-regular, then there is at least one MCM R-module M without free summands, and the non-zero composition $M \longrightarrow M/\mathfrak{m}M \cong k^{(\mu_R(M))} \twoheadrightarrow k$ shows $\delta(k) = 0$. Thus the first three statements are equivalent. Finally, the construction of minimal MCM approximations for CM modules in Proposition 11.15 shows that $\delta(k) = \text{f-rank}(\text{syz}^R_{\dim(R)}(k^\vee)^\vee) = \omega\text{-rank}(\text{syz}^R_{\dim(R)}(k))$, whence (iii) $\iff$ (iv). $\square$

For a moment, let us set $\delta_n = \delta(R/\mathfrak{m}^n)$ for each $n \geqslant 0$. Then the Proposition implies that if R is not regular, then $\delta_0 = 0$. The surjection $R/\mathfrak{m}^{n+1} \twoheadrightarrow R/\mathfrak{m}^n$ gives $\delta_{n+1} \geqslant \delta_n$, and every δ_n is at most 1 by Corollary 11.28. Thus the sequence $\{\delta_n\}$ is non-decreasing, with

$$0 = \delta_0 \leqslant \delta_1 \leqslant \cdots \leqslant \delta_n \leqslant \delta_{n+1} \leqslant \cdots \leqslant 1.$$

If ever $\delta_n = 1$, the sequence stabilizes there. Let us define the *index of* R to be the point at which that stabilization occurs, that is,

$$\text{index}(R) = \min\{\, n \mid \delta(R/\mathfrak{m}^n) = 1\}$$

and set $\text{index}(R) = \infty$ if $\delta(R/\mathfrak{m}^n) = 0$ for every n. Equivalently, $\text{index}(R)$ is the least integer n such that any MCM R-module X mapping onto $R/\mathfrak{m}^n$ has a free direct summand. In these terms, the Proposition says that R is regular if and only if $\text{index}(R) = 1$.

Next we point out that the index of R is finite if R is Gorenstein. Let $\mathbf{x}$ be a system of parameters in the maximal ideal $\mathfrak{m}$. Then $R/(\mathbf{x})$ has finite projective dimension, so $\delta(R/(\mathbf{x})) > 0$ since the MCM approximation is just a free cover (Proposition 11.20). The ideal generated by $\mathbf{x}$ being $\mathfrak{m}$-primary, we have $\mathfrak{m}^n \subseteq (\mathbf{x})$ for some n, and the surjection $R/\mathfrak{m}^n \longrightarrow R/(\mathbf{x})$ gives $\delta_n \geqslant \delta(R/(\mathbf{x})) > 0$. Thus $\text{index}(R) \leqslant n$. In fact, we see that the index of R is bounded above by the *(generalized) Loewy length* of R, defined by

$$\ell(R) = \inf\{\, n \mid \text{there exists a s.o.p. } \mathbf{x} \text{ with } \mathfrak{m}^n \subseteq (\mathbf{x})\}\,.$$

11.38. CONJECTURE (Ding). *Let R be a CM local ring with infinite residue field. Then*

$$\text{index}(R) = \ell(R)\,.$$

This conjecture is known to fail for finite residue fields [**HS97**]. There are some partial results by Ding [**Din92, Din93, Din94**] and by Herzog [**Her94**], who proved it in case R is homogeneous over a field.

In this section we will give Ding's proof that the index of R is finite if and only if R is Gorenstein on the punctured spectrum; moreover, in this case the index is bounded by the Loewy length. This will be Theorem 11.42, to which we come after some preliminaries. Recall that we write $M \mid N$ to indicate M is isomorphic to a direct summand of N.

11.39. LEMMA. *Let $(R, \mathfrak{m})$ be a CM local ring with canonical module ω and let $x \in \mathfrak{m}$ be a non-zerodivisor. Then $\delta(R/(x)) > 0$ if and only if $\omega \mid \mathrm{syz}_1^R(\omega/x\omega)$.*

PROOF. Proposition 11.21 gives the minimal MCM approximation of a module of codepth 1; in the case of $R/(x)$ we see that it is obtained by dualizing a free resolution of
$$(R/(x))^\vee = \mathrm{Ext}_R^1(R/(x), \omega) \cong \omega_{R/(x)} \cong \omega/x\omega\,.$$
It therefore takes the form
$$0 \longrightarrow F^\vee \longrightarrow \mathrm{syz}_1^R(\omega/x\omega)^\vee \longrightarrow R/(x) \longrightarrow 0$$
where F is a free module. Thus $\delta(R/(x)) = \text{f-rank}\left(\mathrm{syz}_1^R(\omega/x\omega)^\vee\right)$ is equal to ω-rank $\left(\mathrm{syz}_1^R(\omega/x\omega)\right)$. □

11.40. LEMMA. *Keep the notation of Lemma 11.39. The following are equivalent for a non-zerodivisor $x \in \mathfrak{m}$:*

(i) $\omega \mid \mathrm{syz}_1^R(\omega/x\omega)$;
(ii) $\mathrm{syz}_1^R(\omega/x\omega) \cong \omega \oplus \mathrm{syz}_1^R(\omega)$;
(iii) the multiplication map $\omega \xrightarrow{x} \omega$ factors through a free module.

PROOF. (i) $\implies$ (ii) Form the pullback of a free cover $F \longrightarrow \omega/x\omega$ and the surjection $\omega \longrightarrow \omega/x\omega$ to obtain a diagram as below.

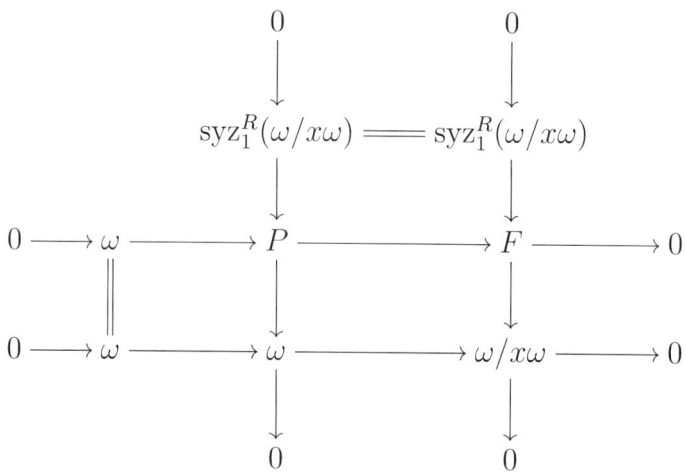

§4. THE INDEX AND APPLICATIONS TO FINITE CM TYPE

The middle row splits, giving a short exact sequence
$$0 \longrightarrow \mathrm{syz}_1^R(\omega/x\omega) \longrightarrow F \oplus \omega \longrightarrow \omega \longrightarrow 0$$
in the middle column. As $\mathrm{Ext}_R^1(\omega,\omega) = 0$, any summand of $\mathrm{syz}_1^R(\omega/x\omega)$ isomorphic to ω must split out as an isomorphism $\omega \longrightarrow \omega$, leaving $\mathrm{syz}_1^R(\omega)$ behind.

(ii) $\implies$ (iii) Letting $F \longrightarrow \omega$ now be a free cover of ω, another pullback gives the diagram

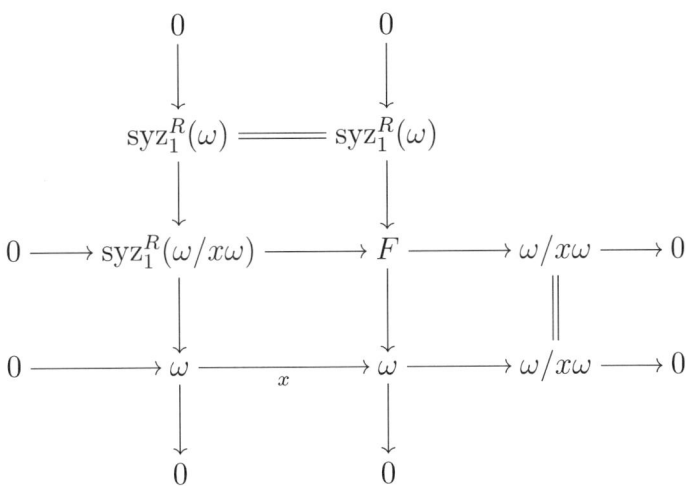

Applying Miyata's Theorem (Theorem 7.1), we find that the left-hand column must split, so that $\omega \xrightarrow{x} \omega$ factors through F.

(iii) $\implies$ (i) If we have a factorization of the multiplication homomorphism $\omega \xrightarrow{x} \omega$ through a free module, say $\omega \longrightarrow G \longrightarrow \omega$, we may pull back in two stages:

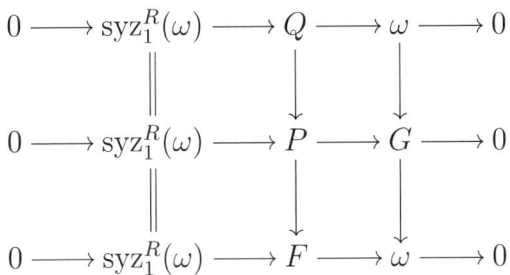

The result is the same as if we had pulled back by $\omega \xrightarrow{x} \omega$ directly, by the functoriality of Ext. Doing so in two stages, however, reveals that the middle row splits as G is free, and so the top row splits as well. This gives $Q \cong \omega \oplus \mathrm{syz}_1^R(\omega)$ and the middle column thus presents Q as the first syzygy of $\mathrm{cok}(\omega \xrightarrow{x} \omega) \cong \omega/x\omega$, giving even property (ii) and in particular (i). $\square$

Putting the lemmas together, we see that $\delta(R/(x)) = 0$ for a non-zerodivisor $x \in \mathfrak{m}$ if and only if x is in the ideal of $\operatorname{End}_R(\omega) \cong R$ consisting of those elements for which the corresponding multiplication factors through a free module. Let us identify this ideal explicitly.

11.41. LEMMA. *Let R be a CM local ring with canonical module ω. The following three ideals of R coincide.*

(i) $\left\{ x \in R \;\middle|\; \omega \xrightarrow{x} \omega \text{ factors through a free module} \right\}$;

(ii) *the trace $\tau_\omega(R)$ of ω in R, which is generated by homomorphic images of ω in R;*

(iii) *the image of the natural map*

$$\alpha \colon \operatorname{Hom}_R(\omega, R) \otimes_R \omega \longrightarrow \operatorname{End}_R(\omega) = R$$

defined by $\alpha(f \otimes a)(b) = f(b) \cdot a$. (Note that this is not *the canonical evaluation homomorphism $\operatorname{ev}(f \otimes a) = f(a)$.)*

PROOF. We prove (i) $\supseteq$ (ii) $\supseteq$ (iii) $\supseteq$ (i).

Let $x \in \tau_\omega(R)$, so that there is a linear functional $f \colon \omega \longrightarrow R$ and an element $a \in \omega$ with $f(a) = x$. Defining $g \colon R \longrightarrow \omega$ by $g(1) = a$, we have a factorization $x = g \circ f \colon \omega \longrightarrow \omega$.

Now if $x \in \operatorname{im} \alpha$, then there exist homomorphisms $f_i \colon \omega \longrightarrow R$ and elements $a_i \in \omega$ such that

$$\alpha\left(\sum_i f_i \otimes a_i\right)(b) = xb$$

for every $b \in \omega$. Define homomorphisms $g_i \colon \omega \longrightarrow R$ by $g_i(b) = \alpha(f_i \otimes b)$ for all $b \in \omega$. Then $\sum_i g_i(a_i) = x$, so that x is contained in the sum of the images of the g_i, hence in the trace ideal.

Finally, suppose we have a commutative diagram

$$\begin{array}{ccc} \omega & \xrightarrow{x} & \omega \\ {\scriptstyle \sum f_i} \searrow & & \nearrow {\scriptstyle \sum g_i} \\ & F & \end{array}$$

with F a free module and $\sum f_i$, $\sum g_i$ the decompositions along an isomorphism $F \cong R^{(n)}$. Then for $a \in \omega$, we have

$$\alpha\left(\sum f_i \otimes g_i(1)\right)(a) = \sum f_i(a) \cdot g_i(1)$$
$$= \sum g_i(f_i(a))$$
$$= xa$$

so that $x \in \operatorname{im} \alpha$. □

We denote by the ideal of Lemma 11.41 by $\tau_\omega(R)$.

From either of the first two descriptions above, we see that $1 \in \tau_\omega(R)$ if and only if R is Gorenstein. It follows that $\tau_\omega(R)$ defines the *Gorenstein locus* of R, that is, a localization $R_\mathfrak{p}$ is Gorenstein if and only if $\tau_\omega(R) \not\subseteq \mathfrak{p}$. In particular, R is Gorenstein on the punctured spectrum if and only if $\tau_\omega(R)$ is $\mathfrak{m}$-primary.

11.42. THEOREM (Ding). *The index of a CM local ring $(R, \mathfrak{m})$ with canonical module ω is finite if and only if R is Gorenstein on the punctured spectrum.*

PROOF. Assume first that R is Gorenstein on the punctured spectrum, so that $\tau_\omega(R)$ is $\mathfrak{m}$-primary. Then there exists a regular sequence $x_1, \ldots, x_d$ in $\tau_\omega(R)$, where $d = \dim R$. We claim by induction on d that $\delta(R/(x_1, \ldots, x_d)) \neq 0$. The case $d = 1$ is immediate from Lemmas 11.39 and 11.40.

Suppose $d > 1$ and X is a MCM R-module with a surjection $X \longrightarrow R/(x_1, \ldots, x_d)$. Tensor with $\overline{R} = R/(x_1)$ to get a surjection $X/x_1 X \longrightarrow \overline{R}/(\overline{x_2}, \ldots, \overline{x_d})$, where overlines indicate passage to $\overline{R}$. Since $\overline{x_2}, \ldots, \overline{x_d}$ are in $\tau_{\overline{\omega}}(\overline{R})$, the inductive hypothesis says that $X/x_1 X$ has an $R/(x_1)$-free direct summand. But then there is a surjection $X \longrightarrow X/x_1 X \longrightarrow \overline{R}$, so that f-rank $X \geqslant \delta(\overline{R}) > 0$, and X has a non-trivial R-free direct summand, showing $\delta(R/(x_1, \ldots, x_d)) > 0$.

Now let us assume that $\tau_\omega(R)$ is not $\mathfrak{m}$-primary. For any power $\mathfrak{m}^n$ of the maximal ideal, we may find a non-zerodivisor $z_n \in \mathfrak{m}^n \setminus \tau_\omega(R)$. By Lemmas 11.39 and 11.40, $\delta(R/(z_n)) = 0$ for every n, and the surjection $R/(z_n) \longrightarrow R/\mathfrak{m}^n$ gives $\delta(R/\mathfrak{m}^n) = 0$ for all n, so that $\text{index}(R) = \infty$. □

As an application of Ding's theorem, we prove that CM local rings of finite CM type are Gorenstein on the punctured spectrum. Of course this follows trivially from Theorem 7.12, since isolated singularities are Gorenstein on the punctured spectrum. This proof is completely independent, however, and may have other applications. It relies upon Guralnick's results in Section §3 of Chapter 1.

11.43. THEOREM. *Let $(R, \mathfrak{m})$ be a CM local ring of finite CM type. Then R has finite index. If in particular R has a canonical module, then R is Gorenstein on the punctured spectrum.*

PROOF. Let $\{M_1, \ldots, M_r\}$ be a complete set of representatives for the isomorphism classes of non-free indecomposable MCM R-modules. By Corollary 1.14, since R is not a direct summand of any M_i, there exist integers n_i, $i = 1, \ldots, r$, such that for $s \geqslant n_i$, $R/\mathfrak{m}^s$ is not a direct summand of $M_i/\mathfrak{m}^s M_i$. Then for $s \geqslant n_i$, there exists no surjection $M_i/\mathfrak{m}^s M_i \longrightarrow R/\mathfrak{m}^s$ by Lemma 1.11. Set $N = \max\{n_i\}$. Let X be any stable MCM R-module, and decompose $X \cong M_1^{(a_1)} \oplus \cdots \oplus M_r^{(a_r)}$. If there were a surjection $X \longrightarrow R/\mathfrak{m}^N$, then (since R is local) one of the summands M_i would map onto $R/\mathfrak{m}^N$, contradicting the choice of N. As X was arbitrary, this shows that $\text{index}(R) < \infty$. $\square$

11.44. REMARK. The foundation of Ding's theorem is identifying the non-zerodivisors x such that $\delta(R/(x)) > 0$. One might also ask about $\delta(\omega/x\omega)$, as well as the corresponding values of the γ-invariant. It's easy to see that the minimal MCM approximation of $\omega/x\omega$ is the short exact sequence $0 \longrightarrow \omega \xrightarrow{x} \omega \longrightarrow \omega/x\omega \longrightarrow 0$, which gives $\delta(\omega/x\omega) = 0$ and $\gamma(\omega/x\omega) = 1$. However, $\gamma(R/(x))$ is much more mysterious. We have $X_{R/(x)} \cong \text{syz}_1^R(\omega/x\omega)^\vee$, so $\gamma(R/(x)) > 0$ if and only if $\text{syz}_1^R(\omega/x\omega)$ has a non-zero free direct summand. We know of no effective criterion for this.

11.45. REMARK. As a final note, we observe that Auslander's criterion for regularity, Proposition 11.37, can be interpreted via the construction of minimal MCM approximations for CM modules in Proposition 11.15. Assume that R is Gorenstein. Then condition (iv) can be written $\delta(\text{syz}_d^R(k)) > 0$, and since $\text{syz}_d^R(k)$ is MCM, this says simply that R is regular if and only if $\text{syz}_d^R(k)$ has a non-trivial free direct summand. This is a special case of a result of Herzog [**Her94**], which generalizes a case of Levin's solution in his thesis [**Lev65**] (see also [**LV68**]) of a conjecture of Kaplansky. Kaplansky's conjecture was that if there exists a finitely generated R-module M such that $\mathfrak{m}M \neq 0$ and $\mathfrak{m}M$ has finite projective dimension, then R is regular; in particular, if $\text{syz}_d^R(R/\mathfrak{m}^n)$ is free for some n then R is regular. Yoshino has conjectured [**Yos98**] that for any positive integers t and n, $\delta(\text{syz}_t^R(R/\mathfrak{m}^n)) > 0$ if and only if R is regular local, and has proven the conjecture when R is Gorenstein and the associated graded ring $\text{gr}_\mathfrak{m}(R)$ has depth at least $d - 1$.

§5. Exercises

11.46. EXERCISE. Prove that the canonical module of a CM local ring is unique up to isomorphism, using the Artinian case and Corollary 1.14.

11.47. EXERCISE. Prove Proposition 11.20: If R is Gorenstein and M is an R-module of finite projective dimension, then the minimal MCM approximation of M is just a minimal free cover.

§5. EXERCISES

11.48. EXERCISE. Let R be a CM local ring with canonical module ω, and let M be a finitely generated R-module of finite injective dimension. Show that M has a finite resolution by copies of ω

$$0 \longrightarrow \omega^{n_t} \longrightarrow \cdots \longrightarrow \omega^{n_1} \longrightarrow \omega^{n_0} \longrightarrow M \longrightarrow 0.$$

11.49. EXERCISE. Let $x \in \mathfrak{m}$ be a non-zerodivisor. Prove that the middle term $X^{R/(x)}$ of the minimal MCM approximation of $R/(x)$ satisfies $X^{R/(x)} \cong \mathrm{syz}_2^R(\omega/x\omega)^\vee$.

11.50. EXERCISE. Let R be CM local and M a finitely generated R-module. Define the *stable MCM trace* of M to be the submodule $\tau(M)$ generated by all homomorphic images $f(X)$, where X is a stable MCM module and $f \in \mathrm{Hom}_R(X, M)$. Show that $\delta(M) = \mu_R(M/\tau(M))$.

11.51. EXERCISE. Let $(R, \mathfrak{m})$ be a local ring. Denote by $\mu^i(\mathfrak{p}, M)$ the number of copies of the injective hull of $R/\mathfrak{p}$ appearing at the ith step of a minimal injective resolution of M. This integer is called the ith *Bass number* of M at $\mathfrak{p}$. It is equal to the vector-space dimension of $\mathrm{Ext}_R^i(R/\mathfrak{p}, M)_\mathfrak{p}$ over the field $(R/\mathfrak{p})_\mathfrak{p}$.

  (i) If $\mu^i(\mathfrak{p}, M) > 0$ and height $\mathfrak{q}/\mathfrak{p} = 1$, prove that $\mu^{i+1}(\mathfrak{q}, M) > 0$.
  (ii) If M has infinite injective dimension, prove that $\mu^i(\mathfrak{m}, M) > 0$ for all $i \geqslant \dim M$. (Hint: go by induction on $\dim M$, the base case being easy. For the inductive step, distinguish two cases: (a) $\mathrm{injdim}_{R_\mathfrak{p}}(M_\mathfrak{p}) = \infty$ for some prime $\mathfrak{p} \neq \mathfrak{m}$, or (b) $\mathrm{injdim}_{R_\mathfrak{p}}(M_\mathfrak{p}) < \infty$ for every $\mathfrak{p} \neq \mathfrak{m}$. In the first case, use the previous part of this exercise; in the second, conclude that $\mathrm{injdim}_R(M) < \infty$.)

In particular, $\mathrm{Ext}_R^{\dim R}(k, Z) \neq 0$ for every finitely generated R-module Z.

11.52. EXERCISE. This exercise gives a proof of the last remaining implication in Proposition 11.14, following [**SS02**]. Let $(R, \mathfrak{m}, k)$ be a CM complete local ring of dimension d with canonical module ω.

  (i) Let M be a MCM R-module with minimal injective resolution $I^\bullet$. Prove that $\mathrm{Ext}_R^d(k, M) = \mathrm{socle}(I^d)$ is an essential submodule of the local cohomology $H_\mathfrak{m}^d(M) = H^d(\Gamma(I^\bullet))$.
  (ii) Let M and N be finitely generated R-modules with M MCM and N having finite injective dimension. Let $f \colon N \longrightarrow M$ be a homomorphism. Prove that the ω-rank of f (that is, the number of direct summands isomorphic to ω common to N and M via f) is equal to the k-dimension of the image of the homomorphism $\mathrm{Ext}_R^d(k, f)$. (Hint: take a MCM approximation of N, and split the middle term $X_N \cong \omega^{(n)}$ according to the image

of $\mathrm{Ext}^d_R(k,f)$. Apply the first part above to the composition $\mathrm{Ext}^d_R(k,\omega^{(n_1)}) \longrightarrow H^d_{\mathfrak{m}}(M)$, then use local duality.)

11.53. EXERCISE. Let R be a Gorenstein local ring (or, more generally, a CM local ring with canonical module ω and satisfying $\tau_\omega(R) \supseteq \mathfrak{m}$) with infinite residue field. Assume that R is not regular. Then
$$e(R) \geqslant \mu_R(\mathfrak{m}) - \dim R - 1 + \mathrm{index}(R).$$
In particular, if R has *minimal multiplicity* $e(R) = \mu_R(\mathfrak{m}) - \dim R + 1$, then $\mathrm{index}(R) = 2$. (Compare with Corollary 6.36.)

CHAPTER 12

Totally Reflexive Modules

Over Gorenstein local rings, MCM modules have a particularly appealing connection with (unbounded) acyclic complexes of finitely generated free modules. This connection is explored in detail in Buchweitz's notes [**Buc86**]. In this chapter we introduce *totally reflexive* modules, which play the same role over arbitrary local rings. The main theorem, Theorem 12.14, which is due to Christensen, Piepmeyer, Striuli, and Takahashi [**CPST08**], states that a local ring with at least one non-free totally reflexive module, but only finite many indecomposable ones, must be a hypersurface singularity of finite CM type. The proof uses a four-term exact sequence (Proposition 12.5) associated to the Auslander transpose $\operatorname{Tr} M$, which we define in the first section.

§1. Stable Hom and Auslander transpose

In this section we introduce two technical tools which will be useful in this chapter and the next. They arise in the context of "algebraic duality," that is, the duality $(-)^* = \operatorname{Hom}_A(-, A)$ into the ring.

12.1. DEFINITION. Let M and N be finitely generated A-modules, where A is a commutative (Noetherian, as always) ring. Denote by $\mathfrak{P}(M, N)$ the submodule of A-homomorphisms from M to N that factor through a projective A-module, and put

$$\underline{\operatorname{Hom}}_A(M, N) = \operatorname{Hom}_A(M, N)/\mathfrak{P}(M, N).$$

We call $\underline{\operatorname{Hom}}_A(M, N)$ the *stable* Hom *module*. We also write $\underline{\operatorname{End}}_A(M)$ for $\underline{\operatorname{Hom}}_A(M, M)$ and refer to it as the *stable endomorphism ring*.

Observe that $\mathfrak{P}(M, M)$ is a two-sided ideal of $\operatorname{End}_A(M)$, so that $\underline{\operatorname{End}}_A(M)$ really is a ring. In particular, it is a quotient of $\operatorname{End}_A(M)$, so the stable endomorphism ring is nc-local if the usual endomorphism ring is.

As with the usual Hom, the stable Hom module $\underline{\operatorname{Hom}}_A(M, N)$ is naturally a left $\underline{\operatorname{End}}_A(N)$-module and a right $\underline{\operatorname{End}}_A(M)$-module. We leave to the reader the straightforward check that these actions are well-defined.

12.2. REMARK. Note that $\mathfrak{P}(M, N)$ is the image of the natural homomorphism
$$\rho_M^N \colon M^* \otimes_A N \longrightarrow \mathrm{Hom}_A(M, N)$$
defined by $\rho(f \otimes y)(x) = f(x)y$ for $f \in M^*$, $y \in N$, and $x \in M$. In particular M is projective if and only if ρ_M^M is surjective.

The other main character of the section is just as easy to define, though we need some more detailed properties from it.

12.3. DEFINITION. Let A be a Noetherian ring and M a finitely generated A-module with projective presentation
$$(12.1) \qquad P_1 \xrightarrow{\varphi} P_0 \longrightarrow M \longrightarrow 0.$$
The *Auslander transpose* $\mathrm{Tr}\, M$ of M is defined by
$$\mathrm{Tr}\, M = \mathrm{cok}(\varphi^* \colon P_0^* \longrightarrow P_1^*),$$
where $(-)^* = \mathrm{Hom}_A(-, A)$. In other words, $\mathrm{Tr}\, M$ is defined by the exactness of the sequence
$$(12.2) \qquad 0 \longrightarrow M^* \longrightarrow P_1^* \xrightarrow{\varphi^*} P_0^* \longrightarrow \mathrm{Tr}\, M \longrightarrow 0.$$

12.4. REMARKS. The Auslander transpose depends, up to projective direct summands, only on M. That is, if $\varphi' \colon P_1' \longrightarrow P_0'$ is another projective presentation of M, then there are projective A-modules Q and Q' such that $\mathrm{cok}(\varphi^*) \oplus Q \cong \mathrm{cok}((\varphi')^*) \oplus Q'$. In particular $\mathrm{Tr}\, M$ is only well-defined up to "stable equivalence." However, we will work with $\mathrm{Tr}\, M$ as if it were well-defined, taking care only to apply in it in situations where the ambiguity will not matter, such as the vanishing of $\mathrm{Ext}_A^i(\mathrm{Tr}\, M, -)$ or $\mathrm{Tor}_i^A(\mathrm{Tr}\, M, -)$ for $i \geqslant 1$.

It is easy to check that $\mathrm{Tr}\, P$ is projective if P is, and that $\mathrm{Tr}(M \oplus N) \cong \mathrm{Tr}\, M \oplus \mathrm{Tr}\, N$ up to projective direct summands. Furthermore, in (12.1) φ^* is a projective presentation of $\mathrm{Tr}\, M$, and $\varphi^{**} = \varphi$ canonically, so we have $\mathrm{Tr}(\mathrm{Tr}\, M) = M$ up to projective summands for every finitely generated A-module M.

When A is a local (or graded) ring, we can give a more apparently intrinsic definition of $\mathrm{Tr}\, M$ by insisting that φ be a minimal presentation, i.e. all the entries of a matrix representing φ lie in the (homogeneous) maximal ideal. However, even then we will not have $\mathrm{Tr}(\mathrm{Tr}\, M) = M$ on the nose in general, since the Auslander transpose of any free module will be zero.

Finally, one can check that $\mathrm{Tr}(-)$ commutes with arbitrary base change. For example, it commutes (up to projective summands, as always) with localization and passing to $A/(x)$ for an arbitrary element $x \in A$.

The Auslander transpose is intimately related to the natural bi-duality homomorphism $\sigma_M \colon M \longrightarrow M^{**}$, defined by
$$\sigma_M(x)(f) = f(x)$$
for $x \in M$ and $f \in M^*$. More generally, we have the following proposition.

12.5. Proposition. *Let M and N be finitely generated A-modules. Then in the exact sequence*
$$0 \longrightarrow \ker \sigma_M^N \longrightarrow M \otimes_A N \xrightarrow{\sigma_M^N} \operatorname{Hom}_A(M^*, N) \longrightarrow \operatorname{cok} \sigma_M^N \longrightarrow 0,$$
in which σ_M^N is defined by $\sigma_M^N(x \otimes y)(f) = f(x)y$ for $x \in M$, $y \in N$, and $f \in M^*$, we have
$$\ker \sigma_M^N \cong \operatorname{Ext}_A^1(\operatorname{Tr} M, N) \quad \text{and} \quad \operatorname{cok} \sigma_M^N \cong \operatorname{Ext}_A^2(\operatorname{Tr} M, N).$$
Moreover
$$\operatorname{Ext}_A^i(\operatorname{Tr} M, N) \cong \operatorname{Ext}_A^{i-2}(M^*, N)$$
for all $i \geqslant 3$. In particular, taking $N = A$ gives an exact sequence
$$0 \longrightarrow \operatorname{Ext}_A^1(\operatorname{Tr} M, A) \longrightarrow M \xrightarrow{\sigma_M} M^{**} \longrightarrow \operatorname{Ext}_A^2(\operatorname{Tr} M, A) \longrightarrow 0$$
and isomorphisms
$$\operatorname{Ext}_A^i(\operatorname{Tr} M, A) \cong \operatorname{Ext}_A^{i-2}(M^*, A)$$
for $i \geqslant 3$. $\qquad\square$

We leave the proof as an exercise. The proposition motivates the following definition.

12.6. Definition. A finitely generated A-module M is said to be *n-torsionless* provided $\operatorname{Ext}_A^i(\operatorname{Tr} M, A) = 0$ for $i = 1, \ldots, n$.

In particular, M is 1-torsionless if and only if $\sigma_M \colon M \longrightarrow M^{**}$ is injective, 2-torsionless if and only if M is reflexive, and n-torsionless for some $n \geqslant 3$ if and only if M is reflexive and $\operatorname{Ext}_A^i(M^*, A) = 0$ for $i = 1, \ldots, n-2$.

12.7. Proposition. *Suppose that a finitely generated A-module M is n-torsionless. Then M is an nth syzygy. The converse holds if $n = 1$.*

Proof. For $n = 0$ there is nothing to prove. For $n = 1$, let $P \longrightarrow M^*$ be a surjection with P projective; then the composition of the injections $M \longrightarrow M^{**}$ and $M^{**} \longrightarrow P^*$ shows that M is a submodule of a projective, whence a first syzygy. Similarly for $n \geqslant 2$, let $P_{n-1} \longrightarrow \cdots P_0 \longrightarrow M^* \longrightarrow 0$ be a projective resolution of M^*. Dualizing and using the definition of n-torsionlessness, we see that
$$0 \longrightarrow M \longrightarrow P_0^* \longrightarrow \cdots \longrightarrow P_{n-1}^*$$

is exact, so M is an nth syzygy. The converse, when $n = 1$, is left as an exercise. □

The converse can fail when $n = 2$ (see Exercise 12.24). On the other hand, we have the following:

12.8. PROPOSITION. *Let R be a CM local ring of dimension d, and let M be a finitely generated R-module. Assume that R is Gorenstein on the punctured spectrum. Then the following are equivalent:*

(i) M is MCM;
(ii) M is a dth syzygy;
(iii) M is d-torsionless, i.e., $\operatorname{Ext}^i_R(\operatorname{Tr} M, R) = 0$ for $i = 1, \ldots, d$.

PROOF. Items (i) and (ii) are equivalent by Corollary A.15, since R is Gorenstein on the punctured spectrum. The implication (iii) $\implies$ (ii) follows from the previous proposition. We have only to prove (i) implies (iii). So assume that M is MCM. The case $d = 0$ is vacuous. For $d = 1$, the four-term exact sequence of Proposition 12.5 and the hypothesis that R is Gorenstein on the punctured spectrum combine to show that $\operatorname{Ext}^1_R(\operatorname{Tr} M, R)$ has finite length. Since $\operatorname{Ext}^1_R(\operatorname{Tr} M, R)$ embeds in M by Proposition 12.5 and M is torsion-free, this implies $\operatorname{Ext}^1_R(\operatorname{Tr} M, R) = 0$.

Now assume that $d \geqslant 2$. Let $P_1 \longrightarrow P_0 \longrightarrow M \longrightarrow 0$ be a free presentation of M, so that

$$0 \longrightarrow M^* \longrightarrow P_0^* \longrightarrow P_1^* \longrightarrow \operatorname{Tr} M \longrightarrow 0$$

is exact. Splice this together with a free resolution of M^* to get a resolution of $\operatorname{Tr} M$

$$G_{d+1} \xrightarrow{\varphi_{d+1}} G_d \xrightarrow{\varphi_d} \cdots \xrightarrow{\varphi_3} G_2 \longrightarrow P_0^* \longrightarrow P_1^* \longrightarrow \operatorname{Tr} M \longrightarrow 0.$$

Dualize, obtaining a complex

$$0 \longrightarrow (\operatorname{Tr} M)^* \longrightarrow P_1 \longrightarrow P_0 \longrightarrow G_2^* \xrightarrow{\varphi_3^*} \cdots \xrightarrow{\varphi_d^*} G_d^* \xrightarrow{\varphi_{d+1}^*} G_{d+1}^*$$

in which $\ker \varphi_3^* \cong M$ since M is reflexive. The truncation of this complex at M

$$(12.3) \qquad 0 \longrightarrow M \longrightarrow G_2^* \xrightarrow{\varphi_3^*} \cdots \xrightarrow{\varphi_d^*} G_d^* \xrightarrow{\varphi_{d+1}^*} G_{d+1}^*$$

is a complex of MCM R-modules, and since R is Gorenstein on the punctured spectrum, the homology $\operatorname{Ext}^{i-2}_R(M^*, R)$ has finite length. The Lemme d'Acyclicité (Exercise 12.22) therefore implies that the complex (12.3) is exact, so that M is a dth syzygy. □

Finally we see how the Auslander transpose and stable Hom interact. Notice that for any A-module M, $\operatorname{Tr} M$ is naturally a module over $\operatorname{End}_A(M)$, since any endomorphism of M lifts to an endomorphism of its projective presentation, thus inducing an endomorphism of $\operatorname{Tr} M$.

12.9. PROPOSITION. *Let A be a commutative ring and M, N two finitely generated A-modules. Then*
$$\underline{\operatorname{Hom}}_A(M, N) \cong \operatorname{Tor}_1^A(\operatorname{Tr} M, N).$$
Furthermore, this isomorphism is natural in both M and N, and is even an isomorphism of $\underline{\operatorname{End}}_A(N)$-$\underline{\operatorname{End}}_A(M)$-bimodules.

PROOF. Let $P_1 \xrightarrow{\varphi} P_0 \longrightarrow M \longrightarrow 0$ be our chosen projective presentation of M. Then we have the exact sequence
$$0 \longrightarrow M^* \longrightarrow P_0^* \xrightarrow{\varphi^*} P_1^* \longrightarrow \operatorname{Tr} M \longrightarrow 0.$$
Tensoring with N yields the complex
$$M^* \otimes_A N \longrightarrow P_0^* \otimes_A N \xrightarrow{\varphi^* \otimes 1_N} P_1^* \otimes_A N \longrightarrow \operatorname{Tr} M \otimes_A N \longrightarrow 0.$$
The homology of this complex at $P_0^* \otimes_A N$ is $\operatorname{Tor}_1^A(\operatorname{Tr} M, N)$. On the other hand, since the P_i are projective A-modules, the natural homomorphisms $\rho_{P_i}^N: P_i^* \otimes_A N \longrightarrow \operatorname{Hom}_A(P_i, N)$ are isomorphisms (Exercise 12.25). It follows that $\ker(\varphi^* \otimes_A 1_N) \cong \operatorname{Hom}_A(M, N)$, so that $\operatorname{Tor}_1^A(\operatorname{Tr} M, N)$ is isomorphic to the quotient of $\operatorname{Hom}_A(M, N)$ by the image of $M^* \otimes_A N \longrightarrow \operatorname{Hom}_A(P_0, N)$, i.e. $\operatorname{Tor}_1^A(\operatorname{Tr} M, N) \cong \underline{\operatorname{Hom}}_A(M, N)$.

We leave the "Furthermore" to the reader. □

§2. Complete resolutions

This section contains two constructions over Gorenstein local rings.

12.10. CONSTRUCTION. Let R be a Gorenstein local ring, and let M be a MCM R-module. Start with a minimal free resolution of M, that is, an exact sequence

(12.4) $\quad \cdots \longrightarrow F_n \longrightarrow \cdots \longrightarrow F_1 \longrightarrow F_0 \longrightarrow M \longrightarrow 0,$

in which each F_i is a free module of minimal rank. Next, we resolve $M^* = \operatorname{Hom}_R(M, R)$ minimally:

(12.5) $\quad \cdots \longrightarrow G_n \longrightarrow \cdots \longrightarrow G_1 \longrightarrow G_0 \longrightarrow M^* \longrightarrow 0.$

By Theorem 11.5, $\operatorname{Ext}^i(M, R) = 0$ for $i > 0$ and M is reflexive. Therefore, upon dualizing (12.5) and setting $F_i = (G_{-1-i})^*$ for $i < 0$, we get an exact sequence

(12.6) $\quad 0 \longrightarrow M \longrightarrow F_{-1} \longrightarrow F_{-2} \longrightarrow \cdots \longrightarrow F_{-n} \longrightarrow \cdots.$

Now we splice (12.4) and (12.6) together, taking the composition $F_0 \longrightarrow M \longrightarrow F_{-1}$ for the map $F_0 \longrightarrow F_{-1}$, and getting an acyclic complex

$$(12.7) \quad F_\bullet : \quad \cdots \longrightarrow F_2 \longrightarrow F_1 \longrightarrow F_0 \longrightarrow F_{-1} \longrightarrow F_2 \longrightarrow \cdots ,$$

in which $M = \operatorname{cok}(F_1 \longrightarrow F_0)$. We call this complex the *complete resolution* of M. We note (see Exercise 12.27) that the dual $F_\bullet^*$ of the complex $F_\bullet$ in (12.7) is acyclic and yields a complete resolution of the MCM module M^*. □

We use these observations to motivate, in the next section, an analogous class of modules over local rings $(R, \mathfrak{m})$ that may not be Gorenstein.

As a bonus application of complete resolutions, we describe the *pitchfork construction* of Auslander and Buchweitz, which gives an independent proof of the existence of MCM approximations over Gorenstein local rings. (Cf. Theorem 11.17.)

12.11. CONSTRUCTION. Let R be a Gorenstein local ring of dimension d, and let M be a finitely generated R-module of codepth t. Let

$$P_\bullet : \quad \cdots \longrightarrow P_n \longrightarrow \cdots \longrightarrow P_1 \longrightarrow P_0 \longrightarrow M \longrightarrow 0$$

be a minimal free resolution of M. Set $C = \operatorname{syz}_t^R(M)$, a MCM module. By the construction above, C has a complete resolution

$$F_\bullet : \quad \cdots \longrightarrow F_2 \longrightarrow F_1 \longrightarrow F_0 \longrightarrow F_{-1} \longrightarrow F_{-2} \longrightarrow \cdots ,$$

which we shift so that $C = \operatorname{cok}(F_{t+1} \longrightarrow F_t)$. Truncating $F_\bullet$ at degree zero, we graft it together with $P_\bullet$ to obtain a commutative pitchfork:

$$\begin{array}{ccccccccc}
 & & P_{t-1} & \longrightarrow & \cdots & \longrightarrow & P_0 & \longrightarrow & M & \longrightarrow 0 \\
 & \nearrow & \uparrow \varphi_{t-1} & & & & \uparrow \varphi_0 & & \uparrow f & \\
\cdots \longrightarrow P_t & & & & & & & & & \\
 & \searrow & \downarrow & & & & \downarrow & & \downarrow & \\
 & & F_{t-1} & \longrightarrow & \cdots & \longrightarrow & F_0 & \longrightarrow & X & \longrightarrow 0
\end{array}$$

We construct the vertical maps φ_i inductively, starting from the equality $\varphi_t : \operatorname{syz}_t^R X = C = \operatorname{syz}_t^R M$. Suppose at the ith stage we have a commutative diagram with exact rows:

$$\begin{array}{ccccccccc}
0 & \longrightarrow & \operatorname{syz}_{i+1}^R M & \longrightarrow & P_i & \longrightarrow & \operatorname{syz}_i^R M & \longrightarrow & 0 \\
 & & \uparrow \overline{\varphi_{i+1}} & & & & & & \\
0 & \longrightarrow & \operatorname{syz}_{i+1}^R X & \longrightarrow & F_i & \longrightarrow & \operatorname{syz}_i^R X & \longrightarrow & 0
\end{array}$$

Since $\operatorname{syz}_{i+1}^R X$ is MCM (being an infinite syzygy), $\operatorname{Ext}_R^1(\operatorname{syz}_{i+1}^R X, R) = 0$, whence the rows of the dualized diagram

$$\begin{array}{ccccccc}
0 & \longrightarrow & (\operatorname{syz}_i^R M)^* & \longrightarrow & P_i^* & \longrightarrow & (\operatorname{syz}_{i+1}^R M)^* \\
& & & & \downarrow \varphi_i & & \downarrow (\overline{\varphi_{i+1}})^* \\
0 & \longrightarrow & (\operatorname{syz}_i^R X)^* & \longrightarrow & F_i^* & \longrightarrow & (\operatorname{syz}_{i+1}^R X)^* & \longrightarrow & 0
\end{array}$$

are also exact. Therefore $(\overline{\varphi_{i+1}})^*$ lifts to $\psi \colon P_i^* \longrightarrow F_i^*$, and re-dualizing $\varphi_i = \psi_i^*$ completes the induction.

We thus obtain a chain map $\varphi_\bullet \colon P_\bullet \longrightarrow F_\bullet$ inducing a homomorphism $f \colon X \longrightarrow M$. We may assume that $\varphi_\bullet$ is surjective in each degree (by adding, if necessary, trivial complexes of free modules), hence in fact split surjective. In particular we assume f is surjective as well. Let $Y = \ker f$, so that

(12.8) $\qquad 0 \longrightarrow Y \longrightarrow X \longrightarrow M \longrightarrow 0$

is exact. The long exact sequence of homology associated to the short exact sequence of complexes

$$0 \longrightarrow \ker \varphi_\bullet \longrightarrow P_\bullet \xrightarrow{\varphi_\bullet} F_\bullet \longrightarrow 0$$

shows that $\ker \varphi_\bullet$ is a complex of projectives with Y its only non-vanishing homology, and that $\ker \varphi_i = 0$ for $i \geqslant t$. It follows that $\ker \varphi_\bullet$ is a finite projective resolution of Y, and that (12.8) is a MCM approximation of M. $\qquad\square$

§3. Totally reflexive modules

12.12. DEFINITION. A doubly-infinite complex

$$F_\bullet \colon \quad \cdots \longrightarrow F_2 \longrightarrow F_1 \longrightarrow F_0 \longrightarrow F_{-1} \longrightarrow F_2 \longrightarrow \cdots$$

over a local ring R is *totally acyclic* provided each F_i is a finitely generated free module and both $(F_\bullet)$ and $(F_\bullet^*)$ are exact. An R-module M is *totally reflexive* [**AM02**] provided $M \cong \operatorname{cok}(F_1 \longrightarrow F_0)$ for some totally acyclic complex $(F_\bullet)$. We say that R has *finite totally reflexive representation type* (*finite TR type* for short) provided there are, up to isomorphism, only finitely many indecomposable totally reflexive modules.

Over a Gorenstein local ring, the totally reflexive modules are exactly the MCM modules, and finite TR type is the same as finite CM type. For a local ring $(R, \mathfrak{m})$, we let $\mathcal{G}(R)$ denote the class of totally reflexive R-modules. (The letter "$\mathcal{G}$" recognizes the fact that these modules are sometimes called "modules of Gorenstein dimension zero" after [**AB69**].) We leave to the reader the proof of the following characterization of totally reflexive modules:

12.13. PROPOSITION. *Let R be a local ring and M a finitely generated R-module. Then M is totally reflexive if and only if the following conditions hold:*

(i) M is reflexive;
(ii) $\operatorname{Ext}_R^i(M, R) = 0$ for each $i > 0$; and
(iii) $\operatorname{Ext}_R^i(M^*, R) = 0$ for each $i > 0$. □

Our goal in this section is the following theorem, due to Christensen, Piepmeyer, Striuli and Takahashi [**CPST08**]:

12.14. THEOREM. *Let $(R, \mathfrak{m}, k)$ be a local ring having at least one non-free totally reflexive module. If R has finite TR type, then R is Gorenstein (and hence has finite CM type).*

This was proved by Takahashi [**Tak05, Tak04b, Tak04a**] in case R is Henselian and has depth at most two. We will give Takahashi's proof for rings of depth zero and then reduce to that case using the approach in [**CPST08**]. We will omit some of the technical details, but include the main ideas of the rather delicate proof. One question that the theorem does not answer, and which is still rather mysterious, is which non-Gorenstein local rings have the property that all of their totally reflexive modules are free. Golod rings [**AM02**, Examples 3.5] and hence, by [**Avr98**, Example 5.2.8], CM local rings with minimal multiplicity (see Conjecture 7.21) have this property.

The key to the proof of Theorem 12.14 is to show that the residue field has an "approximation" by totally reflexive modules. In order to pass to the completion, where we can invoke KRS, we will have to work with more general versions of this concept, and with certain subclasses of $\mathcal{G}$.

12.15. DEFINITION. Let R be a commutative ring and $\mathcal{C}$ a class of finitely generated R-modules with $R \in \mathcal{C}$. Let M be an arbitrary finitely generated R-module, and

(12.9) $\qquad s: \quad 0 \longrightarrow L \xrightarrow{i} C \xrightarrow{p} M \longrightarrow 0$

an exact sequence with $C \in \mathcal{C}$.

(i) We say that *s satisfies $\mathcal{C}$-lifting for M* provided every homomorphism $B \xrightarrow{f} M$, with $B \in \mathcal{C}$, lifts to a homomorphism $B \xrightarrow{g} C$ with $pg = f$; that is, the induced homomorphism

$$\operatorname{Hom}_R(B, C) \longrightarrow \operatorname{Hom}_R(B, M)$$

is a surjection for all $B \in \mathcal{C}$.
  (ii) Say that s is a $\mathcal{C}$-*cover* for M provided it satisfies $\mathcal{C}$-lifting for M and p is right minimal (cf. Definition 11.11), that is, if $\varphi \in \mathrm{End}_R(C)$ and $p\varphi = p$, then φ is an automorphism.
  (iii) Say that s is a $\mathcal{C}$-*approximation* if $\mathrm{Ext}^i_R(B, L) = 0$ for every $B \in \mathcal{C}$ and every $i > 0$.

Part (ii) of the next lemma is known as Wakamatsu's Lemma.

12.16. LEMMA. *Let R be a commutative ring, $\mathcal{C}$ a class of finitely generated R-modules with $R \in \mathcal{C}$, and $s : 0 \longrightarrow L \xrightarrow{i} C \xrightarrow{p} M \longrightarrow 0$ a short exact sequence with $C \in \mathcal{C}$.*
  (i) If s is a $\mathcal{C}$-approximation, then it satisfies $\mathcal{C}$-lifting for M.
  (ii) If $\mathcal{C}$ is closed under extensions and s is a $\mathcal{C}$-cover, then it is a $\mathcal{C}$-approximation.
  (iii) Assume that s satisfies $\mathcal{C}$-lifting for M and that there exists a $\mathcal{C}$-cover for M. Then s is a $\mathcal{C}$-cover if and only if L and C have no non-zero common direct summands under i (cf. Definition 11.10).

PROOF. We leave the proof of item (i) as an exercise. For a proof of Wakamatsu's Lemma, see [**EJ00**, Corollary 7.2.3] or [**Xu96**, 2.1.1]. The proof of (iii) is almost word-for-word the same as the proof of Lemma 11.12 [**CPST08**, Lemma 1.6]. □

Given a class $\mathcal{C}$ of finitely generated R-modules, we let $\mathcal{C}^\perp$ be the class of finitely generated R-modules L such that $\mathrm{Ext}^i_R(C, L) = 0$ for all $C \in \mathcal{C}$ and all $i > 0$.

12.17. DEFINITION. A full subcategory $\mathcal{C}$ of R-mod is a *reflexive subcategory* provided
  (i) $R \in \mathcal{C} \cap \mathcal{C}^\perp$,
  (ii) $M \oplus N \in \mathcal{C} \iff M \in \mathcal{C}$ and $N \in \mathcal{C}$,
  (iii) $M \in \mathcal{C} \implies M^* \in \mathcal{C}$, and
  (iv) $M \in \mathcal{C} \implies \mathrm{syz}_1^R(M) \in \mathcal{C}$.

We say that $\mathcal{C}$ is *closed under extensions* provided, for every short exact sequence
$$0 \longrightarrow C_1 \longrightarrow X \longrightarrow C_2 \longrightarrow 0$$
with $C_i \in \mathcal{C}$, we have $X \in \mathcal{C}$.

It follows from Proposition 12.13 that $\mathcal{G}(R)$ is reflexive and that every reflexive category of modules is contained in $\mathcal{G}(R)$.

To get from the lifting property to approximations and covers, we need the next lemma (cf. [**CPST08**, (2.2) (c) and (2.8)]). The Krull-Remak-Schmidt (KRS) property (Chapter 1, §1) is a key point in the proof of the main theorem and the reason for ascent to the completion.

12.18. LEMMA. *Let $(R, \mathfrak{m}, k)$ be a local ring whose finitely generated modules satisfy KRS, and let $\mathcal{C}$ be a reflexive subcategory of R-mod closed under extensions. These conditions on a finitely generated R-module M are equivalent:*

(i) *There exists a short exact sequence s as in (12.9) with the $\mathcal{C}$-lifting property.*
(ii) *M has a $\mathcal{C}$-cover.*
(iii) *M has a $\mathcal{C}$-approximation.*

PROOF. Assume (i). By discarding direct summands, we may assume that no non-zero direct summand of C is contained in $i(L)$, that is, s is minimal in the sense of Definition 11.10. The proof that (iv) $\implies$ (i) in Lemma 11.12 now shows that s is right minimal, that is, s is a $\mathcal{C}$-cover of M.

The implications (ii) $\implies$ (iii) and (iii) $\implies$ (i) are items (ii) and (i) of Lemma 12.16. □

Here is a connection with finite representation type:

12.19. LEMMA. *Let R be a Noetherian ring and let $\mathcal{C}$ be a class of finitely generated R-modules containing R and closed under direct summands and finite direct sums. Assume that $\mathcal{C}$ contains only finitely many indecomposable modules up to isomorphism. For any finitely generated R-module M, there exists a short exact sequence which satisfies $\mathcal{C}$-lifting for M.*

PROOF. Let $\{C_1, \ldots, C_m\}$ be a set of representatives for the isomorphism classes in $\mathcal{C}$. For each i, let $f_{i1}, \ldots, f_{in}$ be a set of generators for $\mathrm{Hom}_R(C_i, M)$. The map $p \colon C_1^{(n)} \oplus \cdots \oplus C_m^{(n)} \longrightarrow M$, defined in the obvious way using the f_{ij}, is surjective and yields a short exact sequence satisfying the $\mathcal{C}$-lifting property for M. □

We now prove the main theorem for rings of depth zero.

12.20. PROPOSITION. *Let $(R, \mathfrak{m}, k)$ be a Henselian local ring with depth $R = 0$, and let $\mathcal{C}$ be a reflexive subcategory of R-modules closed under extensions. If k has a $\mathcal{C}$-approximation, then either R is Gorenstein or $\mathcal{C}$ contains only free R-modules.*

PROOF. Let $s: 0 \longrightarrow L \xrightarrow{i} C \xrightarrow{p} k \longrightarrow 0$ be a $\mathcal{C}$-approximation of k, and dualize into R, obtaining a four-term exact sequence

$$0 \longrightarrow k^* \xrightarrow{p^*} C^* \xrightarrow{i^*} L^* \longrightarrow \operatorname{Ext}_R^1(k, R) \longrightarrow 0$$

since $\operatorname{Ext}_R^1(C, R) = 0$. Let $Z = \operatorname{im} i^*$, so that

$$t: \quad 0 \longrightarrow k^* \xrightarrow{p^*} C^* \xrightarrow{\theta} Z \longrightarrow 0$$

is exact. Here we have written θ for the map $C^* \longrightarrow Z$, to keep it distinct from i^*.

We claim that t satisfies $\mathcal{C}$-lifting for Z. Assuming the claim for the moment, here is the end of the proof. Since t satisfies $\mathcal{C}$-lifting, either θ is right minimal (so that t is a $\mathcal{C}$-cover) or not. If θ is right minimal, then by Wakamatsu's Lemma (Lemma 12.16 (ii)) t is a $\mathcal{C}$-approximation. Hence $\operatorname{Ext}_R^i(B, k^*) = 0$ for all $B \in \mathcal{C}$ and all $i > 0$. But since R has depth zero, $k^* = \operatorname{Hom}_R(k, R)$ is a finite-dimensional vector space, so this implies that every $B \in \mathcal{C}$ satisfies $\operatorname{Ext}_R^i(B, k) = 0$ for all $i > 0$. Therefore every $B \in \mathcal{C}$ is free. If on the other hand θ is not right minimal, then by Lemma 12.16 (iii) and Lemma 12.18, k^* and C^* have a non-zero common direct summand under p^*. (Here is where we need R to be Henselian.) The only direct summands of k^* are direct sums of copies of k, so we find that $k \mid C^*$. In particular k is TR, whence R is Gorenstein.

Now we prove the claim that t satisfies $\mathcal{C}$-lifting for Z. Let $B \in \mathcal{C}$ be an indecomposable module; the case $B \cong R$ is trivial, so we assume that B is non-free. It suffices to prove that

$$\operatorname{Hom}_R(B, i^*) \colon \operatorname{Hom}_R(B, C^*) \longrightarrow \operatorname{Hom}_R(B, L^*)$$

is a split surjection. Equivalently, by Hom-$\otimes$ adjointness, we may show that $\operatorname{Hom}_R(1_B \otimes i^*, R) \colon (B \otimes_R C)^* \longrightarrow (B \otimes_R L)^*$ is a split surjection. For this it suffices to prove that

$$1_B \otimes i^* \colon B \otimes_R L \longrightarrow B \otimes_R C$$

is a split injection. This is what we will do.

We have a commutative diagram with exact rows

$$\begin{array}{ccccccc}
B \otimes_R L & \xrightarrow{1_B \otimes i} & B \otimes_R C & \xrightarrow{1_B \otimes p} & B \otimes_R k & \longrightarrow & 0 \\
\sigma_B^L \downarrow & & \sigma_B^C \downarrow & & \sigma_B^k \downarrow & & \\
0 \longrightarrow \operatorname{Hom}_R(B^*, L) & \xrightarrow{j} & \operatorname{Hom}_R(B^*, C) & \xrightarrow{q} & \operatorname{Hom}_R(B^*, k) & &
\end{array}$$

in which the vertical maps are the natural homomorphisms

$$\sigma_M^N \colon M \otimes_R N \longrightarrow \operatorname{Hom}_R(M^*, N)$$

defined by $\sigma_M^N(x \otimes y)(f) = f(x)y$. Proposition 12.5 identifies the kernel and cokernel of σ_M^N:

$$\ker \sigma_M^N \cong \operatorname{Ext}_R^1(\operatorname{Tr} M, N), \quad \text{and} \quad \operatorname{cok} \sigma_M^N \cong \operatorname{Ext}_R^2(\operatorname{Tr} M, N).$$

In particular, since $\mathcal{C}$ is closed under syzygies and duals we see that $\operatorname{Tr} B \in \mathcal{C}$, whence $\operatorname{Ext}_R^i(\operatorname{Tr} B, L) = 0$ for all $i > 0$ as the original sequence s is a $\mathcal{C}$-approximation. This implies σ_B^L is an isomorphism. Furthermore, since B is indecomposable and non-free, the image of any homomorphism $f \in B^*$ is contained in $\mathfrak{m}$, and therefore

$$\sigma_B^k(b \otimes \alpha)(f) = f(b)\alpha = 0$$

for any $b \in B$, $\alpha \in k$, and $f \in B^*$. In other words, $\sigma_B^k = 0$.

Now, since $\rho \sigma_B^C = 0$, there exists a homomorphism $g \colon B \otimes_R C \longrightarrow \operatorname{Hom}_R(B^*, L)$ such that $jg = \sigma_B^C$. But then $g \circ (1_B \otimes i) = \sigma_B^L$, an isomorphism, so that $B \otimes i$ is a split injection, as claimed. □

Given a full subcategory $\mathcal{B}$ of R-mod and a local homomorphism $(R, \mathfrak{m}, k) \longrightarrow (S, \mathfrak{n}, \ell)$, we let $S \otimes_R \mathcal{B} = \{S \otimes_R B \mid B \in \mathcal{B}\}$. As in [**CPST08**], we let $\langle S \otimes_R \mathcal{B} \rangle$ denote the smallest class of S-modules that contains $S \otimes_R \mathcal{B}$ and is closed under direct summands and extensions.

PROOF OF THEOREM 12.14. Set $d = \operatorname{depth} R$ and $M = \operatorname{syz}_d^R(k)$. By Lemma 12.19 there is a short exact sequence

$$s \colon \quad 0 \longrightarrow L \xrightarrow{i} C \xrightarrow{p} M \longrightarrow 0$$

with the $\mathcal{G}(R)$-lifting property for M. We now pass to the completion $\widehat{R}$ and observe that the sequence

$$\widehat{s} \colon \quad 0 \longrightarrow \widehat{L} \longrightarrow \widehat{C} \xrightarrow{\widehat{p}} \widehat{M} \longrightarrow 0$$

has the $\widehat{R} \otimes_R \mathcal{G}(R)$-lifting property for $\widehat{M}$. Letting $\mathcal{B} = \operatorname{add}(\widehat{R} \otimes_R \mathcal{G}(R))$, the class of modules that are direct summands of modules in $\widehat{R} \otimes_R \mathcal{G}(R)$ (cf. Definition 2.1), we see that $\widehat{s}$ has the $\mathcal{B}$-lifting property for $\widehat{M}$.

Next we claim that $\widehat{R} \otimes_R \mathcal{G}(R)$ is closed under extensions. To see this, let

$$t \colon \quad 0 \longrightarrow \widehat{G} \longrightarrow V \longrightarrow \widehat{H} \longrightarrow 0$$

be an exact sequence, with $G, H \in \mathcal{G}(R)$. Any R-module W fitting into a short exact sequence $0 \longrightarrow G \longrightarrow W \longrightarrow H \longrightarrow 0$ must be totally reflexive and must satisfy $\mu_R(H) \leqslant \mu_R(G) + \mu_R(H)$. It follows that there are only finitely many such modules W up to isomorphism. By

Theorem 7.11, $\operatorname{Ext}^1_R(H,G)$ has finite length. It follows that $\operatorname{Ext}^1_R(H,G)$ is an $\widehat{R}$-module and that we have natural identifications
$$\operatorname{Ext}^1_R(H,G) \cong \widehat{R} \otimes_R \operatorname{Ext}^1_R(H,G) \cong \operatorname{Ext}^1_{\widehat{R}}(\widehat{H},\widehat{G}).$$
This means that $t = \widehat{u}$ for some exact sequence
$$u: \quad 0 \longrightarrow G \longrightarrow U \longrightarrow H \longrightarrow 0$$
of R-modules. Then $\widehat{U} \cong V$, and the claim is proved. An easy argument now shows that $\mathcal{B}$ is closed under extensions and hence that $\mathcal{B}$ is a reflexive subcategory of $\widehat{R}$-mod.

Let $\mathbf{x} = x_1, \ldots, x_d$ be a regular sequence in $\mathfrak{m}\widehat{R}$, linearly independent modulo $\mathfrak{m}^2\widehat{R}$. Since $\widehat{M}$ has depth d, $\mathbf{x}$ is a regular sequence on $\widehat{M}$ as well. Put $S = \widehat{R}/(\mathbf{x})$, and let $\mathcal{C} = \langle S \otimes_{\widehat{R}} \mathcal{B} \rangle$. One checks (see [**CPST08**, Proposition 2.10]) that $\mathcal{C}$ is a reflexive subcategory of S-mod and that $S \otimes_{\widehat{R}} \widehat{s}$ is exact and has the $\mathcal{C}$-lifting property for $S \otimes_{\widehat{R}} \widehat{M}$. (The key issues here are that $\widehat{R}/(\mathbf{x})$ has finite projective dimension over $\widehat{R}$ and that totally reflexive modules are infinite syzygies. See [**CPST08**] for details.) Next, we need a technical lemma [**CPST08**, Lemma 3.5]:

12.21. LEMMA. *Let $(R, \mathfrak{m}, k)$ be a local ring, and let $\mathbf{x} = x_1, \ldots, x_n$ be a sequence of elements that are linearly independent modulo $\mathfrak{m}^2$. Then $k \mid \operatorname{syz}^R_n(k)/\mathbf{x}\operatorname{syz}^R_n(k)$.* □

Since k is a direct summand of $S \otimes_{\widehat{R}} M = S \otimes_R \operatorname{syz}^R_d(k)$, one can obtain from $S \otimes_{\widehat{R}} \widehat{s}$ an exact sequence
$$0 \longrightarrow X \longrightarrow \overline{C} \xrightarrow{q} k \longrightarrow 0$$
of S-modules with the $\mathcal{C}$-lifting property for k. (Here $\overline{C} = S \otimes_{\widehat{R}} \widehat{C}$ and q is the composition of $1_S \otimes \widehat{p}$ with the projections on k.) (See [**CPST08**, (1.4)].) By Lemma 12.18 there exists a $\mathcal{C}$-approximation for k. Using faithfully flat descent and NAK along the homomorphisms $R \longrightarrow \widehat{R} \longrightarrow S$, we see that $\mathcal{C}$ contains at least one non-free module. Since $\operatorname{depth}(S) = 0$, Proposition 12.20 implies that S is Gorenstein, whence so is R. □

§4. Exercises

12.22. EXERCISE (Lemme d'Acyclicité, [**PS73**]). Let $(A, \mathfrak{m})$ be a local ring and $M_\bullet: 0 \longrightarrow M_s \longrightarrow \cdots \longrightarrow M_0 \longrightarrow 0$ a complex of finitely generated A-modules. Assume that $\operatorname{depth} M_i \geqslant i$ for each i, and that every homology module $H_i(M_\bullet)$ either has finite length or is zero. Then $M_\bullet$ is exact.

12.23. EXERCISE. Prove Proposition 12.5.

12.24. EXERCISE. Let M be a finitely generated A-module. Prove that A is 1-torsionless if and only if A is a first syzygy. Let $R = k[x,y]/(x^2, xy, y^2)$. Prove that the maximal ideal is a second syzygy but is not 2-torsionless.

12.25. EXERCISE. Prove Remark 12.2: there is an exact sequence
$$M^* \otimes_A N \xrightarrow{\rho} \operatorname{Hom}_A(M,N) \longrightarrow \underline{\operatorname{Hom}}_A(M,N) \longrightarrow 0,$$
where ρ sends $f \otimes y$ to the homomorphism $x \mapsto f(x)y$. Prove that ρ is an isomorphism if either M or N is projective. In the special case $M \cong N$, prove that ρ is an isomorphism if and only if M is projective.

12.26. EXERCISE. Let R be a hypersurface and M, N two MCM R-modules. Prove that $\operatorname{Ext}_R^{2i}(M,N) \cong \underline{\operatorname{Hom}}_R(M,N)$ for all $i \geqslant 1$.

12.27. EXERCISE. Let M be a MCM module over a Gorenstein local ring, with complete resolution $F_\bullet$ as in (12.7). Prove that $(F_\bullet^*)$ is an acyclic complex and that $M^* \cong \operatorname{cok}(F_{-2}^* \longrightarrow F_{-1}^*)$.

12.28. EXERCISE. Prove Proposition 12.13.

12.29. EXERCISE. Prove (i) of Lemma 12.16.

CHAPTER 13

Auslander-Reiten Theory

In this chapter we give an introduction to Auslander-Reiten (AR) sequences, also known as almost split sequences, and the Auslander-Reiten quiver. AR sequences are certain short exact sequences which were first introduced in the representation theory of Artin algebras, where they have played a central role. They have since been used fruitfully throughout representation theory. The information contained within the AR sequences is conveniently arranged in the AR quiver, which in some sense gives a picture of the whole category of MCM modules. We illustrate with several examples in §3.

§1. AR sequences

For this section, $(R, \mathfrak{m}, k)$ will be a Henselian CM local ring with canonical module ω.

We begin with the definition.

13.1. DEFINITION. Let M and N be non-zero indecomposable MCM R-modules, and let

(13.1) $$0 \longrightarrow N \xrightarrow{i} E \xrightarrow{p} M \longrightarrow 0$$

be a short exact sequence of R-modules.

(i) We say that (13.1) is an *AR sequence ending in M* if it is non-split, but for every MCM module X and every homomorphism $f \colon X \longrightarrow M$ which is not a split surjection, f factors through p.
(ii) We say that (13.1) is an *AR sequence starting from N* if it is non-split, but for every MCM module Y and every homomorphism $g \colon N \longrightarrow Y$ which is not a split injection, g lifts through i.

We will be concerned almost exclusively with AR sequences ending in a module, and in fact will often call (13.1) an AR sequence *for M*. In fact, the two halves of the definition are equivalent; see Exercise 13.32. We will therefore even allow ourselves to call (13.1) *an AR sequence* without further qualification if it satisfies either condition.

Observe that if (13.1) is an AR sequence, then in particular it is non-split, so that M is not free and N is not isomorphic to the canonical module ω.

As with MCM approximations, we take care of the uniqueness of AR sequences first, then consider existence.

13.2. PROPOSITION. *Suppose that* $0 \longrightarrow N \xrightarrow{i} E \xrightarrow{p} M \longrightarrow 0$ *and* $0 \longrightarrow N' \xrightarrow{i'} E' \xrightarrow{p'} M \longrightarrow 0$ *are two AR sequences for* M. *Then there is a commutative diagram*

$$\begin{array}{ccccccccc} 0 & \longrightarrow & N & \xrightarrow{i} & E & \xrightarrow{p} & M & \longrightarrow & 0 \\ & & \downarrow & & \downarrow & & \parallel & & \\ 0 & \longrightarrow & N' & \xrightarrow{i'} & E' & \xrightarrow{p'} & M & \longrightarrow & 0 \end{array}$$

in which the first and second vertical maps are isomorphisms.

PROOF. Since both sequences are AR sequences for M, neither p nor p' is a split surjection. Therefore each factors through the other, giving a commutative diagram

$$\begin{array}{ccccccccc} 0 & \longrightarrow & N & \xrightarrow{i} & E & \xrightarrow{p} & M & \longrightarrow & 0 \\ & & \psi\downarrow & & \varphi\downarrow & & \parallel & & \\ 0 & \longrightarrow & N' & \xrightarrow{i'} & E' & \xrightarrow{p'} & M & \longrightarrow & 0 \\ & & \psi'\downarrow & & \varphi'\downarrow & & \parallel & & \\ 0 & \longrightarrow & N & \xrightarrow{i} & E & \xrightarrow{p} & M & \longrightarrow & 0 \end{array}$$

with exact rows.

Consider $\psi'\psi \in \operatorname{End}_R(N)$. If $\psi'\psi$ is a unit of this nc-local ring, then $\psi'\psi$ is an isomorphism, so ψ is a split injection. As N and N' are both indecomposable, ψ is an isomorphism, and φ is as well by the Snake Lemma.

If $\psi'\psi$ is not a unit of $\operatorname{End}_R(N)$, then $\sigma := 1_N - \psi'\psi$ is. Define $\tau \colon E \longrightarrow N$ by $\tau(e) = e - \varphi'\varphi(e)$. This has image in N since $p\varphi'\varphi(e) = p(e)$ for all e by the commutativity of the diagram. Now $\tau(i(n)) = \sigma(n)$ for every $n \in N$. Since σ is a unit of $\operatorname{End}_R(N)$, this implies that i is a split surjection, contradicting the assumption that the top row is an AR sequence. $\square$

For existence of AR sequences, we first observe that we will need to impose an additional restriction on M or R.

§1. AR SEQUENCES

13.3. PROPOSITION. *Assume that there exists an AR sequence for M. Then M is locally free on the punctured spectrum of R. In particular, if every indecomposable MCM R-module has an AR sequence, then R has at most an isolated singularity.*

PROOF. Let $\alpha \colon 0 \longrightarrow N \longrightarrow E \longrightarrow M \longrightarrow 0$ be an AR sequence for M. Since α is non-split, M is not free. Let $L = \operatorname{syz}_1^R(M)$, so that there is a short exact sequence
$$0 \longrightarrow L \longrightarrow F \longrightarrow M \longrightarrow 0$$
with F a finitely generated free module. Suppose that $M_\mathfrak{p}$ is not free for some prime ideal $\mathfrak{p} \neq \mathfrak{m}$. Then
$$0 \longrightarrow L_\mathfrak{p} \longrightarrow F_\mathfrak{p} \longrightarrow M_\mathfrak{p} \longrightarrow 0$$
is still non-split, so in particular $\operatorname{Ext}^1_{R_\mathfrak{p}}(M_\mathfrak{p}, L_\mathfrak{p}) = \operatorname{Ext}^1_R(M, L)_\mathfrak{p}$ is non-zero. Choose an indecomposable direct summand K of L such that $\operatorname{Ext}^1_R(M, K)_\mathfrak{p}$ is non-zero, and let $\beta \in \operatorname{Ext}^1_R(M, K)$ be such that $\frac{\beta}{1} \neq 0$ in $\operatorname{Ext}^1_R(M, K)_\mathfrak{p}$. Then the annihilator of β is contained in $\mathfrak{p}$. Let $r \in \mathfrak{m} \setminus \mathfrak{p}$. Then for every $n \geqslant 0$, $r^n \notin \mathfrak{p}$, so that $r^n \beta \neq 0$. In particular $r^n \beta$ is represented by a non-split short exact sequence for all $n \geqslant 0$. Choosing a representative $0 \longrightarrow K \longrightarrow G \longrightarrow M \longrightarrow 0$ for β, and representatives $0 \longrightarrow K \longrightarrow G_n \longrightarrow M \longrightarrow 0$ for each $r^n \beta$ as well, we obtain a commutative diagram

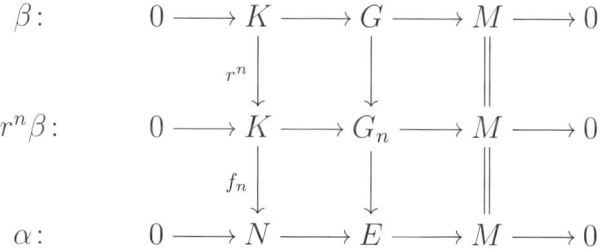

with exact rows. The top half of this diagram is the pushout representing $r^n \beta$ as a multiple of β, while the vertical arrows in the bottom half are provided by the lifting property of AR sequences. Let $f_{n*} \colon \operatorname{Ext}^1_R(M, K) \longrightarrow \operatorname{Ext}^1_R(M, N)$ denote the homomorphism induced by f_n. Then $\alpha = f_{n*}(r^n \beta) = r^n f_{n*}(\beta) \in r^n \operatorname{Ext}^1_R(M, N)$ for every $n \geqslant 0$, and so $\alpha = 0$ by Krull's Intersection Theorem, a contradiction.

The last assertion follows from the first and Lemma 7.9. □

In fact, the converse of Proposition 13.3 holds as well. The proof uses the Auslander transpose $\operatorname{Tr}(-)$ introduced in Chapter 12.

Write $\operatorname{redsyz}_n^R(M)$ for the *reduced* n^{th} syzygy module, i.e. the module obtained by deleting any non-trivial free direct summands from the

n^{th} syzygy module $\mathrm{syz}_n^R(M)$. In particular $\mathrm{redsyz}_0^R(M)$ is gotten from M by deleting any free direct summands.

13.4. PROPOSITION. *Let R be a CM local ring of dimension d and assume that R is Gorenstein on the punctured spectrum. Let M be an indecomposable non-free MCM R-module which is locally free on the punctured spectrum. Then $\mathrm{redsyz}_j^R(\mathrm{Tr}\, M)$ is indecomposable for every $j = 0, \ldots, d$.*

PROOF. Fix a free presentation $P_1 \xrightarrow{\varphi} P_0 \longrightarrow M \longrightarrow 0$ of M, so that $\mathrm{Tr}\, M$ appears in an exact sequence
$$0 \longrightarrow M^* \longrightarrow P_0^* \xrightarrow{\varphi^*} P_1^* \longrightarrow \mathrm{Tr}\, M \longrightarrow 0.$$

First consider the case $j = 0$. It suffices to prove that if $\mathrm{Tr}\, M \cong X \oplus Y$ for R-modules X and Y, then one of X or Y is free. If $\mathrm{Tr}\, M \cong X \oplus Y$, then φ^* can be decomposed as the direct sum of two matrices, that is, φ^* is equivalent to a matrix of the form $[\begin{smallmatrix}\alpha & \\ & \beta\end{smallmatrix}]$ with $X \cong \mathrm{cok}\,\alpha$ and $Y \cong \mathrm{cok}\,\beta$. But then $M = \mathrm{cok}\,\varphi = \mathrm{cok}\,\varphi^{**} \cong \mathrm{cok}(\alpha^*) \oplus \mathrm{cok}(\beta^*)$. This forces one of $\mathrm{cok}(\alpha^*)$ or $\mathrm{cok}(\beta^*)$ to be zero, which means that one of $X \cong \mathrm{cok}\,\alpha$ or $Y \cong \mathrm{cok}\,\beta$ is free.

Next assume that $j = 1$, and let N be the image of $\varphi^* \colon P_1^* \longrightarrow P_0^*$, so that $N \cong \mathrm{redsyz}_1^R(\mathrm{Tr}\, M) \oplus G$ for some finitely generated free module G. Again it suffices to prove that if $N \cong X \oplus Y$, then one of X or Y is free. Let F be a finitely generated free module mapping onto M^*, and let $f \colon F \longrightarrow P_0^*$ be the composition so that we have an exact sequence
$$F \xrightarrow{f} P_0^* \xrightarrow{\varphi^*} P_1^* \longrightarrow \mathrm{Tr}\, M \longrightarrow 0.$$
The dual of this sequence is exact since $\mathrm{Ext}_R^1(\mathrm{Tr}\, M, R) = 0$ by Proposition 12.8, so we obtain the exact sequence
$$P_1^{**} \xrightarrow{\varphi^{**}} P_0^{**} \xrightarrow{f^*} F^*$$
It follows that $M \cong \mathrm{cok}\,\varphi^{**} \cong \mathrm{im}\,f^*$. Now, if $N = \mathrm{cok}\,f$ decomposes as $N \cong X \oplus Y$, then f can be put in block-diagonal form $[\begin{smallmatrix}\alpha & \\ & \beta\end{smallmatrix}]$. It follows that $M \cong \mathrm{im}\,\alpha^* \oplus \mathrm{im}\,\beta^*$, so that one of $\mathrm{im}\,\alpha^*$ or $\mathrm{im}\,\beta^*$ is zero. This implies that one of $X = \mathrm{cok}\,\alpha$ or $Y = \mathrm{cok}\,\beta$ is free.

Now assume that $j \geqslant 2$, and we will show by induction on j that $\mathrm{redsyz}_j^R(\mathrm{Tr}\, M)$ is indecomposable. Note that since $d \geqslant 2$ and R is Gorenstein in codimension one, M is reflexive by Corollary A.13. Thus the case $j = 2$ is clear: if $\mathrm{redsyz}_2^R(\mathrm{Tr}\, M) = \mathrm{redsyz}_0^R(M^*)$ decomposes, then so does $M \cong M^{**}$.

Assume $2 < j \leqslant d$, and that $\mathrm{redsyz}_{j-1}^R(\mathrm{Tr}\, M)$ is indecomposable. Note that Corollary A.13 again implies that both $\mathrm{redsyz}_{j-1}^R(\mathrm{Tr}\, M)$ and

redsyz$_j^R$(Tr M) are reflexive. We have an exact sequence
$$0 \longrightarrow \text{redsyz}_j^R(\text{Tr } M) \oplus G \longrightarrow F \longrightarrow \text{redsyz}_{j-1}^R(\text{Tr } M) \longrightarrow 0,$$
with F and G finitely generated free modules. By Proposition 12.8, we have
$$\text{Ext}_R^1(\text{redsyz}_{j-1}^R(\text{Tr } M), R) = \text{Ext}_R^j(\text{Tr } M, R) = 0,$$
so that the dual sequence
$$0 \longrightarrow (\text{redsyz}_{j-1}^R(\text{Tr } M))^* \longrightarrow F^* \longrightarrow (\text{redsyz}_j^R(\text{Tr } M))^* \oplus G^* \longrightarrow 0$$
is also exact. If redsyz$_j^R$(Tr M) decomposes as $X \oplus Y$ with neither X nor Y free, then syz$_1^R(X^*)$ and syz$_1^R(Y^*)$ appear as direct summands of (redsyz$_{j-1}^R$(Tr M))*. We know that X^* and Y^* are non-zero since both X and Y embed in a free module, and neither X^* nor Y^* is free by the reflexivity of redsyz$_j^R$(Tr M). Thus (redsyz$_{j-1}^R$(Tr M))* is decomposed non-trivially, so that redsyz$_{j-1}^R$(Tr M) is as well, a contradiction. $\square$

Our last preparation before showing the existence of AR sequences is a short sequence of technical lemmas. The first one has the appearance of a spectral sequence, but can be proven by hand just as easily, and we leave it to the reader. See [**CE99**, VI.5.1] if you get stuck.

13.5. LEMMA. *Let A be a commutative ring and X, Y, and Z A-modules. Then the* Hom-*tensor adjointness isomorphism*
$$\text{Hom}_A(X, \text{Hom}_A(Y, Z)) \longrightarrow \text{Hom}_A(X \otimes_A Y, Z)$$
induces homomorphisms
$$\text{Ext}_A^i(X, \text{Hom}_A(Y, Z)) \longrightarrow \text{Hom}_A(\text{Tor}_i^A(X, Y), Z)$$
for every $i \geqslant 0$, which are isomorphisms if Z is injective. $\square$

13.6. LEMMA. *Let $(R, \mathfrak{m}, k)$ be a CM local ring of dimension d with canonical module ω. Let $E = E_R(k)$ be the injective hull of the residue field of R. For any two R-modules X and Y such that Y is MCM and $\text{Tor}_i^R(X, Y)$ has finite length for all $i > 0$, we have*
$$\text{Ext}_R^i(X, \text{Hom}_R(Y, E)) \cong \text{Ext}_R^{i+d}(X, \text{Hom}_R(Y, \omega)).$$

PROOF. Let $I^\bullet : 0 \longrightarrow \omega \longrightarrow I^0 \longrightarrow \cdots \longrightarrow I^d \longrightarrow 0$ be a (finite) injective resolution of ω. Let $\kappa(\mathfrak{p})$ denote the residue field of $R_\mathfrak{p}$ for a prime ideal $\mathfrak{p}$ of R. Since $\text{Ext}_{R_\mathfrak{p}}^i(\kappa(\mathfrak{p}), \omega) = 0$ for $i <$ height $\mathfrak{p}$, and is isomorphic to $\kappa(\mathfrak{p})$ for $i =$ height $\mathfrak{p}$, we see first that $I^d \cong E$, and second (by an easy induction) that $\text{Hom}_R(L, I^j) = 0$ for every $j < d$ and every R-module L of finite length.

Apply $\mathrm{Hom}_R(Y,-)$ to $I^\bullet$. Since Y is MCM, $\mathrm{Ext}^i_R(Y,\omega) = 0$ for $i > 0$, so the result is an exact sequence
(13.2)
$$0 \longrightarrow \mathrm{Hom}_R(Y,\omega) \longrightarrow \mathrm{Hom}_R(Y,I^0) \longrightarrow \cdots \longrightarrow \mathrm{Hom}_R(Y,I^d) \longrightarrow 0$$
Now from Lemma 13.5, we have
$$\mathrm{Ext}^i_R(X, \mathrm{Hom}_R(Y, I^j)) \cong \mathrm{Hom}_R(\mathrm{Tor}^R_i(X,Y), I^j)$$
for every $i,j \geqslant 0$. For $i \geqslant 1$ and $j < d$, however, the right-hand side vanishes since $\mathrm{Tor}^R_i(X,Y)$ has finite length. Thus applying $\mathrm{Hom}_R(X,-)$ to (13.2), we may use the long exact sequence of Ext to find that
$$\mathrm{Ext}^i_R(X, \mathrm{Hom}_R(Y, I^d)) \cong \mathrm{Ext}^i_R(X, \mathrm{Hom}_R(Y, \omega)). \qquad \square$$

Recall that we write $\underline{\mathrm{Hom}}_R(M,N)$ for the stable Hom module (see Definition 12.1).

13.7. PROPOSITION. *Let $(R,\mathfrak{m},k)$ be a CM local ring of dimension d with canonical module ω. Let M and N be finitely generated R-modules with M locally free on the punctured spectrum and N MCM. Then there is an isomorphism*
$$\mathrm{Hom}_R(\underline{\mathrm{Hom}}_R(M,N), E_R(k)) \cong \mathrm{Ext}^1_R(N, (\mathrm{redsyz}^R_d(\mathrm{Tr}\, M))^\vee),$$
where $-^\vee$ as usual denotes $\mathrm{Hom}_R(-,\omega)$. This isomorphism is natural in M and N, and is even an isomorphism of $\underline{\mathrm{End}}_R(N)$-$\underline{\mathrm{End}}_R(M)$-bimodules.

PROOF. Using Proposition 12.9, we substitute $\mathrm{Tor}^R_1(\mathrm{Tr}\, M, N)$ for $\underline{\mathrm{Hom}}_R(M,N)$ on the left-hand side. Applying Lemma 13.5, we see
$$\mathrm{Hom}_R(\underline{\mathrm{Hom}}_R(M,N), E_R(k)) \cong \mathrm{Hom}_R(\mathrm{Tor}^R_1(\mathrm{Tr}\, M, N)), E_R(k)$$
$$\cong \mathrm{Ext}^1_R(\mathrm{Tr}\, M, \mathrm{Hom}_R(N, E_R(k))).$$
By Lemma 13.6, this last is isomorphic to $\mathrm{Ext}^{d+1}_R(\mathrm{Tr}\, M, \mathrm{Hom}_R(N,\omega))$ since $\ell(\mathrm{Tor}^R_i(\mathrm{Tr}\, M, N)) < \infty$ for all $i \geqslant 1$. Take a reduced d^{th} syzygy of $\mathrm{Tr}\, M$, as in Proposition 13.4, to get $\mathrm{Ext}^1_R(\mathrm{redsyz}^R_d(\mathrm{Tr}\, M), N^\vee)$. Finally, canonical duality for the MCM modules $\mathrm{redsyz}^R_d \mathrm{Tr}\, M$ and $N^\vee$ shows that this last module is isomorphic to $\mathrm{Ext}^1_R(N, (\mathrm{redsyz}^R_d \mathrm{Tr}\, M)^\vee)$.

Again we leave the assertion about naturality to the reader. $\square$

For brevity, from now on we write
$$\tau(M) = \mathrm{Hom}_R(\mathrm{redsyz}^R_d \mathrm{Tr}\, M, \omega)$$
and call it the *Auslander-Reiten (AR) translate* of M.

13.8. THEOREM. *Let $(R, \mathfrak{m}, k)$ be a Henselian CM local ring of dimension d and let M be an indecomposable MCM R-module which is locally free on the punctured spectrum. Then there exists an AR sequence for M*

$$\alpha\colon 0 \longrightarrow \tau(M) \longrightarrow E \longrightarrow M \longrightarrow 0\,.$$

Precisely, the $\underline{\mathrm{End}}_R(M)$-module $\mathrm{Ext}^1_R(M, \tau(M))$ has one-dimensional socle, and any representative for a generator for that socle is an AR sequence for M.

PROOF. First observe that $\underline{\mathrm{End}}_R(M)$ is a quotient of the nc-local endomorphism ring $\mathrm{End}_R(M)$, so is again nc-local. Thus the socle of the Matlis dual of $\underline{\mathrm{End}}_R(M)$, that is $\mathrm{Hom}_R(\underline{\mathrm{End}}_R(M), E_R(k))$, is one-dimensional. By Proposition 13.7, the Matlis dual of $\underline{\mathrm{End}}_R(M)$ is isomorphic to $\mathrm{Ext}^1_R(M, \tau(M))$. Let

$$\alpha\colon 0 \longrightarrow \tau(M) \longrightarrow E \longrightarrow M \longrightarrow 0$$

be an extension generating the socle of $\mathrm{Ext}^1_R(M, \tau(M))$.

We know from Proposition 13.4 that $\mathrm{redsyz}_d^R \mathrm{Tr}\, M$ is indecomposable, so its canonical dual $\tau(M)$ is indecomposable as well. It therefore suffices to check the lifting property. Let $f\colon X \longrightarrow M$ be a homomorphism of MCM R-modules. Then pullback along f induces a homomorphism $f^*\colon \mathrm{Ext}^1_R(M, \tau(M)) \longrightarrow \mathrm{Ext}^1_R(X, \tau(M))$. If f does not factor through E, then the image of α in $\mathrm{Ext}^1_R(X, \tau(M))$ is non-zero. Since α generates the socle and α does not go to zero, we see that in fact f^* must be injective. By Proposition 13.7, this injective homomorphism is the same as the one

$$\mathrm{Hom}_R(\underline{\mathrm{End}}_R(M), E_R(k)) \longrightarrow \mathrm{Hom}_R(\underline{\mathrm{Hom}}(X, M), E_R(k))$$

induced by $f\colon X \longrightarrow M$. Since f^* is injective, Matlis duality implies that

$$\underline{\mathrm{Hom}}_R(X, M) \longrightarrow \underline{\mathrm{End}}_R(M)$$

is surjective. In particular, the map $\mathrm{Hom}_R(X, M) \longrightarrow \mathrm{End}_R(M)$ induced by f is surjective. It follows that f is a split surjection, so we are done. □

13.9. COROLLARY. *Let R be a Henselian CM local ring with canonical module, and assume that R is an isolated singularity. Then every indecomposable non-free MCM R-module has an AR sequence.* □

§2. AR quivers

The Auslander-Reiten quiver is a convenient scheme for packaging AR sequences. Up to first approximation, we could define it already: The AR quiver of a Henselian CM local ring with isolated singularity is the directed graph having a vertex $[M]$ for each indecomposable non-free MCM module M, a dotted line joining $[M]$ to $[\tau(M)]$, and an arrow $[X] \longrightarrow [M]$ for each occurrence of X in a direct-sum decomposition of the middle term of the AR sequence for M.

Unfortunately, this first approximation omits the indecomposable free module R. It is also manifestly asymmetrical: it takes into account only the AR sequences ending in a module, and omits those starting from a module. To remedy these defects, as well as for later use (particularly in Chapter 15), we introduce now irreducible homomorphisms between MCM modules, and use them to define the AR quiver. We then reconcile this definition with the naive one above, and check to see what additional information we've gained.

In this section, $(R, \mathfrak{m}, k)$ is a Henselian CM local ring with canonical module ω, and we assume that R has an isolated singularity.

13.10. DEFINITION. Let M and N be MCM R-modules. A homomorphism $\varphi \colon M \longrightarrow N$ is called *irreducible* if it is neither a split injection nor a split surjection, and in any factorization

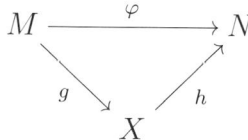

with X a MCM R-module, either g is a split injection or h is a split surjection.

13.11. DEFINITION. Let M and N be MCM R-modules.
(i) Let $\mathrm{rad}(M, N) \subseteq \mathrm{Hom}_R(M, N)$ be the submodule consisting of those homomorphisms $\varphi \colon M \longrightarrow N$ such that, when we decompose $M = \bigoplus_j M_j$ and $N = \bigoplus_i N_i$ into indecomposable modules, and accordingly decompose $\varphi = (\varphi_{ij} \colon M_j \longrightarrow N_i)_{ij}$, no φ_{ij} is an isomorphism.
(ii) Let $\mathrm{rad}^2(M, N) \subseteq \mathrm{Hom}_R(M, N)$ be the submodule of those homomorphisms $\varphi \colon M \longrightarrow N$ for which there is a factorization

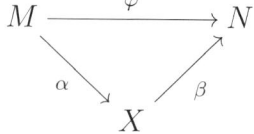

with X MCM, $\alpha \in \operatorname{rad}(M, X)$ and $\beta \in \operatorname{rad}(X, N)$.

13.12. REMARK. Suppose that M and N are indecomposable. If M and N are not isomorphic, then $\operatorname{rad}(M, N)$ is simply $\operatorname{Hom}_R(M, N)$. If, on the other hand, $M \cong N$, then $\operatorname{rad}(M, N) = \mathcal{J}(\operatorname{End}_R(M))$ is the Jacobson radical of the nc-local ring $\operatorname{End}_R(M)$, whence the name. In particular $\mathfrak{m} \operatorname{End}_R(M) \subseteq \operatorname{rad}(M, M)$ by Lemma 1.7.

For any M and N, not necessarily indecomposable, it's clear that the set of irreducible homomorphisms from M to N coincides with $\operatorname{rad}(M, N) \setminus \operatorname{rad}^2(M, N)$. Furthermore we have

$$\mathfrak{m} \operatorname{rad}(M, N) \subseteq \operatorname{rad}^2(M, N)$$

(by Exercise 13.35), so that the following definition makes sense.

13.13. DEFINITION. Let M and N be MCM R-modules, and put
$$\operatorname{Irr}(M, N) = \operatorname{rad}(M, N) / \operatorname{rad}^2(M, N).$$
Denote by $\operatorname{irr}(M, N)$ the k-vector space dimension of $\operatorname{Irr}(M, N)$.

Now we are ready to define the AR quiver of R. We impose an additional hypothesis on R, that the residue field k be algebraically closed.

13.14. DEFINITION. Let $(R, \mathfrak{m}, k)$ be a Henselian CM local ring with a canonical module. Assume that R has an isolated singularity and that k is algebraically closed. The *Auslander-Reiten (AR) quiver* for R is the graph Γ with
- vertices $[M]$ for each indecomposable MCM R-module M;
- r arrows from $[M]$ to $[N]$ if $\operatorname{irr}(M, N) = r$; and
- a dotted (undirected) line between $[M]$ and its AR translate $[\tau(M)]$ for every M.

Without the assumption that k be algebraically closed, we would need to define the AR quiver as a *valued* quiver, as follows. Suppose $[M]$ and $[N]$ are vertices in Γ, and that there is an irreducible homomorphism $M \longrightarrow N$. The abelian group $\operatorname{Irr}(M, N)$ is naturally a $\operatorname{End}_R(N)$-$\operatorname{End}_R(M)$ bimodule, with the left and right actions inherited from those on $\operatorname{Hom}_R(M, N)$. As such, it is annihilated by the radical of each endomorphism ring (see again Exercise 13.35). Let m be the dimension of $\operatorname{Irr}(M, N)$ as a right vector space over $\operatorname{End}_R(M) / \operatorname{rad}(M, M)$, and symmetrically let n be the dimension of $\operatorname{Irr}(M, N)$ as a module over $\operatorname{End}_R(N) / \operatorname{rad}(N, N)$. Then we would draw an arrow from $[M]$ to $[N]$ in Γ, and decorate it with the ordered pair (m, n). In the special case of an algebraically closed field k, $\operatorname{End}_R(M) / \operatorname{rad}(M, M)$ is in fact isomorphic to k for every indecomposable M, so we always have $m = n$.

We now reconcile the definition of the AR quiver with our earlier naive version.

13.15. PROPOSITION. *Let $(R, \mathfrak{m}, k)$ be a Henselian CM local ring with a canonical module, and assume that R has an isolated singularity. Let $0 \longrightarrow N \xrightarrow{i} E \xrightarrow{p} M \longrightarrow 0$ be an AR sequence. Then i and p are irreducible homomorphisms.*

PROOF. We prove only the assertion about p, since the other is exactly dual. First we claim that p is *right minimal*, that is (see Definition 11.11), that whenever $\varphi \colon E \longrightarrow E$ is an endomorphism such that $p\varphi = p$, in fact φ is an automorphism. The proof of this is similar to that of Proposition 13.2: the existence of $\varphi \in \mathrm{End}_R(E)$ such that $p\varphi = p$ defines a commutative diagram

$$\begin{array}{ccccccccc} 0 & \longrightarrow & N & \xrightarrow{i} & E & \xrightarrow{p} & M & \longrightarrow & 0 \\ & & \psi \downarrow & & \varphi \downarrow & & \parallel & & \\ 0 & \longrightarrow & N & \xrightarrow{i} & E & \xrightarrow{p} & M & \longrightarrow & 0 \end{array}$$

of exact sequences, where ψ is the restriction of φ to N. To see that φ is an isomorphism, it suffices by the Snake Lemma to show that ψ is an isomorphism. If not, then (since N is indecomposable and $\mathrm{End}_R(N)$ is therefore nc-local) $1_N - \psi$ is an isomorphism. Then $(1_E - \varphi) \colon E \longrightarrow N$ restricts to an isomorphism on N and therefore splits the AR sequence. This contradiction proves the claim.

We now show p is irreducible. Assume that we have a factorization

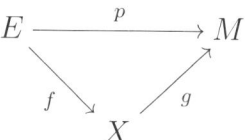

in which g is not a split surjection. The lifting property of AR sequences delivers a homomorphism $u \colon X \longrightarrow E$ such that $g = pu$. Thus we obtain a larger commutative diagram

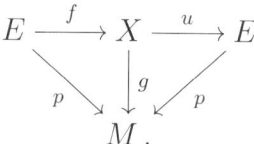

Since p is right minimal by the claim, uf is an automorphism of E. In particular, f is a split injection. □

Recall that we write $A \mid B$ to mean that A is isomorphic to a direct summand of B.

13.16. PROPOSITION. *Let $(R, \mathfrak{m}, k)$ be a Henselian CM local ring with a canonical module, and assume that R has an isolated singularity. Let $0 \longrightarrow N \xrightarrow{i} E \xrightarrow{p} M \longrightarrow 0$ be an AR sequence.*

(i) *A homomorphism $\varphi \colon X \longrightarrow M$ is irreducible if and only if φ is a direct summand of p. Explicitly, this means that $X \mid E$ and φ factors through the inclusion j of X as a direct summand of E, that is, $\varphi = pj$ for a split injection $j \colon X \longrightarrow E$.*

(ii) *A homomorphism $\psi \colon N \longrightarrow Y$ is irreducible if and only if ψ is a direct summand of i. This means $Y \mid E$ and ψ lifts over the projection π of E onto Y, that is, $\psi = \pi i$ for a split surjection $\pi \colon E \longrightarrow Y$.*

PROOF. Again we prove only the first part and leave the dual to the reader.

Assume first that $\varphi \colon X \longrightarrow M$ is irreducible. The lifting property of AR sequences gives a factorization $\varphi = pj$ for some $j \colon X \longrightarrow E$. Since φ is irreducible and p is not a split surjection, j is a split injection.

For the converse, assume that $E \cong X \oplus X'$, and write $p = [\alpha \ \beta] \colon X \oplus X' \longrightarrow M$ along this decomposition. We must show that α is irreducible. First observe that neither α nor β is a split surjection, since p is not. If, now, we have a factorization

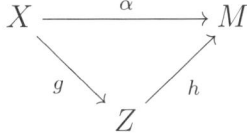

with Z MCM and h not a split surjection, then we obtain a diagram

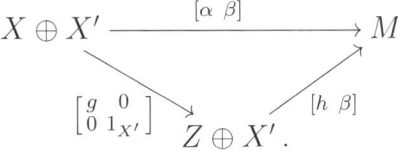

As $p = [\alpha \ \beta]$ is irreducible by Proposition 13.15, and $[h \ \beta]$ is not a split surjection by Exercise 1.23, we find that g is a split injection. $\square$

13.17. COROLLARY. *Let $0 \longrightarrow N \longrightarrow E \longrightarrow M \longrightarrow 0$ be an AR sequence. Then for any indecomposable MCM R-module X, $\operatorname{irr}(N, X) = \operatorname{irr}(X, M)$ is the multiplicity of X in the decomposition of E as a direct sum of indecomposables.* $\square$

Now we deal with $[R]$.

13.18. PROPOSITION. *Let $(R, \mathfrak{m})$ be a Henselian CM local ring with a canonical module, and assume that R has an isolated singularity. Let $0 \longrightarrow Y \longrightarrow X \xrightarrow{q} \mathfrak{m} \longrightarrow 0$ be the minimal MCM approximation of the maximal ideal $\mathfrak{m}$. (If $\dim R \leqslant 1$, we take $X = \mathfrak{m}$ and $Y = 0$.) Then a homomorphism $\varphi \colon M \longrightarrow R$ with M MCM is irreducible if and only if φ is a direct summand of q. In other words, φ is irreducible if and only if $M \mid X$ and φ factors through the inclusion of M as a direct summand of X, that is, $\varphi = qj$ for some split injection $j \colon M \longrightarrow X$.*

PROOF. Assume that $\varphi \colon M \longrightarrow R$ is irreducible. Since φ is not a split surjection, the image of φ is contained in $\mathfrak{m}$. We can therefore lift φ to factor through q, obtaining a factorization $M \xrightarrow{j} X \xrightarrow{q} \mathfrak{m}$. This factorization composes with the inclusion of $\mathfrak{m}$ into R to give a factorization of $\varphi \colon M \xrightarrow{j} X \longrightarrow R$. Since φ is irreducible and $X \longrightarrow R$ is not surjective, j is a split injection. □

13.19. REMARK. Putting Propositions 13.16 and 13.18 together, we find in particular that the AR quiver is *locally finite*, i.e. each vertex has only finitely many arrows incident to it. The local structure of the quiver is

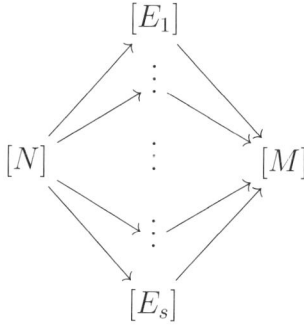

where $N = \tau(M)$ and $E = \bigoplus_{i=1}^{s} E_i$ is the middle term of the AR sequence ending in M.

§3. Examples

13.20. EXAMPLE. We can compute the AR quiver for a power series ring $R = k[\![x_1, \ldots, x_d]\!]$ directly. It has a single vertex, $[R]$, and the irreducible homomorphisms $R \longrightarrow R$ are by Propositions 13.18 and 11.20 the direct summands of $R^{(d)} \xrightarrow{[x_1, \ldots, x_d]} R$, the beginning of the Koszul resolution of $\mathfrak{m} = (x_1, \ldots, x_d)$. Thus $\mathrm{irr}(R, R) = d$ and

is the AR quiver. Alternatively, one can observe that $\mathfrak{m} = \operatorname{rad}(R,R) = \mathcal{J}(\operatorname{End}_R(R))$, while $\mathfrak{m}^2 = \operatorname{rad}^2(R,R)$, and $\dim_k(\mathfrak{m}/\mathfrak{m}^2) = d$.

13.21. EXAMPLE. We can also compute directly the AR quiver for the two-dimensional (A_1) singularity $k[\![x,y,z]\!]/(xz-y^2)$, though this one is less trivial. By Example 5.25, there is a single non-free indecomposable MCM module, namely the ideal

$$I = (x,y)R \cong \operatorname{cok}\left(\begin{bmatrix} y & -x \\ -z & y \end{bmatrix}, \begin{bmatrix} y & x \\ z & y \end{bmatrix}\right).$$

We compute $\operatorname{Irr}(I,I)$ from the definition: we have $\operatorname{Hom}_R(I,I) \cong R$ since R is integrally closed, so that $\operatorname{rad}(I,I) = \mathfrak{m}$, the maximal ideal (x,y,z). Furthermore, for any element $f \in \mathfrak{m}$, the endomorphism of I given by multiplication by f factors through $R^{(2)}$. Indeed, I is isomorphic to the submodule of $R^{(2)}$ generated by the column vectors $\binom{y}{x}$ and $\binom{z}{y}$. If $f = ax + by + cz$, then the diagram

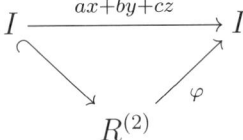

commutes, where φ is defined by $\varphi(e_1) = \binom{ax+cz}{cy}$ and $\varphi(e_2) = \binom{bz}{ax+by}$. Therefore $\operatorname{rad}^2(I,I) = \mathfrak{m} = \operatorname{rad}(I,I)$ and $\operatorname{Irr}(I,I) = 0$. (See Exercise 16.9 for another approach to this calculation.)

It follows that in the AR sequence ending in I,

$$0 \longrightarrow \tau(I) \longrightarrow E \longrightarrow I \longrightarrow 0,$$

E has no direct summands isomorphic to I, so is necessarily free. Since $\tau(I) = (\operatorname{redsyz}_2^R(\operatorname{Tr} I))^\vee = (I^*)^\vee = I$, the AR sequence is of the form

$$0 \longrightarrow I \longrightarrow R^{(2)} \longrightarrow I \longrightarrow 0,$$

and is the beginning of the free resolution of I. We conclude that the AR quiver of R is

$$[R] \rightleftarrows [I].$$

The direct approach of Example 13.21 is impractical in general, but we can use the material of Chapters 5 and 6 to compute the AR quivers of the complete Kleinian singularities (A_n), (D_n), (E_6), (E_7), and (E_8) of Table 6.24. They are isomorphic to the McKay-Gabriel quivers of the associated finite subgroups of $\operatorname{SL}(2,k)$.

Recall the setup and definition of the McKay-Gabriel quiver in dimension two. Let k be a field and $V = ku + kv$ a two-dimensional k-vector space. Let $G \subseteq \operatorname{GL}(V) \cong \operatorname{GL}(2,k)$ be a finite group with order invertible in k, and assume that G acts on V with no non-trivial pseudo-reflections. In this situation the categories of projective modules over the skew group ring $S\#G$ and the MCM R-modules are equivalent by Proposition 5.18, Corollaries 5.20 and 6.4 and Theorem 6.3. Explicitly, the functor defined by $P \mapsto P^G$ gives an equivalence between the finitely generated projective $S\#G$-modules and $\operatorname{add}_R(S)$, the R-direct summands of S. Since $\dim V = 2$, these are all the MCM R-modules by Theorem 6.3.

Writing $V_0 = k, V_1, \ldots, V_d$ for a complete set of non-isomorphic irreducible representations of G, we set

$$P_j = S \otimes_k V_j \qquad \text{and} \qquad M_j = (S \otimes_k V_j)^G$$

for $j = 0, \ldots, d$. Then $P_0 = S$, $P_1, \ldots, P_d$ are the indecomposable finitely generated projective $S\#G$-modules, and $M_0 = R$, $M_1, \ldots, M_d$ are the indecomposable MCM R-modules.

The McKay-Gabriel quiver Γ for G (see Definitions 5.21 and 5.22 and Theorem 5.23) has as vertices the indecomposable projective $S\#G$-modules $P_0, \ldots, P_d$. For each i and j, we draw m_{ij} arrows $P_i \longrightarrow P_j$ if V_i appears with multiplicity m_{ij} in the irreducible direct-sum decomposition of $V \otimes_k V_j$.

13.22. PROPOSITION. *With notation as above, the McKay-Gabriel quiver is isomorphic to the AR quiver of $R = S^G$. (We ignore the AR translate τ.)*

PROOF. First observe that R is a two-dimensional normal domain, whence an isolated singularity, so that AR quiver of R is defined.

It follows from Corollaries 5.20 and 6.4 and Theorem 6.3, as in the discussion above, that the equivalence of categories defined by

$$P_j = S \otimes_k V_j \quad \mapsto \quad M_j = (S \otimes_k V_j)^G$$

induces a bijection between the vertices of the McKay-Gabriel quiver and those of the AR quiver. It remains to determine the arrows.

Consider the Koszul complex over S

$$0 \longrightarrow S \otimes_k \bigwedge^2 V \longrightarrow S \otimes_k V \longrightarrow S \longrightarrow k \longrightarrow 0 \,,$$

which is also an exact sequence of $S\#G$-modules, and tensor with V_j to obtain
(13.3)
$$0 \longrightarrow S \otimes_k \left(\bigwedge^2 V \otimes_k V_j \right) \longrightarrow S \otimes_k (V \otimes_k V_j) \longrightarrow P_j \longrightarrow V_j \longrightarrow 0 \,.$$

Since $\bigwedge^2 V \otimes_k (\bigwedge^2 V)^* \cong k$, we see that $\bigwedge^2 V \otimes_k V_j$ is an indecomposable $k[G]$-module, so that $S \otimes_k (\bigwedge^2 V \otimes_k V_j)$ is an indecomposable projective $S\#G$-module. Take fixed points; since each V_j is simple, we have $V_j^G = 0$ for all $j \neq 0$, and $V_0^G = k^G = k$. We obtain exact sequences of R-modules
(13.4)
$$0 \longrightarrow \left(S \otimes_k \left(\bigwedge^2 V \otimes_k V_j\right)\right)^G \longrightarrow (S \otimes_k (V \otimes_k V_j))^G \xrightarrow{p_j} M_j \longrightarrow 0$$
for each $j \neq 0$, and

(13.5) $$0 \longrightarrow \left(S \otimes_k \bigwedge^2 V\right)^G \longrightarrow (S \otimes_k V)^G \xrightarrow{p_0} R \longrightarrow k \longrightarrow 0$$

for $j = 0$.

We now claim that (13.4) is the AR sequence ending in M_j for all $j = 1, \ldots, d$, while the map p_0 in (13.5) is the minimal MCM approximation of the maximal ideal of R. It will then follow from Propositions 13.16 and 13.18 that the number of arrows $[M_i] \longrightarrow [M_j]$ in the AR quiver is equal to the multiplicity of M_i in a direct-sum decomposition of $(S \otimes_k (V \otimes_k V_j))^G$, which is equal to the multiplicity of V_i in the direct-sum decomposition of $V \otimes_k V_j$.

First assume that $j \neq 0$. We saw earlier that $S \otimes_k (\bigwedge^2 V \otimes_k V_j)$ is an indecomposable projective $S\#G$-module, whence its submodule of fixed points $(S \otimes_k (\bigwedge^2 V \otimes_k V_j))^G$ is an indecomposable MCM R-module. Since the sequence (13.3) is not split, p_j is non-split as well. Assume that X is a MCM R-module and $f \colon X \longrightarrow M_j$ is a homomorphism that is not a split surjection. There then exists a homomorphism of projective $S\#G$-modules $\widetilde{f} \colon \widetilde{X} \longrightarrow P_j = S \otimes_k V_j$, also not a split surjection, such that $\widetilde{X}^G = X$ and $\widetilde{f}^G = f$. This fits into a diagram

$$\begin{array}{c} \widetilde{X} \\ \downarrow \widetilde{f} \\ S \otimes_k (V \otimes_k V_j) \xrightarrow{\widetilde{p}_j} S \otimes_k V_j \longrightarrow V_j \longrightarrow 0 \,. \end{array}$$

Since the image of $f \colon X \longrightarrow M_j$ is contained in that of
$$p_j \colon (S \otimes_k (V \otimes_k V_j))^G \longrightarrow M_j,$$
the image of $\widetilde{f}$ is contained in that of $\widetilde{p}_j$. But $\widetilde{X}$ is projective, so there exists $\widetilde{g} \colon \widetilde{X} \longrightarrow S \otimes_k (V \otimes_k V_j)$ such that $\widetilde{f} = \widetilde{p}_j \widetilde{g}$. Set $g = \widetilde{g}^G$; then $f = p_j g$, proving the claim in this case.

For $j = 0$, the argument is essentially the same; if $f\colon X \longrightarrow \mathfrak{m}$ is any homomorphism from a MCM R-module X to the maximal ideal of R, then the composition $X \longrightarrow \mathfrak{m} \longrightarrow R$ lifts to a homomorphism $\widetilde{f}\colon \widetilde{X} \longrightarrow S$ of projective $S\#G$-modules. The image of $\widetilde{f}$ is contained in the image of $\widetilde{p}_0\colon S\otimes_k V \longrightarrow S$, so again there exists $\widetilde{g}\colon \widetilde{X} \longrightarrow S\otimes_k V$ making the obvious diagram commute, and f factors through p_0. $\square$

It follows from Proposition 13.22 and §3 of Chapter 6 that the AR quivers for the Kleinian singularities (A_n), (D_n), (E_6), (E_7), and (E_8) are (after replacing pairs of opposing arrows by undirected edges) the corresponding extended ADE diagrams listed in Table 6.24. Indeed, we need not even worry about the Auslander-Reiten translate τ: since R is Gorenstein of dimension two. $\tau(X) = (\mathrm{redsyz}_R^d(\mathrm{Tr}\,X))^\vee \cong X$ for every MCM X.

Glancing back at Example 5.25, we can write down a few more AR quivers. For instance, let $R = k[\![u^5, u^2v, uv^3, v^5]\!]$, the fixed ring of the cyclic group of order 5 generated by $\mathrm{diag}(\zeta_5, \zeta_5^3)$. The AR quiver looks like

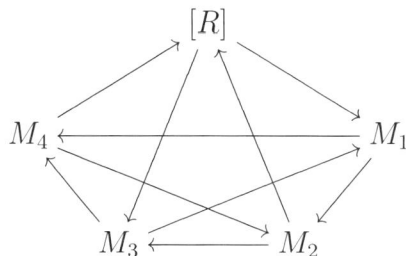

where

$$M_1 = R(u^4, uv, v^3) \cong (u^5, u^2v, uv^3)$$
$$M_2 = R(u^3, v) \cong (u^5, u^2v)$$
$$M_3 = R(u^2, uv^2, v^4) \cong (u^5, u^4v^2, u^3v^4)$$
$$M_4 = R(u, v^2) \cong (u^5, u^4v^2).$$

For another example, let $R = k[\![u^8, u^3v, uv^3, v^8]\!]$. The AR quiver is

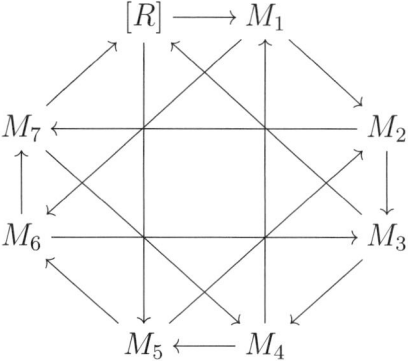

where this time

$$M_1 = R(u^7, u^2v, v^3) \cong (u^8, u^3v, uv^3)$$
$$M_2 = R(u^6, uv, v^6) \cong (u^8, u^3v, u^2v^6)$$
$$M_3 = R(u^5, v) \cong (u^8, u^3v)$$
$$M_4 = R(u^4, u^2v^2, v^4) \cong (u^8, u^6v^2, u^4v^4)$$
$$M_5 = R(u^3, uv^2, v^7) \cong (u^8, u^6v^2, u^5v^7)$$
$$M_6 = R(u^2, u^5v, v^2) \cong (u^2v^6, u^5v^7, v^8)$$
$$M_7 = R(u, v^5) \cong (uv^3, v^8).$$

Before leaving the case of dimension two, we briefly describe how to compute the AR quiver for an arbitrary two-dimensional normal domain which is not necessarily a ring of invariants. The short exact sequence (13.5)

$$0 \longrightarrow \left(S \otimes_k \bigwedge^2 V\right)^G \longrightarrow (S \otimes_k V)^G \xrightarrow{p_0} R \longrightarrow k \longrightarrow 0$$

appearing in the proof of Proposition 13.22 is called the *fundamental sequence for* R, and contains within it all the information carried by the entire AR quiver, as the proof of Proposition 13.22 shows. There is an analog of this sequence for general two-dimensional normal domains.

Assume that $(R, \mathfrak{m}, k)$ is a complete local normal domain of dimension 2. Let ω be the canonical module for R. Then we know that $\mathrm{Ext}^2_R(k, \omega) = k$, so there is up to isomorphism a unique four-term exact sequence of the form

$$0 \longrightarrow \omega \xrightarrow{a} E \xrightarrow{b} R \longrightarrow k \longrightarrow 0$$

representing a non-zero element of $\operatorname{Ext}_R^2(k,\omega)$. This is known as the *fundamental sequence for R*. The module E is easily seen to be MCM of rank 2.

Let $f\colon X \longrightarrow R$ be a homomorphism of MCM R-modules which is not a split surjection. Then the image of f is contained in $\mathfrak{m} = \operatorname{im} b$, and since $\operatorname{Ext}_R^1(X,\omega) = 0$, the pullback diagram

$$\begin{array}{ccccccccc} 0 & \longrightarrow & \omega & \longrightarrow & Q & \longrightarrow & X & \longrightarrow & 0 \\ & & \| & & \downarrow & & \downarrow f & & \\ 0 & \longrightarrow & \omega & \xrightarrow{a} & E & \xrightarrow{b} & R & & \end{array}$$

has split-exact top row. It follows that f factors through $b\colon E \longrightarrow R$, so that b is a minimal MCM approximation of the maximal ideal $\mathfrak{m}$.

More is true. Recall from Exercise 6.48 that for R-modules M and N, the reflexive product $M \cdot N$ is defined by $M \cdot N = (M \otimes_R N)^{\vee\vee}$, where $-^\vee$ denotes the canonical dual. See [**Aus86b**] for a proof of the following result.

13.23. THEOREM (Auslander). *Let $(R,\mathfrak{m},k)$ be a two-dimensional complete local normal domain with canonical module ω. Let*

$$0 \longrightarrow \omega \longrightarrow E \longrightarrow R \longrightarrow k \longrightarrow 0$$

be the fundamental sequence for R, and let M be an indecomposable non-free MCM R-module. Then the induced sequence

(13.6) $$0 \longrightarrow \omega \cdot M \longrightarrow E \cdot M \longrightarrow M \longrightarrow 0$$

is exact. If (13.6) is non-split, then it is the AR sequence ending in M. In particular, if rank M *is a unit in R, then (13.6) is non-split, so is an AR sequence. The converse is true if k is algebraically closed.* □

Let us return to the ADE singularities. The AR quivers for the one-dimensional ADE hypersurface singularities can also be obtained from those in dimension two, together with the explicit matrix factorizations for the indecomposable MCM modules listed in §4 of Chapter 6.

For example, consider the one-dimensional (E_6) singularity $R = k[\![x,y]\!]/(x^3 + y^4)$, where k is a field of characteristic not 2, 3, or 5. Let $R^\# = k[\![x,y,z]\!]/(x^3 + y^4 + z^2)$ be the double branched cover. The matrix factorizations for the indecomposable MCM $R^\#$-modules are all of the form $(zI_n - \varphi, zI_n + \varphi)$, where φ is one of the matrices $\varphi_1, \varphi_2, \varphi_3, \varphi_3^\vee, \varphi_4$, or $\varphi_4^\vee$ of 9.22. Flatting those matrix factorizations, i.e. killing z, amounts to ignoring z entirely and focusing simply on the φ_j. When we do this, certain of the matrix factorizations split into non-isomorphic pairs (as indicated by the anti-diagonal block format of the matrices), while certain other pairs of matrix factorizations collapse into a single isomorphism class.

Specifically, we can see that φ_1 splits into two non-equivalent matrices

$$\left(\begin{bmatrix} x & y^3 \\ y & -x^2 \end{bmatrix}, \begin{bmatrix} x^2 & y^3 \\ y & -x \end{bmatrix}\right)$$

forming a matrix factorization, and φ_2 splits similarly into the matrix factorization

$$\left(\begin{bmatrix} x & 0 & y^2 \\ y & x & 0 \\ 0 & 0 & x \end{bmatrix}, \begin{bmatrix} x^2 & y^3 & -xy^2 \\ -xy & x^2 & y^3 \\ y^2 & -xy & x^2 \end{bmatrix}\right).$$

On the other hand, over R,

$$\varphi_3 = \begin{bmatrix} iy^2 & 0 & -x^2 & 0 \\ 0 & iy^2 & -xy & -x^2 \\ x & 0 & -iy^2 & 0 \\ -y & x & 0 & -iy^2 \end{bmatrix} \text{ and } \varphi_3^\vee = \begin{bmatrix} -iy^2 & 0 & -x^2 & 0 \\ 0 & -iy^2 & -xy & -x^2 \\ x & 0 & iy^2 & 0 \\ -y & x & 0 & iy^2 \end{bmatrix}$$

have isomorphic cokernels, as do

$$\varphi_4 = \begin{bmatrix} iy^2 & -x^2 \\ x & -iy^2 \end{bmatrix} \text{ and } \varphi_4^\vee = \begin{bmatrix} -iy^2 & -x^2 \\ x & iy^2 \end{bmatrix}.$$

Therefore R has 6 non-isomorphic non-free indecomposable MCM modules, namely

$$M_{1a} = \operatorname{cok} \begin{bmatrix} x & y^3 \\ y & -x^2 \end{bmatrix}, \qquad M_{1b} = \operatorname{cok} \begin{bmatrix} x^2 & y^3 \\ y & -x \end{bmatrix},$$

$$M_{2a} = \operatorname{cok} \begin{bmatrix} x & 0 & y^2 \\ y & x & 0 \\ 0 & 0 & x \end{bmatrix} \qquad M_{2b} = \operatorname{cok} \begin{bmatrix} x^2 & y^3 & -xy^2 \\ -xy & x^2 & y^3 \\ y^2 & -xy & x^2 \end{bmatrix}$$

$$M_3 = \operatorname{cok} \varphi_3 = \operatorname{cok} \varphi_3^\vee$$
$$M_4 = \operatorname{cok} \varphi_4 = \operatorname{cok} \varphi_4^\vee.$$

Since each of these modules is self-dual and the AR translate τ is given by $(\operatorname{redsyz}_1^R(-^*))^*$, we have $\tau(M_{1a}) = M_{1b}$, $\tau(M_{2a}) = M_{2b}$, and vice versa, while τ fixes M_3 and M_4. One can compute the irreducible homomorphisms among these modules and obtain the AR quiver

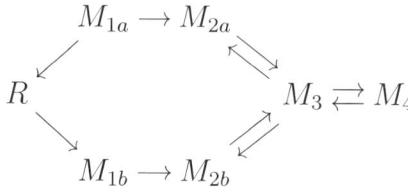

where τ is given by reflection across the horizontal axis.

For completeness, we draw the AR quivers for all the one-dimensional ADE singularities below.

13.24. The extended Coxeter-Dynkin diagram $(\widetilde{A}_n)$ has $n+1$ nodes. The splitting/collapsing behavior of the matrix factorizations depends on the parity of n. When $n = 2m$ is even, we find

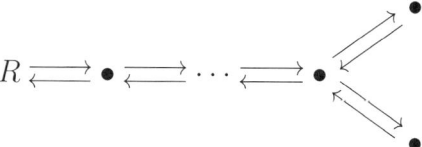

with $m+1$ vertices. The AR translate τ is the identity. When $n = 2m+1$ is odd, the quiver is

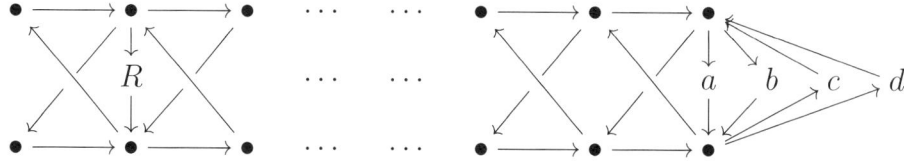

with $m+2$ vertices. Here τ is reflection across the horizontal axis.

13.25. The extended diagram $(\widetilde{D}_n)$ also has $n+1$ nodes, and again the quiver depends on the parity of n. When $n = 2m$ is even, every non-free MCM module splits, and the quiver looks like

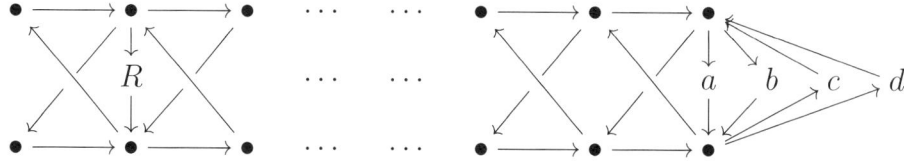

with $4m+1$ vertices. The translate τ is given by reflection in the horizontal axis for those vertices not on the axis, swaps a and d, and swaps b and c. When $n = 2m+1$ is odd, the two "legs" at the opposite end of the (D_n) diagram from the free module collapse into a single

module, giving the quiver

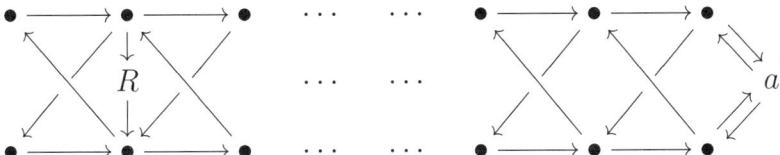

with $4m$ vertices. Again, τ is reflection across the horizontal axis.

13.26. We saw above the the quiver for the one-dimensional (E_6) singularity has the form

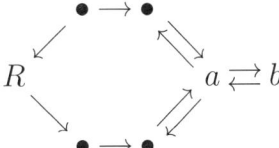

with 7 vertices and τ given by reflection across the horizontal axis.

13.27. In the (E_7) case, every non-free indecomposable splits when flatted, giving 15 vertices in the AR quiver for the one-dimensional singularity.

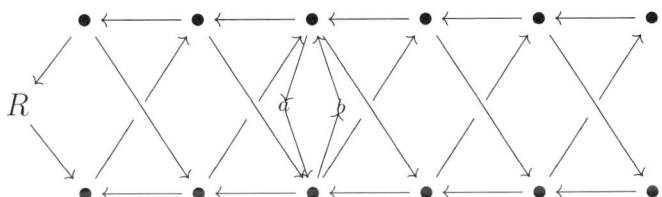

The translate is reflection across the horizontal axis for every vertex except a and b, which are interchanged by τ.

13.28. For the (E_8) singularity, once again every non-free indecomposable splits when flatted.

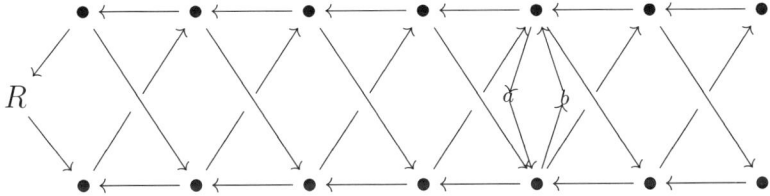

Here there are 17 vertices; the translate is reflection across the horizontal axis and interchanges a and b.

13.29. EXAMPLE. Let $A = k[\![t^3, t^4, t^5]\!]$. Then A is a finite birational extension of the (E_6) singularity $R = k[\![x, y]\!]/(x^3 + y^4) \cong k[\![t^3, t^4]\!]$, so has finite CM type by Theorem 4.13. In fact, A is isomorphic to the endomorphism ring of the maximal ideal of R. By Lemma 4.9 every indecomposable MCM R-module other than R itself is actually a MCM A-module, and $\operatorname{Hom}_R(M, N) = \operatorname{Hom}_A(M, N)$ for all non-free MCM R-modules M and N. Thus the AR quiver for A is obtained from the one for R by erasing $[R]$ and all the arrows into and out of $[R]$. As R-modules, $A \cong (t^4, t^6)$, so the quiver is the one below.

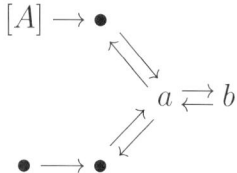

§4. Exercises

13.30. EXERCISE. Prove that a short exact sequence $0 \longrightarrow N \longrightarrow E \longrightarrow M \longrightarrow 0$ is split if and only if every homomorphism $X \longrightarrow M$ factors through E, or equivalently the sequence has the lifting property with respect to all modules.

13.31. EXERCISE. Let $R = D/(t^n)$, where (D, t) is a complete DVR. Then the indecomposable finitely generated R-modules are

$$D/(t), \ D/(t^2), \ \ldots, \ D/(t^n) = R.$$

Compute the AR sequences for each of the indecomposables, directly from the definition. (Hint: start with $n = 2$.)

13.32. EXERCISE. Prove, mimicking the proof of Proposition 13.2, that (13.1) is an AR sequence ending in M if and only if it is an AR sequence starting from N. (Hint: Given $\psi \colon N \longrightarrow Y$, it suffices to show that the short exact sequence obtained from the pushout is split. If not, use the lifting property to obtain an endomorphism α of N such that either α is an isomorphism and splits ψ, or $\alpha - 1_N$ is an isomorphism and splits (13.1).)

13.33. EXERCISE. Assume that $0 \longrightarrow N \xrightarrow{i} E \xrightarrow{p} M \longrightarrow 0$ is a non-split short exact sequence of MCM modules satisfying the lifting property to be an AR sequence ending in M. Prove that M is indecomposable.

13.34. EXERCISE. This exercise shows that if R is an Artinian local ring and M is an indecomposable R-module with an AR sequence
$$0 \longrightarrow N \longrightarrow E \longrightarrow M \longrightarrow 0,$$
then $N \cong (\operatorname{Tr} M)^\vee$.

(a) Let $P_1 \longrightarrow P_0 \longrightarrow X \longrightarrow 0$ be an exact sequence with P_0 and P_1 finitely generated projective, and let Z be an arbitrary finitely generated R-module. Use the proof of Proposition 12.9 to show that the kernel and cokernel of the natural map $\operatorname{Hom}_R(P_0, Z) \longrightarrow \operatorname{Hom}_R(P_1, Z)$ are $\operatorname{Hom}_R(X, Z)$ and $\operatorname{Tr} X \otimes_R Z$, respectively, and conclude that we have an equality of lengths
$$\ell(\operatorname{Hom}_R(X, Z)) - \ell(\operatorname{Hom}_R(Z, (\operatorname{Tr} X)^\vee))$$
$$= \ell(\operatorname{Hom}_R(P_0, Z)) - \ell(\operatorname{Hom}_R(P_1, Z)).$$

(b) Let $\sigma \colon 0 \longrightarrow A \xrightarrow{f} B \xrightarrow{g} C \longrightarrow 0$ be an exact sequence of finitely generated R-modules, and define the *defects* of σ on an R-module X by
$$\sigma_*(X) = \operatorname{cok}[\operatorname{Hom}_R(B, X) \longrightarrow \operatorname{Hom}_R(A, X)]$$
$$\sigma^*(X) = \operatorname{cok}[\operatorname{Hom}_R(X, B) \longrightarrow \operatorname{Hom}_R(X, C)].$$
Show that $\ell(\sigma^*(X)) = \ell(\sigma_*((\operatorname{Tr} X)^\vee))$ for every X. Conclude that the following two conditions are equivalent:
   (i) every homomorphism $X \longrightarrow C$ factors through g;
   (ii) every homomorphism $A \longrightarrow (\operatorname{Tr} X)^\vee$ factors through f.

(c) Prove that if $0 \longrightarrow N \longrightarrow E \longrightarrow M \longrightarrow 0$ is an AR sequence for M, then $N \cong (\operatorname{Tr} M)^\vee$. (Hint: let $h \colon N \longrightarrow Y$ be given with Y indecomposable and not isomorphic to $(\operatorname{Tr} M)^\vee$. Apply the previous part to $X = \operatorname{Tr}(Y^\vee)$.)

13.35. EXERCISE. Prove that $\operatorname{rad}(M, N)/\operatorname{rad}^2(M, N)$ is annihilated by the maximal ideal $\mathfrak{m}$, so is a finite-dimensional k-vector space. Your proof will actually show that the quotient is annihilated by the radical of $\operatorname{End}_R(M)$ (acting on the right) and the radical of $\operatorname{End}_R(N)$ (acting on the left).

13.36. EXERCISE ([**Eis95**, A.3.22]). If $\sigma \colon A \longrightarrow B \longrightarrow C \longrightarrow 0$ is an exact sequence, prove that (there exists a choice of $\operatorname{Tr} M$ such that) the sequence
$$0 \longrightarrow \operatorname{Hom}_R(\operatorname{Tr} M, A) \longrightarrow \operatorname{Hom}_R(\operatorname{Tr} M, B) \longrightarrow \operatorname{Hom}_R(\operatorname{Tr} M, C)$$
$$\longrightarrow M \otimes_R A \longrightarrow M \otimes_R B \longrightarrow M \otimes_R C \longrightarrow 0$$

is exact. In other words, Tr can be thought of as measuring the non-exactness of $M \otimes_R -$ and, if we set $N = \operatorname{Tr} M$, of $\operatorname{Hom}_R(N, -)$.

13.37. EXERCISE. Recall that an inclusion of modules $A \subset B$ is *pure* if $M \otimes_R A \longrightarrow M \otimes_R B$ is injective for all R-modules M. If σ is as in the previous exercise with $A \longrightarrow B$ pure, then prove that

$$0 \longrightarrow \operatorname{Hom}_R(N, A) \longrightarrow \operatorname{Hom}_R(N, B) \longrightarrow \operatorname{Hom}_R(N, C) \longrightarrow 0$$

is exact for every finitely presented module N. Conclude that if C is finitely presented, then σ splits. (See Exercise 7.23 for a different proof.)

CHAPTER 14

Countable Cohen-Macaulay Type

We shift directions now, and focus on a representation type mentioned in passing in earlier chapters: countable type.

14.1. DEFINITION. A Cohen-Macaulay local ring $(R, \mathfrak{m})$ is said to have *countable Cohen-Macaulay type* if it admits only countably many isomorphism classes of maximal Cohen-Macaulay modules.

(By Theorem 2.2, it is equivalent to assume that there are only countably many *indecomposable* MCM modules, up to isomorphism.)

The property of countable type has received less attention than finite type, and correspondingly less is known about it. There is however an analogue of Auslander's theorem (Theorem 14.3), as well as a complete classification (Theorem 14.16) of complete hypersurface singularities over an uncountable field with countable CM type, due to Buchweitz-Greuel-Schreyer [**BGS87**]. This has recently been revisited by Burban-Drozd [**BD08, BD10**]; we present here their approach, which echoes nicely the material in Chapter 4. They use a construction similar to the conductor square to prove that the (A_∞) and (D_∞) hypersurface singularities $k[\![x,y,z]\!]/(xy)$ and $k[\![x,y,z]\!]/(x^2y-z^2)$ have countable type. The material of Chapters 8 and 9 can then be used to show that, in any dimension, the higher-dimensional (A_∞) and (D_∞) singularities are the only hypersurfaces with countably infinite CM type. Apart from these results, there are a few examples due to Schreyer (see Section §4), but much remains to be done.

§1. Structure

The main structural result on CM local rings of countable CM type was conjectured by Schreyer in 1987 [**Sch87**, Section 7]. He predicted that an analytic local ring R over the complex numbers having countable type has at most a one-dimensional singular locus, that is, $R_\mathfrak{p}$ is regular for all $\mathfrak{p} \in \mathrm{Spec}(R)$ with $\dim(R/\mathfrak{p}) > 1$. In this section we prove Schreyer's conjecture more generally for all CM local rings satisfying a souped-up version of prime avoidance [**Bur72**, Lemma 3]; see also [**SV85**]. In particular, this property holds if either the ring is complete or the residue field is uncountable. Some assumption of uncountability is necessary to avoid the degenerate case of a countable ring, which has only countably many isomorphism classes of finitely generated modules!

14.2. LEMMA (Countable Prime Avoidance). *Let A be a Noetherian ring satisfying either of these conditions.*
   (i) A is complete local, or
   (ii) there is an uncountable set of elements $\{u_\lambda\}_{\lambda \in \Lambda}$ of A such that $u_\lambda - u_\mu$ is a unit of A for every $\lambda \neq \mu$.

Let $\{\mathfrak{p}_i\}_{i=1}^\infty$ be a countable set of prime ideals of R, and I an ideal with $I \subseteq \bigcup_{i=1}^\infty \mathfrak{p}_i$. Then $I \subseteq \mathfrak{p}_i$ for some i.

Notice that the second condition is satisfied if, for example, $(A, \mathfrak{m})$ is local with $A/\mathfrak{m}$ uncountable. In fact the proof will show that when (ii) is verified the ideals $\mathfrak{p}_i$ need not even be prime.

We postpone the proof to the end of this section.

14.3. THEOREM. *Let $(R, \mathfrak{m})$ be an excellent CM local ring of countable CM type. Assume that R satisfies countable prime avoidance. Then the singular locus of R has dimension at most one.*

PROOF. Set $d = \dim R$, and assume that the singular locus of R has dimension greater than one. Since R is excellent, $\operatorname{Sing} R$ is a closed subset of $\operatorname{Spec}(R)$, defined by an ideal J such that $\dim(R/J) \geqslant 2$. Consider the set Ω consisting of prime ideals $\mathfrak{p} \neq \mathfrak{m}$ such that $\mathfrak{p} = \operatorname{Ann}_R(\operatorname{Ext}_R^i(M, N))$ for some $i \geqslant 1$ and MCM R-modules M, N. Then of course Ω is a countable set, and each $\mathfrak{p} \in \Omega$ contains J. Applying countable prime avoidance, we find an element $r \in \mathfrak{m} \setminus \bigcup_{\mathfrak{p} \in \Omega} \mathfrak{p}$. Choose a minimal prime $\mathfrak{q}$ of $J + (r)$; since $\dim(R/J) \geqslant 2$ we have $\mathfrak{q} \neq \mathfrak{m}$, and $\mathfrak{q} \notin \Omega$.

Set $M = \operatorname{syz}_d^R(R/\mathfrak{q})$ and $N = \operatorname{syz}_{d+1}^R(R/\mathfrak{q})$, and consider the ideal $\mathfrak{a} = \operatorname{Ann}_R\left(\operatorname{Ext}_R^1(M, N)\right)$. Clearly $\mathfrak{q}$ is contained in $\mathfrak{a}$, as $\operatorname{Ext}_R^1(M, N) \cong \operatorname{Ext}_R^{d+1}(R/\mathfrak{q}, N)$. Since $\mathfrak{q}$ contains the defining ideal J, the localization $R_\mathfrak{q}$ is not regular, so the residue field $R_\mathfrak{q}/\mathfrak{q} R_\mathfrak{q}$ has infinite projective dimension and $\operatorname{Ext}_R^1(M, N)_\mathfrak{q} \neq 0$. Therefore $\mathfrak{a} \subseteq \mathfrak{q}$, and we see that $\mathfrak{q} \in \Omega$, a contradiction. □

14.4. REMARKS. With a suitable assumption of prime avoidance for sets of cardinality $\aleph_n$, the same proof shows that if R has at most "$\aleph_{m-1}$ CM type," then the singular locus of R has dimension at most m.

Theorem 14.3 implies that for an excellent CM local ring of countable CM type, satisfying countable prime avoidance, there are at most finitely many non-maximal prime ideals $\mathfrak{p}_1, \ldots, \mathfrak{p}_n$ such that $R_{\mathfrak{p}_i}$ is not a regular local ring. Each of these localizations has dimension $d-1$. Naturally, one would like to know more about these $R_{\mathfrak{p}_i}$. Peeking ahead at the examples later on in this chapter, we find that in each of them, every $R_{\mathfrak{p}_i}$ has finite CM type! Whether or not this holds in general is still an open question. The next result gives partial information: at least each $R_{\mathfrak{p}_i}$ has countable type. It is a nice application of MCM approximations (Chapter 11).

14.5. THEOREM. *Let $(R, \mathfrak{m})$ be a CM local ring with a canonical module. If R has countable CM type, then $R_{\mathfrak{p}}$ has countable CM type for every $\mathfrak{p} \in \operatorname{Spec}(R)$.*

PROOF. Let $\mathfrak{p} \in \operatorname{Spec}(R)$ and suppose that $\{M^\alpha\}$ is an uncountable family of finitely generated R-modules such that the localized modules $\{M^\alpha_{\mathfrak{p}}\}$ are non-isomorphic MCM $R_{\mathfrak{p}}$-modules. For each α there is by Theorem 11.17 a MCM approximation of M^α

$$(14.1) \qquad \chi^\alpha: \quad 0 \longrightarrow Y^\alpha \longrightarrow X^\alpha \longrightarrow M^\alpha \longrightarrow 0$$

with X^α MCM and $\operatorname{injdim}_R Y^\alpha < \infty$. Since there are only countably many non-isomorphic MCM modules, there must be uncountably many short exact sequences

$$(14.2) \qquad \chi^\beta: \quad 0 \longrightarrow Y^\beta \longrightarrow X \longrightarrow M^\beta \longrightarrow 0$$

where X is a fixed MCM module.

Localize at $\mathfrak{p}$; since $M^\beta_{\mathfrak{p}}$ is MCM over $R_{\mathfrak{p}}$ and $Y^\beta_{\mathfrak{p}}$ has finite injective dimension, $\operatorname{Ext}^1_R(M^\beta, Y^\beta)_{\mathfrak{p}} \cong \operatorname{Ext}^1_{R_{\mathfrak{p}}}(M^\beta_{\mathfrak{p}}, Y^\beta_{\mathfrak{p}}) = 0$ by Proposition 11.3. In particular, the extension χ^β splits when localized at $\mathfrak{p}$. This implies that $M^\beta_{\mathfrak{p}} \mid X_{\mathfrak{p}}$ for uncountably many β, which cannot happen by Theorem 2.2. $\square$

The results above, together with the examples in §4, suggest a reasonable question:

14.6. QUESTION. *Let R be a complete local Cohen-Macaulay ring of dimension at least one, and assume that R has an isolated singularity. If R has countable CM type, must it have finite CM type?*

Here is the proof we omitted earlier. We follow [**SV85**] closely.

PROOF OF COUNTABLE PRIME AVOIDANCE. First, we consider a complete local ring $(A, \mathfrak{m})$. Suppose that $I \not\subseteq \mathfrak{p}_i$ for each i, but that $I \subseteq \bigcup_i \mathfrak{p}_i$. Obviously $I \subseteq \mathfrak{m}$. Since A is Noetherian, all chains in $\operatorname{Spec}(A)$ are finite, so we may replace each chain by its maximal element to assume that there are no inclusions among the $\mathfrak{p}_i$. Note that A is complete with respect to the I-adic topology [**Mat89**, Ex. 8.2].

Construct a Cauchy sequence in A as follows. Choose $x_1 \in I \setminus \mathfrak{p}_1$, and suppose inductively that we have chosen $x_1, \ldots, x_r$ to satisfy

(a) $x_j \notin \mathfrak{p}_i$, and
(b) $x_i - x_j \in I^i \cap \mathfrak{p}_i$

for all $i \leqslant j \leqslant r$. If $x_r \notin \mathfrak{p}_{r+1}$, put $x_{r+1} = x_r$. Otherwise, take $y_{r+1} \in (I^r \cap \mathfrak{p}_1 \cap \cdots \cap \mathfrak{p}_r) \setminus \mathfrak{p}_{r+1}$ (this is possible since there are no containments among the $\mathfrak{p}_i$) and set $x_{r+1} = x_r + y_{r+1}$. In either case, we have

(c) $x_{r+1} \notin \mathfrak{p}_i$ for $i \leqslant r+1$, and
(d) $x_{r+1} - x_r \in \mathfrak{m}^{r+1} \cap \mathfrak{p}_1 \cap \cdots \cap p_r$, so that if $i < r+1$ then $x_i - x_{r+1} \in \mathfrak{m}^i \cap \mathfrak{p}_i$.

By condition (d), $\{x_1, x_2, \ldots\}$ is a Cauchy sequence, so converges to $x \in A$. Since $x_i - x_s \in \mathfrak{p}_i$ for all $i \leqslant s$, and $x_s \longrightarrow x$, we obtain $x_i - x \in \mathfrak{p}_i$ for all i, since $\mathfrak{p}_i$ is closed in the I-adic topology [**Mat89**, Thm. 8.14]. Therefore $x \notin \mathfrak{p}_i$ for all i, but $x \in I$, a contradiction.

Now let $\{u_\lambda\}_{\lambda \in \Lambda}$ be an uncountable family of elements of A as in (ii) of Lemma 14.2. Take generators $a_1, \ldots, a_k$ for the ideal I, and for each $\lambda \in \Lambda$ set

$$z_\lambda = a_1 + u_\lambda a_2 + u_\lambda^2 a_3 + \cdots + u_\lambda^{k-1} a_k,$$

an element of I. Since $\{\mathfrak{p}_i\}$ is countable, and $I \subseteq \bigcup_i \mathfrak{p}_i$, there exist some $j \geqslant 1$ and uncountably many $\lambda \in \Lambda$ such that $z_\lambda \in \mathfrak{p}_j$. In particular there are distinct elements $\lambda_1, \ldots, \lambda_k$ such that $z_{\lambda_i} \in \mathfrak{p}_j$ for $i = 1, \ldots, k$.

The $k \times k$ Vandermonde matrix

$$P = \left(u_{\lambda_i}^{j-1}\right)_{i,j}$$

has determinant $\prod_{i \neq j}(u_{\lambda_i} - u_{\lambda_j})$, so is invertible. But

$$P \begin{pmatrix} a_1 & \cdots & a_k \end{pmatrix}^T = \begin{pmatrix} z_{\lambda_1} & \cdots & z_{\lambda_k} \end{pmatrix}^T,$$

so

$$\begin{pmatrix} a_1 & \cdots & a_k \end{pmatrix}^T = P^{-1} \begin{pmatrix} z_{\lambda_1} & \cdots & z_{\lambda_k} \end{pmatrix}^T,$$

which implies $I = (a_1, \ldots, a_k) \subseteq \mathfrak{p}_j$. □

§2. Burban-Drozd triples

Our goal in this section and the next is to classify the complete equicharacteristic hypersurface singularities of countable CM type in characteristic other than 2. They are the "natural limits" (A_∞) and (D_∞) of the (A_n) and (D_n) singularities. This classification is originally due to Buchweitz-Greuel-Schreyer [**BGS87**]; they construct all the indecomposable MCM modules over the one-dimensional (A_∞) and (D_∞) hypersurface singularities, and use the property of countable simplicity (Definition 9.1) to show that no other one-dimensional hypersurfaces have countable type. They then use the double branched cover construction of Chapter 8 to obtain the result in all dimensions.

We modify this approach by describing a special case of some recent results of Burban and Drozd [**BD10**], which allow us to construct all the indecomposable MCM modules over the *surface* singularities rather than over the curves. In addition to its satisfying parallels with our treatment of hypersurfaces of finite CM type in Chapters 6 and 9, this method is also pleasantly akin to the "conductor square" construction in Chapter 4. It also allows us to write down, in a manner analogous to §4 of Chapter 9, a complete list of the indecomposable matrix factorizations over the two-dimensional (A_∞) and (D_∞) hypersurfaces.

14.7. NOTATION. Throughout this section we consider a reduced, CM, complete local ring $(R, \mathfrak{m})$ of dimension 2 which is *not* normal. (The assumption that R is reduced is no imposition, thanks to Theorem 14.3.) We will impose further assumptions later on, cf. 14.12. Since normality is equivalent to both (R_1) and (S_2) by Proposition A.9, this means that R is not regular in codimension one. Let S be the integral closure of R in its total quotient ring. Since R is complete and reduced, S is a finitely generated R-module (Theorem 4.6), and is a direct product of complete local normal domains, each of which is CM.

Let $\mathfrak{c} = (R :_R S) = \mathrm{Hom}_R(S, R)$ be the conductor ideal as in Chapter 4, the largest common ideal of R and S. Set $\overline{R} = R/\mathfrak{c}$ and $\overline{S} = S/\mathfrak{c}$.

14.8. LEMMA. *In the notation above, the following properties hold.*
 (i) *The conductor ideal $\mathfrak{c}$ is a MCM module over both R and S.*
 (ii) *The quotients $\overline{R}$ and $\overline{S}$ are one-dimensional CM (possibly non-reduced) rings with $\overline{R} \subseteq \overline{S}$.*
 (iii) *The diagram*

$$\begin{array}{ccc} R & \hookrightarrow & S \\ \downarrow & & \downarrow \\ \overline{R} & \hookrightarrow & \overline{S} \end{array}$$

is a pullback diagram of ring homomorphisms.

PROOF. Since $\mathfrak{c} = \mathrm{Hom}_R(S, R)$, Exercise 5.37 implies that $\mathfrak{c}$ has depth 2 when considered as an R-module. Since $R \subseteq S$ is a finite extension, $\mathfrak{c}$ is also MCM over S.

The conductor $\mathfrak{c}$ defines the non-normal locus of $\operatorname{Spec} R$. Since for a height-one prime $\mathfrak{p}$ of R, $R_\mathfrak{p}$ is normal if and only if it is regular, and R is not regular in codimension one, we see that $\mathfrak{c}$ has height at most one in R. On the other hand, R is reduced, so its localizations at minimal primes are fields, and it follows that $\mathfrak{c}$ has height exactly one in R. Thus $\dim \overline{R} = 1$, and, since $\overline{R} \hookrightarrow \overline{S}$ is integral, $\overline{S}$ is one-dimensional as well. Since $\mathfrak{c}$ has depth 2, the quotients $\overline{R}$ and $\overline{S}$ have depth 1 by the Depth Lemma.

The third statement is easy to check. □

Recall from the exercises to Chapter 6 that the reflexive product $N \cdot M = (N \otimes_R M)^{\vee\vee}$ of two R-modules M and N is a MCM R-module, where $-^\vee = \operatorname{Hom}_R(-, \omega_R)$. In the special case $N = S$, the reflexive product $S \cdot M$ inherits an S-module structure and so is a MCM S-module. Recall also that for any (not necessarily reflexive) S-module X, there is an exact sequence (Exercise 14.31)

$$(14.3) \qquad 0 \longrightarrow \operatorname{tor}(X) \longrightarrow X \longrightarrow X^{\vee\vee} \longrightarrow L \longrightarrow 0,$$

where $\operatorname{tor}(X)$ denotes the torsion submodule of X and L is an S-module of finite length.

Let M be a MCM R-module. Set $\overline{M} = M/\mathfrak{c}M$ and $\overline{S \cdot M} = (S \cdot M)/\mathfrak{c}(S \cdot M)$, modules over $\overline{R}$ and $\overline{S}$, respectively. By Exercise 14.30, applied to $\overline{R}$ and to $R_\mathfrak{p}$, respectively, we have $\overline{M}^{\vee\vee} \cong \overline{M}/\operatorname{tor}(\overline{M})$ and $(S \cdot M)_\mathfrak{p} \cong (S_\mathfrak{p} \otimes_{R_\mathfrak{p}} M_\mathfrak{p})/\operatorname{tor}(S_\mathfrak{p} \otimes_{R_\mathfrak{p}} M_\mathfrak{p})$.

Finally, let A and B be the total quotient rings of $\overline{R}$ and $\overline{S}$, respectively. We are thus faced with a commutative diagram of ring homomorphisms

$$(14.4) \qquad \begin{array}{ccc} R & \hookrightarrow & S \\ \downarrow & & \downarrow \\ \overline{R} & \hookrightarrow & \overline{S} \\ \downarrow & & \downarrow \\ A & \hookrightarrow & B \end{array}$$

in which the top square is a pullback. Furthermore, the bottom row is an Artinian pair in the sense of Chapter 3, and a MCM R-module yields a module over the Artinian pair, as we now show.

14.9. LEMMA. *Keep the notation established so far, and let M be a MCM R-module.*

(i) *We have $B = A \otimes_R S$, that is, $B = \{\text{non-zerodivisors}\}^{-1} \overline{S}$. In particular B is a finitely generated A-module.*

(ii) The natural homomorphism of B-modules

$$\theta_M \colon B \otimes_A (A \otimes_{\overline{R}} \overline{M}) \xrightarrow{\cong} B \otimes_S (S \otimes_R M) \longrightarrow B \otimes_S (S \cdot M)$$

is surjective.

(iii) The natural homomorphism of A-modules

$$A \otimes_{\overline{R}} \overline{M} \longrightarrow B \otimes_A (A \otimes_{\overline{R}} \overline{M}) \xrightarrow{\theta_M} B \otimes_S (S \cdot M)$$

is injective.

PROOF. For the first statement, set $C = U^{-1}\overline{S}$. Any $b \in B$ can be written $b = \frac{c}{v}$ where $c \in C$ and v is a non-zerodivisor of S. Since C is Artinian, there is an integer n such that $Cv^n = Cv^{n+1}$, say $v^n = dv^{n+1}$. Then $v^n(1 - dv) = 0$ so that $dv = 1$ in B. This shows that $b = dc \in C$.

The exact sequence (14.3), with $N = S \otimes_R M$, shows that the cokernel of the natural homomorphism $S \otimes_R M \longrightarrow S \cdot M$ has finite length. Hence that cokernel vanishes when we tensor with B and θ_M is surjective.

To prove (iii), set $N = (S \otimes_R M)/\operatorname{tor}(S \otimes_R M)$. Then the natural map $M \longrightarrow N$ sending $x \in M$ to $\overline{1 \otimes x}$ is injective. It follows that the restriction $\mathfrak{c}M \longrightarrow \mathfrak{c}N$ is also injective. In fact, it is also surjective: for any $a \in \mathfrak{c}$, $s \in S$, and $x \in M$, we have

$$s(\overline{s \otimes x}) = \overline{as \otimes x} = \overline{1 \otimes asx}$$

in the image of $\mathfrak{c}M$, since $as \in \mathfrak{c}$.

Since N is torsion-free, we have an exact sequence

$$0 \longrightarrow N \longrightarrow N^{\vee\vee} \longrightarrow L \longrightarrow 0$$

where the duals $(-)^\vee$ are computed over S and L is an S-module of finite length. It follows that the cokernel of the restriction $\mathfrak{c}N \hookrightarrow \mathfrak{c}N^{\vee\vee}$ also has finite length. Consider the composition $g \colon M \longrightarrow N \longrightarrow N^{\vee\vee}$ and the induced diagram

$$\begin{array}{ccccccccc}
0 & \longrightarrow & \mathfrak{c}M & \longrightarrow & M & \longrightarrow & \overline{M} & \longrightarrow & 0 \\
& & \downarrow f & & \downarrow g & & \downarrow h & & \\
0 & \longrightarrow & \mathfrak{c}N^{\vee\vee} & \longrightarrow & N^{\vee\vee} & \longrightarrow & \overline{N^{\vee\vee}} & \longrightarrow & 0
\end{array}$$

with exact rows, where f is the restriction of g to $\mathfrak{c}M$. Since g is injective and the cokernel of f has finite length, the Snake Lemma implies that $\ker h$ has finite length as well. Thus $A \otimes_{\overline{R}} h \colon A \otimes_{\overline{R}} \overline{M} \longrightarrow A \otimes_{\overline{R}} \overline{N^{\vee\vee}}$ is injective. Finally we observe that $A \otimes_{\overline{R}} h$ is the natural homomorphism in (iii), since $(S \cdot M)_\mathfrak{p} \cong (S_\mathfrak{p} \otimes_{R_\mathfrak{p}} M_\mathfrak{p})/\operatorname{tor}(S_\mathfrak{p} \otimes_{R_\mathfrak{p}} M_\mathfrak{p})$ for all primes $\mathfrak{p}$ minimal over $\mathfrak{c}$. $\square$

14.10. DEFINITION. Keeping all the notation introduced so far in this section, consider the following *category of Burban-Drozd triples* $\mathrm{BD}(R)$. The objects of $\mathrm{BD}(R)$ are triples (N, V, θ), where
- N is a MCM S-module,
- V is a finitely generated A-module, and
- $\theta \colon B \otimes_A V \longrightarrow B \otimes_S N$ is a surjective homomorphism between B-modules such that the composition
$$V \longrightarrow B \otimes_A V \xrightarrow{\theta} B \otimes_S N$$
is injective.

The induced map of A-modules $V \longrightarrow B \otimes_S N$ is called a *gluing map*.

A morphism between two triples (N, V, θ) and (N', V', θ') is a pair (f, F) such that $f \colon V \longrightarrow V'$ is a homomorphism of A-modules and $F \colon N \longrightarrow N'$ is a homomorphism of S-modules combining to make the diagram

$$\begin{array}{ccc} B \otimes_A V & \xrightarrow{\theta} & B \otimes_S N \\ {\scriptstyle 1 \otimes f} \downarrow & & \downarrow {\scriptstyle 1 \otimes F} \\ B \otimes_A V' & \xrightarrow{\theta'} & B \otimes_S N' \end{array}$$

commutative.

The category of Burban-Drozd triples is finer than the category of modules over the Artinian pair $A \hookrightarrow B$, since the homomorphism F above must be defined over S rather than just over B. In particular, an isomorphism of pairs $(f, F) \colon (V, N) \longrightarrow (V', N')$ includes as part of its data an isomorphism of S-modules $F \colon N \longrightarrow N'$, of which there are fewer than there are isomorphisms of B-modules $B \otimes_S N \longrightarrow B \otimes_S N'$.

14.11. THEOREM (Burban-Drozd). *Let R be a reduced CM complete local ring of dimension 2 which is not an isolated singularity. Let $\mathbb{F}$ be the functor from MCM R-modules to $\mathrm{BD}(R)$ defined on objects by*
$$\mathbb{F}(M) = (S \cdot M, A \otimes_R M, \theta_M).$$
Then $\mathbb{F}$ is an equivalence of categories.

Lemma 14.9 shows that the functor $\mathbb{F}$ is well-defined. The proof that it is an equivalence is somewhat technical. For the applications we have in mind, a more restricted version suffices.

14.12. ASSUMPTIONS. We continue to assume that R is a reduced, CM, complete local ring of dimension two and that $S \neq R$ is its integral closure in the total quotient ring. Let $\mathfrak{c}$ be the conductor and $\overline{R} = R/\mathfrak{c}$, $\overline{S} = S/\mathfrak{c}$. We impose two additional assumptions.

(i) Assume that S is a *regular ring*. Since R is Henselian, this is equivalent to S being a direct product of regular local rings. Every MCM S-module is thus projective.

(ii) Assume that $\overline{R} = R/\mathfrak{c}$ is also a *regular local ring*, that is, a DVR. It follows that $\overline{S}$ is a free $\overline{R}$-module, and even more, that a finitely generated $\overline{S}$ module is MCM if and only if it is free over $\overline{R}$. Also, the total quotient ring A of $\overline{R}$ is a field.

Under these simplifying assumptions, we define a category of modified Burban-Drozd triples $\mathrm{BD}'(R)$.

14.13. DEFINITION. Keep the assumptions established in 14.12. A *modified Burban-Drozd triple* $(N, X, \widetilde{\theta})$ consists of the following data:
- N is a finitely generated projective S-module;
- $X \cong \overline{R}^{(n)}$ is a free $\overline{R}$ module of finite rank; and
- $\widetilde{\theta} \colon X \longrightarrow \overline{N} = N \otimes_S \overline{S}$ is a *split* injection of $\overline{R}$-modules such that in the induced commutative square

$$\begin{array}{ccc} A \otimes_{\overline{R}} X & \xrightarrow{A \otimes_{\overline{R}} \widetilde{\theta}} & A \otimes_{\overline{R}} \overline{N} = (A \otimes_R S) \otimes_S \overline{N} \\ \downarrow & & \| \\ B \otimes_{\overline{R}} X & \longrightarrow & B \otimes_S \overline{N} \end{array}$$

the lower horizontal arrow is a split surjection. (The right-hand vertical arrow comes from Lemma 14.9(i).)

A morphism between modified triples $(N, X, \widetilde{\theta})$ and $(N', X', \widetilde{\theta}')$ is a pair $(f \colon X \longrightarrow X',\ F \colon N \longrightarrow N')$ such that $f \colon X \longrightarrow X'$ is a homomorphism between free $\overline{R}$-modules and $F \colon N \longrightarrow N'$ is a homomorphism of S-modules fitting into a commutative diagram

$$\begin{array}{ccc} B \otimes_{\overline{R}} X & \longrightarrow & B \otimes_S N \\ {\scriptstyle 1 \otimes f} \downarrow & & \downarrow {\scriptstyle 1 \otimes F} \\ B \otimes_{\overline{R}} X' & \longrightarrow & B \otimes_S N' \end{array}$$

where the horizontal arrows are induced by $\widetilde{\theta}$ and $\widetilde{\theta}'$, respectively.

Observe that if $(N, X, \widetilde{\theta})$ is a modified Burban-Drozd triple, then $(N,\ A \otimes_{\overline{R}} X, B \otimes_{\overline{R}} \widetilde{\theta})$ is a Burban-Drozd triple.

14.14. LEMMA. *Assume the hypotheses of 14.12, and let M be a MCM R-module. Then*

$$\mathcal{F}(M) = \left(S \cdot M,\ \overline{M}^{\vee\vee},\ \widetilde{\theta}_M \right)$$

is a modified Burban-Drozd triple, where $\widetilde{\theta}_M \colon \overline{M}^{\vee\vee} \longrightarrow \overline{S \cdot M}$ is the natural map.

PROOF. Since S is a regular ring of dimension 2, the reflexive S-module $S \cdot M$ is in fact projective. Furthermore, the natural homomorphism of R-modules $M \longrightarrow S \cdot M$ is obtained by applying $\operatorname{Hom}_R(-, M)$ to the short exact sequence $0 \longrightarrow \mathfrak{c} \longrightarrow R \longrightarrow \overline{R} \longrightarrow 0$. In particular, we have the short exact sequence

$$(14.5) \qquad 0 \longrightarrow M \longrightarrow S \cdot M \longrightarrow E \longrightarrow 0,$$

where $E = \operatorname{Ext}^1_R(\overline{R}, M)$. Since E is annihilated by $\mathfrak{c}$, it is naturally a $\overline{R}$-module, and has depth one over R by the Depth Lemma applied to (14.5). Since $\overline{R}$ is a DVR by assumption, this implies that E is a free $\overline{R}$-module. The induced exact sequence of $\overline{R}$-modules

$$\overline{M} \longrightarrow \overline{S \cdot M} \longrightarrow E \longrightarrow 0,$$

where overlines indicate passage modulo $\mathfrak{c}$, is thus split exact on the right.

The projective $\overline{S}$-module $\overline{S \cdot M}$ is torsion-free over $\overline{R}$, so there is a commutative diagram

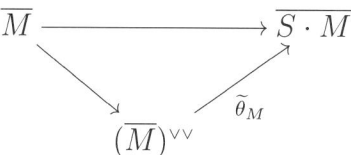

where as usual $-^{\vee}$ is canonical duality over R. Since $\overline{M}^{\vee\vee} \cong \overline{M}/\operatorname{tor}(\overline{M})$ by Exercise 14.31, we have $A \otimes_{\overline{R}} \overline{M} = A \otimes_{\overline{R}} (\overline{M})^{\vee\vee}$, so that

$$A \otimes_{\overline{R}} \widetilde{\theta}_M \colon A \otimes_{\overline{R}} \overline{M}^{\vee\vee} \longrightarrow A \otimes_{\overline{R}} \overline{S \cdot M} = B \otimes_{\overline{S}} \overline{S \cdot M}$$

is injective by Lemma 14.9(iii). This shows that the kernel of $\widetilde{\theta}_M$ is torsion, hence zero as $\overline{M}^{\vee\vee}$ is torsion-free. We therefore have the split-exact sequence of $\overline{R}$-modules

$$0 \longrightarrow \overline{M}^{\vee\vee} \xrightarrow{\widetilde{\theta}_M} \overline{S \cdot M} \longrightarrow E \longrightarrow 0.$$

In the induced commutative diagram

$$\begin{array}{c}
A \otimes_{\overline{R}} \overline{M}^{\vee\vee} \xrightarrow{A \otimes_{\overline{R}} \widetilde{\theta}_M} A \otimes_{\overline{R}} \overline{S \cdot M} = B \otimes_S N \\
\searrow \qquad \nearrow \\
B \otimes_{\overline{R}} \overline{M}^{\vee\vee}
\end{array}$$

the northeasterly arrow is surjective by Lemma 14.9(ii), and is even split surjective since $B \otimes_S N$ is projective over B. □

We now define a functor $\mathcal{G}$ from $\mathrm{BD}'(R)$ to MCM R-modules which is inverse to $\mathcal{F}$ on objects, still under the assumptions 14.12. Let $(N, X, \widetilde{\theta})$ be an object of $\mathrm{BD}'(R)$. Let $\pi \colon N \longrightarrow \overline{N} = N/\mathfrak{c}N$ be the natural projection, and define M by the pullback diagram

(14.6)
$$\begin{array}{ccc} M & \longrightarrow & N \\ \downarrow & & \downarrow \pi \\ X & \xrightarrow{\widetilde{\theta}} & \overline{N} \end{array}$$

of R-modules. Since $\widetilde{\theta}$ is a split injection of torsion-free modules over the DVR $\overline{R}$, its cokernel is an R-module of depth 1. This cokernel is isomorphic to the cokernel of $M \longrightarrow N$, and it follows that $\mathrm{depth}_R M = 2$, so that M is a MCM R-module. Define

$$\mathcal{G}(N, X, \widetilde{\theta}) = M \,.$$

14.15. THEOREM. *The functors $\mathcal{F}$ and $\mathcal{G}$ are inverses on objects, namely, for a MCM R-module M and a modified Burban-Drozd triple $(N, X, \widetilde{\theta})$, we have*

$$\mathcal{G}\mathcal{F}(M) \cong M$$

and

$$\mathcal{F}\mathcal{G}(N, X, \widetilde{\theta}) \cong (N, X, \widetilde{\theta}) \,.$$

PROOF. For the first assertion, it suffices to show that

$$\begin{array}{ccc} M & \longrightarrow & S \cdot M \\ \downarrow & & \downarrow \pi \\ (\overline{M})^{\vee\vee} & \xrightarrow{\widetilde{\theta}} & \overline{S \cdot M} \end{array}$$

is a pullback diagram. We have already seen that the homomorphisms $M \longrightarrow S \cdot M$ and $(\overline{M})^{\vee\vee} \longrightarrow \overline{S \cdot M}$ have the same cokernel, identified as $\mathrm{Ext}^1_R(\overline{R}, M)$. It follows from the Snake Lemma that

$$\ker \left(M \longrightarrow (\overline{M})^{\vee\vee} \right) \cong \ker \left(S \cdot M \longrightarrow \overline{S \cdot M} \right) \,.$$

From this it follows easily that M is the pullback of the diagram above.

For the converse, let $(N, X, \widetilde{\theta})$ be an object of $\mathrm{BD}'(R)$ and let M be defined by the pullback (14.6). Then $\mathrm{cok}\,(M \longrightarrow N)$ is isomorphic to $\operatorname{cok}\left(\widetilde{\theta}\colon X \longrightarrow \overline{N}\right)$, and is in particular an $\overline{R}$-module. The Snake Lemma applied to the diagram

$$\begin{array}{ccccccccc} 0 & \longrightarrow & \mathfrak{c}M & \longrightarrow & M & \longrightarrow & \overline{M} & \longrightarrow & 0 \\ & & \downarrow & & \downarrow & & \downarrow & & \\ 0 & \longrightarrow & \mathfrak{c}N & \longrightarrow & N & \longrightarrow & \overline{N} & \longrightarrow & 0 \end{array}$$

gives an exact sequence

$$0 \longrightarrow \ker(\overline{M} \longrightarrow \overline{N}) \longrightarrow \operatorname{cok}(\mathfrak{c}M \longrightarrow \mathfrak{c}N) \longrightarrow \operatorname{cok}(M \longrightarrow N).$$

This shows that $\operatorname{cok}(\mathfrak{c}M \longrightarrow \mathfrak{c}N)$ is annihilated by $\mathfrak{c}^2$, so in particular is a torsion R-module. Now the commutative diagram

$$\begin{array}{ccccccccc} 0 & \longrightarrow & \mathfrak{c}M & \longrightarrow & M & \longrightarrow & \overline{M} & \longrightarrow & 0 \\ & & \downarrow & & \| & & \downarrow & & \\ 0 & \longrightarrow & \mathfrak{c}N & \longrightarrow & M & \longrightarrow & X & \longrightarrow & 0 \end{array}$$

implies that $\overline{M} \longrightarrow X$ is surjective with torsion kernel. Therefore $X \cong \overline{M}/\operatorname{tor}(\overline{M}) \cong (\overline{M})^{\vee\vee}$.

The inclusion $M \hookrightarrow N$ induces a homomorphism $S \cdot M \longrightarrow N$ of reflexive S-modules, so in particular of reflexive R-modules. It suffices by Exercise 14.39 to prove that this is an isomorphism in codimension 1 in R, that is, $(S \cdot M)_\mathfrak{p} \longrightarrow N_\mathfrak{p}$ is an isomorphism for all height-one primes $\mathfrak{p} \in \operatorname{Spec} R$. Over $R_\mathfrak{p}$, the localization of (14.6) is still a pullback diagram.

$$\begin{array}{ccccc} M_\mathfrak{p} & \longrightarrow & (S \cdot M)_\mathfrak{p} & \longrightarrow & N_\mathfrak{p} \\ \downarrow & & & & \downarrow \\ (\overline{M})_\mathfrak{p} & & \longrightarrow & & \overline{N}_\mathfrak{p} \end{array}$$

Since $(S \cdot M)_\mathfrak{p} \cong (S_\mathfrak{p} \otimes_{R_\mathfrak{p}} M_\mathfrak{p})/\operatorname{tor}(S_\mathfrak{p} \otimes_{R_\mathfrak{p}} M_\mathfrak{p})$ and the bottom line is a module over the Artinian pair $A \hookrightarrow B$, we can use the machinery of Chapter 4 to see that $(S \cdot M)_\mathfrak{p} \cong N_\mathfrak{p}$. □

§3. Hypersurfaces of countable CM type

We now apply Theorem 14.15 to obtain the complete classification of indecomposable MCM modules over the two-dimensional (A_∞) and (D_∞) complete hypersurface singularities, and show in particular that (A_∞) and (D_∞) have countably infinite CM type in all dimensions. Then we will establish the following result of Knörrer and Buchweitz-Greuel-Schreyer:

14.16. THEOREM. *Assume that k is an uncountable algebraically closed field of characteristic different from 2, and let R be the hypersurface $k[\![x, y, x_2, \ldots, x_d]\!]/(f)$, where $0 \neq f \in (x, y, x_2, \ldots, x_d)^2$. These are equivalent:*

  (i) *R has countably infinite CM type;*
  (ii) *R is a countably simple singularity which is not simple, i.e. there is a countably infinite set of ideals L of $k[\![x, y, x_2, \ldots, x_d]\!]$ such that $f \in L^2$; and*
  (iii) *$R \cong k[\![x, y, x_2, \ldots, x_d]\!]/(g + x_2^2 + \cdots + x_d^2)$, where $g \in k[\![x, y]\!]$ is one of the following:*
  (A_∞) $g = x^2$; *or*
  (D_∞) $g = x^2 y$.

Observe that the equations defining the (A_∞) and (D_∞) hypersurface singularities are natural limiting cases of the (A_n) and (D_n) equations as $n \longrightarrow \infty$, since high powers of the variables are small in the $\mathfrak{m}$-adic topology. As we shall see, the same is true of the matrix factorizations over these singularities.

By Knörrer's Theorem 8.20, we may reduce the proof of the implication (iii) $\implies$ (i) of Theorem 14.16 to the case of dimension $d = 2$. Thus we prove in Propositions 14.17 and 14.19 that the hypersurface singularities defined by $x^2 + z^2$ and $x^2 y + z^2$, respectively, have countably infinite CM type.

14.17. PROPOSITION. *Let $R = k[\![x, y, z]\!]/(x^2 + z^2)$ be an (A_∞) hypersurface singularity with k an algebraically closed field of characteristic other than 2. Let $i \in k$ be a square root of -1. Let M be an indecomposable non-free MCM R-module. Then M is isomorphic to $\mathrm{cok}(zI - \varphi, zI + \varphi)$, where φ is one of the following matrices over $k[\![x, y]\!]$:*

- (ix) *or* $(-ix)$; *or*
- $\begin{pmatrix} -ix & y^j \\ 0 & ix \end{pmatrix}$ *for some $j \geq 1$.*

In particular R has countable CM type.

PROOF. For simplicity in the proof we replace x by ix to assume that
$$R = k[\![x, y, z]\!]/(z^2 - x^2).$$
The integral closure S of R is then
$$S = R/(z - x) \times R/(z + x)$$

with the normalization homomorphism $\nu \colon R \longrightarrow S = S_1 \times S_2$ given by the diagonal embedding $\nu(r) = (\overline{r}, \overline{r})$. In particular, S is a regular ring.

Put another way, S is the R-submodule of the total quotient ring generated by the orthogonal idempotents

$$e_1 = \frac{z+x}{2z} \in S_1 \quad \text{and} \quad e_2 = \frac{z-x}{2z} \in S_2,$$

which are the identity elements of S_1 and S_2 respectively. In these terms, $\nu(r) = r(e_1 + e_2)$ for $r \in R$.

The conductor of R in S is the ideal $\mathfrak{c} = (x,z)R = (x,z)S$, so that

$$\overline{R} = k[\![x,y,z]\!]/(x,z) \cong k[\![y]\!]$$

is a DVR, and $\overline{S} \cong \overline{R} \times \overline{R}$ is a direct product of two copies of $\overline{R}$. The inclusion $\overline{\nu} \colon \overline{R} \longrightarrow \overline{S}$ is again diagonal, $\overline{\nu}(\overline{r}) = (\overline{r}, \overline{r})$. Finally, the quotient field A of $\overline{R}$ is $k(\!(y)\!)$, which embeds diagonally into $B = k(\!(y)\!) \times k(\!(y)\!)$. Thus all the assumptions of 14.12 are verified, and we may apply Theorem 14.15.

Let $(N, X, \widetilde{\theta})$ be an object of $\mathrm{BD}'(R)$, so that $N \cong S_1^{(p)} \oplus S_2^{(q)}$ for some $p, q \geqslant 0$, while $X \cong \overline{R}^{(n)}$ for some n and $\widetilde{\theta} \colon X \longrightarrow \overline{N}$ is a split injection. The gluing morphism $\theta \colon B \otimes_{\overline{R}} X \longrightarrow B \otimes_S N$ is thus a linear transformation of A-vector spaces $B^{(n)} \longrightarrow B^{(p)} \oplus B^{(q)}$. More precisely, $\widetilde{\theta}$ defines a pair of matrices

$$(\theta_1, \theta_2) \in M_{p \times n}(A) \times M_{q \times n}(A)$$

representing an embedding

$$\theta = \begin{pmatrix} \theta_1 \\ \theta_2 \end{pmatrix} \colon A^{(n)} \longrightarrow B^{(p)} \oplus B^{(q)}$$

such that θ is injective (has full column rank) and both θ_1 and θ_2 are surjective (full row rank). Thus in particular we have $\max(p,q) \leqslant n \leqslant p + q$.

Two pairs of matrices (θ_1, θ_2) and (θ'_1, θ'_2) define isomorphic modified Burban-Drozd triples if and only if there exist isomorphisms

$$f \colon A^{(n)} \longrightarrow A^{(n)}$$
$$F_1 \colon S_1^{(p)} \longrightarrow S_1^{(p)}$$
$$F_2 \colon S_2^{(q)} \longrightarrow S_2^{(q)}$$

such that as homomorphisms $B^{(n)} \longrightarrow B^{(p)}$ and $B^{(n)} \longrightarrow B^{(q)}$ we have

$$\theta'_1 = F_1^{-1} \theta_1 f$$
$$\theta'_2 = F_2^{-1} \theta_2 f.$$

§3. HYPERSURFACES OF COUNTABLE CM TYPE

See Exercise 14.37 for a guided proof of the next lemma.

14.18. LEMMA. *The indecomposable objects of* $\mathrm{BD}'(R)$ *are*

(i) $(S_1, \overline{R}, ((1), \emptyset))$ *and* $(S_2, \overline{R}, (\emptyset, (1)))$
(ii) $(S_1 \times S_2, \overline{R}, ((1), (1)))$
(iii) $(S_1 \times S_2, \overline{R}, ((1), (y^j)))$ *and* $(S_1 \times S_2, \overline{R}, ((y^j), (1)))$ *for some* $j \geqslant 1$.

Now we derive the matrix factorizations corresponding to the modified Burban-Drozd triples listed above. The pullback diagram corresponding to the triple $(S_1, \overline{R}, ((1), \emptyset))$

$$\begin{array}{ccc} M & \longrightarrow & S_1 \\ \downarrow & & \downarrow \\ \overline{R} & \xrightarrow[\binom{1}{\emptyset}]{} & \overline{S_1} \end{array}$$

clearly gives $M \cong S_1 = \operatorname{cok}(z - x, z + x)$, the first component of the integral closure. Similarly, the modified triple $(S_2, \overline{R}, (\emptyset, (1)))$ yields $M \cong S_2 = \operatorname{cok}(z + x, z - x)$.

The diagonal map $((1),(1)): \overline{R} \longrightarrow \overline{S}_1 \times \overline{S}_2$ obviously defines the free module R. By symmetry, it suffices now to consider the modified Burban-Drozd triple $(S_1 \times S_2, \overline{R}, ((1), (y^j)))$. The pullback diagram

$$\begin{array}{ccc} M & \longrightarrow & S_1 \times S_2 \\ \downarrow & & \downarrow \\ \overline{R} & \xrightarrow[\binom{1}{y^j}]{} & \overline{S}_1 \times \overline{S}_2 \end{array}$$

defines M as the module of ordered triples of polynomials

$$(f(y), g_1(x, y, x), g_2(x, y, -x)) \in \overline{R} \times S_1 \times S_2$$

such that $f - g_1 \in \mathfrak{c}S_1$ and $y^j f - g_2 \in \mathfrak{c}S_2$. This is equal to the R-submodule of S generated by $\mathfrak{c} = (x, z) = (z + x, z - x)$ and $e_1 + y^j e_2$, where again $e_1 = (z + x)/2z$ and $e_2 = (z - x)/2z$ are idempotent. Multiplying by the non-zerodivisor $(2z)^j = ((z - x) + (z + x))^j$ to knock the generators down into R, we find

$$(x, z, e_1 + y^j e_2)S \cong (2z)^j \left(z + x, z - x, \frac{z+x}{2z} + y^j \frac{z-x}{2z} \right)$$

$$= \left((z+x)^{j+1}, (z-x)^{j+1}, (2z)^j \left(\frac{z+x}{2z} \right)^j + y^j (2z)^j \left(\frac{z-x}{2z} \right)^j \right)$$

$$= \left((z+x)^{j+1}, (z-x)^{j+1}, (z+x)^j + (z-x)^j y^j \right)$$

$$= \left((z-x)^j, (z+x)^j + (z-x)^j y^j \right).$$

The matrix factorization

$$\left(\begin{pmatrix} z+x & y^j \\ 0 & z-x \end{pmatrix}, \begin{pmatrix} z-x & -y^j \\ 0 & z+x \end{pmatrix} \right)$$

provides a minimal free resolution of this ideal and finishes the proof. □

As an aside, we note that the restriction on the characteristic of k could be removed by working instead with the hypersurface defined by xz instead of $x^2 + z^2$. In characteristic not two, of course the hypersurface singularities are isomorphic, and $k[\![x, y, z]\!]/(xz)$ can be shown to have countable type in all characteristics.

14.19. PROPOSITION. *Let $R = k[\![x, y, z]\!]/(x^2 y + z^2)$ be a (D_∞) hypersurface singularity, where k is a field of arbitrary characteristic. Let M be an indecomposable non-free MCM R-module. Then M is isomorphic to $\mathrm{cok}(zI - \varphi, zI + \varphi)$ for φ one of the following matrices over $k[\![x, y]\!]$.*

- $\begin{pmatrix} 0 & -y \\ x^2 & 0 \end{pmatrix}$
- $\begin{pmatrix} 0 & -xy \\ x & 0 \end{pmatrix}$
- $\begin{pmatrix} -xy & 0 \\ -y^{j+1} & xy \\ x & 0 \\ y^j & -x \end{pmatrix}$ *for some $j \geq 1$;*
- $\begin{pmatrix} -xy & 0 \\ -y^j & x \\ x & 0 \\ y^j & -xy \end{pmatrix}$ *for some $j \geq 1$.*

In particular R has countable CM type.

PROOF. In this case, the integral closure of R is obtained by adjoining the element $t = \frac{z}{x}$ of the quotient field, so $S = R\left[\frac{z}{x}\right]$. The

§3. HYPERSURFACES OF COUNTABLE CM TYPE

maximal ideal of R is then $(x,y,z)R = (x,t^2,tx)R$ and that of S is $(x,t)S$. In particular, S is a regular local ring. The conductor is now $\mathfrak{c} = (x,z)R = (x,tx)S = xS$, so that $\overline{R} = R/(x,z) \cong k[\![t^2]\!]$ and $\overline{S} = S/(x) \cong k[\![t]\!]$ are both DVRs, with $\overline{\nu}\colon \overline{R} \longrightarrow \overline{S}$ the obvious inclusion. The Artinian pair $A = k((t^2)) \longrightarrow B = k((t))$ is thus a field extension of degree 2.

Let $(N, X, \widetilde{\theta})$ be an object of $\mathrm{BD}'(R)$. The integral closure S being regular local, $N \cong S^{(n)}$ is a free S-module, while $X \cong \overline{R}^{(m)}$ is a free $\overline{R}$-module. The gluing map $\theta \colon B \otimes_A V \cong B^{(m)} \longrightarrow B^{(n)} \cong B \otimes_S N$ is thus simply an $n \times m$ matrix over B with full row rank. The condition that the composition $A^{(m)} \longrightarrow B^{(n)}$ be injective amounts to writing $\theta = \theta_0 + t\theta_1$ and requiring $\binom{\theta_0}{\theta_1} \colon A^{(m)} \longrightarrow A^{(2n)} \cong B^{(n)}$ to have full column rank as a matrix over A. In particular we have $n \leqslant m \leqslant 2n$.

Two $n \times m$ matrices θ, θ' over B define isomorphic modified Burban-Drozd triples if and only if there exist isomorphisms

$$f \colon A^{(m)} \longrightarrow A^{(m)} \quad \text{and} \quad F \colon S^{(n)} \longrightarrow S^{(n)}$$

such that, when considered as matrices over B, we have

$$\theta' = F^{-1}\theta f.$$

In other words, we are allowed to perform row operations over $\overline{S} = k[\![t]\!]$ and column operations over $A = k((t^2))$.

14.20. LEMMA. *The indecomposable objects of* $\mathrm{BD}'(R)$ *are*

(i) $\left(S, \overline{R}, (1)\right)$
(ii) $\left(S, \overline{R}, (t)\right)$
(iii) $\left(S, \overline{R}^{(2)}, (1\ t)\right)$
(iv) $\left(S^{(2)}, \overline{R}^{(2)}, \left(\begin{smallmatrix} 1 & t \\ t^m & 0 \end{smallmatrix}\right)\right)$ *for some* $m \geqslant 1$.

We leave the proof of Lemma 14.20 as Exercise 14.38.

The MCM R-module corresponding to $(S, \overline{R}, (1))$ is given by the pullback

$$\begin{array}{ccc} M & \longrightarrow & S \\ \downarrow & & \downarrow \\ \overline{R} & \longrightarrow & \overline{S} \end{array}$$

where the bottom line is the given inclusion of $A = k((t^2))$ into $B = k((t))$, so is clearly the free module R. In $(S, \overline{R}, (t))$, the natural inclusion is replaced by multiplication by t. The pullback M is the R-submodule of S generated by $\mathfrak{c} = (x, z)$ and $t = \frac{z}{x}$. Multiplying through

by the non-zerodivisor x, we find
$$M \cong (x^2, xz, z)R$$
$$= (x^2, z)R$$
$$\cong \operatorname{cok}\left(\begin{pmatrix} z & y \\ -x^2 & z \end{pmatrix}, \begin{pmatrix} z & -y \\ x^2 & z \end{pmatrix}\right).$$

The modified Burban-Drozd triple $(S, \overline{R}, (1\ t))$ is defined by the isomorphism $\theta \colon A \xrightarrow{(1\ t)} B$, so corresponds to the integral closure S, which has matrix factorization
$$\left(\begin{pmatrix} z & xy \\ -x & z \end{pmatrix}, \begin{pmatrix} z & -xy \\ x & z \end{pmatrix}\right).$$

Finally, let $m \geqslant 1$ and let M be the R-module defined by the pullback diagram
$$\begin{array}{ccc} M & \longrightarrow & S^2 \\ \downarrow & & \downarrow \\ \overline{R}^{(2)} & \xrightarrow{\begin{pmatrix} 1 & t \\ t^m & 0 \end{pmatrix}} & \overline{S}^{(2)} \end{array}.$$

Then M is the R-submodule of $S^{(2)}$ generated by $\mathfrak{c}S^{(2)}$ and the elements
$$\begin{pmatrix} 1 \\ t^m \end{pmatrix}, \begin{pmatrix} t \\ 0 \end{pmatrix}.$$

Substitute $t = \frac{z}{x}$ to see that the generators are therefore
$$\begin{pmatrix} x \\ 0 \end{pmatrix}, \begin{pmatrix} z \\ 0 \end{pmatrix}, \begin{pmatrix} 0 \\ x \end{pmatrix}, \begin{pmatrix} 0 \\ z \end{pmatrix}, \begin{pmatrix} 1 \\ z^m/x^m \end{pmatrix}, \quad \text{and} \quad \begin{pmatrix} z/x \\ 0 \end{pmatrix}.$$

Notice that the second generator is a multiple of the last. Multiplication by x on the first component and x^m on the second is injective on S^2, so M_1 is isomorphic to the module generated by
$$\begin{pmatrix} x^2 \\ 0 \end{pmatrix}, \begin{pmatrix} 0 \\ x^{m+1} \end{pmatrix}, \begin{pmatrix} 0 \\ x^m z \end{pmatrix}, \begin{pmatrix} x \\ z^m \end{pmatrix}, \quad \text{and} \quad \begin{pmatrix} z \\ 0 \end{pmatrix}.$$

Observe that
$$\begin{pmatrix} x^2 \\ 0 \end{pmatrix} = x \begin{pmatrix} x \\ z^m \end{pmatrix} - \begin{pmatrix} 0 \\ xz^m \end{pmatrix},$$

so we may replace the first generator by $\begin{pmatrix} 0 \\ xz^m \end{pmatrix}$, getting
$$M = \left\langle \begin{pmatrix} 0 \\ xz^m \end{pmatrix}, \begin{pmatrix} 0 \\ x^{m+1} \end{pmatrix}, \begin{pmatrix} 0 \\ x^m z \end{pmatrix}, \begin{pmatrix} x \\ z^m \end{pmatrix}, \begin{pmatrix} z \\ 0 \end{pmatrix} \right\rangle.$$

At this point we distinguish two cases. If $m = 2j$ is even, then using the relation $xy^2 = -z^2$ in R,
$$xz^m = xz^{2j} = xx^{2j}y^j = x^{m+1}y^j$$
up to sign, so the first generator is a multiple of the second. If $m = 2j+1$ is odd, then
$$xz^m = xz^{2j+1} = xx^{2j}y^j z = x^{m+1}y^j z$$
again up to sign, so that again the first generator is a multiple of the second. In either case, M is generated by
$$\left\langle \begin{pmatrix} x \\ z^m \end{pmatrix}, \begin{pmatrix} 0 \\ x^m z \end{pmatrix}, \begin{pmatrix} z \\ 0 \end{pmatrix}, \begin{pmatrix} 0 \\ x^{m+1} \end{pmatrix}, \right\rangle.$$
Now it's easy to check that when m is odd, say $m = 2j+1$ for some $j \geqslant 0$,
$$M \cong \operatorname{cok} \left(\begin{pmatrix} z & & -xy & 0 \\ & z & -y^{j+1} & x \\ x & 0 & z & \\ y^{j+1} & -xy & & z \end{pmatrix}, \begin{pmatrix} z & & xy & 0 \\ & z & y^{j+1} & -x \\ -x & 0 & & z \\ -y^{j+1} & xy & & z \end{pmatrix} \right)$$
and in case m is even, say $m = 2j$ for some $j \geqslant 1$,
$$M \cong \operatorname{cok} \left(\begin{pmatrix} z & & -xy & 0 \\ & z & -y^{j+1} & xy \\ x & 0 & z & \\ y^j & -x & & z \end{pmatrix}, \begin{pmatrix} z & & xy & \\ & z & y^{j+1} & -xy \\ -x & 0 & & z \\ -y^j & x & & z \end{pmatrix} \right)$$
(after a permutation of the generators). $\square$

Together with Theorem 8.20, Propositions 14.17 and 14.19 show that the (A_∞) and (D_∞) hypersurface singularities have countable CM type in all dimensions. To show that these are the only ones and complete the proof of Theorem 14.16, we need the following classification of countably simple singularities (the proof of which is considerably simpler than the corresponding classification for simple singularities on pages 145–150).

14.21. THEOREM. *Let k be an algebraically closed field of characteristic different from 2, and let $R = k[\![x,y]\!]/(f)$ be a one-dimensional complete hypersurface singularity over k. Assume k is uncountable. If R is a countably simple but not simple singularity, then either $R \cong k[\![x,y]\!]/(x^2)$ or $R \cong k[\![x,y]\!]/(x^2 y)$.*

PROOF. By Lemma 9.3, we see that $\operatorname{e}(R) \leqslant 3$ and $f \notin (\alpha, \beta^2)^3$ for every $\alpha, \beta \in (x, y)$. If in addition R is reduced, then by Remark 9.13 it is a simple singularity. Hence we may assume that in the irreducible factorization $f = uf_1^{e_1} \cdots f_r^{e_r}$, with u a unit and the f_i distinct irreducibles, we have $e_i \geqslant 2$ for at least one i. Say $e_1 \geqslant 2$. Since f is not divisible by any cube (by Lemma 9.3(iia)) we must have $e_1 = 2$. Since the multiplicity of R is at most 3, we must have $r \leqslant 2$ and that each f_i has non-zero linear term. Make the linear change of variable sending $\sqrt{u}f_1$ to x, so that now $f = x^2 f_2^{e_2}$ with $e_2 = 0$ or 1. Now if $e_2 = 0$ we have $f = x^2$, while if $e_2 = 1$ we make the change of variables sending f_2 to y, so that $f = x^2 y$. □

Now we finish the proof of the main theorem.

PROOF OF THEOREM 14.16. If $R = k[\![x, y, x_2, \ldots, x_d]\!]/(f)$ is of countably infinite CM type, then R is a countably simple singularity but not simple by Theorem 9.2.

To prove that countably simplicity implies one of the forms listed in item (iii), we may, as in the proof of Theorem 9.8, reduce to the case of dimension one, where Theorem 14.21 finishes.

Finally, Propositions 14.17 and 14.19 show that the (A_∞) and (D_∞) singularities have countably infinite CM type, completing the proof. □

We remarked above that the equations defining the (A_∞) and (D_∞) hypersurface singularities, and even their matrix factorizations, are "natural limits" of the cases (A_n) and (D_n). This suggests the following question.

14.22. QUESTION. *Must every CM local ring of countable CM type be a "natural limit" of a "series of singularities" of finite CM type? For those that are, are the indecomposable MCM modules "limits" of MCM modules over singularities in the series?*

To address the question, of course, the first order of business must be to give meaning to the phrases in quotation marks. This is problematic, as Arnold remarked [**Arn81**]: "Although the series undoubtedly exist, it is not at all clear what a series of singularities is."

§4. Other examples

Besides the hypersurface examples of the last section, very few nontrivial examples of countable CM type are known. In this section we present a few, taken from Schreyer's survey article [**Sch87**].

In dimension one, we have the following example, which will return triumphantly in Chapter 17.

14.23. EXAMPLE. Let k be an arbitrary field, and consider the one-dimensional (D_∞) hypersurface singularity $R = k[\![x,y]\!]/(x^2 y)$ over k.

§4. OTHER EXAMPLES

Set $E = \operatorname{End}_R(\mathfrak{m})$, where $\mathfrak{m} = (x,y)$ is the maximal ideal. Then we claim that
$$E \cong k[\![x,y,z]\!]/(yz, x^2 - xz, xz - z^2)$$
$$\cong k[\![a,b,c]\!]/(ab, ac, c^2).$$

In particular E is local, and it follows from Proposition 4.14 that R has countable CM type.

That the two alleged presentations of E are isomorphic is a simple matter of a linear change of variables:
$$a = z, \qquad b = y, \qquad c = x - z.$$

To show that in fact E is isomorphic to $A = k[\![x,y,z]\!]/(yz, x^2 - xz, xz - z^2)$, note that the element $x + y \in R$ is a non-zerodivisor, and that the fraction $z := \frac{x^2}{x+y}$ is in $\operatorname{End}_R(\mathfrak{m})$ but not in R. Now, $E = \operatorname{Hom}_R(\mathfrak{m}, R)$ since $\mathfrak{m}$ does not have a non-zero free direct summand, and it follows by duality over the Gorenstein ring R that $E/R \cong \operatorname{Ext}^1_R(R/\mathfrak{m}, R) \cong k$. Therefore $E = R[z]$. Since
$$z^2 = \frac{x^2(x+y)^2}{(x+y)^2} = x^2 \in \mathfrak{m},$$

E is local. One verifies the relations $yz = 0$ and $x^2 = xz = z^2$ in E. Thus we have a surjective homomorphism of R-algebras $A \longrightarrow E$. Since R is a subring of E, and the inclusion $R \hookrightarrow E$ factors through A, we see that R is also a subring of A, and that the surjection $A \longrightarrow E$ fixes R.

The induced homomorphism $A/R \longrightarrow E/R$ is still surjective, and in fact is bijective since A/R is simple as well. It follows from the Five Lemma that $A \longrightarrow E$ is an isomorphism.

By Lemma 4.9, the indecomposable MCM E-modules are precisely the non-free indecomposable MCM R-modules. These turn out to be exactly the cokernels of the following matrices over $R = k[\![x,y]\!]/(x^2 y)$:

$$(y); \quad (x^2); \quad (x); \quad (xy)$$

$$\begin{pmatrix} x & \\ y^j & -x \end{pmatrix}; \quad \begin{pmatrix} xy & \\ y^j & -xy \end{pmatrix}; \quad \begin{pmatrix} x & \\ y^j & -xy \end{pmatrix}; \quad \begin{pmatrix} xy & \\ y^j & -x \end{pmatrix}$$

for $j \geqslant 1$. To see this, note that, for each of the $R^\#$-modules M of Proposition 14.19, $M^\flat$ decomposes as a direct sum of two modules. These are the cokernels of the matrices on the list above. One can argue directly that each of these modules is indecomposable. (In characteristic different from two, indecomposability follows from Corollary 8.19.) By Proposition 8.18, the list is complete.

For two-dimensional examples, we note that Herzog's result Proposition 6.2 implies the following.

14.24. PROPOSITION. *Let S be a two-dimensional CM complete local ring which is Gorenstein in codimension one. (For example, S could be one of the two-dimensional (A_∞) and (D_∞) hypersurface singularities.) Let G be a finite group with order invertible in S, acting by linear changes of variables on S. Set $R = S^G$. If S has countable CM type, then R has countable CM type.* □

14.25. EXAMPLE. Fix an integer $r \geqslant 2$. R be the two-dimensional (A_∞) hypersurface $R = k[\![x,y,z]\!]/(xy)$, and let the cyclic group $\mathbb{Z}/r\mathbb{Z}$ act on R, the generator sending (x,y,z) to $(x, \zeta_r y, \zeta_r z)$, where ζ_r is a primitive r^{th} root of unity. (Here k is an algebraically closed field of characteristic prime to r.) The invariant subring is generated by $x, y^r, y^{r-1}z, \ldots, z^r$ (see Exercise 5.35), and is thus isomorphic to the quotient of $k[\![t_0, t_1, \ldots, t_r, x]\!]$ by the 2×2 minors of

$$\begin{pmatrix} t_0 & \cdots & t_{r-1} & 0 \\ t_1 & \cdots & t_r & x \end{pmatrix}.$$

14.26. EXAMPLE. Fix an odd integer $r = 2m+1$, and let R be the two-dimensional (D_∞) hypersurface $k[\![x,y,z]\!]/(x^2y - z^2)$, where k is a field with characteristic prime to r. Let $r = 2m+1$ be an odd positive integer, and let $\mathbb{Z}/r\mathbb{Z}$ act on R by the action sending $(x,y,z) \mapsto (\zeta_r^2 x, \zeta_r^{-1} y, \zeta_r^{m+2} z)$.

The ring of invariants is complicated to describe in general; see Exercise 5.36. If $m = 1$, it is generated by x^3, xy^2, y^3, z and hence is isomorphic to

$$k[\![a,b,c,z]\!] \Big/ I_2 \begin{pmatrix} a & z^2 & b \\ z^2 & b & c \end{pmatrix}.$$

If $m = 2$, there are 7 generating invariants

$$x^5, x^3y, x^3z, xy^2, xyz, y^5, y^4z,$$

and 15 relations among them. When $m = 4$, the greatest common divisor of $m+2$ and $2m+1$ is no longer 1, and things get really weird.

14.27. REMARK. As Schreyer points out, the phenomenon observed in Question 14.22 repeats here. The one-dimensional example E is obtained as a limit of the endomorphism rings of the maximal ideals of the (D_n) hypersurface singularities:

$$\operatorname{End}_{D_n}(\mathfrak{m}) \cong k[\![x,y,z]\!]/I_n,$$

where I_n is the ideal of 2×2 minors of $\begin{pmatrix} y & x-z & 0 \\ x-z & y^n & z \end{pmatrix}$.

Similarly, for Example 14.25 we may take the limit of the quotients of $k[\![t_0, t_1, \ldots, t_{r+1}]\!]$ by the 2×2 minors of

$$\begin{pmatrix} t_0 & \cdots & t_{r-1} & t_r^n \\ t_1 & \cdots & t_r & t_{r+1} \end{pmatrix},$$

and for Example 14.26 with $m = 1$, we take the quotient of $k[\![a, b, c, d]\!]$ by the 2×2 minors of

$$\begin{pmatrix} d^2 + a^n & c & b \\ b & d^2 & a \end{pmatrix}.$$

As assured by Theorem 7.19, both of these are invariant rings of a finite group acting on power series, the first for a cyclic group action $\mathcal{C}_{nr-n+1,n}$, and the second by a binary dihedral $\mathcal{D}_{2+3n, 2+2n}$ (cf. [**Sch87**, **Rie81**]).

These examples add some strength to Question 14.22. We also mention the related question, also first asked by Schreyer in [**Sch87**]:

14.28. QUESTION. *Is every CM local ring of countable CM type a quotient of one of the (A_∞) or (D_∞) hypersurface singularities by a finite group action?*

Burban and Drozd have recently announced a negative answer to this question [**BD10**]. Namely, set

$$A_{m,n} = k[\![x_1, x_2, y_1, y_2, z]\!] \Big/ (x_1 y_1, x_1 y_2, x_2 y_1, x_2 y_2, x_1 z - x_2^n, y_1 z - y_2^m).$$

Then $A_{m,n}$ has countable CM type for every $n, m \geq 0$. For $n = m$ this ring is isomorphic to a ring of invariants of the (A_∞) hypersurface, but for $m \neq n$ it is not.

§5. Exercises

14.29. EXERCISE. Let $R = \mathbb{Q}[x, y, z]_{(x,y,z)}/(x^2)$. The completion $\widehat{R} = \mathbb{Q}[\![x, y, z]\!]/(x^2)$ has a two-dimensional singular locus and therefore has uncountable CM type. Show that only countably many indecomposable $\widehat{R}$-modules are used in direct-sum decompositions of modules of the form $\widehat{R} \otimes_R M$, for MCM R-modules M. Thus the set $\mathcal{U}$ in the proof of Theorem 10.1 is properly contained in the set of all MCM $\widehat{R}$-modules.

14.30. EXERCISE. Let R be a one-dimensional CM local ring with canonical module ω, and let M be a finitely generated R-module. Prove that $M^{\vee\vee} \cong M/\operatorname{tor}(M)$.

14.31. EXERCISE. Let R be a two-dimensional local ring which is Gorenstein on the punctured spectrum. Let M be a finitely generated R-module. Prove that there is an exact sequence
$$0 \longrightarrow \operatorname{tor}(M) \longrightarrow M \xrightarrow{\sigma_M} M^{**} \longrightarrow L \longrightarrow 0 \,,$$
where σ_M is the biduality homomorphism, defined by $\sigma_M(m)(f) = f(m)$, and L is a module of finite length. If R is CM with canonical module ω, prove that there is also an exact sequence
$$0 \longrightarrow \operatorname{tor}(M) \longrightarrow M \xrightarrow{\tau_M} M^{\vee\vee} \longrightarrow L' \longrightarrow 0 \,,$$
where τ_M is again the corresponding evaluation homomorphism and L' also has finite length.

14.32. EXERCISE. Let R be a reduced local ring satisfying Serre's condition (S_2) and let M and N be two finitely generated R-modules. Assume that N is reflexive. Prove that
$$\operatorname{Hom}_R(M,N) = \operatorname{Hom}_R(M^{**},N) \,.$$
If in addition R is CM with canonical module ω, then
$$\operatorname{Hom}_R(M,N) = \operatorname{Hom}_R(M^{\vee\vee},N) \,.$$
(Hint: first reduce to the torsion-free case.)

14.33. EXERCISE. Let R be a reduced CM two-dimensional local ring with canonical module ω. Assume that R is Gorenstein in codimension one. Prove that there is a natural isomorphism $M^{\vee\vee} \longrightarrow M^{**}$.

14.34. EXERCISE. Let R be a reduced Noetherian ring and assume that the integral closure S is a finitely generated R-module. Let $\mathfrak{c}$ be the conductor. Prove that $S = \operatorname{End}_R(\mathfrak{c})$.

14.35. EXERCISE. Let R and S be as in 14.7 and let N be a finitely generated S-module. Prove that
$$\operatorname{Hom}_S(\operatorname{Hom}_S(N,S),S) \cong \operatorname{Hom}_R(\operatorname{Hom}_R(N,R),R) \,.$$

14.36. EXERCISE. Let R be a CM local ring and M a reflexive R-module which is locally free in codimension one. Let N be an arbitrary finitely generated R-module, and let $M \cdot N$ denote the reflexive product of M and N (cf. Exercise 6.48). Show that $M \cdot N \cong \operatorname{Hom}_R(M^*,N)$. Conclude that $S \cdot N \cong \operatorname{Hom}_R(\mathfrak{c},N)$ in the setup of 14.7.

14.37. EXERCISE. Prove Lemma 14.18, that the modified Burban-Drozd triples listed there constitute a full set of representatives for the indecomposables of $\operatorname{BD}'(R)$, along the following lines.

- The listed forms are pairwise non-isomorphic and cannot be further decomposed.

- Every object of $\mathrm{BD}'(R)$ splits into direct summands with either $n = p = q$ or $n = p+q$. (Consider the complement of $(\ker \theta_1) + \ker(\theta_2)$ in $A^{(n)}$.)
- In the case $n = p + q$, the object further splits into direct summands with either $n = p$ or $n = q$. Any triple with $n = p$ or $n = q$ can be completely diagonalized, giving one of the factors of the integral closure.

14.38. EXERCISE. Prove Lemma 14.20, that the modified Burban-Drozd triples listed there form a full set of representatives for the indecomposables of $\mathrm{BD}'(R)$, along the following lines.
- The listed forms are pairwise non-isomorphic and cannot be further decomposed.
- The $m \times n$ matrix θ can be reduced (using the allowable moves: row operations over $\overline{S}$ and column operations over A) to the block form
$$\begin{pmatrix} t^{d_1} I_{s_1} & A_{1,2} & \cdots & A_{1,\nu} & A_{1,\nu+1} \\ & t^{d_2} I_{s_2} & \cdots & A_{2,\nu} & A_{2,\nu+1} \\ & & \ddots & \vdots & \vdots \\ & & & t^{d_\nu} I_{s_\nu} & A_{\nu,\nu+1} \end{pmatrix}$$
where
  - $d_1 < d_2 < \cdots < d_\nu$ and $d_1 = 0$ or 1.
  - Each entry of $A_{i,j}$ has order in t at least d_i+1 for $1 \leqslant i \leqslant \nu$ and $1 \leqslant j \leqslant \nu + 1$.
  - Each entry of $A_{i,j}$ has order in t at most d_j for $1 \leqslant i \leqslant \nu$ and $1 \leqslant j \leqslant \nu$.
- If $A_{1,j} = 0$ for all $j = 2, \ldots, \nu + 1$, then either (1) or (t) is a direct summand of θ and we are done by induction on the number of rows.
- If $A_{1,j} \neq 0$ for some $j \leqslant \nu$, write $A_{1,j} = t^{d_1} B_{1,j}$ for some matrix $B_{1,j}$ with entries in $k[\![t]\!]$. Show that we may assume $B_{1,j}$ has entries in $k[\![t^2]\!]$, and then diagonalize over $k[\![t^2]\!]$ to assume $B_{1,j} = \begin{pmatrix} I_{s'} & 0 \\ 0 & 0 \end{pmatrix}$. If $s' = 0$, return to the previous step, while if $s' > 0$, split off one of
$$\begin{pmatrix} 1 & t \\ 0 & t^{d_j} \end{pmatrix} \quad \text{or} \quad \begin{pmatrix} t & t^2 \\ 0 & t^{d_j} \end{pmatrix}.$$
- Consider two cases for each of the above matrices: $d_j = 1$ versus $d_j \neq 1$ in the first matrix, and $d_j = 2$ versus $d_j \neq 2$ in the second. Split off one of the forms listed in Lemma 14.20 in each case.

- Finally, if $A_{1,j} = 0$ for all $j = 2, \ldots, \nu$ but $A_{1,\nu+1} \neq 0$, then one of (1), (t), $(1\ t)$, or $(t\ t^2) \sim (1\ t)$ is a direct summand of θ.

14.39. EXERCISE. Generalize Lemma 5.11 as follows. Let R be a reduced ring satisfying Serre's condition (S_2), and let $f \colon M \longrightarrow N$ be a homomorphism of R-modules, each of which satisfies (S_2). Then f is an isomorphism if and only if $f_\mathfrak{p} \colon M_\mathfrak{p} \longrightarrow N_\mathfrak{p}$ is an isomorphism for every height-one prime $\mathfrak{p}$ of R.

CHAPTER 15

The Brauer-Thrall Conjectures

In a brief abstract published in the 1941 Bulletin of the AMS, Brauer announced that he had found sufficient conditions for a finite-dimensional algebra A over a field k to have infinitely many non-isomorphic indecomposable finitely generated modules [**Bra41**]. Some years later, Thrall claimed similar results [**Thr47**]: he wrote that Brauer had in fact given three conditions, each sufficient to ensure that A has indecomposable modules of arbitrarily high k-dimension, and he gave a fourth sufficient condition. These were stated in terms of the so-called "Cartan invariants" [**ANT44**, p. 106] of the rings A, $A/\mathcal{J}(A)$, $A/\mathcal{J}(A)^2$, etc. Neither Brauer nor Thrall ever published the details of their work, leaving it to Thrall's student Jans to publish them. Jans attributes the following conjectures [**Jan57**] to both Brauer and Thrall. Say that a finite-dimensional k-algebra A has *bounded representation type* if the k-dimensions of indecomposable finitely generated A-modules are bounded, and *strongly unbounded representation type* if A has infinitely many pairwise non-isomorphic modules of k-dimension n for infinitely many n.

15.1. CONJECTURE (Brauer-Thrall Conjectures). *Let A be a finite-dimensional algebra over a field k.*
  I. *If A has bounded representation type then A actually has finite representation type.*
  II. *Assume that k is infinite. If A has unbounded representation type, then A has strongly unbounded representation type.*

Under mild hypotheses, both of these conjectures are now theorems. Brauer-Thrall I was proved by Roĭter [**Roĭ68**], while Brauer-Thrall II for perfect fields k is due to Nazarova and Roĭter [**NR73**]. See [**Rin80**] or [**Gus82**] for some history on these results. (It's perhaps interesting to note that Auslander gave a proof of Roĭter's theorem for arbitrary Artinian rings [**Aus74**]—with length standing in for k-dimension—and that this is where "almost split sequences" made their first appearance.)

We import the definition of bounded type to the context of MCM modules almost verbatim. Recall that the multiplicity of a finitely generated module M over a local ring R is denoted $e(M)$.

15.2. DEFINITION. We say that a CM local ring R has *bounded CM type* provided there is a bound on the multiplicities of the indecomposable MCM R-modules.

If an R-module M has constant rank r, then it is known that $\mathrm{e}(M) = r\,\mathrm{e}(R)$ (see Appendix A, §2). Thus for modules with constant rank, a bound on multiplicities is equivalent to a bound on ranks.

The first example showing that that bounded and finite type are not equivalent in the context of MCM modules, that is, that Brauer-Thrall I fails, was given by Dieterich in 1980 [**Die81**]: Let k be a field of characteristic 2, let $A = k[\![x]\!]$, and let G be the two-element group. Then the group algebra AG has bounded but infinite CM type. Indeed, note that $AG \cong k[\![x,y]\!]/(y^2)$ (via the map sending the generator of the group to $y-1$). Thus AG has multiplicity 2 but is not reduced, whence AG has bounded but infinite CM type by Theorem 4.18. In fact, as we saw in Chapter 14, $k[\![x,y]\!]/(y^2)$ has (countably) infinite CM type for every field k.

Theorem 4.10 says, in part, that if an analytically unramified local ring $(R, \mathfrak{m}, k)$ of dimension one with infinite residue field k fails to have finite CM type, then R has $|k|$ indecomposable MCM modules of every rank n. Thus, for these rings, finite CM type and bounded CM type are equivalent, just as for finite-dimensional algebras, and moreover Brauer-Thrall II even holds for these rings. In this chapter we present the proof, due independently to Dieterich [**Die87**] and Yoshino [**Yos87**], of Brauer-Thrall I for all complete, equicharacteristic, CM isolated singularities over a perfect field (Theorem 15.20) and show how to use the results of the previous chapters to weaken the hypothesis of completeness to that of excellence. We also give a new proof (independent of the one in Chapter 4) that Brauer-Thrall II holds for complete one-dimensional reduced rings with algebraically closed residue field (Theorem 15.27). The latter result uses Smalø's "inductive step" (Theorem 15.26) for building infinitely many indecomposables in a higher rank from infinitely many in a lower one. As another application of Smalø's theorem we observe that Brauer-Thrall II holds for rings of uncountable CM type.

§1. The Harada-Sai lemma

We will reduce the proof of the first Brauer-Thrall conjecture to a statement about modules of finite length, namely the Harada-Sai Lemma 15.4. In this section we give Eisenbud and de la Peña's 1998 proof [**EdlP98**] of Harada-Sai, and in the next section we show how to extend it to MCM modules. The Lemma gives an upper bound on the lengths of non-zero paths in the Auslander-Reiten quiver. To state it, we make a definition.

15.3. DEFINITION. Let R be a commutative ring and let
$$(15.1) \qquad M_1 \xrightarrow{f_1} M_2 \xrightarrow{f_2} \cdots \xrightarrow{f_{s-1}} M_s$$
be a sequence of homomorphisms between R-modules. We say (15.1) is a *Harada-Sai sequence* if
  (i) each M_i is indecomposable of finite length;
  (ii) no f_i is an isomorphism; and
  (iii) the composition $f_{s-1} f_{s-2} \cdots f_1$ is non-zero.

Fitting's Lemma (Exercise 1.25) implies that, in the special case where $M_i = M$ and $f_i = f$ are constant for all i, the longest possible Harada-Sai sequence has length $\ell(M) - 1$, where as usual $\ell(M)$ denotes the length of M. In general, the Harada-Sai Lemma gives a bound on the length of a Harada-Sai sequence in terms of the lengths of the modules.

15.4. LEMMA. *Let (15.1) be a Harada-Sai sequence such that the length of each M_i is bounded above by b. Then $s \leqslant 2^b - 1$.*

In fact we will prove a more precise statement, which determines exactly which sequences of lengths $\ell(M_i)$ are possible in a Harada-Sai sequence.

15.5. DEFINITION. The *length sequence* of a sequence (15.1) of modules of finite length is the integer sequence
$$\underline{\lambda} = (\ell(M_1), \ell(M_2), \ldots, \ell(M_s)).$$
We define *special* integer sequences as follows:
$$\lambda^{(1)} = (1)$$
$$\lambda^{(2)} = (2, 1, 2)$$
$$\lambda^{(3)} = (3, 2, 3, 1, 3, 2, 3)$$
and, in general, $\lambda^{(b)}$ is obtained by inserting b at the beginning, the end, and between every two entries of $\lambda^{(b-1)}$. Alternatively,
$$\lambda^{(b+1)} = (\lambda^{(b)} + \mathbf{1}, 1, \lambda^{(b)} + \mathbf{1}),$$
where $\mathbf{1}$ is the sequence of all 1s. Notice that $\lambda^{(b)}$ is a list of $2^b - 1$ integers.

We say that one integer sequence λ of length n *embeds* in another integer sequence μ of length m if there is a strictly increasing function $\sigma \colon \{1, \ldots, n\} \longrightarrow \{1, \ldots, m\}$ such that $\lambda_i = \mu_{\sigma(i)}$.

Lemma 15.4 follows from the next result.

15.6. THEOREM. *There is a Harada-Sai sequence with length sequence $\underline{\lambda}$ if and only if $\underline{\lambda}$ embeds in $\lambda^{(b)}$ for some b.*

PROOF. First let

$$(15.2) \qquad M_1 \xrightarrow{f_1} M_2 \xrightarrow{f_2} \cdots \xrightarrow{f_{s-1}} M_s$$

be a Harada-Sai sequence with length sequence $\underline{\lambda} = (\lambda_1, \ldots, \lambda_s)$. Set $b = \max\{\lambda_i\}$. If $b = 1$, then each M_i is simple. As the composition is non-zero and no f_i is an isomorphism, the length of the sequence must be 1. Thus $\underline{\lambda} = (1)$ embeds in $\lambda^{(1)} = (1)$. Suppose then that $b > 1$.

If two consecutive entries of $\underline{\lambda}$ are equal, say $\lambda_i = \lambda_{i+1}$, then we may insert some indecomposable summand of $\operatorname{im}(f_i)$ between M_i and M_{i+1}, chosen so that the composition is still non-zero. This gives a new Harada-Sai sequence, one step longer. Thus we may assume that no two consecutive λ_i are equal.

Observe that no composition of two consecutive f_j is an isomorphism; indeed, this would force both to split, contradicting the indecomposability of M_j.

Let $\underline{\lambda}'$ be the integer sequence gotten from $\underline{\lambda}$ by deleting every occurrence of b. Then $\underline{\lambda}'$ is the length sequence of the Harada-Sai sequence obtained by "collapsing" (15.2): for each M_i having length equal to b, delete M_i and replace the pair of homomorphisms f_i and f_{i+1} by the composition $f_{i+1}f_i \colon M_{i-1} \longrightarrow M_{i+1}$. By induction $\underline{\lambda}'$ embeds into $\lambda^{(b-1)}$. Since every second element of $\lambda^{(b)}$ is b and the b's in $\underline{\lambda}$ never repeat, this can be extended to an embedding $\underline{\lambda} \longrightarrow \lambda^{(b)}$.

To prove the other direction, it suffices by the same "collapsing" argument to show that there is a Harada-Sai sequence with $\lambda^{(b)}$ for its length sequence. For this we refer to [**EdlP98**], where Eisenbud and de la Peña construct such sequences over the ring $k[x,y]/(xy)$. □

§2. Faithful systems of parameters

The goal of this section is to prove an analog of the Harada-Sai Lemma 15.4 for MCM modules. We will reduce to the case of finite length modules by passing to the quotient by a particularly nice regular sequence: one that preserves indecomposability, non-isomorphism, and even non-split short exact sequences of MCM modules.

Throughout, $(R, \mathfrak{m}, k)$ is a CM local ring of dimension d. We will need to impose additional restrictions later on; see Theorem 15.19 for the full list.

§2. FAITHFUL SYSTEMS OF PARAMETERS

15.7. DEFINITION. Let $\mathbf{x} = x_1, \ldots, x_d$ be a system of parameters for R. We say $\mathbf{x}$ is a *faithful* system of parameters if for every pair M, N of finitely generated R-modules with M MCM, $\mathbf{x}\operatorname{Ext}_R^1(M,N) = 0$.

In what follows, we write $\mathbf{x}^2$ for the system of parameters $x_1^2, \ldots, x_d^2$. Here is the basic property of faithful systems of parameters that makes them well suited to our purposes. It's interesting to observe the similarity of this statement to that of Guralnick's Lemma 1.11. The statement could even be given the same form: a commutative rectangle consisting of two squares, the bottom of which also commutes, though the top square might not.

15.8. PROPOSITION. *Let $\mathbf{x} = x_1, \ldots x_d$ be a faithful system of parameters, and let M and N be MCM R-modules. For every homomorphism $\varphi \colon M/\mathbf{x}^2 M \longrightarrow N/\mathbf{x}^2 N$, there exists $\widetilde{\varphi} \in \operatorname{Hom}_R(M, N)$ such that φ and $\widetilde{\varphi}$ induce the same homomorphism $M/\mathbf{x} M \longrightarrow N/\mathbf{x} N$.*

PROOF. Our goal is the case $i = 0$ of the following statement: there exists a homomorphism
$$\varphi_i \colon M/\left(x_1^2, \ldots x_i^2\right) M \longrightarrow N/\left(x_1^2, \ldots, x_i^2\right) N$$
such that $\varphi_i \otimes_R R/(\mathbf{x}) = \varphi \otimes_R R/(\mathbf{x})$. We prove this by descending induction on i, taking $\varphi_d = \varphi$ for the base case $i = d$.

Assume that φ_{i+1} has been constructed. Then it suffices to find a homomorphism $\varphi_i \colon M/\left(x_1^2, \ldots x_i^2\right) M \longrightarrow N/\left(x_1^2, \ldots, x_i^2\right) N$ with the following stronger property:
$$\varphi_i \otimes_R R/\left(x_1^2, \ldots, x_i^2, x_{i+1}\right) = \varphi \otimes_R R/\left(x_1^2, \ldots, x_i^2, x_{i+1}\right),$$
for then of course killing $x_1, \ldots, x_i, x_{i+2}, \ldots, x_d$ we obtain $\varphi_i \otimes_R R/(\mathbf{x}) = \varphi \otimes_R R/(\mathbf{x})$.

Set $\mathbf{y}_i = x_1^2, \ldots, x_i^2$ and $\mathbf{z}_i = x_1^2, \ldots, x_i^2, x_{i+1}$. Then we have a commutative diagram with exact rows (as N is MCM and x_{i+1} is an R-regular element)

$$\begin{array}{ccccccccc}
0 & \longrightarrow & N/\mathbf{y}_i N & \xrightarrow{x_{i+1}^2} & N/\mathbf{y}_i N & \longrightarrow & N/\mathbf{y}_{i+1} N & \longrightarrow & 0 \\
& & \downarrow{\scriptstyle x_{i+1}} & & \| & & \downarrow & & \\
0 & \longrightarrow & N/\mathbf{y}_i N & \xrightarrow{x_{i+1}} & N/\mathbf{y}_i N & \longrightarrow & N/\mathbf{z}_i N & \longrightarrow & 0.
\end{array}$$

Apply $\operatorname{Hom}_R(M, -)$ to obtain a commutative exact diagram:

$$\begin{array}{ccccc}
\operatorname{Hom}_R(M, N/\mathbf{y}_i N) & \longrightarrow & \operatorname{Hom}_R(M, N/\mathbf{y}_{i+1} N) & \longrightarrow & \operatorname{Ext}_R^1(M, N/\mathbf{y}_i N) \\
\| & & \downarrow & & \downarrow{\scriptstyle x_{i+1}} \\
\operatorname{Hom}_R(M, N/\mathbf{y}_i N) & \longrightarrow & \operatorname{Hom}_R(M, N/\mathbf{z}_i N) & \longrightarrow & \operatorname{Ext}_R^1(M, N/\mathbf{y}_i N)
\end{array}$$

By the definition of a faithful system of parameters, the right-hand vertical map is zero. We have φ_{i+1} living in $\operatorname{Hom}_R(M, N/\mathbf{y}_{i+1}N)$ in the middle of the top row, and an easy diagram chase delivers φ_i in the top-left corner such that $\varphi_i \otimes_R R/(\mathbf{z}_i) \cong \varphi_{i+1} \otimes_R R/(\mathbf{z}_i)$. □

Here are the main consequences of Proposition 15.8. The first and third corollaries are sometimes called "Maranda's Theorem," having first been proven by Maranda [**Mar53**] in the case of the group ring of a finite group over the ring of p-adic integers, and extended by Higman [**Hig60**] to arbitrary orders over complete discrete valuation rings.

15.9. COROLLARY. *Let $\mathbf{x}$ be a faithful system of parameters for R, and let M and N be MCM R-modules. Suppose there is an isomorphism $\varphi \colon M/\mathbf{x}^2 M \longrightarrow N/\mathbf{x}^2 N$. Then there is an isomorphism $\widetilde{\varphi} \colon M \longrightarrow N$ such that $\widetilde{\varphi} \otimes_R R/(\mathbf{x}) = \varphi \otimes_R R/(\mathbf{x})$.*

PROOF. Proposition 15.8 gives us the homomorphism $\widetilde{\varphi}$; it remains to see that $\widetilde{\varphi}$ is an isomorphism. Since $\widetilde{\varphi}$ is surjective modulo $\mathbf{x}^2$, it is at least surjective by NAK. Similarly, applying the Proposition to φ^{-1}, we find that there is a surjection $\widetilde{\varphi^{-1}} \colon N \longrightarrow M$. By Exercise 4.26, the surjections $\widetilde{\varphi^{-1}}\widetilde{\varphi}$ and $\widetilde{\varphi}\widetilde{\varphi^{-1}}$ are both isomorphisms, so $\widetilde{\varphi}$ and $\widetilde{\varphi^{-1}}$ are as well. □

15.10. COROLLARY. *Let $\mathbf{x}$ be a faithful system of parameters for R, and let $s \colon 0 \longrightarrow N \xrightarrow{i} E \xrightarrow{p} M \longrightarrow 0$ be a short exact sequence of MCM modules. Then s is non-split if and only if $s \otimes_R R/(\mathbf{x}^2)$ is non-split.*

PROOF. Sufficiency is clear: a splitting for s immediately gives a splitting for $s \otimes_R R/(\mathbf{x}^2)$. For the other direction, suppose $\overline{p} = p \otimes_R R/(\mathbf{x}^2)$ is a split epimorphism. Then there exists $\varphi \colon M/\mathbf{x}^2 M \longrightarrow E/\mathbf{x}^2 E$ such that $\overline{p}\varphi$ is the identity on $M/\mathbf{x}^2 M$. Let $\widetilde{\varphi} \colon M \longrightarrow E$ be the lifting guaranteed by Proposition 15.8. Then $(p\widetilde{\varphi}) \otimes_R R/(\mathbf{x})$ is the identity on $M/\mathbf{x}M$, so $p\widetilde{\varphi}$ is an isomorphism. Thus s is split. □

15.11. COROLLARY. *Assume that R is Henselian. Let $\mathbf{x}$ be a faithful system of parameters for R, and let M be a MCM R-module. Then M is indecomposable if and only if $M/\mathbf{x}^2 M$ is indecomposable.*

PROOF. Again, we have only to prove one direction: if M decomposes non-trivially, then so must $M/\mathbf{x}^2 M$ by NAK. For the other direction, assume that M is indecomposable. Then $\operatorname{End}_R(M)$ is a nc-local ring since R is Henselian (see Chapter 1). We have a commutative

diagram

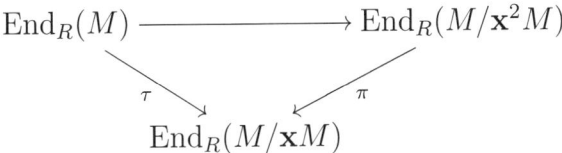

where each map is the natural one induced by tensoring with $R/(\mathbf{x})$ or $R/(\mathbf{x}^2)$. Let $e \in \mathrm{End}_R(M/\mathbf{x}^2 M)$ be an idempotent; we'll show that e is either 0 or 1, so that $M/\mathbf{x}^2 M$ is indecomposable. The image $\pi(e)$ of e in $\mathrm{End}_R(M/\mathbf{x}M)$ is still idempotent, and is contained in $\tau(\mathrm{End}_R(M))$ by Proposition 15.8. Since $\mathrm{End}_R(M)$ is nc-local, so is its homomorphic image $\tau(\mathrm{End}_R(M))$, so $\pi(e)$ is either 0 or 1.

If $\pi(e) = 0$, then $e \otimes_R R/(\mathbf{x}) = 0$, so that $e(M/\mathbf{x}^2 M) \subseteq \mathbf{x}(M/\mathbf{x}^2 M)$. But e is idempotent, so that $\mathrm{im}(e) = \mathrm{im}(e^2) \subseteq \mathrm{im}(\mathbf{x}^2) = 0$ and so $e = 0$. If $\pi(e) = 1$, then the same argument applies to $1 - e$, giving $e = 1$. □

To address the existence of faithful systems of parameters, consider a couple of general lemmas. We leave the proof of the first as an exercise. The second is an easy special case of [**Wan94**, Lemma 5.10].

15.12. LEMMA. *Let Γ be a ring, I an ideal of Γ, and $\Lambda = \Gamma/I$. Then $\mathrm{Ann}_\Gamma I$ annihilates $\mathrm{Ext}^1_\Gamma(\Lambda, K)$ for every Γ-module K.* □

15.13. LEMMA. *Let Γ be a ring, I an ideal of Γ, and $\Lambda = \Gamma/I$. Let*

$$(15.3) \qquad L \xrightarrow{\varphi} M \xrightarrow{\psi} N$$

be an exact sequence of Γ-modules. Then the homology H of the complex

$$(15.4) \qquad \mathrm{Hom}_\Gamma(\Lambda, L) \xrightarrow{\varphi_*} \mathrm{Hom}_\Gamma(\Lambda, M) \xrightarrow{\psi_*} \mathrm{Hom}_\Gamma(\Lambda, N)$$

is annihilated by $\mathrm{Ann}_\Gamma I$.

PROOF. Let $K = \ker \varphi$ and $X = \mathrm{im}\,\varphi$, and let $\eta \colon L \longrightarrow X$ be the surjection induced by φ. Then applying $\mathrm{Hom}_\Gamma(\Lambda, -)$, we see that the cohomology in the middle of (15.4) is equal to the cokernel of

$$\mathrm{Hom}_\Gamma(\Lambda, \eta) \colon \mathrm{Hom}_\Gamma(\Lambda, L) \longrightarrow \mathrm{Hom}_\Gamma(\Lambda, X)\,.$$

This cokernel is also a submodule of $\mathrm{Ext}^1_\Gamma(\Lambda, K)$, so we are done by the previous lemma. □

We will apply Lemma 15.13 to the *homological different* $\mathfrak{H}_T(R)$ of a homomorphism $T \longrightarrow R$, where R is as above a CM local ring and T is a regular local ring. Recall (from e.g. Appendix B) that if $A \longrightarrow B$ is a homomorphism of commutative rings, we let $\mu \colon B \otimes_A B \longrightarrow B$ be the diagonal map defined by $\mu(b \otimes b') = bb'$, and we set $\mathcal{J} = \ker \mu$. The homological different $\mathfrak{H}_A(B)$ is then defined to be

$$\mathfrak{H}_A(B) = \mu(\operatorname{Ann}_{B \otimes_A B} \mathcal{J}).$$

Notice that for any two B-modules M and N, $\operatorname{Hom}_A(M, N)$ is naturally a $B \otimes_A B$-module via the rule $[\varphi(b \otimes b')](m) = \varphi(bm)b'$ for any $\varphi \in \operatorname{Hom}_A(M, N)$, $m \in M$, and $b, b' \in B$. Since for any $B \otimes_A B$-module X, $\operatorname{Hom}_{B \otimes_A B}(R, X)$ is the submodule of X annihilated by $\mathcal{J}$, and $\mathcal{J}$ is generated by elements of the form $b \otimes 1 - 1 \otimes b$, we see that

$$\operatorname{Hom}_{B \otimes_A B}(B, \operatorname{Hom}_A(M, N)) \cong \operatorname{Hom}_B(M, N)$$

for all M, N. This is in particular an isomorphism of $B \otimes_A B$-modules, where the action of $B \otimes_A B$ on $\operatorname{Hom}_B(M, N)$ is via μ.

15.14. PROPOSITION. *Let R be a CM local ring and assume $T \subseteq R$ is a regular local ring such that R is a finitely generated T-module. Then $\mathfrak{H}_T(R)$ annihilates $\operatorname{Ext}^1_R(M, N)$ for every MCM R-module M and arbitrary R-module N.*

PROOF. Let $0 \longrightarrow N \longrightarrow I^0 \longrightarrow I^1 \longrightarrow I^2 \longrightarrow \cdots$ by an injective resolution of N over R. Since M is MCM over R, it is finitely generated and free over T, and the complex

$$\operatorname{Hom}_T(M, I^0) \xrightarrow{\varphi} \operatorname{Hom}_T(M, I^1) \xrightarrow{\psi} \operatorname{Hom}_T(M, I^2)$$

is exact. Apply $\operatorname{Hom}_{R \otimes_T R}(R, -)$; by the discussion above the result is

$$\operatorname{Hom}_R(M, I^0) \xrightarrow{\varphi_*} \operatorname{Hom}_R(M, I^1) \xrightarrow{\psi_*} \operatorname{Hom}_R(M, I^2).$$

The homology H of this complex is naturally $\operatorname{Ext}^1_R(M, N)$, and is by Lemma 15.13 annihilated by $\operatorname{Ann}_{R \otimes_T R} \mathcal{J}$. Since the $R \otimes_T R$-module structure on these Hom modules is via μ, we see that the homological different $\mathfrak{H}_T(R) = \mu(\operatorname{Ann}_{R \otimes_T R} \mathcal{J})$ annihilates $\operatorname{Ext}^1_R(M, N)$. $\square$

Put $\mathfrak{H}(R) = \sum_T \mathfrak{H}_T(R)$, where the sum is over all regular local subrings T of R such that R is a finitely generated T-module. It follows immediately from Proposition 15.14 that $\mathfrak{H}(R)$ annihilates $\operatorname{Ext}^1_R(M, N)$ whenever M is MCM.

Let us now introduce a more classical ideal, the Jacobian ideal. Let T be a Noetherian ring and R a finitely generated T-algebra. Then R has a presentation $R = T[x_1, \ldots, x_n]/(f_1, \ldots, f_m)$ for some n and m. The *Jacobian ideal of R over T* is the ideal $J_T(R)$ in R generated by the maximal minors of the Jacobian matrix $(\partial f_i/\partial x_j)_{ij}$. We set $J(R) = \sum_T J_T(R)$, where again the sum is over all regular subrings T of R over which R is module-finite.

One can see ([**Wan94**, Prop. 5.8] or Exercise 15.34) that $J_T(R) \subseteq \mathfrak{H}_T(R)$ for every T, so that $J(R) \subseteq \mathfrak{H}(R)$. Thus we have

15.15. COROLLARY. *Let R be a CM local ring and let $J(R)$ be the Jacobian ideal of R. Then J annihilates $\mathrm{Ext}^1_R(M,N)$ for every pair of R-modules M, N with M MCM.* □

There are two problems with this result. The first is the question of whether any regular local subrings T as in the definition of $J(R)$ actually exist. Luckily, Cohen's Structure Theorems assure us that when R is complete and contains its residue field k, there exist plenty of regular local rings $T = k[\![x_1, \ldots, x_d]\!]$ over which R is module-finite.

The second problem is that $J(R)$ may be trivial if the residue field is not perfect.

15.16. REMARK. If R is a hypersurface $R = k[x_1, \ldots, x_d]/(f(x))$, then $J(R)$ is the ideal of R generated by the partial derivatives $\partial f / \partial x_i$ of f. If k is not perfect, this ideal can be zero. For example, suppose that k is an imperfect field of characteristic p, and let $\alpha \in k \setminus k^p$. Put $R = k[x, y]/(x^p - \alpha y^p)$. Then $J(R) = 0$. Note that R is a one-dimensional domain, so is an isolated singularity. Thus in particular J does not define the singular locus of R.

To address this second problem, we appeal to Nagata's Jacobian criterion for smoothness of complete local rings [**GD64**, 22.7.2] (see also [**Wan94**, Props. 4.4 and 4.5]).

15.17. THEOREM. *Let $(R, \mathfrak{m}, k)$ be an equidimensional complete local ring containing its residue field k. Assume that k is perfect. Then the Jacobian ideal $J(R)$ of R defines the singular locus: for a prime ideal $\mathfrak{p}$, $R_\mathfrak{p}$ is a regular local ring if and only if $J(R) \not\subseteq \mathfrak{p}$.* □

This immediately gives existence of faithful systems of parameters, and our extension of the Harada-Sai Lemma to MCM modules. We leave the details of the proof of existence as an exercise (Exercise 15.35).

15.18. THEOREM (Yoshino). *Let $(R, \mathfrak{m}, k)$ be a complete CM local ring containing its residue field k. Assume that k is perfect and that R has an isolated singularity. Then R admits a faithful system of parameters.* □

15.19. THEOREM (Harada-Sai for MCM modules). *Let R be a complete equicharacteristic CM local ring with perfect residue field and an isolated singularity. Let $\mathbf{x}$ be a faithful system of parameters for R. Let M_0, M_1, $\ldots$, M_{2^n} be indecomposable MCM R-modules, and let $f_i \colon M_i \longrightarrow M_{i+1}$ be homomorphisms that are not isomorphisms. If $\ell(M_i/\mathbf{x}^2 M_i) \leqslant n$ for all $i = 0, \ldots, 2^n$, then $f_{2^n-1} \cdots f_2 f_1 \otimes_R R/(\mathbf{x}^2) = 0$.*

PROOF. Set $\widetilde{M_i} = M_i/\mathbf{x}^2 M_i$ and $\widetilde{f_i} = f_i \otimes_R R/(\mathbf{x}^2)$. Then $\widetilde{M_0} \xrightarrow{\widetilde{f_0}} \cdots \xrightarrow{\widetilde{f_{2^n-1}}} \widetilde{M_{2^n}}$ is a sequence of indecomposable $R/(x^2)$-modules, each of length at most n, in which no $\widetilde{f_i}$ is an isomorphism. It is too long to be a Harada-Sai sequence, however, so we conclude $f_{2^n} \cdots f_2 f_1 \otimes_R R/(\mathbf{x}^2) = 0$. □

§3. Proof of Brauer-Thrall I

In this section we prove the following theorem, proved in the complete case by Dieterich [**Die87**] and Yoshino [**Yos87**] independently. See also [**PR90**] and [**Wan94**]. Our proof follows [**Yos90**] closely.

15.20. THEOREM (Dieterich, Yoshino). *Let $(R, \mathfrak{m}, k)$ be an excellent equicharacteristic CM local ring with perfect residue field k. Then R has finite CM type if and only if R has bounded CM type and at most an isolated singularity.*

Of course one direction of the theorem follows immediately from Auslander's Theorem 7.12, and requires no hypotheses on R other than Cohen-Macaulayness. The content of the theorem is that bounded type and isolated singularity together imply finite type.

We begin by considering the case where R is complete and the residue field k is algebraically closed, and at the end of the section we show how to relax these restrictions. When k is algebraically closed and R is complete and has at most an isolated singularity, we have access to the Auslander-Reiten quiver of R, as well as to faithful systems of parameters. In this case, we will prove

15.21. THEOREM. *Let $(R, \mathfrak{m}, k)$ be a complete equicharacteristic CM local ring with algebraically closed residue field k. Assume that R has at most an isolated singularity. Let Γ be the AR quiver of R and Γ° a non-empty connected component of Γ. If Γ° has bounded multiplicities, that is, there exists an integer B such that $e(M) \leqslant B$ for all $[M] \in \Gamma^\circ$, then $\Gamma = \Gamma^\circ$ and Γ is finite. In particular R has finite CM type.*

Let us be precise about what it means for Γ° to be a connected component. We take it to mean that Γ° is *closed under irreducible homomorphisms*, meaning that if $X \longrightarrow Y$ is an irreducible homomorphism between indecomposable MCM modules, then $[X] \in \Gamma^\circ$ if and only if $[Y] \in \Gamma^\circ$.

Here is the strategy of the proof. Assume that Γ° is a connected component of Γ with bounded multiplicities. We will show that for any $[M]$ and $[N]$ in Γ, if either of $[M]$ or $[N]$ is in Γ° then there is a path from $[M]$ to $[N]$ in Γ, and furthermore that such a path can be chosen to have bounded length. To do this, we assume no such path exists and derive a contradiction to the Harada-Sai Lemma 15.19.

We fix notation as in Theorem 15.21, so that $(R, \mathfrak{m}, k)$ is a complete equicharacteristic CM local ring with algebraically closed residue field k and with an isolated singularity. Let Γ be the AR quiver of R. By Theorem 15.18 there exists a faithful system of parameters $\mathbf{x}$ for R. We say that a homomorphism $\varphi\colon M \longrightarrow N$ between R-modules is non-trivial modulo $\mathbf{x}^2$ if $\varphi \otimes_R R/(\mathbf{x}^2) \neq 0$. Abusing notation slightly, we also say that a path in Γ is non-trivial modulo $\mathbf{x}^2$ if the corresponding composition of irreducible maps is non-trivial modulo $\mathbf{x}^2$.

15.22. LEMMA. *Fix a non-negative integer n. Let M and N be indecomposable MCM R-modules and $\varphi\colon M \longrightarrow N$ a homomorphism which is non-trivial modulo $\mathbf{x}^2$. Assume that there is no directed path in Γ from $[M]$ to $[N]$ of length $< n$ which is non-trivial modulo $\mathbf{x}^2$. Then the following two statements hold.*

(i) *There is a sequence of homomorphisms*
$$M = M_0 \xrightarrow{f_1} M_1 \xrightarrow{f_2} \cdots \xrightarrow{f_n} M_n \xrightarrow{g} N$$
with each M_i indecomposable, each f_i irreducible, and the composition $gf_n \cdots f_1$ non-trivial modulo $\mathbf{x}^2$.

(ii) *There is a sequence of homomorphisms*
$$M \xrightarrow{h} N_n \xrightarrow{g_n} N_{n-1} \xrightarrow{g_{n-1}} \cdots \xrightarrow{g_1} N_0 = N$$
with each N_i indecomposable, each g_i irreducible, and the composition $g_1 \cdots g_n h$ non-trivial modulo $\mathbf{x}^2$.

PROOF. We prove part (ii); the other half is similar.

If $n = 0$, then we may simply take $h = \varphi\colon M \longrightarrow N$. Assume therefore that $n > 0$, there is no directed path of length $< n$ from $[M]$ to $[N]$ which is non-trivial modulo $\mathbf{x}^2$, and that we have constructed a sequence of homomorphisms
$$M \xrightarrow{h} N_{n-1} \xrightarrow{g_{n-1}} \cdots \xrightarrow{g_1} N_0 = N$$
with each N_i indecomposable, each g_i irreducible, and the composition $g_1 \cdots g_{n-1} h$ non-trivial modulo $\mathbf{x}^2$. We wish to insert an indecomposable module N_n into the sequence, extending it by one step. There are two cases, according to whether or not N_{n-1} is free.

If N_{n-1} is not free, then there is an AR sequence $0 \longrightarrow \tau(N_{n-1}) \xrightarrow{i} E \xrightarrow{p} N_{n-1} \longrightarrow 0$ ending in N_{n-1}. Since there is no path from $[M]$ to $[N]$ of length $n-1$, we see that h is not an isomorphism, so is not a split surjection since M and N_{n-1} are both indecomposable. Therefore h factors through E, say as $M \xrightarrow{\alpha} E \xrightarrow{p} N_{n-1}$. Write E as a direct sum of indecomposable MCM modules $E = \bigoplus_{i=1}^r E_i$, and decompose α and q accordingly, $M \xrightarrow{\alpha_i} E \xrightarrow{p_i} N_{n-1}$. Each p_i is irreducible by Proposition 13.16, and there must exist at least one i such that $g_1 \cdots g_{n-1} p_i \alpha_i$ is non-trivial modulo $\mathbf{x}^2$. Set $N_n = E_i$ and $g_n = p_i$, extending the sequence one step.

If N_{n-1} is free, then $N_{n-1} \cong R$, and the image of M is contained in $\mathfrak{m}$ since h is not an isomorphism. Let $0 \longrightarrow Y \xrightarrow{i} X \xrightarrow{p} \mathfrak{m} \longrightarrow 0$ be a minimal MCM approximation of $\mathfrak{m}$. (If $\dim R \leqslant 1$, we take $X = \mathfrak{m}$ and $Y = 0$.) The homomorphism $h \colon M \longrightarrow \mathfrak{m}$ factors through X as $M \xrightarrow{\alpha} X \xrightarrow{p} \mathfrak{m}$. Decompose $X = \bigoplus_{i=1}^r X_i$ where each X_i is indecomposable, and write $p = \sum_{i=1}^r p_i$, where $p_i \colon X_i \longrightarrow \mathfrak{m}$. By Proposition 13.18, each composition $X_i \xrightarrow{p_i} \mathfrak{m} \hookrightarrow R$ is an irreducible homomorphism, and again we may choose i so that the composition $g_1 \cdots g_{n-1} p_i \alpha_i$ is non-trivial modulo $\mathbf{x}^2$. □

15.23. LEMMA. *Let Γ° be a connected component of the AR quiver Γ, and assume that $\ell(M/\mathbf{x}^2 M) \leqslant m$ for every $[M]$ in Γ°. Let $\varphi \colon M \longrightarrow N$ be a homomorphism between indecomposable MCM R-modules which is non-trivial modulo $\mathbf{x}^2$, and assume that either $[M]$ or $[N]$ is in Γ°. Then there is a directed path of length $< 2^m$ from $[M]$ to $[N]$ in Γ which is non-trivial modulo $\mathbf{x}^2$. In particular, both $[M]$ and $[N]$ are in Γ° if either one is.*

PROOF. Set $n = 2^m$, and assume that $[N]$ is in Γ°. If there is no directed path of length $< n$ from $[M]$ to $[N]$, then by Lemma 15.22 there is a sequence of homomorphisms

$$M \xrightarrow{h} N_n \xrightarrow{g_n} N_{n-1} \xrightarrow{g_{n-1}} \cdots \xrightarrow{g_1} N_0 = N$$

with each N_i indecomposable, each g_i irreducible, and the composition $g_1 \cdots g_n h$ non-trivial modulo $\mathbf{x}^2$. Since Γ° is connected, each $[N_i]$ is in Γ°, so that $\ell(N_i/\mathbf{x}^2 N_i) \leqslant m$ for each i. By the Harada-Sai Lemma 15.19, $g_1 \cdots g_n$ is trivial modulo $\mathbf{x}^2$, a contradiction.

A symmetric argument using the other half of Lemma 15.22 takes care of the case where $[M]$ is in Γ°. □

We are now ready for the proof of Brauer-Thrall I in the complete case. Keep notation as in the statement of Theorem 15.21.

PROOF OF THEOREM 15.21. We have $\mathrm{e}(M) \leqslant B$ for every $[M]$ in Γ°. Choose t large enough that $\mathfrak{m}^t \subseteq (\mathbf{x}^2)$, where $\mathbf{x}$ is the faithful system of parameters guaranteed by Theorem 15.18. Then (see Theorem A.21) $\ell(M/\mathbf{x}^2 M) \leqslant t^{\dim R} B$ for every $[M]$ in Γ°. Set $m = t^{\dim R} B$.

Let M be any indecomposable MCM module such that $[M]$ is in Γ°. By NAK, there is an element $z \in M \setminus \mathbf{x}^2 M$. Define $\varphi \colon R \longrightarrow M$ by $\varphi(1) = z$; then φ is non-trivial modulo $\mathbf{x}^2$. By Lemma 15.23, $[R]$ is in Γ°, and is connected to $[M]$ by a path of length $< 2^m$ in Γ°.

Now let $[N]$ be arbitrary in Γ. The same argument shows that there is a homomorphism $\psi \colon R \longrightarrow N$ which is non-trivial modulo $\mathbf{x}^2$, whence $[N]$ is in Γ° as well, connected to $[R]$ by a path of length $< 2^m$. Thus $\Gamma = \Gamma^\circ$, and since Γ is a locally finite graph (Remark 13.19) of finite diameter, Γ is finite. □

To complete the proof of Theorem 15.20, we need to know that for R an excellent isolated singularity with perfect residue field, the hypotheses ascend along a gonflement making the residue field algebraically closed, and thence to the completion, and the conclusion descends back down to R. We have verified most of these details in previous chapters, and all that remains is to assemble the pieces.

PROOF OF BRAUER-THRALL I (THEOREM 15.20). Let R be as in the statement of the theorem, so that R is excellent and equicharacteristic, with perfect residue field. If R has finite CM type, then R has at most an isolated singularity by Theorem 7.12, and of course R has bounded CM type.

For the converse, we may assume that $\dim(R) > 0$, since by Theorem 3.3 bounded and finite CM type are equivalent for Artinian rings. Then R is reduced, by Proposition A.8. Let b bound the multiplicities of the indecomposable MCM R-modules. Choose a gonflement $(R, \mathfrak{m}, k) \longrightarrow (S, \mathfrak{n}, K)$, where K is the algebraic closure of k. Consider the flat local homomorphisms

(15.5) $$R \longrightarrow S \longrightarrow S^{\mathrm{h}} \longrightarrow \widehat{S}.$$

By Propositions 10.15 and 10.7, S^{h} is excellent and has at most an isolated singularity, and now Proposition 10.9 implies that $\widehat{S}$ has at most an isolated singularity.

Let N be an arbitrary indecomposable MCM $\widehat{S}$-module. Using Propositions 10.5, 10.7 and 10.15, and Corollary 10.11, we see that N is weakly extended from a MCM R-module M, say $N \oplus X \cong \widehat{S} \otimes_R M$. Write $M = V_1 \oplus \cdots \oplus V_t$, where the V_i are indecomposable. Then $N \oplus X \cong (\widehat{S} \otimes_R V_1) \oplus \cdots \oplus (\widehat{S} \otimes_R V_t)$. By KRS, $N \mid \widehat{S} \otimes_R V_i$ for some i and hence $e_S(N) \leqslant e_S(\widehat{S} \otimes V_i)$. But $e_S(\widehat{S} \otimes_R V_i) = e_R(V_i)$ by Exercise 10.24. We have shown that b bounds the multiplicities of the indecomposable MCM $\widehat{S}$-modules. Thus S has bounded CM type, and hence (Theorem 15.21) finite CM type. Finally, Theorem 10.1 shows that R has finite CM type. □

One cannot completely remove the hypothesis of excellence in Theorem 15.20. For example, let S be any one-dimensional analytically ramified local domain. It is known [**Mat73**, pp. 138–139] that there is a one-dimensional local domain R between S and its quotient field such that $e(R) = 2$ and $\widehat{R}$ is not reduced. Then R has bounded but infinite CM type by Theorem 4.18, and of course R has an isolated singularity.

§4. Brauer-Thrall II

Let $(R, \mathfrak{m}, k)$ be a complete local ring with isolated singularity and with algebraically closed residue field k. The second Brauer-Thrall conjecture, transplanted to the context of MCM modules, states that if R has infinite CM type then there is an infinite sequence of positive integers $n_1 < n_2 < n_3 < \ldots$ with the following property: for each i there are infinitely many non-isomorphic indecomposable MCM R-modules of multiplicity n_i.

Dieterich [**Die87**] verified Brauer-Thrall II for hypersurface singularities $k[\![x_0, \ldots, x_d]\!]/(f)$ where $\operatorname{char}(k) \neq 2$. Popescu and Roczen generalized Dieterich's results to excellent Henselian local rings [**PR90**] and to characteristic two in [**PR91**].

In Chapter 4 we proved a strong version of Brauer-Thrall II for one-dimensional rings. Here we give a less computational proof (with mild restrictions). This proof uses an inductive step, due to Smalø [**Sma80**], for concluding, from the existence of infinitely many indecomposable modules of a given multiplicity, infinitely many of a higher multiplicity. This inductive step works in any dimension.

Smalø's theorem also confirms Brauer-Thrall II for isolated singularities of uncountable CM type, as we point out at the end of the section. Smalø's result is quite general, and we feel it deserves to be better known.

We need two lemmas to control the growth of multiplicity as one walks through an AR quiver. The first is a general fact about Betti numbers [**Avr98**, Lemma 4.2.7].

15.24. LEMMA. *Let $(R, \mathfrak{m}, k)$ be a CM local ring of dimension d and multiplicity e, and let M be a finitely generated R-module. Then*
$$\mu_R(\operatorname{syz}^R_{n+1}(M)) \leqslant (e-1)\mu_R(\operatorname{syz}^R_n(M))$$
for all $n > d - \operatorname{depth} M$.

PROOF. We may replace M by $\mathrm{syz}^R_{d-\mathrm{depth}\,M}(M)$ to assume that M is MCM. We may also assume that the residue field k is infinite, by passing if necessary to an elementary gonflement $R' = R[t]_{\mathfrak{m}[t]}$, which preserves the multiplicity of R and number of generators of syzygies of M. In this case (see Appendix A, §2), there exists an R-regular and M-regular sequence $\mathbf{x} = x_1, \ldots, x_d$ such that $\mathrm{e}(R) = \mathrm{e}(R/(\mathbf{x})) = \ell(R/(\mathbf{x}))$, and we have $\mu_R(\mathrm{syz}^R_n(M)) = \mu_{R/(\mathbf{x})}(\mathrm{syz}^{R/(\mathbf{x})}_n(M \otimes_R R/(\mathbf{x})))$. We are thus reduced to the case where R is Artinian of length e.

In a minimal free resolution $F_\bullet$ of M, we have $\mathrm{syz}^R_{n+1}(M) \subseteq \mathfrak{m} F_n$, so that

$$(e-1)\mu_R(\mathrm{syz}^R_n(M)) = \ell(\mathfrak{m} F_n)$$
$$\geqslant \ell(\mathrm{syz}^R_{n+1}(M))$$
$$\geqslant \mu_R(\mathrm{syz}^R_{n+1}(M)),$$

for all $n \geqslant 1$. $\square$

15.25. LEMMA. *Let $(R, \mathfrak{m})$ be a complete CM local ring with algebraically closed residue field, and assume that R has an isolated singularity. Then there exists a constant $c = c(R)$ such that if $X \longrightarrow Y$ is an irreducible homomorphism of MCM R-modules, then $\mathrm{e}(X) \leqslant c\,\mathrm{e}(Y)$ and $\mathrm{e}(Y) \leqslant c\,\mathrm{e}(X)$.*

PROOF. Recall from Chapter 13 that the Auslander-Reiten translate τ is given by $\tau(M) = \mathrm{Hom}_R(\mathrm{redsyz}^R_d \mathrm{Tr}\, M, \omega)$, where ω is the canonical module for R and $d = \dim(R)$. We first claim that

$$(15.6) \qquad \mathrm{e}(\tau(M)) \leqslant e(e-1)^{d+1}\,\mathrm{e}(M),$$

where $e = \mathrm{e}(R)$ is the multiplicity of R. To see this, it suffices to prove the inequality for $\mathrm{e}(\mathrm{syz}^R_d(\mathrm{Tr}\, M))$, since redsyz is a direct summand of syz and dualizing into the canonical module preserves multiplicity. By Lemma 15.24, we have only to prove that $\mathrm{e}(\mathrm{Tr}\, M) \leqslant e(e-1)\,\mathrm{e}(M)$. Let $F_1 \longrightarrow F_0 \longrightarrow M \longrightarrow 0$ be a minimal free presentation of M, so that $F_0^* \longrightarrow F_1^* \longrightarrow \mathrm{Tr}\, M \longrightarrow 0$ is a free presentation of $\mathrm{Tr}\, M$. Then

$$\mathrm{e}(\mathrm{Tr}\, M) \leqslant \mathrm{e}(F_1^*) = e\,\mu_R(\mathrm{syz}^R_1(M)) \leqslant e(e-1)\mu_R(M) \leqslant e(e-1)\,\mathrm{e}(M),$$

finishing the claim.

Now to the proof of the lemma. We may assume that X and Y are indecomposable. First suppose that Y is not free. Then there is an AR sequence

$$0 \longrightarrow \tau(Y) \longrightarrow E \longrightarrow Y \longrightarrow 0$$

ending in Y, and X is a direct summand of E by Proposition 13.16. Then
$$\begin{aligned}\mathrm{e}(E) &= \mathrm{e}(\tau(Y)) + \mathrm{e}(Y) \\ &\leqslant [e(e-1)^{d+1} + 1]\,\mathrm{e}(Y)\end{aligned}$$
so $\mathrm{e}(X) \leqslant [e(e-1)^{d+1} + 1]\,\mathrm{e}(Y)$.

Now suppose that Y is free, so that $Y \cong R$. Then X is a direct summand of the minimal MCM approximation E of the maximal ideal $\mathfrak{m}$ by Proposition 13.16, so $\mathrm{e}(X) \leqslant \mathrm{e}(E)$, and in particular $\mathrm{e}(X)$ is bounded in terms of $\mathrm{e}(R)$.

The other inequality is similar. $\square$

15.26. THEOREM (Smalø). *Let $(R, \mathfrak{m})$ be a complete CM local ring with algebraically closed residue field, and assume that R has an isolated singularity. Assume that $\{M_i \mid i \in I\}$ is an infinite family of pairwise non-isomorphic indecomposable MCM R-modules of multiplicity b. Then there exists an integer $b' > b$, a positive integer t, and a subset $J \subseteq I$ with $|J| = |I|$ such that there is a family $\{N_j \mid j \in J\}$ of pairwise non-isomorphic indecomposable MCM R-modules of multiplicity b'. Furthermore there exist non-zero homomorphisms $M_j \longrightarrow N_j$, each of which is a composition of t irreducible homomorphisms.*

PROOF. Set $s = 2^b - 1$. First observe that since the AR quiver of R is locally finite, there are at most finitely many M_i such that there is a chain of strictly fewer than s irreducible homomorphism starting at M_i and ending at the canonical module ω. Deleting these indices i, we obtain $J' \subseteq I$.

Each M_i is MCM, so $\mathrm{Hom}_R(M_i, \omega)$ is non-zero for each remaining M_i. By NAK, there exists $\varphi \in \mathrm{Hom}_R(M_i, \omega)$ which is non-trivial modulo $\mathbf{x}^2$. Hence by Lemma 15.22 there is a sequence of homomorphisms
$$M_i = N_{i,0} \xrightarrow{f_{i,1}} N_{i,1} \xrightarrow{f_{i,2}} \cdots \longrightarrow N_{i,s-1} \xrightarrow{f_{i,s}} N_{i,s} \xrightarrow{g} \omega$$
with each $N_{i,j}$ indecomposable, each $f_{i,j}$ irreducible, and the composition $g_i f_{i,s} \cdots f_{i,1}$ non-trivial modulo $\mathbf{x}^2$.

By the Harada-Sai Lemma 15.19, not all the $N_{i,j}$ can have multiplicity less than or equal to b. So there exists $J'' \subseteq J'$, of the same cardinality, and $t \leqslant s$ such that $\mathrm{e}(N_{i,t}) > b$ for all i.

Applying Lemma 15.25 to the irreducible homomorphisms connecting M_i to $N_{i,t}$, we find that
$$b < \mathrm{e}(N_{i,t}) \leqslant c^t\,\mathrm{e}(N_{i,0}) = c^t b$$

for some constant c depending only on R. There are thus only finitely many possibilities for $\mathrm{e}(N_{i,t})$ as i ranges over J'', and we take $J''' \subseteq J''$ such that $\mathrm{e}(N_{i,t}) = b' > b$ for all $i \in J'''$.

There may be some repetitions among the isomorphism classes of the $N_{i,t}$. However, for any indecomposable MCM module N, there are only finitely many M with chains of irreducible homomorphisms of length t from M to N, so each isomorphism class of $N_{i,t}$ occurs only finitely many times. Pruning away these repetitions, we finally obtain $J = J'''' \subseteq I$ as desired. $\square$

In Theorem 4.10 we proved a strong form of Brauer-Thrall II for the case of a one-dimensional analytically unramified local ring $(R, \mathfrak{m}, k)$. Here we indicate how one can use Smalø's theorem to give a much less computational proof of strongly unbounded CM type when R is complete and k is algebraically closed.

15.27. THEOREM. *Let $(R, \mathfrak{m}, k)$ be a complete reduced CM local ring of dimension one with algebraically closed residue field k. Suppose that R does not satisfy the Drozd-Roĭter conditions (DR1) and (DR2) of Chapter 4. Then, for infinitely many positive integers n, there exist $|k|$ pairwise non-isomorphic indecomposable MCM R-modules of multiplicity n.*

PROOF. It will suffice to show that R has infinitely many non-isomorphic faithful ideals, for then an easy argument like that in Exercise 4.32 produces infinitely many non-isomorphic ideals that are indecomposable as R-modules, and, consequently, infinitely many of some fixed multiplicity. By Construction 4.1 it will suffice to produce an infinite family of pairwise non-isomorphic modules $V_t \hookrightarrow \overline{R}/\mathfrak{c}$ over the Artinian pair $(R/\mathfrak{c} \hookrightarrow \overline{R}/\mathfrak{c})$. We follow the argument in the proof of [**Wie89**, Proposition 4.2]. Using Lemmas 3.10 and 3.11 and Proposition 3.12, we can pass to the Artinian pair $\mathbf{A} := (k \hookrightarrow D)$, where either (i) $\dim_k(D) \geqslant 4$ or (ii) $D \cong k[x,y]/(x^2, xy, y^2)$. It is easy to see that if U and V are distinct rings between k and D then the $\mathbf{A}$-modules $U \hookrightarrow D$ and $V \hookrightarrow D$ are non-isomorphic. Therefore we may assume that there are only finitely many intermediate rings. The usual proof of the primitive element theorem then provides an element $\alpha \in D$ such that $D = k[\alpha]$. This rules out (ii), so we may assume that $\dim_k(D) \geqslant 4$.

For each $t \in k$, let I_t be the k-subspace of D spanned by 1 and $\alpha + t\alpha^2$. By Exercise 15.37, there are infinitely many non-isomorphic $\mathbf{A}$-modules $I_t \hookrightarrow D$ as t varies over k. $\square$

In higher dimensions, one cannot hope to prove the base case of Brauer-Thrall II by constructing an infinite family of MCM ideals. At least for hypersurfaces, there are lower bounds on the ranks of stable MCM modules (Corollary 15.29 below). These bounds depend on the following theorem of Bruns [**Bru81**, Corollary 2]:

15.28. THEOREM (Bruns). *Let R be a commutative Noetherian ring and M a finitely generated R-module which is free of constant rank r. Let N be a second syzygy of M, and set $s = \operatorname{rank} N$. If M is not free, then the codimension of the non-free locus of M is $\leqslant r + s + 1$.* □

15.29. COROLLARY. *Let $(R, \mathfrak{m})$ be a hypersurface ring, and suppose that the singular locus of R is contained in a closed set of codimension c. Let M be a non-zero stable MCM module of constant rank r. Then $r \geqslant \frac{1}{2}(c - 1)$.*

PROOF. Since R is a hypersurface and M is stable MCM, Proposition 8.6 says that the second syzygy of M is isomorphic to M. Moreover, the non-free locus of M is contained in the singular locus of R by the Auslander-Buchsbaum formula. Therefore c is less than or equal to the codimension of the non-free locus of M. The inequality in Theorem 15.28 now gives the inequality $c \leqslant 2r + 1$. □

15.30. COROLLARY. *Let $(R, \mathfrak{m})$ be a hypersurface ring, and assume R is an isolated singularity of dimension d. Let M be a non-zero stable MCM module of constant rank r. Then $r \geqslant \frac{1}{2}(d - 1)$.* □

This bound is probably much too low. In fact, Buchweitz, Greuel and Schreyer [**BGS87**] conjecture that $r \geqslant 2^{d-2}$ for isolated hypersurface singularities. Still, the bound given in the corollary rules out MCM ideals once the dimension exceeds three.

Suppose, for example, that $R = \mathbb{C}[\![x_0, x_1, x_2, x_3, x_4]\!]/(x_0^4 + x_1^5 + x_2^2 + x_3^2 + x_4^2)$. This has uncountable CM type, by the results of Chapters 9 and 14. Every MCM ideal of R, however, is principal, by the Corollary above. On the other hand, since R has uncountable CM type there must be *some* positive integer r for which there are uncountably many indecomposable MCM modules of rank r. Of course this works whenever we have uncountable CM type:

15.31. PROPOSITION. *Let $(R, \mathfrak{m}, k)$ be a CM local ring with uncountable CM type. Then Brauer-Thrall II holds for R.* □

Using the structure theorem for hypersurfaces of countable CM type, we can recover Dieterich's theorem [**Die87**], as long as the ground field is uncountable:

15.32. THEOREM (Dieterich). *Let $(R, \mathfrak{m}, k)$ be a complete hypersurface singularity which is an isolated singularity. Assume k is uncountable, algebraically closed, and of characteristic different from two. If R has infinite CM type, then Brauer-Thrall II holds for R.*

PROOF. Since R is an isolated singularity and does not have finite CM type, Theorem 14.16 ensures that R has uncountable CM type. □

§5. Exercises

15.33. EXERCISE. Prove Lemma 15.12: For any ring Γ and any quotient ring $\Lambda = \Gamma/I$, the annihilator $\mathrm{Ann}_\Gamma I$ annihilates $\mathrm{Ext}^1_\Gamma(\Lambda, K)$ for every Γ-module K.

15.34. EXERCISE. Let R be a Noetherian ring and T a subring over which R is finitely generated as an algebra. Prove that $J_T(R) \subseteq \mathfrak{H}_T(R)$.

15.35. EXERCISE. Fill in the details of the proof of Theorem 15.18: show by induction on j that there exist regular local subrings $T_1, \ldots, T_j$ and elements $x_i \in J_{T_i}(R)$ such that $x_1, \ldots, x_j$ is part of a system of parameters. For the inductive step, use prime avoidance.

15.36. EXERCISE. Suppose that $\mathbf{x} = x_1, \ldots, x_d$ is a faithful system of parameters in a local ring R. Prove that R has at most an isolated singularity.

15.37. EXERCISE. Let k be an infinite field and D a k-algebra with $4 \leqslant d := \dim_k(D) < \infty$. Assume there is an element $\alpha \in D$ such that $D = k[\alpha]$. For $t \in k$, let $I_t = k + k(\alpha + t\alpha^2)$, and consider the $(k \hookrightarrow D)$-modules $I_t \hookrightarrow D$. For fixed $t \in k$, show that there are at most two elements $u \in k$ for which $I_u \hookrightarrow D$ and $I_t \hookrightarrow D$ are isomorphic as $(k \hookrightarrow D)$-modules. (It is helpful to treat the cases $d = 4$ and $d > 4$ separately.)

CHAPTER 16

Finite CM Type in Higher Dimensions

The results of Chapters 3, 4, and 7 give clear descriptions of the CM local rings of finite CM type in small dimension. For dimension greater than two, much less is known. Gorenstein rings of finite CM type are characterized by Theorem 9.15, but there are only two non-Gorenstein examples of dimension greater than two in the literature. In this chapter we describe these examples, and also present the theorem of Eisenbud and Herzog that these examples, together with those of the previous chapters, encompass all the *homogeneous* CM rings of finite CM type.

§1. Two examples

We give in this section the two known examples of non-Gorenstein Cohen-Macaulay local rings of finite CM type in dimension at least 3. They are taken from work of Auslander and Reiten [**AR89**]. We also quote two theorems from [**AR89**] to the effect that each example is the only one of its kind.

First we strengthen Brauer-Thrall I, Theorem 15.21, slightly for non-Gorenstein rings.

Let $(R, \mathfrak{m}, k)$ be a complete equicharacteristic CM local ring with algebraically closed residue field k, and assume that R has an isolated singularity. It follows from Theorem 15.21 that if $\mathcal{C} = \{M_1, \ldots, M_r\}$ is a finite set of indecomposable MCM R-modules which is closed under irreducible homomorphisms (i.e. for an irreducible homomorphism $X \longrightarrow Y$ between indecomposable MCM modules, we have $X \in \mathcal{C}$ if and only if $Y \in \mathcal{C}$), then R has finite CM type and $\mathcal{C}$ contains all the indecomposables. When R is not Gorenstein, a slightly weaker condition suffices. Say that a set $\mathcal{C}$ of indecomposable MCM modules is *closed under AR sequences* if for each indecomposable non-free module $M \in \mathcal{C}$, and each indecomposable module $N \in \mathcal{C}$ not isomorphic to the canonical module ω, all indecomposable summands of the terms in the AR sequences $0 \longrightarrow \tau(M) \longrightarrow E \longrightarrow M \longrightarrow 0$ and $0 \longrightarrow N \longrightarrow E' \longrightarrow \tau^{-1}(N) \longrightarrow 0$ are in $\mathcal{C}$.

16.1. PROPOSITION. *Let $(R, \mathfrak{m}, k)$ be a complete equicharacteristic CM local ring with algebraically closed residue field k and with an isolated singularity. Assume that R is not Gorenstein. If $\mathcal{C}$ is a set of indecomposable MCM R-modules which contains R and the canonical module ω, and is closed under AR sequences, then $\mathcal{C}$ is closed under irreducible homomorphisms. If in addition $\mathcal{C}$ is finite, then R has finite CM type.*

PROOF. The last sentence follows from the ones before by Theorem 15.21.

Let $f \colon X \longrightarrow Y$ be an irreducible homomorphism with X and Y indecomposable MCM R-modules. Assume that $Y \in \mathcal{C}$; the other case is dual. We may assume that $X \not\cong \omega$. If $Y \not\cong R$, then there is an AR sequence ending in Y: $0 \longrightarrow \tau Y \longrightarrow E \xrightarrow{p} Y \longrightarrow 0$. By Proposition 13.16, f is a component of p, so in particular $X \mid E$. Since $\mathcal{C}$ is closed under AR sequences, $X \in \mathcal{C}$.

If $Y \cong R$, then $Y \not\cong \omega$. There is thus an AR sequence beginning in Y: $0 \longrightarrow Y \longrightarrow E \longrightarrow \tau^{-1} Y \longrightarrow 0$. By Exercise 16.8, $f \colon X \longrightarrow Y$ induces an irreducible homomorphism $\tau^{-1} f \colon \tau^{-1} X \longrightarrow \tau^{-1} Y$. Since $\tau^{-1} Y \not\cong R$, the first case implies that $\tau^{-1} X \in \mathcal{C}$, whence $X \in \mathcal{C}$ and we are done. □

16.2. EXAMPLE. Let $S = k[\![x, y, z, u, v]\!]$ and put $R = S/(yv - zu, yu - xv, xz - y^2)$, where k is an algebraically closed field of characteristic different from 2. Then R has finite CM type.

Define matrices over S

$$\psi = \begin{bmatrix} yv - zu & yu - xv & xz - y^2 \end{bmatrix} \quad \text{and} \quad \varphi = \begin{bmatrix} x & y \\ y & z \\ u & v \end{bmatrix},$$

so that the entries of ψ are the 2×2 minors of φ, and we have $R = \operatorname{cok} \psi$. Then the S-free resolution of R is a Hilbert-Burch type resolution

$$0 \longrightarrow S^{(2)} \xrightarrow{\varphi} S^{(3)} \xrightarrow{\psi} S \longrightarrow R \longrightarrow 0.$$

In particular, R has depth 3. The regular sequence x, v, $z - u$ is a system of parameters, so R is CM. Since the characteristic of k is not 2, the Jacobian criterion (Theorem 15.17) implies that R is an isolated singularity, whence in particular a normal domain.

The canonical module $\omega = \operatorname{Ext}_S^2(R, S)$ is presented over R by the transpose of the matrix φ. This is easily checked to be isomorphic to the ideal (u, v). (The natural map from ω to $(-v, u)$ is surjective and has kernel of rank zero, so is an isomorphism.)

§1. TWO EXAMPLES

Set $I = (x, y, u)$ and $J = (x, y, z)$. Each is an ideal of height one in R, with quotient a power series ring of dimension 2, so is a MCM R-module. We also have $I \cong \operatorname{redsyz}_1^R(\omega)$ and $J \cong I^\vee = \operatorname{Hom}_R(I, \omega)$. By Exercise 16.10, we have $I \cong \omega^*$ as well. Therefore, in the class group $\operatorname{Cl}(R)$, we have $[\omega] = -[I]$.

Let us compute the AR translates
$$\tau(-) = \operatorname{redsyz}_3^R(\operatorname{Tr}(-))^\vee \cong (\operatorname{redsyz}_1^R(-^*))^\vee,$$
by (12.2). We have
$$\tau(I) = \operatorname{redsyz}_1^R(I^*)^\vee \cong \operatorname{redsyz}_1^R(\omega)^\vee \cong I^\vee = J.$$
Similarly
$$\tau(\omega) = \operatorname{redsyz}_1^R(\omega^*)^\vee \cong \operatorname{redsyz}_1^R(I)^\vee.$$
Set $M = \operatorname{redsyz}_1^R(I)^\vee$, a MCM R-module of rank 2 which is indecomposable by Proposition 13.4, so that $\tau(\omega) = M$. To finish the AR translates, note that J^* is isomorphic to the ideal (x, u^2) and that $J \cong \operatorname{redsyz}_1^R((x, u^2))$, whence $\tau(J) = J^\vee \cong I$. Finally $\tau(M) = (I^*)^\vee = R$.

The syzygy module $M^\vee = \operatorname{redsyz}_1^R((x, y, u))$ is generated by the following six elements of $R^{(3)}$: the Koszul relations
$$z_1 = \begin{pmatrix} 0 \\ -u \\ y \end{pmatrix}, \quad z_2 = \begin{pmatrix} -u \\ 0 \\ x \end{pmatrix}, \quad z_3 = \begin{pmatrix} -y \\ x \\ 0 \end{pmatrix},$$
and the three additional relations
$$z_4 = \begin{pmatrix} 0 \\ -v \\ z \end{pmatrix}, \quad z_5 = \begin{pmatrix} -v \\ 0 \\ y \end{pmatrix}, \quad z_6 = \begin{pmatrix} -z \\ y \\ 0 \end{pmatrix}.$$

Define homomorphisms $f \colon \omega \longrightarrow M^\vee$, $g \colon \omega \longrightarrow M^\vee$, and $h \colon I \longrightarrow M^\vee$ by
$$f(u) = z_1, \qquad f(v) = z_4,$$
$$g(u) = z_2, \qquad g(v) = z_5,$$
$$h(x) = z_3, \quad h(y) = z_6, \quad h(u) = \begin{pmatrix} -v & u & 0 \end{pmatrix}^T$$

One checks easily that f, g, and h are well-defined, and that the sum $(f, g, h) \colon \omega^{(2)} \oplus I \longrightarrow M^\vee$ is surjective. Letting H be the kernel, we have, in the divisor class group $\operatorname{Cl}(R)$,
$$[H] = 2[\omega] \oplus I - [M^\vee] = -[I] - [\operatorname{redsyz}_1^R(I)] = -[I] + [I] = 0.$$

Moreover, $[H]$ has rank 1 and therefore is isomorphic to R. Thus we have a non-split short exact sequence

(16.1) $\qquad 0 \longrightarrow R \longrightarrow \omega^{(2)} \oplus I \longrightarrow M^\vee \longrightarrow 0.$

Dualizing gives another non-split short exact sequence

(16.2) $$0 \longrightarrow M \longrightarrow R^{(2)} \oplus J \longrightarrow \omega \longrightarrow 0\,.$$

To show that these are both AR sequences, it suffices to prove that the stable endomorphism ring $\underline{\mathrm{End}}_R(\omega)$ is isomorphic to k. Indeed, in that case Proposition 13.7 reads

$$\mathrm{Ext}^1_R(\omega, \tau(\omega)) \cong \mathrm{Hom}_R(\underline{\mathrm{End}}_R(\omega), E_R(k)) = k$$

so that any non-split extension in $\mathrm{Ext}^1_R(\omega, M)$ represents the AR sequence.

Since $\omega I' = (x, y, z, u, v)$ for $I' = (\frac{x}{u}, \frac{y}{u}, 1) \cong I$, Exercise 16.9 implies that multiplication by any $r \in (x, y, z, u, v)$ factors through a free module. Thus $\underline{\mathrm{End}}_R(\omega) = k$, and both (16.1) and (16.2) are AR sequences.

For the AR sequence ending in J, note that we already have an arrow $M \longrightarrow J$ in the AR quiver. The AR sequence ending in J has rank-one modules on both ends, so the middle term has rank two. The middle term is thus isomorphic to M, and we have the AR sequence

$$0 \longrightarrow I \longrightarrow M \longrightarrow J \longrightarrow 0\,.$$

We also have an arrow $I \longrightarrow M^\vee$, so we must have $M \cong M^\vee$. Finally, the AR quiver is already known to contain arrows $J \longrightarrow \omega$ and $R \longrightarrow I$, so the AR sequence ending in I is

$$0 \longrightarrow J \longrightarrow R \oplus \omega \longrightarrow I \longrightarrow 0\,.$$

The set $\{R, \omega, I, J, M\}$ is thus closed under AR sequences, so that R has finite CM type by Proposition 16.1, and the AR quiver looks like the one below.

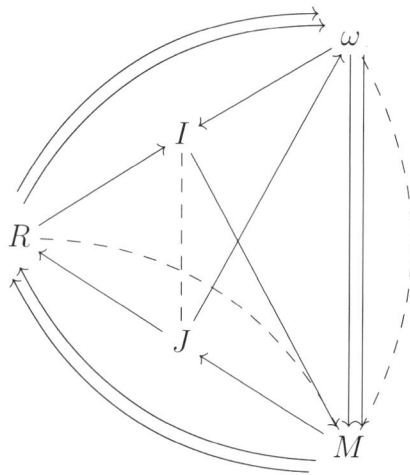

The ring of Example 16.2 is an example of a *scroll*. Let $(m_1, \ldots, m_r)$ be positive integers with $m_1 \geqslant m_2 \geqslant \cdots \geqslant m_r$, and consider a matrix of indeterminates

$$X = \begin{bmatrix} x_{1,0} & \cdots & x_{1,m_1-1} \\ x_{1,1} & \cdots & x_{1,m_1} \end{bmatrix} \cdots \begin{bmatrix} x_{r,0} & \cdots & x_{r,m_r-1} \\ x_{r,1} & \cdots & x_{r,m_r} \end{bmatrix}$$

over an infinite field k. The quotient ring $R = k[\![x_{ij}]\!]/I_2(X)$ by the ideal of 2×2 minors of X is called a (complete) *scroll of type* $(m_1, \ldots, m_r)$. Replacing the power series ring $k[\![x_{ij}]\!]$ by the polynomial ring $k[x_{ij}]$ gives the *graded* scrolls. In either case, R is a CM normal domain of dimension $r + 1$ and has an isolated singularity.

If $r = 1$, then the complete scroll R is isomorphic to the invariant ring $k[\![u^m, u^{m-1}v, \ldots, v^m]\!]$, so has finite CM type by Theorem 6.3. If R has type $(1,1)$, then $R \cong k[\![x, y, z, w]\!]/(xy - zw)$ is a three-dimensional (A_1) hypersurface singularity, so again has finite CM type. The ring of Example 16.2 is of type $(2, 1)$. These are the only examples with finite CM type:

16.3. THEOREM (Auslander-Reiten). *Let R be a complete or graded scroll of type different from (m), $(1, 1)$, and $(2, 1)$. Then R has infinite CM type.* □

Auslander and Reiten prove Theorem 16.3 by constructing an infinite family of rank-two graded MCM modules over the graded scrolls $k[x_{ij}]/I_2(X)$. For example, assume that R is a graded scroll of type $(n, 1)$, with $n \geqslant 3$. Then R is the quotient of the polynomial ring $k[x_0, \ldots, x_n, u, v]$ by the 2×2 minors of the matrix

$$\begin{pmatrix} x_0 & \cdots & x_{n-1} & u \\ x_1 & \cdots & x_n & v \end{pmatrix}$$

so is a three-dimensional normal domain. Set $A = (x_0^2, x_0x_1, \ldots, x_0x_n)$ and $B = (x_0^2, x_0x_1, \ldots, x_0x_{n-1}, x_0u)$. Then both A and B are indecomposable MCM R-modules of rank one. For $\lambda \in k$, let M_λ be the submodule of $R^{(2)}$ generated by the vectors $a_j = (x_0x_{j-1}, 0)$ for $1 \leqslant j \leqslant n$, $a_{n+1} = (x_0u, 0)$, $a_j = (0, x_0x_{j-n-2})$ for $n + 2 \leqslant j \leqslant 2n$, $a_{2n+1} = (x_0u, x_0x_{n-1})$, and $a_{2n+2} = (x_1u + \lambda x_0v, x_0x_n)$. Then we have a natural inclusion $B \xrightarrow{\beta} M_\lambda$ and a surjection $M_\lambda \xrightarrow{\alpha} A$.

Auslander and Reiten show that for each λ the sequence $0 \longrightarrow B \xrightarrow{\beta} M_\lambda \xrightarrow{\alpha} A \longrightarrow 0$ is exact, so that M_λ is MCM. Furthermore an isomorphism $f \colon M_\lambda \longrightarrow M_\mu$ induces an isomorphism between the corresponding extensions, which forces $\lambda = \mu$. The modules $\{M_\lambda\}_{\lambda \in k}$ thus form an infinite family of rank-two MCM modules.

The case of type $(m_1, m_2, \ldots, m_r)$ with $m_2 + \cdots + m_r \geq 2$ is handled similarly.

Here is the other example of this section.

16.4. EXAMPLE. Let R be the invariant ring of Example 5.24, so that k is a field of characteristic different from 2, $S = k[\![x, y, z]\!]$, G is the cyclic group of order 2 with the generator acting on $V = kx \oplus ky \oplus kz$ by negating each variable, and $R = S^G = k[\![x^2, xy, xz, y^2, yz, z^2]\!]$. Then R has finite CM type.

Let k denote the trivial representation of G, and k_- the other irreducible representation. Note that $V \cong k_-^{(3)}$. The Koszul complex

$$0 \longrightarrow S \otimes_k \bigwedge^3 V \longrightarrow S \otimes_k \bigwedge^2 V \longrightarrow S \otimes_k V \longrightarrow S \longrightarrow k \longrightarrow 0$$

resolves k both over S and over the skew group ring $S\#G$. Replacing V by $k_-^{(3)}$ and writing $S_- = S \otimes_k k_-$, we get

$$(16.3) \qquad 0 \longrightarrow S_- \longrightarrow S^{(3)} \longrightarrow S_-^{(3)} \longrightarrow S \longrightarrow k \longrightarrow 0.$$

Tensor with k_- to obtain an exact sequence

$$(16.4) \qquad 0 \longrightarrow S \longrightarrow S_-^{(3)} \longrightarrow S^{(3)} \longrightarrow S_- \longrightarrow k_- \longrightarrow 0.$$

As in Example 5.24, we find that the fixed submodule S_-^G is the R-submodule of S generated by (x, y, z). In particular, we have $S \cong S^G \oplus S_-^G$ as R-modules. Since S is Gorenstein, we must have $\operatorname{Hom}_R(S, \omega) \cong S$, where ω is the canonical module for R. In particular $R \oplus S_-^G \cong \omega \oplus (S_-^G)^\vee$. As R is not Gorenstein, this implies that $\omega \cong S_-^G$.

Applying $(-)^G$ to the resolution of k_- gives an exact sequence of R-modules

$$0 \longrightarrow R \longrightarrow \omega^{(3)} \longrightarrow R^{(3)} \longrightarrow \omega \longrightarrow 0.$$

Set $M = \operatorname{redsyz}_1^R(\omega)$, the kernel in the middle of this sequence, so that we have two short exact sequences

$$0 \longrightarrow R \longrightarrow \omega^{(3)} \longrightarrow M \longrightarrow 0,$$
$$0 \longrightarrow M \longrightarrow R^{(3)} \longrightarrow \omega \longrightarrow 0.$$

By the symmetry of the Koszul complex, the canonical dual of the first sequence is isomorphic to the second, so that $M^\vee \cong M$. Furthermore the square of the fractional ideal $\omega = (x, y, z)R$ is isomorphic to the maximal ideal of R, so that $\omega^* = \omega$. These allow us to compute the AR translates

$$\tau(\omega) = (\operatorname{redsyz}_1^R(\omega^*))^\vee \cong M^\vee \cong M$$

and
$$\tau(M) = (\text{redsyz}_1^R(M^*))^\vee \cong \omega^\vee = R.$$
As in the previous example, the fact that $\omega^2 = (x^2, xy, xz, y^2, yz, z^2)$ is the maximal ideal of R implies that $\underline{\text{End}}_R(\omega) \cong k$, so that $\text{Ext}_R^1(\omega, M)$ is one-dimensional and
$$0 \longrightarrow M \longrightarrow R^{(3)} \longrightarrow \omega \longrightarrow 0$$
is the AR sequence for ω. Dualizing gives
$$0 \longrightarrow R \longrightarrow \omega^{(3)} \longrightarrow M \longrightarrow 0,$$
the AR sequence for M.

The set of MCM modules $\{R, \omega, M\}$ is thus closed under AR sequences, so that R has finite CM type by Proposition 16.1, and the AR quiver is below.

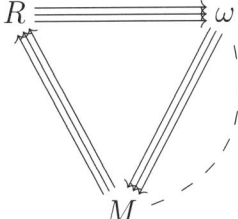

As with the earlier example, the ring of Example 16.4 is the only one of its kind with finite CM type. The proof is more involved than in the earlier case; see [**AR89**].

16.5. THEOREM (Auslander-Reiten). *Let $S = k[\![x_1, \ldots, x_n]\!]$, where k is an algebraically closed field and $n \geq 3$. Let G be a finite non-trivial group acting faithfully on S, such that $|G|$ is invertible in k. Then the invariant ring $R = S^G$ is of finite CM type if and only if $n = 3$ and G is the group of order two, where the generator sends each variable to its negative.* □

§2. Classification for homogeneous CM rings

Together with the results of previous chapters, the examples of the previous section exhaust the known CM complete local rings of finite CM type. There is no complete classification known. For homogeneous CM rings, however, there is such a classification, due to Eisenbud and Herzog [**EH88**].

Let k be an algebraically closed field of characteristic zero, and let $R = \bigoplus_{n=0}^\infty R_n$ be a positively graded k-algebra, generated in degree one and with $R_0 = k$. We call such an R a *homogeneous* ring. We further say that a CM homogeneous ring R has *finite CM type* if, up to isomorphism and shifts of the grading, there are only finitely many graded MCM R-modules.

16.6. THEOREM (Eisenbud-Herzog). *Let R be a CM homogeneous ring. Then R has finite CM type if and only if R is isomorphic to one of the rings in the following list.*
   (i) $k[x_0, \ldots, x_n]$ for some $n \geqslant 0$;
   (ii) $k[x_0, \ldots, x_n]/(x_0^2 + \cdots x_n^2)$ for some $n \geqslant 0$;
   (iii) $k[x]/(x^m)$ for some $m \geqslant 1$;
   (iv) $k[x,y]/(xy(x+y))$, a graded (D_4) hypersurface singularity;
   (v) $k[x,y,z]/(xy, yz, xz)$;
   (vi) $k[x_0, \ldots, x_m]/I_2 \begin{bmatrix} x_0 & \cdots & x_{m-1} \\ x_1 & \cdots & x_m \end{bmatrix}$ for some $m \geqslant 1$, a graded scroll of type (m);
   (vii) $k[x,y,z,u,v]/I_2 \begin{bmatrix} x & y & u \\ y & z & v \end{bmatrix}$, a graded scroll of type $(2,1)$; and
   (viii) $k[x,y,z,u,v,w]/I_2(A)$, where A is the generic symmetric 3×3 matrix
$$A = \begin{bmatrix} x & y & z \\ y & u & v \\ z & v & w \end{bmatrix}.$$

The rings in (vii) and (viii) are homogeneous versions of the rings in Examples 16.2 and 16.4. In particular
$$k[x,y,z,u,v,w]/I_2(A) \cong k[x^2, xy, xz, y^2, yz, z^2]$$
as non-homogeneous rings, though the ring on the right is not generated in degree one.

The classification follows from verifying Conjecture 7.21 for CM homogeneous domains:

16.7. THEOREM. *Let R be a CM homogeneous domain of finite CM type. Then for any maximal regular sequence $\mathbf{x}$ of elements of degree 1 in R, the quotient $R' = R/(x)$ satisfies $\dim_k R'_n \leqslant 1$ for all $n \geqslant 2$. In particular, if R is not Gorenstein then R has minimal multiplicity, i.e.*
$$e(R) = \mu_R(\mathfrak{m}) - \dim R + 1,$$
where $\mathfrak{m} = \bigoplus_{n=1}^{\infty} R_n$ is the irrelevant maximal ideal of R.

Artinian homogeneous rings satisfying the condition $\dim_k R'_n \leqslant 1$ for all $n \geqslant 2$ are called *stretched* [**Sal79**]. Equivalently, $R_{e-t} \neq 0$, where $e = e(R)$ is the multiplicity and $t = \dim_k R_1$ is the embedding dimension of R.

PROOF OF THEOREM 16.6 ASSUMING THEOREM 16.7. Assume R is not the polynomial ring. If R has dimension zero, then by Theorem 3.3 it is a principal ideal ring, so isomorphic to $k[x]/(x^m)$ for some

$m \geqslant 1$. If $\dim R = 1$, then by the graded version of Theorem 4.13 R is a finite birational extension of an ADE hypersurface singularity; the only homogeneous rings among these are (ii) with $n = 1$, (iv), and (v).

Now assume that R has dimension at least 2. If R is Gorenstein, then by Theorem 9.16 it is an ADE hypersurface singularity of multiplicity 2. Since R is homogeneous, this implies that R is the (A_1) hypersurface of (ii). Thus we may assume that R is not Gorenstein.

By Theorem 7.12, R is an isolated singularity. In particular, by Serre's criterion (Proposition A.9) R is a normal domain and has minimal multiplicity by Theorem 16.7. The homogeneous domains of minimal multiplicity are classified by the Del Pezzo-Bertini Theorem (see for example [**EH87**]). The ones with isolated singularities are

(a) (A_1) hypersurface rings $k[x_0, \ldots, x_n]/(x_0^2 + \cdots + x_n^2)$;
(b) graded scrolls of arbitrary type $(m_1, \ldots, m_r)$; and
(c) the ring of (viii).

Each ring in the first and third classes has finite CM type, while the only graded scrolls of finite CM type are those of type (m), $(1,1)$ (also an (A_1) hypersurface), and $(2,1)$ by Theorem 16.3. □

We sketch the proof that CM homogeneous rings of finite CM type are stretched. Let $\mathbf{x} = x_1, \ldots, x_d$ be a maximal regular sequence of elements of degree one in R, set $R' = R/(\mathbf{x})$, and assume that $\dim_k R'_c \geqslant 2$ for some $c \geqslant 2$. For each $u \in R'_c$, let $L_u = R'/(u)$, and set $M_u = \text{redsyz}_d^R(L_u)$. Then each M_u is a graded MCM R-module. Eisenbud and Herzog show:

(a) there is an upper bound on the ranks of the M_u;
(b) each M_u has a unique (up to scalar multiple) generator f_u of degree d, and all other generators have degree $> d$; and
(c) if we denote by $\overline{f_u}$ the image of f_u in $M_u/\mathbf{x}M_u$, then $\text{Ann}_R(\overline{f_u}) = \text{Ann}_R(L_u)$.

See the Exercises for proofs of these assertions. Assuming them, we can show that R is of infinite CM type. Indeed, if $M_u \cong M_{u'}$ for $u, u' \in R'_c$, then $M_u/\mathbf{x}M_u \cong M_{u'}/\mathbf{x}M_{u'}$, via an isomorphism taking $R\overline{f_u}$ to $R\overline{f_{u'}}$. It follows that the annihilators of L_u and $L_{u'}$ are equal, and in particular $(u) = (u')$ as ideals of R'. But since $\dim_k R'_c \geqslant 2$, there are infinitely many ideals of the form (u) for $u \in R'_c$. The bound on the ranks of the M_u thus implies that R has infinite CM type.

§3. Exercises

16.8. EXERCISE. In the notation of Proposition 16.1, let $0 \longrightarrow N \longrightarrow E \longrightarrow \tau^{-1}N \longrightarrow 0$ be an AR sequence and $f: N \longrightarrow Z$ a homomorphism between indecomposable MCM modules with $Z \not\cong \omega$. Prove that there is an induced homomorphism $\tau^{-1}f\colon \tau^{-1}N \longrightarrow \tau^{-1}Z$, and that $\tau^{-1}f$ is irreducible if f is.

16.9. EXERCISE. Let R be a Noetherian domain and I an ideal of R. Assume that there is a fractional ideal J of R such that $IJ \subseteq R$. Show that multiplication by an element $r \in IJ$, as a homomorphism $I \longrightarrow I$, factors through a free R-module.

16.10. EXERCISE. Let $L = (a,b)$ be a two-generated ideal of a ring R. Assume that L contains a non-zerodivisor, and let $L^{-1} = \{\alpha \in K \mid \alpha L \subseteq R\}$, where K is the total quotient ring. Prove that there is a short exact sequence $0 \longrightarrow L^{-1} \xrightarrow{\begin{bmatrix} b \\ -a \end{bmatrix}} R^{(2)} \xrightarrow{[a\ b]} L \longrightarrow 0$, so that L^{-1} is isomorphic to $\operatorname{syz}_1^R(L)$. Prove that $L^{-1} \cong L^*$.

16.11. EXERCISE. In the setup of the proof of Theorem 16.7, prove item (a), that the ranks of the modules M_u are bounded, by showing that the lengths of L_u are bounded and that
$$\operatorname{rank}(\operatorname{redsyz}_R^d(L)) \leqslant \ell(L)\operatorname{rank}(\operatorname{redsyz}_d^R(k))$$
for a module L of finite length.

16.12. EXERCISE. Prove (b) from the proof of Theorem 16.7 by constructing a comparison map between the R-free resolution $F_\bullet$ of L_u and the Koszul complex $K_\bullet$ on the regular sequence $\mathbf{x}$. Show by induction on i that the induced maps $K_i/\mathfrak{m}_R K_i \longrightarrow F_i/\mathfrak{m}_R F_i$ are all injective, so that $K_\bullet$ is a direct summand of $F_\bullet$. Finally, show that the minimal generators of the quotient F_i/K_i are all in degrees $> i$.

16.13. EXERCISE. Continuing the notation of Exercise 16.12, prove that the kernel of the map $M_u/\mathbf{x}M_u \longrightarrow F_{d-1}/\mathbf{x}F_{d-1}$ is isomorphic to the dth Koszul homology of $\mathbf{x}$ on L_u, which is L_u. Conclude that the generator $\overline{f_u}$ of $K_d/\mathbf{x}K_d$ has annihilator equal to that of L_u.

CHAPTER 17

Bounded CM Type

In this chapter we classify the complete equicharacteristic hypersurface rings of bounded CM type with residue field of characteristic not equal to 2. It is an astounding coincidence that the answer turns out to be precisely the same as in Chapter 14: The hypersurface rings of bounded but infinite type are the (A_∞) and (D_∞) hypersurface singularities in all positive dimensions. Note that the families of ideals showing (countable) non-simplicity in Lemma 9.3 for certain classes of hypersurface rings do not give rise to indecomposable modules of large rank; thus there does not seem to be a way to use the results of Chapter 14 to demonstrate unbounded CM type directly.

We also classify the one-dimensional complete CM local rings containing an infinite field and having bounded CM type. There is only one additional isomorphism type, which we have seen already in Example 14.23. The explicit classification, together with the results of Chapter 2, allows us to show that bounded type descends from the completion in dimension one.

§1. Hypersurface rings

To classify the complete hypersurface rings of bounded CM type, we must use Knörrer's results from Chapter 8 to reduce the problem to the case of dimension one. Recall (Definition 15.2) that bounded CM type was defined in terms of multiplicities of indecomposable MCM modules. It will be more convenient in what follows to find bounds on the minimal number of generators of MCM modules; luckily, this is the same as bounding their multiplicity. We leave the proof of this fact as an exercise (Exercise 17.12).

17.1. LEMMA. *Let A be a CM local ring. Then R has bounded CM type if and only if there is an integer b such that $\mu_R(M) \leqslant b$ for each indecomposable MCM R-module M.* □

17.2. PROPOSITION. *Let $R = S/(f)$ be a complete equicharacteristic hypersurface singularity, where $S = k[\![x_0, \ldots x_d]\!]$ and f is a non-zero non-unit of S. Set $R^\# = S[\![z]\!]/(f + z^2)$, the double branched cover of R.*

(i) If $R^\sharp$ has bounded CM type, then R has bounded CM type.

(ii) If the characteristic of k is not 2, then the converse holds as well. More precisely, if $\mu_R(M) \leqslant B$ for each indecomposable MCM R-module, then $\mu_{R^\sharp}(N) \leqslant 2B$ for each indecomposable MCM $R^\sharp$-module N.

PROOF. Assume that $R^\sharp$ has bounded CM type, and let B bound the minimal number of generators of MCM $R^\sharp$-modules. Let M be an indecomposable non-free MCM R-module. Then by Proposition 8.15 $M^{\sharp\flat} \cong M \oplus \mathrm{syz}_1^R(M)$, so M is a direct summand of $M^{\sharp\flat}$. Decompose $M^\sharp$ into indecomposable MCM $R^\sharp$-modules, $M^\sharp \cong N_1 \oplus \cdots \oplus N_t$. Then $M^{\sharp\flat} \cong N_1^\flat \oplus \cdots \oplus N_t^\flat$, and by KRS M is a direct summand of some $N_j^\flat$. Since $\mu_R(N_j^\flat) = \mu_{R^\sharp}(N_j)$ for each j, we have $\mu_R(M) \leqslant B$.

For the converse, assume that $\mu_R(M) \leqslant B$ for every indecomposable MCM R-module M, and let N be an indecomposable non-free MCM $R^\sharp$-module. By Proposition 8.18, $N^{\flat\sharp} \cong N \oplus \mathrm{syz}_1^{R^\sharp}(N)$. Decompose $N^\flat$ into indecomposable MCM R-modules, $N^\flat \cong M_1 \oplus \cdots \oplus M_s$. Then $N^{\flat\sharp} \cong M_1^\sharp \oplus \cdots \oplus M_s^\sharp$. By KRS again, N is a direct summand of some $M_j^\sharp$. It will suffice to show that $\mu_{R^\sharp}(M_j^\sharp) \leqslant 2B$ for each j.

By (ii) of Lemma 8.17, M_j is stable, and we have
$$\mu_{R^\sharp}(M_j^\sharp) = \mu_R(M_j^{\sharp\flat}) = \mu_R(M_j) + \mu_R(\mathrm{syz}_1^R(M))$$
by Proposition 8.15. But since M_j is a MCM R-module, all of its Betti numbers are equal to $\mu_R(M_j)$ by Proposition 8.6. Thus $\mu_R(M_j^{\sharp\flat}) = 2\mu_R(M_j) \leqslant 2B$. If, on the other hand, $M_j = R$, then $M_j^\sharp \cong R^\sharp$, and so $\mu_{R^\sharp}(M_j^\sharp) = 1$. □

Our next concern is to show that a hypersurface ring of bounded CM type has multiplicity at most two, as long as the dimension is at least two. This is a corollary of the following impressive theorem due to Kawasaki [**Kaw96**, Theorem 4.1], proven originally in the graded case by Herzog and Sanders [**HS88**]. (A similar result was obtained by Dieterich [**Die87**] using a theorem on the structure of the AR quiver of a complete isolated hypersurface singularity.) Recall that an *abstract hypersurface ring* is a Noetherian local ring $(A, \mathfrak{m})$ such that the $\mathfrak{m}$-adic completion $\widehat{A}$ is isomorphic to $B/(f)$ for some regular local ring B and non-unit f.

17.3. THEOREM. *Let $(A, \mathfrak{m})$ be an abstract hypersurface ring of dimension d. Assume that the multiplicity $e = \mathrm{e}(A)$ is greater than 2. Then for each $n > e$, the module $\mathrm{syz}_{d+1}^A(A/\mathfrak{m}^n)$ is indecomposable and*
$$\mu_R\left(\mathrm{syz}_{d+1}^A(A/\mathfrak{m}^n)\right) \geqslant \binom{d+n-1}{d-1}.$$

We omit the proof. Putting Kawasaki's theorem together with Herzog's Theorem 9.15, we have the following result.

17.4. PROPOSITION. *Let $(R, \mathfrak{m}, k)$ be a Gorenstein local ring of dimension at least two, and assume that R has bounded CM type. Then R is an abstract hypersurface ring of multiplicity at most* 2. □

When the hypersurface ring in Proposition 17.4 is complete and has an algebraically closed coefficient field of characteristic other then 2, we can show by the same arguments as in Chapter 9 that it is an iterated double branched cover of a one-dimensional hypersurface ring of bounded type.

17.5. THEOREM. *Let k be an algebraically closed field of characteristic not equal to 2, and let $R = k[\![x_0, \ldots, x_d]\!]/(f)$, where f is a non-zero non-unit of the formal power series ring and $d \geqslant 2$. Then R has bounded CM type if and only if $R \cong k[\![x, \ldots, x_d]\!]/(g + x_2^2 + \cdots + x_d^2)$ for some non-zero $g \in k[\![x_0, x_1]\!]$ such that $k[\![x_0, x_1]\!]/(g)$ has bounded CM type.* □

Notice that assumption that g be non-zero is essential, in view of Proposition 3.4.

§2. Dimension one

The results of the previous section reduce the problem of classifying hypersurface rings of bounded CM type to dimension one. In this section we will deal with those one-dimensional hypersurface rings, as well as the case of non-hypersurface rings of dimension one.

Our problem breaks down according to the multiplicity of the ring. Recall from Theorem 4.18 that over a one-dimensional CM local ring of multiplicity 2 or less, every MCM R-module is isomorphic to a direct sum of ideals of R, whence R has bounded CM type. If on the other hand R has multiplicity 4 or more, then by Proposition 4.3 R has an overring S with $\mu_R(S) \geqslant 4$, and then we may apply Theorem 4.2 to obtain an indecomposable MCM module of constant rank n for every $n \geqslant 1$.

Now we address the troublesome case of multiplicity three for complete equicharacteristic hypersurface rings. Let $R = k[\![x, y]\!]/(f)$, where k is a field and $f \in (x, y)^3 \setminus (x, y)^4$. If R is reduced, we know by Theorem 4.10 that R has bounded CM type if and only if R has finite CM type, that is, if and only if R satisfies the condition

(DR2) $\frac{\mathfrak{m}\overline{R}+R}{R}$ is cyclic as an R-module.

Hence we focus on the case where R is not reduced. Our strategy will be to build finite birational extensions S of R satisfying the hypothesis of Theorem 4.2: that $\mu_R\left(\frac{\mathfrak{m}S+R}{R}\right) \geq 2$, or equivalently $\mu_R\left(\frac{\mathfrak{m}S}{\mathfrak{m}}\right) \geq 2$.

17.6. THEOREM. *Let $R = k[\![x, y]\!]/(f)$, where k is a field and f is a non-zero non-unit of the formal power series ring $k[\![x, y]\!]$. Assume that*
  (i) $\mathrm{e}(R) = 3$;
  (ii) R is not reduced; and
  (iii) $R \not\cong k[\![x, y]\!]/(x^2 y)$.
For each positive integer n, R has an indecomposable MCM module of constant rank n.

PROOF. We know f has order 3 and that its factorization into irreducibles has a repeated factor. Thus, up to a unit, we have either $f = g^3$ or $f = g^2 h$, where g and h are irreducible elements of $k[\![x, y]\!]$ of order 1, and, in the second case, g and h are relatively prime. After a k-linear change of variables we may assume that $g = x$.

In the second case, if the leading form of h is not a constant multiple of x, then by another change of variable [**ZS75**, Cor. 2, p. 137] we may assume that $h = y$. This is the case we have ruled out in (iii).

Suppose now that the leading form of h is a constant multiple of x. By Corollary 9.6 to the Weierstrass Preparation Theorem, there exist a unit u and a non-unit power series $q \in k[\![y]\!]$ such that $h = u(x+q)$. Moreover, $q \in y^2 k[\![y]\!]$ (since the leading form of h is a constant multiple of x). In summary, there are two cases to consider:
  (a) $f = x^3$.
  (b) $f = x^2(x+q)$ for some $0 \neq q \in y^2 k[\![y]\!]$.

Let $\mathfrak{m} = (x, y)$ be the maximal ideal of R. We must show that R has a finite birational extension S such that $\mu_R(S) = 3$ and $\mathfrak{m}S/\mathfrak{m}$ is not cyclic as an R-module. We sketch the arguments, leaving the details to the reader.

In case (a) we put $S = R\left[\frac{x}{y^2}\right] = R + R\frac{x}{y^2} + R\frac{x^2}{y^4}$. Clearly $\mu_R(S) = 3$, and one checks that $\frac{\mathfrak{m}S}{\mathfrak{m}^2 S + \mathfrak{m}}$ is two-dimensional over $R/\mathfrak{m}$.

Assume now that we are in case (b). One can argue by descending induction that it suffices to consider the case where q has order 2 in $k[\![y]\!]$. Put $u = \frac{x}{y^2}$, $v = \frac{x^2 + qx}{y^5}$, and $S = R[u, v]$. Once again this can be seen to satisfy the assumptions of Theorem 4.2, and this finishes the proof. □

The argument in the proof of Theorem 17.6 does not apply to the (D_∞) hypersurface singularity $R = k[\![x, y]\!]/(x^2 y) \cong k[\![u, v]\!]/(u^2 v - u^3)$. One adjoins the idempotent $\frac{u^2}{v^2}$ to obtain a ring isomorphic to $k[\![v]\!] \times$

$k[\![u,v]\!]/(u^2)$, whose integral closure is $k[\![v]\!] \times \bigcup_{n=1}^{\infty} R\left[\frac{u}{v^n}\right]$. From this information one can easily check that $\mathfrak{m}S/\mathfrak{m}$ is a cyclic R-module for every finite birational extension S of $(R, \mathfrak{m})$, so we cannot apply Theorem 4.2. However, the calculations in Chapter 14 do indeed verify that the one-dimensional (A_∞) and (D_∞) hypersurface rings have bounded type. Combining this with Theorem 17.6, we have a complete classification of the complete one-dimensional equicharacteristic hypersurface rings of bounded CM type.

17.7. THEOREM. *Let k be an arbitrary field, and let $R = k[\![x,y]\!]/(f)$ be a complete hypersurface ring of dimension one, where f is a non-zero non-unit. Then R has bounded but infinite CM type if and only if R is isomorphic either to the (A_∞) singularity or to the (D_∞) singularity.*

Further, if R has unbounded CM type, then R has, for each positive integer r, an indecomposable MCM module of constant rank r. □

Turning now to the non-hypersurface situation in multiplicity 3, we have the following structural result for the relevant rings.

17.8. LEMMA. *Let $(R, \mathfrak{m}, k)$ be a one-dimensional local CM ring with k infinite, and suppose $e(R) = \mu_R(\mathfrak{m}) = 3$. Let N be the nilradical of R. Then:*
  (i) $N^2 = 0$.
  (ii) $\mu_R(N) \leqslant 2$.
  (iii) *If $\mu_R(N) = 2$, then $\mathfrak{m}$ is generated by three elements u, v, w such that $\mathfrak{m}^2 = \mathfrak{m}u$ and $N = Rv + Rw$.*
  (iv) *If $\mu_R(N) = 1$, then $\mathfrak{m}$ is generated by three elements u, v, w such that $\mathfrak{m}^2 = \mathfrak{m}u$, $N = Rw$, and $vw = w^2 = 0$.*

PROOF. Since the residue field of R is infinite, we can find a minimal reduction for $\mathfrak{m}$, that is, a non-zerodivisor $u \in \mathfrak{m}$ such that $\mathfrak{m}^{n+1} = u\mathfrak{m}^n$ for all $n \gg 0$. Now, using the formula

(17.1) $$\mu_R(J) \leqslant e(R) - e(R/J)$$

for an ideal J of height 0 in a one-dimensional CM local ring R (Theorem A.29(ii)), it is straightforward to show (i) and (ii). The other two assertions are easy as well. □

17.9. THEOREM. *Let k be an infinite field. The following is a complete list, up to k-isomorphism, of the complete, equicharacteristic, CM local rings of dimension one with bounded but infinite CM type and with residue field k:*
  (i) *the (A_∞) hypersurface singularity $k[\![x,y]\!]/(x^2)$;*
  (ii) *the (D_∞) hypersurface singularity $k[\![x,y]\!]/(x^2y)$;*

(iii) the endomorphism ring E of the maximal ideal of the (D_∞) singularity, which satisfies

$$E \cong k[\![x,y,z]\!]/(yz, x^2-xz, xz-z^2) \cong k[\![a,b,c]\!]/(ab, ac, c^2)\,.$$

Moreover, if $(R, \mathfrak{m}, k)$ is a one-dimensional, complete, equicharacteristic CM local ring and R does not *have bounded CM type, then R has, for each positive integer r, $|k|$ pairwise non-isomorphic indecomposable MCM modules of constant rank r.*

PROOF. The (A_∞) and (D_∞) hypersurface rings have bounded but infinite CM type by the calculations in Chapter 14. In Example 14.23, we showed that E has the presentations asserted above, and that E has countable CM type. More precisely, we used Lemma 4.9 to see that the indecomposable MCM E-modules are precisely the non-free indecomposable MCM modules over the (D_∞) hypersurface ring, whence E has bounded but infinite CM type as well.

To prove that the list is complete and to prove the "Moreover" statement, assume now that $(R, \mathfrak{m}, k)$ is a one-dimensional, complete, equicharacteristic CM local ring with k infinite, and that R has infinite CM type but does *not* have indecomposable MCM modules of arbitrarily large constant rank. We will show that R is isomorphic to one of the rings in the statement of the Theorem. As above, we proceed by building finite birational extensions of R to which we may apply Theorem 4.2.

If R is a hypersurface ring, Theorem 17.7 tells us that R is isomorphic to either $k[\![x,y]\!]/(x^2)$ or $k[\![x,y]\!]/(x^2y)$. Thus we assume that $\mu_R(\mathfrak{m}) \geqslant 3$. But $e(R) \leqslant 3$ by Theorem 4.2 and we know by Exercise 11.53 that $e(R) \geqslant \mu_R(\mathfrak{m}) - \dim R + 1$. Therefore we may assume that $e(R) = \mu_R(\mathfrak{m}) = 3$. Thus we are in the situation of Lemma 17.8. Moreover, we may assume that R is not reduced, else we are done by Theorem 4.10, so R has non-trivial nilradical N.

If N requires two generators, then by Lemma 17.8(iii), we can find elements u, v, w in R such that $\mathfrak{m} = Ru + Rv + Rw$, u is a minimal reduction of $\mathfrak{m}$ with $\mathfrak{m}^2 = \mathfrak{m}u$, and $N = Rv + Rw$. Put $S = R\left[\frac{v}{u^2}, \frac{w}{u^2}\right]$. It is easy to verify (by clearing denominators) that $\{1, \frac{v}{u^2}, \frac{w}{u^2}\}$ is a minimal generating set for S as an R-module, and that the images of $\frac{v}{u}$ and $\frac{w}{u}$ form a minimal generating set for $\frac{\mathfrak{m}S}{\mathfrak{m}}$. Thus R has indecomposables of arbitrarily large constant rank by Theorem 4.2 and our basic assumption is violated.

We may therefore assume that N is principal. This is the hard case of the proof; we sketch the argument, and point to [**LW05**] for the details. Using Lemma 17.8(iv), we once again find elements u, v, w in

R such that $\mathfrak{m} = Ru + Rv + Rw$, u is again a minimal reduction of $\mathfrak{m}$ with $\mathfrak{m}^2 = \mathfrak{m}u$, and $N = Rw$ with $vw = w^2 = 0$.

Since $v^2 \in \mathfrak{m}u \subset Ru$, we see that R/Ru is a three-dimensional k-algebra. Further, since $\bigcap_n (Ru^n) = 0$, it follows that R is finitely generated (and free) as a module over the discrete valuation ring $D = k[\![u]\!]$. One checks that $R = D + Dv + Dw$ (and therefore $\{1, v, w\}$ is a basis for R as a D-module).

In order to understand the structure of R we must analyze the equation that puts v^2 into $u\mathfrak{m}$. Thus we write $v^2 = u^r(\alpha u + \beta v + \gamma w)$, where $r \geqslant 1$ and $\alpha, \beta, \gamma \in D$. Since u is a non-zerodivisor and $vw = w^2 = 0$, we see immediately that $\alpha = 0$. Thus we have
$$v^2 = u^r(\beta v + \gamma w),$$
with β and γ in D. Moreover, at least one of β and γ must be a unit of D.

If $r \geqslant 2$, put $S = R[\frac{v}{u^2}, \frac{w}{u^2}]$. This finite birational extension forces R to have indecomposables of arbitrarily large constant rank by Theorem 4.2, so we must have $v^2 = u(\beta v + \gamma w)$ with $\beta, \gamma \in D$ and at least one of β, γ a unit of D. We will produce a hypersurface subring $A = D[\![g]\!]$ of R such that $R = \operatorname{End}_A(\mathfrak{m}_A)$. We will then show that $A \cong k[\![x, y]\!]/(x^2 y)$ and the proof will be complete.

In the case where γ is not a unit, set $A = D[v + w]$. Then one can show that A is a local ring with maximal ideal $\mathfrak{m}_A = Au + A(v+w)$, and that R is a finite birational extension of A. Since $v(v+w) = (v+w)^2$ and $w(v+w) = 0$, we see that v and w are in $\operatorname{End}_A(\mathfrak{m}_A)$. Since $\operatorname{End}_A(\mathfrak{m}_A)/A$ is simple (as A is Gorenstein), it follows that $R = \operatorname{End}_A(\mathfrak{m}_A)$.

If on the other hand γ is a unit of D, we put $A = D[v] \subseteq R$. Then A is a local ring with maximal ideal $\mathfrak{m}_A = Au + Av$. (The relevant equation this time is $v^3 = u\beta v^2$.) We have $uw = \gamma^{-1} v^2 - \gamma^{-1} \beta uv \in \mathfrak{m}_A$. As in the first case, we conclude that $R = \operatorname{End}_A(\mathfrak{m}_A)$.

By Lemma 4.9, A has infinite CM type but does not have indecomposable MCM modules of arbitrarily large constant rank. Moreover, A cannot have multiplicity 2, since it has a finite birational extension of multiplicity greater than 2. By Theorem 17.7, $A \cong k[\![x, y]\!]/(x^2 y)$, as desired. $\square$

§3. Descent in dimension one

In this section we use the classification theorem in the previous section, together with the results on extended modules in Chapter 2, to show that bounded CM type passes to and from the completion of an equicharacteristic one-dimensional CM local ring $(R, \mathfrak{m}, k)$ with k infinite. Contrary to the situation in Chapter 10, we do not assume that R is excellent with an isolated singularity; indeed, in dimension one the latter assumption would make $\widehat{R}$ reduced, in which case finite and bounded CM type are equivalent by Theorem 4.10. We do, however, insist that k be infinite, in order to use the crucial fact from §2 that failure of bounded CM type implies the existence of indecomposable MCM modules of unbounded *constant* rank and also to use the explicit matrices worked out in Proposition 14.19 and Example 14.23 for the indecomposable MCM modules over $k[\![x,y]\!]/(x^2 y)$.

17.10. THEOREM. *Let $(R, \mathfrak{m}, k)$ be a one-dimensional equicharacteristic CM local ring with completion $\widehat{R}$. Assume that k is infinite. Then R has bounded CM type if and only if $\widehat{R}$ has bounded CM type. If R has unbounded CM type, then R has, for each r, an indecomposable MCM module of constant rank r.*

PROOF. Assume that $\widehat{R}$ does not have bounded CM type. Fix a positive integer r. By Theorem 17.9 we know that $\widehat{R}$ has an indecomposable MCM module M of constant rank r. By Corollary 2.8 there is a finitely generated R-module N, necessarily MCM and with constant rank r, such that $\widehat{N} \cong M$. Obviously N too must be indecomposable.

Assume from now on that $\widehat{R}$ has bounded CM type. If $\widehat{R}$ has *finite* CM type, the same holds for R by Theorem 10.1. Therefore we assume that $\widehat{R}$ has infinite CM type. Then $\widehat{R}$ is isomorphic to one of the three rings of Theorem 17.9.

If $\widehat{R} \cong k[\![x,y]\!]/(x^2)$, then $\mathrm{e}(R) = \mathrm{e}(\widehat{R}) = 2$, and R has bounded CM type by Theorem 4.18. Suppose for the moment that we have verified bounded CM type for any local ring S whose completion is isomorphic to $E = k[\![x,y,z]\!]/(yz, x^2 - xz, xz - z^2)$. If, now, $\widehat{R} \cong k[\![x,y]\!]/(x^2 y)$, put $S = \mathrm{End}_R(\mathfrak{m})$. Then $\widehat{S} \cong E$, whence S has bounded CM type. Therefore so has R, by Lemma 4.9. Thus we assume that $\widehat{R} \cong E$.

Our plan is to examine each of the indecomposable non-free E-modules and then use Corollary 2.8 to determine exactly which MCM E-modules are extended from R. As we saw in Example 14.23, those indecomposable MCM modules are the cokernels of the following matrices over $T = k[\![x,y]\!]/(x^2 y)$:

$$
\begin{array}{cccc}
(y) & (x^2) & (x) & (xy) \\
\alpha = \begin{pmatrix} x & \\ y^j & -x \end{pmatrix} & & \beta = \begin{pmatrix} xy & \\ y^j & -xy \end{pmatrix} & \\
\gamma = \begin{pmatrix} x & \\ y^j & -xy \end{pmatrix} & & \delta = \begin{pmatrix} xy & \\ y^j & -x \end{pmatrix} &
\end{array}
$$

§3. DESCENT IN DIMENSION ONE

where j is a positive integer. Let $\mathfrak{P} = (x)$ and $\mathfrak{Q} = (y)$ be the two minimal prime ideals of T. Note that $T_\mathfrak{P} \cong k((y))[x]/(x^2)$ and $T_\mathfrak{Q} \cong k((x))$. With the exception of $U := \mathrm{cok}(x)$ and $V := \mathrm{cok}(xy)$, each of the E-modules listed above is locally free at both $\mathfrak{P}$ and $\mathfrak{Q}$. The ranks are given in the following table.

φ	$\mathrm{rank}_\mathfrak{P}\,\mathrm{cok}\,\varphi$	$\mathrm{rank}_\mathfrak{Q}\,\mathrm{cok}\,\varphi$
(x^2)	1	0
(y)	0	1
α	1	0
β	1	2
γ	1	1
δ	1	1

Let M be a MCM $\widehat{R}$-module, and write

(17.2)
$$M \cong \left(\bigoplus_{i=1}^{a} A_i\right) \oplus \left(\bigoplus_{j=1}^{b} B_j\right) \oplus \left(\bigoplus_{k=1}^{c} C_k\right) \oplus \left(\bigoplus_{l=1}^{d} D_l\right) \oplus U^{(e)} \oplus V^{(f)},$$

where the A_i, B_j, C_k, D_l are indecomposable generically free modules, of ranks $(1,0)$, $(0,1)$, $(1,1)$, $(1,2)$ respectively, and again $U = \mathrm{cok}(x)$ and $V = \mathrm{cok}(xy)$.

Suppose first that R is a domain. Then M is extended if and only if $a = b+d$ and $e = f = 0$. Now the indecomposable MCM R-modules are those whose completions have (a,b,c,d,e,f) minimal and non-trivial with respect to these relations. (We are implicitly using Corollary 1.15 here.) One checks that the possibilities are $(0,0,1,0,0,0)$, $(1,1,0,0,0,0)$, and $(1,0,0,1,0,0)$, and we conclude that the indecomposable MCM R-modules have rank 1 or 2.

Next suppose that R is reduced but not a domain. Then R has exactly two minimal prime ideals, and we see from Corollary 2.8 that every generically free $\widehat{R}$-module is extended from R; however, neither U nor V can be a direct summand of an extended module. In this case, the indecomposable MCM R-modules are generically free, with ranks $(1,0)$, $(0,1)$, $(1,1)$ and $(1,2)$ at the minimal prime ideals.

Finally, we assume that R is not reduced. We must now consider the two modules U and V that are not generically free. We will see that $U = \mathrm{cok}(x)$ is always extended and that $V = \mathrm{cok}(xy)$ is extended if and only if R has two minimal prime ideals. Note that $U \cong Txy = Exy$ (the nilradical of $E = \widehat{R}$), while $V \cong Tx = Ex$.

The nilradical N of R is of course contained in the nilradical Exy of $\widehat{R}$. Moreover, since $Exy \cong E/(x,z)$ is a faithful module over

$E/(x, z) \cong k[\![y]\!]$, every non-zero submodule of Exy is isomorphic to Exy. In particular, $N\widehat{R} \cong Exy$. This shows that U is extended.

Next we deal with V. The kernel of the surjective map $Ex \longrightarrow Exy$, given by multiplication by y, is Ex^2. Thus we have a short exact sequence

(17.3) $$0 \longrightarrow Ex^2 \longrightarrow V \xrightarrow{y} U \longrightarrow 0.$$

Observe that $Ex^2 = Tx^2 \cong \operatorname{cok}(y)$ is generically free of rank $(0, 1)$. Let K be the common total quotient ring of T and $\widehat{R}$. Then $K \otimes_E Ex^2$ is a projective K-module, and as K is Gorenstein, (17.3) splits when tensored up to K. In particular, this gives

$$K \otimes_E V \cong \left(K \otimes_E Ex^2\right) \oplus \left(K \otimes_E U\right).$$

If, now, R has two minimal primes, then every generically free $\widehat{R}$-module is extended, by Corollary 2.8. In particular Ex^2 is extended, and by Lemma 2.7 so is V. Thus every indecomposable MCM $\widehat{R}$-module is extended, and R has bounded CM type.

If, on the other hand, R has just one minimal prime ideal, then the module M in (17.2) is extended if and only if $a = b + d + f$. The $\widehat{R}$-modules corresponding to indecomposable MCM R-modules are therefore U, $V \oplus W$, where W is some generically free module of rank $(0, 1)$, and the modules of constant rank 1 and 2 described above. $\square$

What if k is finite? If R (as in Theorem 17.10) has bounded but infinite CM type, we can take an elementary gonflement $(R, \mathfrak{m}, k) \longrightarrow (S, \mathfrak{n}, \ell)$ with ℓ infinite. We get the homomorphisms $R \longrightarrow S \longrightarrow S^{\mathrm{h}} \longrightarrow \widehat{S}$ of (15.5). Using Proposition 10.6, we see as before that $\widehat{S}$ must have bounded but infinite CM type and therefore be isomorphic to one of the three rings listed in Theorem 17.9. Conversely, if $\widehat{S}$ has unbounded CM type, we know that S has, for each $r \geqslant 1$, an indecomposable MCM module of constant rank r. Unfortunately, we don't know whether or not these modules are extended from R, or even whether or not R must have unbounded CM type.

Proving descent of bounded CM type in general seems quite difficult. Part of the difficulty lies in the fact that, in general, there is no bound on the number of indecomposable MCM $\widehat{R}$-modules required to decompose the completion of an indecomposable MCM R-module. Thus the argument of Theorem 10.1, while sufficient for showing descent of finite CM type, is not enough for bounded CM type.

Here is an example to illustrate. Recall that for a two-dimensional normal domain, the divisor class group essentially controls which modules are extended to the completion. Precisely (Proposition 2.15), if

R and $\widehat{R}$ are both normal domains, then a torsion-free $\widehat{R}$-module N is extended from R if and only if $\operatorname{cl}(N)$ is in the image of the natural map on divisor class groups $\operatorname{Cl}(R) \longrightarrow \operatorname{Cl}(\widehat{R})$.

17.11. EXAMPLE. Let R be a complete local two-dimensional normal domain containing a field, and assume that the divisor class group $\operatorname{Cl}(R)$ has an element α of infinite order. For example, one might take the ring of Lemma 2.16.

By Heitmann's theorem [**Hei93**], there is a unique factorization domain A contained in R such that $\widehat{A} = R$. Choose, for each integer n, a divisorial ideal I_n corresponding to $n\alpha \in \operatorname{Cl}(\widehat{A})$. For each $n \geqslant 1$, let $M_n = I_n \oplus N_n$, where N_n is the direct sum of n copies of I_{-1}. Then M_n has trivial divisor class and therefore is extended from A by Proposition 2.15. However, no non-trivial proper direct summand of M_n has trivial divisor class, and it follows that M_n (a direct sum of $n+1$ indecomposable $\widehat{A}$-modules) is extended from an indecomposable MCM A-module.

It is important to note that the example above does not give a counterexample to descent of bounded CM type, but merely points out one difficulty in studying descent.

§4. Exercises

17.12. EXERCISE. Let A be a local ring. Prove that there is an upper bound on the multiplicities of the indecomposable MCM A-modules if and only if there is a bound on their minimal numbers of generators. (See Corollary A.24.)

17.13. EXERCISE. Complete the proof of Theorem 17.6.

17.14. EXERCISE. Show that the argument of Theorem 17.6 does not apply to $R = k[\![u,v]\!]/(u^2v - v^3)$, since $\mathfrak{m}S/\mathfrak{m}$ is a cyclic R-module for every finite birational extension S of R.

17.15. EXERCISE. Finish the proof of Lemma 17.8.

APPENDIX A

Basics and Background

Here we collect some basic definitions and results that are necessary but somewhat peripheral to the main themes of the book. Some of the results are stated without proof; for these, one can find proofs in [**Mat89**]. We refer to [**Mat89**] also for any unexplained terminology.

§1. Depth, syzygies, and Serre's conditions

Throughout this section we let $(R, \mathfrak{m}, k)$ be a local ring.

A.1. DEFINITION. Let M be a finitely generated R-module. The *depth* of M is given by
$$\operatorname{depth}_R M = \inf \{n \mid \operatorname{Ext}_R^n(k, M) \neq 0\}.$$

Note that $\operatorname{depth}_R 0 = \inf(\emptyset) = \infty$. Conversely, non-zero modules have finite depth:

A.2. PROPOSITION. *Let M be a finitely generated R-module. If $M \neq 0$, then*
  (i) $\operatorname{depth}_R M < \infty$.
  (ii) $\operatorname{depth}_R M = \sup \{n \mid \exists\ M\text{-regular sequence } (x_1, \ldots, x_n) \subset \mathfrak{m}\}$.
  (iii) Every maximal M-regular sequence in $\mathfrak{m}$ has length n.
  (iv) $\operatorname{depth}_R M \leqslant \dim(R/\mathfrak{p})$ for every $\mathfrak{p} \in \operatorname{Ass} M$. In particular, we have $\operatorname{depth} M \leqslant \dim M \leqslant \dim R$.
  (v) If $(S, \mathfrak{n}) \longrightarrow (R, \mathfrak{m})$ is a local homomorphism and R is finitely generated as an S-module, then $\operatorname{depth}_S M = \operatorname{depth}_R M$.
  (vi) If $\mathfrak{p} \in \operatorname{Spec} R$, then $\operatorname{depth}_{R_\mathfrak{p}}(M_\mathfrak{p}) = 0 \iff \mathfrak{p} \in \operatorname{Ass} M$. □

When the base ring R is clear, or when, e.g. as in item (v) it is irrelevant, we often omit the subscript and write "$\operatorname{depth} M$".

Depth is closely related to the absence of *torsion*.

A.3. DEFINITION. Let A be any commutative ring and M an A-module. Say that M is *torsion-free* if every non-zerodivisor in A is a non-zerodivisor on M. Equivalently, the natural map $M \longrightarrow K \otimes_A M$, where K is the total quotient ring, is injective. At the other extreme, a module M is *torsion* provided each element of M is annihilated by some non-zerodivisor of R. Equivalently, $K \otimes_A M = 0$. The set of elements that are annihilated by non-zerodivisors is called the *torsion submodule* of M and is denoted by $\operatorname{tors}(M)$. Clearly $M/\operatorname{tors}(M)$ is torsion-free.

The next result is called the *Depth Lemma*. It follows easily from the long exact sequence of Ext.

A.4. LEMMA. *Let* $0 \longrightarrow U \longrightarrow V \longrightarrow W \longrightarrow 0$ *be a short exact sequence of finitely generated R-modules.*
  (i) If $\operatorname{depth} W < \operatorname{depth} V$, *then* $\operatorname{depth} U = \operatorname{depth} W + 1$.
  (ii) $\operatorname{depth} U \geqslant \min\{\operatorname{depth} V, \operatorname{depth} W\}$.
  (iii) $\operatorname{depth} V \geqslant \min\{\operatorname{depth} U, \operatorname{depth} W\}$. □

See [**Mat89**, Thm. 19.1] for a proof of the next result, called the Auslander-Buchsbaum formula. We write $\operatorname{pd}_R M$ for the projective dimension of an R-module M.

A.5. THEOREM (Auslander-Buchsbaum Formula). *Let M be an R-module of finite projective dimension. Then*
$$\operatorname{depth} M + \operatorname{pd}_R M = \operatorname{depth} R.$$

We often use a simple consequence of the Auslander-Buchsbaum formula: if M is a MCM module over a regular local ring, then M is free.

A.6. DEFINITION. Let M be a finitely generated module over a local ring $(R, \mathfrak{m})$, and let n be a non-negative integer. Then M *Serre's condition* (S_n) provided
$$\operatorname{depth}_{R_\mathfrak{p}}(M_\mathfrak{p}) \geqslant \min\{n, \dim(R_\mathfrak{p})\} \text{ for every } \mathfrak{p} \in \operatorname{Spec} R.$$

A.7. WARNING. Our terminology differs from that of EGA [**GD65**, Definition 5.7.2] and Bruns-Herzog [**BH93**, Section 2.1]. Where we have "$\dim(R_\mathfrak{p})$" those authors have "$\dim(M_\mathfrak{p})$". Notice, for example, that by the EGA definition every finite length module would satisfy (S_n) for all n, while this is certainly not the case with the definition we use. Of course, the two conditions agree for the ring itself.

The (S_n) conditions allow characterizations of reducedness and normality.

A.8. PROPOSITION. *The ring R is reduced if and only if the following hold.*
  (i) R satisfies (S_1), and
  (ii) $R_\mathfrak{p}$ is a field for every minimal prime ideal $\mathfrak{p}$. □

§1. DEPTH, SYZYGIES, AND SERRE'S CONDITIONS

A.9. PROPOSITION (Serre's criterion). *These are equivalent.*

(i) *R is a normal domain.*
(ii) *R satisfies (S_2), and $R_\mathfrak{p}$ is a regular local ring for each prime ideal $\mathfrak{p}$ of height at most one.* □

We will say that a finitely generated module M over a ring A is an rth *syzygy* (of N), provided there is an exact sequence

$$(A.1) \qquad 0 \longrightarrow M \longrightarrow F_{r-1} \longrightarrow \cdots \longrightarrow F_0 \longrightarrow N \longrightarrow 0,$$

where N is a finitely generated module and each F_i is a finitely generated projective A-module.

Syzygies are uniquely defined up to projective direct summands by Schanuel's Lemma:

A.10. LEMMA (Schanuel's Lemma). *Let M_1 and M_2 be r^{th} syzygies of a finitely generated N over a Noetherian ring A. Then there are finitely generated projective A-modules G_1 and G_2 such that $M_1 \oplus G_2 \cong M_2 \oplus G_1$.*

If R is local and each F_i is chosen minimally, then the resolution is essentially unique. In particular, the syzygies are unique up to isomorphism, and we let $\mathrm{syz}_r^R(N)$ denote the rth syzygy with respect to a minimal resolution. We define $\mathrm{redsyz}_r^R(N)$ to be the *reduced r^{th} syzygy*, obtained from $\mathrm{syz}_r^R(N)$ by deleting all non-zero free direct summands.

Serre's conditions are closely related to the property of being an nth syzygy. We explain this now using the following result, which is proved, but not quite stated correctly, in [**EG85**]. (Compare with Proposition 12.7.)

Recall that maximal Cohen-Macaulay modules over Gorenstein local rings are reflexive (by, for example, Theorem 11.5).

A.11. THEOREM. *Let M be a finitely generated R-module satisfying Serre's condition (S_n), where $n \geqslant 1$. Assume*

(i) *R satisfies (S_{n-1}), and*
(ii) *$R_\mathfrak{p}$ is Gorenstein for every prime $\mathfrak{p}$ with $\dim(R_\mathfrak{p}) \leqslant n-1$.*

Then there is an exact sequence

$$(A.2) \qquad 0 \longrightarrow M \xrightarrow{\alpha} F \longrightarrow N \longrightarrow 0,$$

in which F is a finitely generated free module and N satisfies (S_{n-1}).

PROOF. We start with an exact sequence

$$(A.3) \qquad 0 \longrightarrow K \longrightarrow G \longrightarrow M^* \longrightarrow 0,$$

where G is a finitely generated free module and $M^* = \operatorname{Hom}_R(M, R)$. Put $F = G^*$, and dualize (A.3), getting an exact sequence

(A.4) $\quad 0 \longrightarrow M^{**} \xrightarrow{\beta} F \longrightarrow K^* \longrightarrow \operatorname{Ext}^1_R(M^*, R) \longrightarrow 0.$

Let $\sigma \colon M \longrightarrow M^{**}$ be the canonical map, let $\alpha = \beta\sigma$, and put $N = \operatorname{cok} \alpha$.

To verify exactness of (A.2), we just have to show that σ is one-to-one. Supposing, by way of contradiction, that $L = \ker(\sigma)$ is non-zero, we choose $\mathfrak{p} \in \operatorname{Ass} L$. Then depth $L_\mathfrak{p} = 0$. Given any minimal prime $\mathfrak{q}$, we know $R_\mathfrak{q}$ is a zero-dimensional Gorenstein ring (since $n \geqslant 1$), and $M_\mathfrak{q}$ is a MCM $R_\mathfrak{q}$-module, whence $\sigma_\mathfrak{q}$ is an isomorphism. Thus $L_\mathfrak{q} = 0$ for each minimal prime $\mathfrak{q}$. In particular, $\dim(R_\mathfrak{p}) \geqslant 1$, so $\operatorname{depth}(M_\mathfrak{p}) \geqslant 1$. But this contradicts the fact that $\operatorname{depth}(L_\mathfrak{p}) = 0$.

Let $\mathfrak{p}$ be a prime of height h. If $h \leqslant n - 1$, we need to show that $N_\mathfrak{p}$ is MCM. Since $R_\mathfrak{p}$ is Gorenstein and $M_\mathfrak{p}$ is MCM, the canonical map $\sigma_\mathfrak{p}$ is an isomorphism. Also, $M^*_\mathfrak{p}$ is MCM, so $\operatorname{Ext}^1_{R_\mathfrak{p}}(M^*_\mathfrak{p}, R_\mathfrak{p}) = 0$. The upshot of all of this is that $N_\mathfrak{p} \cong K^*_\mathfrak{p}$. Now (A.3) shows that $K_\mathfrak{p}$ is MCM, and therefore so is its dual $K^*_\mathfrak{p}$.

To complete the proof that N satisfies (S_{n-1}), we assume now that $h \geqslant n$. We need to show that $\operatorname{depth}_{R_\mathfrak{p}}(N_\mathfrak{p}) \geqslant n - 1$. Suppose on the contrary that $\operatorname{depth}_{R_\mathfrak{p}}(N_\mathfrak{p}) < n - 1$. Since $\operatorname{depth}_{R_\mathfrak{p}}(F_\mathfrak{p}) \geqslant n - 1$, the Depth Lemma A.4, applied to (A.2), shows that

$$\operatorname{depth}_{R_\mathfrak{p}}(M_\mathfrak{p}) = 1 + \operatorname{depth}_{R_\mathfrak{p}}(N_\mathfrak{p}) < n,$$

a contradiction. □

A.12. COROLLARY. *Let $(R, \mathfrak{m})$ be a local ring, M a finitely generated R-module and n a positive integer. Assume R satisfies Serre's condition (S_n) and $R_\mathfrak{p}$ is Gorenstein for each prime $\mathfrak{p}$ of height at most $n - 1$. These are equivalent.*

(i) M is an nth syzygy.
(ii) M satisfies (S_n).

PROOF. (i) $\implies$ (ii) by the Depth Lemma, and (ii) $\implies$ (i) by Theorem A.11. □

A.13. COROLLARY. *Let $(R, \mathfrak{m})$ be a local ring that satisfies (S_2) and is Gorenstein in codimension one. These are equivalent for a finitely generated R-module M.*

(i) M is reflexive.
(ii) M satisfies (S_2).
(iii) M is a second syzygy.

A.14. COROLLARY. *Let R be a local normal domain and let M be a finitely generated R-module. If M is MCM, then M is reflexive. The converse holds if R has dimension two.* □

A.15. COROLLARY. *Let $(R, \mathfrak{m})$ be a CM local ring of dimension d, and assume that $R_\mathfrak{p}$ is Gorenstein for every prime ideal $\mathfrak{p} \neq \mathfrak{m}$. These are equivalent, for a finitely generated R-module M.*

(i) M is MCM.
(ii) M is a dth syzygy. □

A.16. REMARK. The hypothesis that R be Gorenstein on the punctured spectrum cannot be weakened, at least when R has a canonical module (or, more generally, a *Gorenstein module* [**Sha70**], that is, a finitely generated module whose completion is a direct sum of copies of the canonical module $\omega_{\widehat{R}}$). Let $(R, \mathfrak{m})$ be a d-dimensional CM local ring having a canonical module ω. If ω is a dth syzygy, then R is Gorenstein on the punctured spectrum. To see this, we build an exact sequence

$$(A.5) \qquad 0 \longrightarrow \omega \longrightarrow F \longrightarrow M \longrightarrow 0,$$

where F is free and M is a $(d-1)^{\text{st}}$ syzygy. Now let $\mathfrak{p}$ be any non-maximal prime ideal. Since $M_\mathfrak{p}$ is MCM and $\omega_\mathfrak{p}$ is a canonical module for $R_\mathfrak{p}$, (A.5) splits when localized at $\mathfrak{p}$ (apply Proposition 11.3). But then $\omega_\mathfrak{p}$ is free, and it follows that $R_\mathfrak{p}$ is Gorenstein. (We thank Bernd Ulrich for showing us this argument (cf. also [**LW00**, Lemma 1.4]).)

§2. Multiplicity and rank

In this section we gather the definitions and basic results on multiplicity and rank that are used in the body of the text. See Chapter 14 of [**Mat89**] for proofs.

Throughout we let $(R, \mathfrak{m}, k)$ be a local ring of dimension d, let I be an $\mathfrak{m}$-primary ideal of R, and let M be a finitely generated R-module.

A.17. DEFINITION. The *multiplicity* of I on M is defined by

$$\mathrm{e}_R(I, M) = \lim_{n \to \infty} \frac{d!}{n^d} \ell_R(M/I^n M),$$

where $\ell_R(-)$ denotes length as an R-module. In particular we set $\mathrm{e}_R(M) = \mathrm{e}_R(\mathfrak{m}, M)$ and call it the *multiplicity of M*. Finally, we denote $\mathrm{e}(R) = \mathrm{e}_R(R)$ and call it the *multiplicity of the ring R*.

It is standard that the Hilbert function $n \mapsto \ell_R(M/I^{n+1}M)$ is eventually given by a polynomial in n of degree equal to $\dim(M)$. Thus $\frac{\mathrm{e}_R(I,M)}{d!}$ is the coefficient of n^d in this Hilbert polynomial, and

$e_R(I, M) \neq 0$ if and only if $\dim(M) = d$. In particular if $d = 0$ then $e_R(I, M) = \ell_R(M)$ for any I.

It follows immediately from the definition that if $I \subseteq J$ are two $\mathfrak{m}$-primary ideals, then $e_R(I, M) \geqslant e_R(J, M)$. One case where equality holds is particularly useful.

A.18. DEFINITION. Let $I \subseteq J$ be ideals of R (not necessarily $\mathfrak{m}$-primary). We say I is a *reduction* of J if
$$I^{n+1} = JI^n$$
for some $n \geqslant 1$. Equivalently, $I^{n+k} = J^k I^n$ for all $n \gg 0$ and all $k \geqslant 1$.

The proof of the next result is a short calculation from the definitions.

A.19. PROPOSITION. *Let $I \subseteq J$ be $\mathfrak{m}$-primary ideals of R such that I is a reduction of J. Then $e_R(I, M) = e_R(J, M)$.* □

Reductions are often better-behaved ideals. In particular, under a mild assumption there is a reduction which is generated by a system of parameters. (Recall that a system of parameters consists of $d = \dim(R)$ elements generating an $\mathfrak{m}$-primary ideal.) See [**Mat89**, Theorem 14.14] for a proof.

A.20. THEOREM. *Assume that the residue field k is infinite. Then there exists a system of parameters $x_1, \ldots, x_d$ contained in I such that $(x_1, \ldots, x_d)$ is a reduction of I. Indeed, if I is generated by $a_1, \ldots, a_t$, then the x_i may be taken to be "sufficiently general" linear combinations $x_i = \sum r_{ij} a_j$ (for $r_{ij} \in R$ avoiding the common zeros of a finite list of polynomials). Such a reduction is called a* minimal reduction. □

The restriction on the residue field is rarely an obstacle in practice. For many questions, the general case can be reduced to this one by passing to a gonflement $R' = R[x]_{\mathfrak{m}R[x]}$ (see Chapter 10, §3). Since $R \longrightarrow R'$ is faithfully flat, it is easy to check that the association $I \mapsto IR'$ preserves containment, height, number of generators, and colength if I is $\mathfrak{m}$-primary. It thus preserves multiplicities. Since the residue field of R' is $R'/\mathfrak{m}R' = (R[x]/\mathfrak{m}R[x])_{\mathfrak{m}R[x]}$, the quotient field of $(R/\mathfrak{m})[x]$, it is an infinite field.

Theorem A.20 reduces many computations of multiplicity to the case of ideals generated by systems of parameters. The next results relate multiplicities over R to multiplicities calculated modulo a system of parameters.

A.21. THEOREM. *Let* $\mathbf{x} = x_1, \ldots, x_d$ *be a system of parameters contained in* I, *and set* $\overline{R} = R/(\mathbf{x})$, $\overline{I} = I/(\mathbf{x})$, *and* $\overline{M} = M/\mathbf{x}M$. *If* $x_i \in I^{s_i}$ *for each* i, *then*
$$\ell_R(\overline{M}) = \mathrm{e}_{\overline{R}}(\overline{I}, \overline{M}) \geqslant s_1 \cdots s_d \, \mathrm{e}_R(I, M).$$
In particular, if $x_i \in \mathfrak{m}^s$ *for all* i, *then* $\ell_R(\overline{M}) \geqslant s^d \, \mathrm{e}_R(M)$. □

A.22. COROLLARY. *Let* $(S, \mathfrak{n})$ *be a regular local ring and* $f \in S$ *a non-zero non-unit. Then the multiplicity of the hypersurface ring* $R = S/(f)$ *is the largest integer* s *such that* $f \in \mathfrak{n}^s$. □

The behavior of multiplicity for ideals generated by systems of parameters is most satisfactory in the Cohen-Macaulay case. This is [**Mat89**, Theorems 14.10 and 14.11].

A.23. THEOREM. *Let* $\mathbf{x}$ *be a system of parameters for* R. *Then*
$$\ell_R(M/\mathbf{x}M) \geqslant \mathrm{e}_R((\mathbf{x}), M),$$
and if $\mathbf{x}$ *is a regular sequence on* M *then equality holds.* □

Denote by $\mu_R(M)$ the minimal number of generators required for M.

A.24. COROLLARY. *Let* $(R, \mathfrak{m}, k)$ *be a local ring and* M *a MCM* R-*module. Then* $\mu_R(M) \leqslant \mathrm{e}_R(M)$.

PROOF. We may assume, by passing to a gonflement, that k is infinite. Using Theorem A.20, we obtain a system of parameters $\mathbf{x}$ such that $(\mathbf{x})$ is a reduction of $\mathfrak{m}$. Then $\mathrm{e}(M) = \mathrm{e}((\mathbf{x}), M) = \ell_R(M/\mathbf{x}M)$ by Proposition A.19 and Theorem A.23. Finally, we have $\ell_R(M/\mathbf{x}M) \geqslant \ell_R(M/\mathfrak{m}M) = \mu_R(M/\mathfrak{m}M)$, and $\mu_R(M/\mathfrak{m}M) = \mu_R(M)$ by NAK. □

Maximal Cohen-Macaulay modules M for which $\mu_R(M) = \mathrm{e}_R(M)$ are said to be *maximally generated* [**BHU87**] and are known also as *Ulrich* modules [**HK87**]. It is unknown whether or not every local CM ring has an Ulrich module.

Here are a few more basic facts on multiplicity.

A.25. PROPOSITION. *Let* $0 \longrightarrow M' \longrightarrow M \longrightarrow M'' \longrightarrow 0$ *be a short exact sequence of finitely generated* R-*modules. Then*
$$\mathrm{e}_R(I, M) = \mathrm{e}_R(I, M') + \mathrm{e}_R(I, M'').$$
□

A.26. PROPOSITION (*"Associativity Formula"*). *We have*
$$\mathrm{e}_R(I, M) = \sum_{\mathfrak{p}} \ell_{R_{\mathfrak{p}}}(M_{\mathfrak{p}}) \cdot \mathrm{e}_{R/\mathfrak{p}}(\overline{I}, R/\mathfrak{p}),$$

where the sum is over all minimal primes $\mathfrak{p}$ of R such that $\dim(R/\mathfrak{p}) = d$, and $\overline{I}$ denotes the image of I in $R/\mathfrak{p}$. If in particular R is a domain with quotient field K, then $e_R(I, M) = \dim_K(K \otimes_R M)$. □

The quantity $\dim_K(K \otimes_R M)$ in Proposition A.26 is known as the *rank* of M. We extend this notion as follows.

A.27. DEFINITION. Let A be a Noetherian ring and N a finitely generated A-module. Denote by K the total quotient ring of A, obtained by inverting the complement of the associated primes of A. Say that N has *constant rank* provided $K \otimes_A N$ is a free K-module. If $K \otimes_A N \cong K^{(r)}$ (equivalently, $N_\mathfrak{p} \cong A_\mathfrak{p}^{(r)}$ for every $\mathfrak{p} \in \operatorname{Ass} A$), we say that N has constant rank r.

The following useful fact follows directly from Proposition A.26.

A.28. PROPOSITION. *Let M be a finitely generated R-module with constant rank r. Then $e_R(M) = e(R) \cdot r$.* □

In dimension one, the multiplicity of R carries a great deal of structural information. The second statement of the next result goes back to Akizuki [**Aki37**].

A.29. THEOREM. *Assume $d = 1$.*
  (i) *The multiplicity of R is the minimal number of generators required for high powers of $\mathfrak{m}$.*
  (ii) *If R is Cohen-Macaulay, then $e(R)$ is the sharp bound on $\mu_R(I)$ as I runs over all ideals of R. Moreover, for every ideal I of R we have the inequality*
$$\mu_R(I) \leqslant e(R) - e(R/I).$$
  (iii) *If R is Cohen-Macaulay, then $e(R)$ is the sharp bound on $\mu_R(S)$ for S a finite birational extension of R.*
  (iv) *If R is reduced and the integral closure $\overline{R}$ is finitely generated over R, then $e(R) = \mu_R(\overline{R})$.*

PROOF. Since $\dim(R) = 1$, we have $\ell_R(R/\mathfrak{m}^{n+1}) = en - p$ for $n \gg 0$ and some $p \in \mathbb{Z}$. Since also $\ell_R(R/\mathfrak{m}^{n+1}) = \ell_R(R/\mathfrak{m}^n) + \mu_R(\mathfrak{m}^n)$, part (i) follows.

For (ii), it will suffice, by (i), to prove the inequality. Noting that every non-zero ideal of R is an MCM R-module, we have, from Corollary A.24
$$\mu_R(I) \leqslant e_R(I) = e(R) - e_R(R/I),$$
by additivity of multiplicity along exact sequences.

The bound in (ii) is sharp by (i). Every finite birational extension of R is isomorphic as an R-module to an ideal of R (clear denominators), and is therefore generated by at most $e(R)$ elements. Proposition 4.3 shows that the bound is sharp.

For (iv) we observe that $\overline{R}$ is a principal ideal ring, so there exists a non-zerodivisor $x \in R$ such that $\mathfrak{m}\overline{R} = x\overline{R}$. Thus $\mu_R(\overline{R}) = \ell_R(\overline{R}/\mathfrak{m}\overline{R}) = \ell_R(\overline{R}/x\overline{R})$. By Theorem A.23 and Proposition A.19 this is equal to $e_R((x), \overline{R}) = e_R(\mathfrak{m}, \overline{R})$. Since $\overline{R}$ is a birational extension of R, it has constant rank 1, so $e_R(\mathfrak{m}, \overline{R}) = e(R)$ by Proposition A.28. □

§3. Henselian rings

We gather here a few equivalent conditions for a local ring to be Henselian. Condition (v) is the definition used in Chapter 1; condition (i) is one of the classical formulations.

A.30. THEOREM. *Let $(R, \mathfrak{m}, k)$ be a local ring. These are equivalent.*

(i) *For every monic polynomial f in $R[x]$ and every factorization $\overline{f} = g_0 h_0$ of its image in $k[x]$, where g_0 and h_0 are relatively prime monic polynomials, there exist monic polynomials $g, h \in R[x]$ such that $g \equiv g_0 \bmod \mathfrak{m}$, $h \equiv h_0 \bmod \mathfrak{m}$, and $f = gh$.*

(ii) *Every commutative module-finite R-algebra which is an integral domain is local.*

(iii) *Every commutative module-finite R-algebra is a direct product of local rings.*

(iv) *Every module-finite R-algebra of the form $R[x]/(f)$, where f is a monic polynomial, is a direct product of local rings.*

(v) *For every module-finite R-algebra Λ (not necessarily commutative) with Jacobson radical $\mathcal{J}(\Lambda)$, each idempotent of $\Lambda/\mathcal{J}(\Lambda)$ lifts to an idempotent of Λ.*

PROOF. We prove (i) $\Longrightarrow$ (ii) $\Longrightarrow$ (iii) $\Longrightarrow$ (v) $\Longrightarrow$ (iv) $\Longrightarrow$ (i).

(i) $\Longrightarrow$ (ii): Let D be a domain that is module-finite over R, and suppose D is not local. Then there exist non-units α and β of D such that $\alpha + \beta = 1$. Set $S = R[\alpha] \subseteq D$; then α and β are still non-units of S. Since in particular they are not in $\mathfrak{m}S$ by Lemma 1.7, it follows that $S/\mathfrak{m}S = k[\overline{\alpha}]$ is a non-local finite-dimensional k-algebra. Thus the minimal polynomial $p(x) \in k[x]$ for $\overline{\alpha}$ is not just a power of a single irreducible polynomial. Let $f \in R[x]$ be a monic polynomial of least degree with $f(\alpha) = 0$. Then p divides $\overline{f} \in k[x]$, so that the irreducible factorization of $\overline{f}$ involves at least two distinct monic irreducible factors. Therefore we may write $\overline{f} = g_0 h_0$, where g_0 and

h_0 are monic polynomials of positive degree satisfying $\gcd(g_0, h_0) = 1$. Lifting this factorization to $R[x]$, we have $f = gh$. By the minimality of $\deg f$, we have $g(\alpha) \neq 0 \neq h(\alpha)$, but $g(\alpha)h(\alpha) = f(\alpha) = 0$ in D, a contradiction.

(ii) $\implies$ (iii): Let S be a commutative, module-finite R-algebra. Then S is semilocal, say with maximal ideals $\mathfrak{m}_1, \ldots, \mathfrak{m}_t$. Set $X_i = \{\mathfrak{p} \in \operatorname{Spec} S \mid \mathfrak{p} \subseteq \mathfrak{m}_i\}$ for $i = 1, \ldots, t$. By applying (ii) to each of the domains $S/\mathfrak{p}$, as $\mathfrak{p}$ runs over $\operatorname{Spec} S$, we see that the sets X_i are pairwise disjoint. Moreover, letting $\mathfrak{p}_{ij}$, $j = 1, \ldots, s_i$, be the minimal prime ideals contained in each $\mathfrak{m}_i$, we see that $X_i = V(\mathfrak{p}_{i1}) \cup \cdots \cup V(\mathfrak{p}_{is_i})$, a closed set. Thus $\operatorname{Spec} S$ is a disjoint union of open-and-closed sets, and (iii) follows.

(iii) $\implies$ (v): Let Λ be a module-finite R-algebra, not necessarily commutative, and let $e \in \Lambda/\mathcal{J}(\Lambda)$ satisfy $e^2 = e$. Let $\alpha \in \Lambda$ be any lifting of e, and set $S = R[\alpha]$. One checks that $\mathcal{J}(S) = S \cap \mathcal{J}(\Lambda)$, so that $\alpha^2 - \alpha \in \mathcal{J}(S)$. As S is a direct product of local rings, the same is true of $S/\mathcal{J}(S)$. The idempotent $\overline{\alpha} \in S/\mathcal{J}(S)$ is therefore a sum of some subset of the primitive idempotents of $S/\mathcal{J}(S)$. Each of these primitive idempotents clearly lifts to S, so α lifts to an idempotent of S, which lifts e as well.

(v) $\implies$ (iv): Suppose $S = R[x]/(f)$ with f a monic polynomial. Then $S/\mathfrak{m}S = k[x]/(\overline{f})$ is a direct product of local finite-dimensional k-algebras. Since $\mathfrak{m}S \subseteq \mathcal{J}(S)$ by Lemma 1.7, we have $S/\mathfrak{m}S \twoheadrightarrow S/\mathcal{J}(S)$, so $S/\mathcal{J}(S) \cong \overline{T}_1 \times \cdots \times \overline{T}_n$ is also a direct product of local rings $\overline{T}_i$. The primitive idempotents of this decomposition lift to idempotents $e_1, \ldots, e_n$ of S, giving a decomposition $S = T_1 \times \cdots \times T_n$ with $T_i = e_i S$. Since each $\overline{T}_i = T_i/\mathcal{J}(S)T_i$ is local so is each T_i.

(iv) $\implies$ (i): Let $f \in R[x]$ be monic and let $\overline{f} = g_0 h_0$ be a factorization of the image $\overline{f} \in k[x]$ into relatively prime monic polynomials. Set $S = R[x]/(f)$, a direct product $S_1 \times \cdots \times S_n$ of local rings by assumption. Then $S/\mathfrak{m}S = k[x]/(\overline{f}) \cong k[x]/(g_0) \times k[x]/(h_0)$ by the Chinese Remainder Theorem, and also $S/\mathfrak{m}S = S_1/\mathfrak{m}S_1 \times \cdots \times S_n/\mathfrak{m}S_n$. After reordering the factors S_i if necessary, we may assume that $k[x]/(g_0) = S_1/\mathfrak{m}S_1 \times \cdots \times S_l/\mathfrak{m}S_l$ and $k[x]/(h_0) = S_{l+1}/\mathfrak{m}S_{l+1} \times \cdots \times S_n/\mathfrak{m}S_n$ for some l with $1 < l < n$. Set $A = S_1 \times \cdots \times S_l$, a free R-module of rank $\deg g_0$. Let $t \in A$ denote the image of $x \in S$, and let $g \in R[T]$ be the characteristic polynomial of the R-linear operator $A \longrightarrow A$ given by multiplication by t. Note that $\overline{g} = g_0$ in $k[x]$. Now $g(t) = 0$ by the Cayley-Hamilton Theorem, so we have a surjective homomorphism $R[x]/(g) \longrightarrow A$, which is in fact an isomorphism by NAK. The map

$R[x] \longrightarrow A$ factors through S by construction, so we may write $f = gh$ for some monic $h \in R[x]$. □

A.31. COROLLARY. *Let R be a Henselian local ring, let $\alpha \in R$ be a unit, and let n be a positive integer prime to $\mathrm{char}(k)$. If $\overline{\alpha}$ has an nth root in k, then α has an nth root in R.*

PROOF. Let $f = x^n - \alpha \in R[x]$, and let β be a root of $x^n - \overline{\alpha} \in k[x]$. Write $x^n - \overline{\alpha} = (x - \beta)h(x)$. The hypotheses imply that $x^n - \overline{\alpha}$ has n distinct roots, so $x - \beta$ and $h(x)$ are relatively prime. Since R is Henselian, we get $\widetilde{\beta} \in R^\times$ and $\widetilde{h} \in R[x]$ such that $x^n - \alpha = (x - \widetilde{\beta})\widetilde{h}$. Then $\widetilde{\beta}^n = \alpha$. □

A.32. REMARK. For completeness we mention a few more equivalent conditions. The proof of these equivalences is beyond our scope. Let $(R, \mathfrak{m}, k)$ be a local ring. Recall (Definition 10.2) that a *pointed étale neighborhood* of R is a flat local R-algebra $(S, \mathfrak{n})$, essentially of finite type, such that $\mathfrak{m}S = \mathfrak{n}$ and $S/\mathfrak{n} = k$. There is a structure theory for such extensions [**Ive73**, III.2]: S is a pointed étale neighborhood of R if and only if $S \cong (R[x]/(f))_\mathfrak{p}$, where f is a monic polynomial, $\mathfrak{p}$ is a maximal ideal of $R[x]/(f)$ satisfying $f' \notin \mathfrak{p}$, and $S/\mathfrak{p}S = R/\mathfrak{m}$.

The following conditions are then equivalent to the ones in Theorem A.30.

(v) *If $R \longrightarrow S$ is a pointed étale neighborhood, then $R \cong S$.*

(vi) *For every monic polynomial $f \in R[x]$ and every $\alpha \in R$ such that $f(\alpha) \in \mathfrak{m}$ and $f'(\alpha) \notin \mathfrak{m}$, there exists $r \in R$ such that $r \equiv \alpha \bmod \mathfrak{m}$ and $f(r) = 0$.*

(vii) *For every system of polynomials $f_1, \ldots, f_n \in R[x_1, \ldots, x_n]$ and every $(\alpha_1, \ldots, \alpha_n) \in R^{(n)}$ such that $f_i(\alpha_1, \ldots, \alpha_n) \in \mathfrak{m}$ and the Jacobian determinant $\det\left[\frac{\partial f_i}{\partial x_j}(\alpha_1, \ldots, \alpha_n)\right]$ is a unit, there exist ring elements $r_1, \ldots, r_n$ such that $r_i \equiv \alpha_i \bmod \mathfrak{m}$ and $f_i(r_1, \ldots, r_n) = 0$ for all $i = 1, \ldots, n$.*

Condition (vi) ("simple roots lift from k to R") is also sometimes used as the definition of Henselianness.

APPENDIX B

Ramification Theory

This appendix contains the basic results we need in the body of the text on unramified and étale ring homomorphisms, as well as the ramification behavior of prime ideals in integral extensions. We also include proofs of the theorem on the purity of the branch locus (Theorem B.12) and results relating ramification to pseudo-reflections in finite groups of linear ring automorphisms.

§1. Unramified homomorphisms

Recall that a ring homomorphism $A \longrightarrow B$ is said to be *of finite type* if B is a finitely generated A-algebra, that is, $B \cong A[x_1, \ldots, x_n]/I$ for some polynomial variables $x_1, \ldots, x_n$ and an ideal I. We say $A \longrightarrow B$ is *essentially of finite type* if B is a localization (at an arbitrary multiplicatively closed set) of an A-algebra of finite type.

B.1. DEFINITION. Assume that $(A, \mathfrak{m}, k) \longrightarrow (B, \mathfrak{n}, \ell)$ is a local homomorphism of local rings. We say that $A \longrightarrow B$ is an *unramified local homomorphism* provided

(i) $\mathfrak{m}B = \mathfrak{n}$,
(ii) $B/\mathfrak{m}B$ is a finite separable field extension of $A/\mathfrak{m}$, and
(iii) B is essentially of finite type over A.

If in addition $A \longrightarrow B$ is flat, we say it is *étale*.

B.2. REMARKS. Let $A \longrightarrow B$ be a local homomorphism between local rings. Let $\widehat{A}$ and $\widehat{B}$ be the $\mathfrak{m}$-adic and $\mathfrak{n}$-adic completions of A and B, respectively. It is straightforward to check that $A \longrightarrow B$ is unramified, respectively étale, if and only if $\widehat{A} \longrightarrow \widehat{B}$ is so.

If $A \longrightarrow B$ is an unramified local homomorphism, then $\widehat{B}$ is a finitely generated $\widehat{A}$-module. Indeed, it follows from the complete version of NAK ([**Mat89**, Theorem 8.4] or [**Eis95**, Exercises 7.2 and 7.4]) that any $k = \widehat{A}/\widehat{\mathfrak{m}}$-vector space basis for $\ell = \widehat{B}/\widehat{\mathfrak{n}}$ lifts to a set of $\widehat{A}$-module generators for $\widehat{B}$. If, in particular, there is no residue field growth (for instance, if k is separably or algebraically closed), then $\widehat{A} \longrightarrow \widehat{B}$ is surjective.

If $A \longrightarrow B$ is étale, then $\widehat{B}$ is a finitely generated flat $\widehat{A}$-module, whence $\widehat{B} \cong \widehat{A}^{(n)}$ for some n. If in this case $k = \ell$, then $\widehat{B} = \widehat{A}$.

It's easy to check that if $A \longrightarrow B$ is étale, then A and B share the same Krull dimension and the same depth. Furthermore, A is regular if and only if B is regular. For further permanence results along these lines, we need to globalize the definition.

B.3. DEFINITION. Let A and B be Noetherian rings, and $A \longrightarrow B$ a homomorphism essentially of finite type. Let $\mathfrak{q} \in \operatorname{Spec} B$ and set $\mathfrak{p} = A \cap \mathfrak{q}$. We say that $A \longrightarrow B$ is *unramified at* $\mathfrak{q}$ (or also $\mathfrak{q}$ *is unramified over* A) if and only if the induced map $A_\mathfrak{p} \longrightarrow B_\mathfrak{q}$ is an unramified local homomorphism of local rings. Similarly, $A \longrightarrow B$ is *étale at* $\mathfrak{q}$ if and only if $A_\mathfrak{p} \longrightarrow B_\mathfrak{q}$ is an étale local homomorphism. Finally, $A \longrightarrow B$ is *unramified*, respectively *étale*, if it is unramified, respectively étale, at every prime ideal $\mathfrak{q} \in \operatorname{Spec} B$.

Here is an easy transitivity property of unramified primes.

B.4. LEMMA. *Let $A \longrightarrow B \longrightarrow C$ be homomorphisms, essentially of finite type, of Noetherian rings. Let $\mathfrak{r} \in \operatorname{Spec} C$ and set $\mathfrak{q} = B \cap \mathfrak{r}$.*
  (i) *If $\mathfrak{r}$ is unramified over B and $\mathfrak{q}$ is unramified over A, then $\mathfrak{r}$ is unramified over A.*
  (ii) *If $\mathfrak{r}$ is unramified over A, then $\mathfrak{r}$ is unramified over B.*

It is clear that a local homomorphism $(A, \mathfrak{m}) \longrightarrow (B, \mathfrak{n})$ essentially of finite type is an unramified local homomorphism if and only if $\mathfrak{n}$ is unramified over A. However, it's not at all clear that an unramified local homomorphism is unramified in the sense of Definition B.3. To reconcile these definitions, we must show that being unramified is preserved under localization. The easiest way to do this is to give an alternative description, following [**AB59**].

B.5. DEFINITION. Let $A \longrightarrow B$ be a homomorphism of Noetherian rings. Define the *diagonal map* $\mu \colon B \otimes_A B \longrightarrow B$ by $\mu(b \otimes b') = bb'$ for all $b, b' \in B$, and set $\mathcal{J} = \ker \mu$. Thus we have a short exact sequence of $B \otimes_A B$-modules

(B.1) $$0 \longrightarrow \mathcal{J} \longrightarrow B \otimes_A B \xrightarrow{\mu} B \longrightarrow 0.$$

B.6. REMARKS.
  (i) The ideal $\mathcal{J}$ is generated by all elements of the form $b \otimes 1 - 1 \otimes b$, where $b \in B$. Indeed, if $\mu\left(\sum_j b_j \otimes b'_j\right) = 0$, then $\sum_j b_j b'_j = 0$, so that
  $$\sum_j b_j \otimes b'_j = \sum_j (1 \otimes b'_j)(b_j \otimes 1 - 1 \otimes b_j).$$

(ii) The ring $B \otimes_A B$, also called the *enveloping algebra* of the A-algebra B, has two A-module structures, one on each side. Thus $\mathcal{J}$ also has two different B-structures. However, these two module structures coincide modulo $\mathcal{J}^2$. The reason is that

$$\mathcal{J}/\mathcal{J}^2 = ((B \otimes_A B)/\mathcal{J}) \otimes_{B \otimes_A B} \mathcal{J}$$

is a $(B \otimes_A B)/\mathcal{J}$-module, and $(B \otimes_A B)/\mathcal{J} = B$. In particular, $\mathcal{J}/\mathcal{J}^2$ has an unambiguous B-module structure.

(iii) The B-module $\mathcal{J}/\mathcal{J}^2$ is also known as the *module of (relative) Kähler differentials of B over A*, denoted $\Omega_{B/A}$ [**Eis95**, Chapter 16]. It is the universal module of A-linear derivations on B, in the sense that the map $\delta \colon B \longrightarrow \mathcal{J}/\mathcal{J}^2$ sending b to $b \otimes 1 - 1 \otimes b$ is an A-linear derivation (satisfies the Leibniz rule), and given any A-linear derivation $\epsilon \colon B \longrightarrow M$, there exists a unique B-linear homomorphism $\mathcal{J}/\mathcal{J}^2 \longrightarrow M$ making the obvious diagram commute. In particular we have $\mathrm{Der}_A(B, M) \cong \mathrm{Hom}_B(\mathcal{J}/\mathcal{J}^2, M)$ for every B-module M. Though it is very important for a deeper study of unramified maps, will not need this interpretation in this book.

(iv) If $A \longrightarrow B$ is assumed to be essentially of finite type, $\mathcal{J}$ is a finitely generated $B \otimes_A B$-module. To see this, first observe that the question reduces at once to the case where B is of finite type over A. In that case, if $x_1, \ldots, x_n$ are A-algebra generators for B, one checks that the elements $x_i \otimes 1 - 1 \otimes x_i$, for $i = 1, \ldots, n$, generate $\mathcal{J}$. It follows that if $A \longrightarrow B$ is essentially of finite type then $\mathcal{J}/\mathcal{J}^2$ is a finitely generated B-module.

(v) The term "diagonal map" comes from the geometry. Suppose $f \colon A \hookrightarrow B$ is an integral extension of integral domains which are finitely generated algebras over an algebraically closed field k. Then there is a corresponding surjective map of irreducible varieties $f^\# \colon Y \longrightarrow X$, where X is the maximal ideal spectrum of A and Y is that of B. In this case, the maximal ideal spectrum of $B \otimes_A B$ is the fiber product

$$Y \times_X Y = \left\{ (y_1, y_2) \in Y \times Y \mid f^\#(y_1) = f^\#(y_2) \right\}.$$

The map $\mu \colon B \otimes_A B \longrightarrow B$ corresponds to the diagonal embedding $\mu^\# \colon Y \longrightarrow Y \times_X Y$ taking y to (y, y). In these terms, $\mathcal{J}$ is the ideal of functions on $Y \times_X Y$ vanishing on the diagonal.

B.7. LEMMA. *Let $A \longrightarrow B$ be a homomorphism of Noetherian rings. Then the following conditions are equivalent.*

(i) *B is a projective $B \otimes_A B$-module.*

(ii) The exact sequence $0 \longrightarrow \mathcal{J} \longrightarrow B \otimes_A B \xrightarrow{\mu} B \longrightarrow 0$ splits as $B \otimes_A B$-modules.

(iii) $\mu\left(\operatorname{Ann}_{B \otimes_A B}(\mathcal{J})\right) = B$.

If $\mathcal{J}/\mathcal{J}^2$ is a finitely generated B-module (for example, if $A \longrightarrow B$ is essentially of finite type), then these are equivalent to

(iv) $\mathcal{J}$ is generated by an idempotent.

(v) $\mathcal{J}/\mathcal{J}^2 = 0$.

PROOF. (i) $\iff$ (ii) is clear.

(ii) $\iff$ (iii): The map $\mu \colon B \otimes_A B \longrightarrow B$ splits over $B \otimes_A B$ if and only if the induced homomorphism

$$\operatorname{Hom}_{B \otimes_A B}(B, \mu) \colon \operatorname{Hom}_{B \otimes_A B}(B, B \otimes_A B) \longrightarrow \operatorname{Hom}_{B \otimes_A B}(B, B)$$

is surjective. However, the isomorphism $B \cong (B \otimes_A B)/\mathcal{J}$ shows that we have $\operatorname{Hom}_{B \otimes_A B}(B, B \otimes_A B) \cong \operatorname{Ann}_{B \otimes_A B}(\mathcal{J})$, so that μ splits if and only if $\operatorname{Hom}_{B \otimes_A B}(B, \mu)$ is surjective, if and only if $\mu\left(\operatorname{Ann}_{B \otimes_A B}(\mathcal{J})\right) = B$.

The final two statements are always equivalent for a finitely generated ideal. Assume (iv), so that there exists $z \in \mathcal{J}$ with $xz = x$ for every $x \in \mathcal{J}$. Define $q \colon B \otimes_A B \longrightarrow \mathcal{J}$ by $q(x) = xz$. Then for $x \in \mathcal{J}$, we have $q(x) = x$, so that the sequence splits. Conversely, any splitting q of the map $\mathcal{J} \longrightarrow B \otimes_A B$ yields an idempotent $z = q(1)$, so (ii) and (iv) are equivalent. $\square$

The proof[1] of the next result is too long for us to include here, even though it is the foundation for the theory. See for example [**Eis95**, Corollary 16.16].

B.8. PROPOSITION. *Suppose that A is a field and B is an A-algebra essentially of finite type. Then the equivalent conditions of Lemma B.7 hold if and only if B is a direct product of a finite number of fields, each finite and separable over A.*

The condition in the Proposition that B be a direct product of a finite number of fields, each finite and separable over A, is sometimes called a "(classically) separable algebra" in the literature. Equivalently, $K \otimes_A B$ is a reduced ring for every field extension K of A.

We now relate the equivalent conditions of Lemma B.7 to the definitions at the beginning of the Appendix.

---

[1]Sketch: In the special case where A and B are both fields, one can show that if B is projective over $B \otimes_A B$ then $A \longrightarrow B$ is necessarily module-finite. Then a separability idempotent $z \in \mathcal{J}$ is given as follows: let $\alpha \in B$ be a primitive element, with minimal polynomial $f(x) = (x - \alpha) \sum_{i=0}^{n-1} b_i x^i$. Then $z = \left(1 \otimes \frac{1}{f'(\alpha)}\right) \sum_{i=0}^{n-1} \alpha^i \otimes b_i$ is idempotent.

B.9. PROPOSITION. *Let $A \longrightarrow B$ be a homomorphism, essentially of finite type, of Noetherian rings. The following statements are equivalent.*

(i) *The exact sequence $0 \longrightarrow \mathcal{J} \longrightarrow B \otimes_A B \xrightarrow{\mu} B \longrightarrow 0$ splits as $B \otimes_A B$-modules.*

(ii) *B is unramified over A.*

(iii) *Every maximal ideal of B is unramified over A.*

PROOF. (i) $\Longrightarrow$ (ii): Let $\mathfrak{q} \in \operatorname{Spec} B$, and let $\mathfrak{p} = A \cap \mathfrak{q}$ be its contraction to A. It is enough to show that $B_\mathfrak{p}/\mathfrak{p}B_\mathfrak{p}$ is unramified over the field $A_\mathfrak{p}/\mathfrak{p}A_\mathfrak{p}$, i.e. is a finite direct product of finite separable field extensions. By Proposition B.8, it suffices to show that $B_\mathfrak{p}/\mathfrak{p}B_\mathfrak{p}$ is a projective module over $B_\mathfrak{p}/\mathfrak{p}B_\mathfrak{p} \otimes_{A_\mathfrak{p}/\mathfrak{p}A_\mathfrak{p}} B_\mathfrak{p}/\mathfrak{p}B_\mathfrak{p}$. Let $p \colon B \longrightarrow B \otimes_A B$ be a splitting for μ, so that $\mu p = 1_B$. Set $y = p(1)$. Then $\mu(y) = 1$ and $y \ker \mu = 0$; in fact, the existence of an element y satisfying these two conditions is easily seen to be equivalent to the existence of a splitting of μ. Consider the diagram

$$\begin{array}{ccccc}
B \otimes_A B & \xrightarrow{f} & B_\mathfrak{p} \otimes_{A_\mathfrak{p}} B_\mathfrak{p} & \xrightarrow{g} & B_\mathfrak{p}/\mathfrak{p}B_\mathfrak{p} \otimes_{A_\mathfrak{p}} B_\mathfrak{p}/\mathfrak{p}B_\mathfrak{p} \\
\mu \downarrow & & \downarrow \mu' & & \downarrow \mu'' \\
B & \longrightarrow & B_\mathfrak{p} & \longrightarrow & B_\mathfrak{p}/\mathfrak{p}B_\mathfrak{p}
\end{array}$$

in which the horizontal arrows are the natural ones and the vertical arrows are the respective diagonal maps. Put $y'' = gf(y)$. Then $\mu''(y'') = 1$ and $y \ker(\mu'') = 0$, so that μ'' splits. Since the top-right ring is also $B_\mathfrak{p}/\mathfrak{p}B_\mathfrak{p} \otimes_{A_\mathfrak{p}/\mathfrak{p}A_\mathfrak{p}} B_\mathfrak{p}/\mathfrak{p}B_\mathfrak{p}$, this shows that $A_\mathfrak{p}/\mathfrak{p}A_\mathfrak{p} \longrightarrow B_\mathfrak{p}/\mathfrak{p}B_\mathfrak{p}$ is unramified.

(ii) $\Longrightarrow$ (iii) is obvious.

(iii) $\Longrightarrow$ (i): Since $\mathcal{J}$ is a finitely generated B-module, it suffices to assume that $A \longrightarrow B$ is an unramified local homomorphism of local rings and show that $\mathcal{J} = \mathcal{J}^2$. Once again we reduce to the case where A is a field and B is a separable A-algebra. In this case Proposition B.8 implies that $\mathcal{J} = \mathcal{J}^2$. $\square$

B.10. REMARKS. This proposition reconciles the two definitions of unramifiedness given at the beginning of the Appendix, since it implies that unramifiedness is preserved by localization. This has some very satisfactory consequences. One can now use the characterizations of reducedness and normality in terms of the conditions (R_n) and (S_n) to see that if $A \longrightarrow B$ is étale, then A is reduced, respectively normal, if and only if B is so. Note that this fact would be *false* without the hypothesis that $A \longrightarrow B$ is essentially of finite type. Indeed, the natural completion homomorphism $A \longrightarrow \widehat{A}$ satisfies (i) and (ii) of Definition B.1, and is of course flat, but there are examples of completion not preserving reducedness or normality.

Proposition B.9 also allows us to expand our use of language, saying that a prime ideal $\mathfrak{p} \in \operatorname{Spec} A$ is *unramified in B* if the localization $A_\mathfrak{p} \longrightarrow B_\mathfrak{p}$ is unramified, that is, every prime ideal of B lying over $\mathfrak{p}$ is unramified.

We now define the *homological different* of the A-algebra B. It is the ideal of B
$$\mathfrak{H}_A(B) = \mu\left(\operatorname{Ann}_{B \otimes_A B}(\mathcal{J})\right),$$
where $\mu \colon B \otimes_A B \longrightarrow B$ is again the diagonal map. The homological different defines the *branch locus* of $A \longrightarrow B$, that is, the primes of B which are ramified over A, as we now show.

B.11. THEOREM. *Let $A \longrightarrow B$ be a homomorphism, essentially of finite type, of Noetherian rings. A prime ideal $\mathfrak{q} \in \operatorname{Spec} B$ is unramified over A if and only if $\mathfrak{q}$ does not contain $\mathfrak{H}_A(B)$.*

PROOF. This fact follows from Proposition B.9 and condition (iii) of Lemma B.7, together with the observation that formation of $\mathcal{J}$ commutes with localization at $\mathfrak{q}$ and $A \cap \mathfrak{q}$. Precisely, let $\mathfrak{q} \in \operatorname{Spec} B$ and set $\mathfrak{p} = A \cap \mathfrak{q}$. Let S be the multiplicatively closed set of simple tensors $u \otimes v$, where u and v range over $B \setminus \mathfrak{q}$. Then $(B \otimes_A B)_S \cong B_\mathfrak{q} \otimes_A B_\mathfrak{q} \cong B_\mathfrak{q} \otimes_{A_\mathfrak{p}} B_\mathfrak{q}$ and the kernel of the map $\widetilde{\mu} \colon B_\mathfrak{q} \otimes_{A_\mathfrak{p}} B_\mathfrak{q} \longrightarrow B_\mathfrak{q}$ coincides with $(\ker \mu)_S$. □

§2. Purity of the branch locus

Turn now to the theorem on the purity of the branch locus. The proof we give, following Auslander-Buchsbaum [**AB59**] and Auslander [**Aus62**], is somewhat lengthy.

For the rest of this Appendix, we will be mainly concerned with finite integral extensions $A \longrightarrow B$ of Noetherian domains. In particular they will be of finite type. Recall that for a finite integral extension, we have the "lying over" and "going up" properties; if in addition A is normal, then we also have "going down" [**Mat89**, Theorems 9.3 and 9.4]. In particular, in this case we have height $\mathfrak{q} = \operatorname{height}(A \cap \mathfrak{q})$ for $\mathfrak{q} \in \operatorname{Spec} B$ ([**Mat89**, 9.8, 9.9]).

Recall also that since a normal domain satisfies Serre's condition (S_2), the associated primes of a principal ideal all have height one (Proposition A.9). In other words, principal ideals have pure height one.

B.12. THEOREM (Purity of the Branch Locus). *Let A be a regular ring and $A \hookrightarrow B$ a module-finite ring extension with B normal. Then $\mathfrak{H}_A(B)$ is an ideal of pure height one in B. In particular, if $A \longrightarrow B$ is unramified in codimension one, then $A \longrightarrow B$ is unramified.*

First we observe that the condition "unramified in codimension one" can be interpreted in terms of the sequence (B.1).

Assume $A \longrightarrow B$ is a module-finite extension of Noetherian normal domains. We write $B \cdot B$ for the reflexive product $(B \otimes_A B)^{**}$, where $-^* = \operatorname{Hom}_B(-, B)$. Since the B-module B is reflexive, and any homomorphism from $B \otimes_A B$ to a reflexive B-module factors through $B \cdot B$, we see that $\mu \colon B \otimes_A B \longrightarrow B$ factors as $B \otimes_A B \longrightarrow B \cdot B \xrightarrow{\mu^{**}} B$.

B.13. PROPOSITION. *A module-finite extension of Noetherian normal domains $A \longrightarrow B$ is unramified in codimension one if and only if μ^{**} is a split surjection of $B \otimes_A B$-modules.*

PROOF. If μ^{**} is a split surjection, then $\mu_{\mathfrak{p}}^{**}$ is a split surjection for all primes $\mathfrak{q}$ of height one in B. For these primes, however, $\mu_{\mathfrak{q}}^{**} = \mu_{\mathfrak{q}}$ since $B_{\mathfrak{q}} \otimes_{A_{\mathfrak{p}}} B_{\mathfrak{q}} = (B \otimes_A B)_{\mathfrak{q}}$ is a reflexive module over the DVR $B_{\mathfrak{q}}$, where $\mathfrak{p} = A \cap \mathfrak{q}$. Thus μ splits locally at every height-one prime of B, so $A \longrightarrow B$ is unramified in codimension one.

Now assume $A \longrightarrow B$ is unramified in codimension one. Let K be the quotient field of A and L the quotient field of B. Since $A \longrightarrow B$ is unramified at the zero ideal, $K \longrightarrow L$ is unramified, equivalently, a finite separable field extension. In particular, the diagonal map $\eta \colon L \otimes_K L \longrightarrow L$ is a split epimorphism of $L \otimes_K L$-modules.

Since $B \cdot B$ is B-reflexive, it is in particular torsion-free, and so $B \cdot B$ is a submodule of $L \otimes_K L$. We therefore have a commutative diagram of short exact sequences

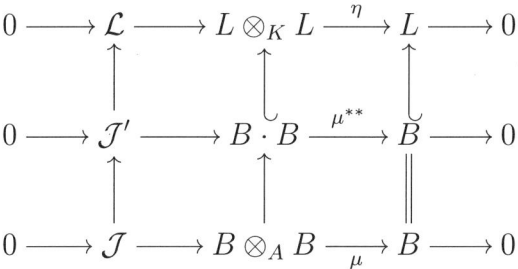

in which the left-hand modules are by definition the kernels, and in which the top row splits over $L \otimes_K L$ since L/K is separable. Let $\epsilon \colon L \longrightarrow L \otimes_K L$ be a splitting, and let ζ be the restriction of ϵ to B. It will suffice to show that $\zeta(B) \subseteq B \cdot B$, for then ζ will be the splitting of μ^{**} we need. For a height-one prime ideal $\mathfrak{q}$ of B, with $\mathfrak{p} = A \cap \mathfrak{q}$, we do have $\zeta_{\mathfrak{q}}(B_{\mathfrak{q}}) \subseteq (B_{\mathfrak{q}} \otimes_{A_{\mathfrak{p}}} B_{\mathfrak{q}})^{**} = (B \otimes_A B)_{\mathfrak{q}}$, since $A \longrightarrow B$ is unramified in codimension one. But $\mathrm{im}(\zeta) = \bigcap_{\text{height } \mathfrak{q}=1} \mathrm{im}(\zeta_{\mathfrak{q}})$ and $B = \bigcap_{\text{height } \mathfrak{q}=1} B_{\mathfrak{q}}$ as B is normal, so the image of ζ is contained in $B \cdot B$ and ζ is a splitting for μ^{**}. $\square$

Following Auslander and Buchsbaum, we shall first prove Theorem B.12 in the special case where B is a finitely generated projective A-module. In this case the homological different coincides with the Dedekind different from number theory, which we describe now.

Let $A \longrightarrow B$ be a module-finite extension of normal domains. Let K and L be the quotient fields of A and B, respectively. We assume that $K \longrightarrow L$ is a separable extension. (In the situation of Theorem B.12, this follows from the hypothesis.) In this case the trace form $(x, y) \mapsto \mathrm{Tr}_{L/K}(xy)$ is a non-degenerate pairing $L \otimes_K L \longrightarrow L$, and since $A \longrightarrow B$ is integral and A is integrally closed in K we have $\mathrm{Tr}_{L/K}(B) \subseteq A$. Set

$$\mathfrak{C}_A(B) = \left\{ x \in L \mid \mathrm{Tr}_{L/K}(xB) \subseteq A \right\},$$

and call it the *Dedekind complementary module for B/A*. It is a fractional ideal of B.

We set $\mathfrak{D}_A(B) = (\mathfrak{C}_A(B))^{-1}$, the inverse of the fractional ideal $\mathfrak{C}_A(B)$. This is the *Dedekind different* of B/A. Since $B \subseteq \mathfrak{C}_A(B)$, we have that $\mathfrak{D}_A(B) \subseteq B$ and $\mathfrak{D}_A(B)$ is an ideal of B. It is even a reflexive ideal since it is the inverse of a fractional ideal.

The following theorem is attributed to Noether ([**Noe50**], posthumous) and Auslander-Buchsbaum.

B.14. THEOREM. *Let $A \longrightarrow B$ be a module-finite extension of Noetherian normal domains which induces a separable extension of quotient fields. We have $\mathfrak{H}_A(B) \subseteq \mathfrak{D}_A(B)$, and if B is projective as an A-module then $\mathfrak{H}_A(B) = \mathfrak{D}_A(B)$.*

PROOF. Let K and L be the respective quotient fields of A and B as in the discussion above. Set $L^* = \mathrm{Hom}_K(L, K)$ and $B^* = \mathrm{Hom}_A(B, A)$. Define

$$\sigma_L \colon L \otimes_K L \longrightarrow \mathrm{Hom}_L(L^*, L)$$

by $\sigma_L(x \otimes y)(f) = xf(y)$. Then σ_L restricts to a similarly defined map $\sigma_B \colon B \otimes_A B \longrightarrow \mathrm{Hom}_B(B^*, B)$. It's straightforward to show that σ_B is an isomorphism if B is projective over A; in particular, σ_L is an isomorphism. Its inverse is defined by $(\sigma_L)^{-1}(f) = \sum_j f(x_j^*) \otimes x_j$, where $\{x_j\}$ and $\{x_j^*\}$ are dual bases for L and L^* over K.

Consider the diagram

$$\begin{array}{ccccccc}
\operatorname{Hom}_B(B^*, B) & \xrightarrow{i_B} & \operatorname{Hom}_A(B^*, B) & \xleftarrow{\sigma_B} & B \otimes_A B & \xrightarrow{\mu_B} & B \\
\downarrow & & \downarrow & & \downarrow & & \downarrow \\
\operatorname{Hom}_L(L^*, L) & \xrightarrow[i_L]{} & \operatorname{Hom}_K(L^*, L) & \xleftarrow[\cong]{\sigma_L} & L \otimes_K L & \xrightarrow{\mu_L} & L
\end{array}$$

in which μ_B and μ_L are the respective diagonal maps, i_B and i_L are inclusions, and the vertical arrows are all induced from the inclusion of B into L. Now $\operatorname{Tr}_{L/K}(x) = \sum_j x_j^*(xx_j)$, so if $f \in \operatorname{Hom}_L(L^*, L)$ then we have $f(\operatorname{Tr}_{L/K}) = \sum_j x_j f(x_j^*)$. Thus the composition of the entire bottom row, left to right, is given by

$$\mu_L\,(\sigma_L)^{-1}\,i_L(f) = \mu_L\left(\sum_j f(x_j^*) \otimes x\right) = \sum_j f(x_j^*) x_j = f(\operatorname{Tr}_{L/K})\,.$$

It follows that the image of $\operatorname{Hom}_B(B^*, B)$ in L is $\mathfrak{D}_A(B)$.

The module $\operatorname{Hom}_A(B^*, B)$ is naturally a $B \otimes_A B$-module via the rule $((b \otimes b')\,(f))\,(g) = bf(g \circ b')$, where the b' on the right represents the map on B given by multiplication by that element. Thus σ_B is a $B \otimes_A B$-module homomorphism. An element $\operatorname{Hom}_A(B^*, B)$ is in the image of i_B if and only if it is a B-module homomorphism, i.e. $(b \otimes 1)\,(f) = (1 \otimes b)\,(f)$ for every $b \in B$. This is exactly saying that f annihilates $\mathcal{J} = \ker \mu_B$. Thus implies that $\sigma_B\left(\operatorname{Ann}_{B \otimes_A B}(\mathcal{J})\right) \subseteq \operatorname{im} i_B$. It follows that $\mathfrak{H}_A(B) = \mu_B\left(\operatorname{Ann}_{B \otimes_A B}(\mathcal{J})\right) \subseteq \mathfrak{D}_A(B)$.

Finally, if B is projective as an A-module then σ_B is an isomorphism and $\sigma_B\left(\operatorname{Ann}_{B \otimes_A B}(\mathcal{J})\right)$ is equal to the image of i_B. Thus $\mathfrak{H}_A(B) = \mathfrak{D}_A(B)$. $\square$

Next we show that $\mathfrak{D}_A(B)$ has pure height one, so in case they are equal $\mathfrak{H}_A(B)$ does as well. We need a general fact about modules over normal domains.

B.15. PROPOSITION. *Let A be a Noetherian normal domain. Let*

$$0 \longrightarrow M \longrightarrow N \longrightarrow T \longrightarrow 0$$

be a short exact sequence of non-zero finitely generated A-modules in which M is reflexive and T is torsion. Then $\operatorname{Ann}_A(T)$ is an ideal of pure height one in A.

PROOF. This is similar to Lemma 5.11. Let $\mathfrak{p}$ be a prime ideal minimal over the annihilator of T. Then in particular $\mathfrak{p}$ is an associated prime of T, so that depth $T_\mathfrak{p} = 0$. Since M is reflexive, it satisfies (S_2), so that if $\mathfrak{p}$ has height two or more then $M_\mathfrak{p}$ has depth at least two. This contradicts the Depth Lemma.

B.16. COROLLARY. *Let $A \longrightarrow B$ be a module-finite extension of normal domains. Assume that the induced extension of quotient fields is separable. If $\mathfrak{D}_A(B) \neq B$, then $\mathfrak{D}_A(B)$ is an ideal of pure height one in B. Consequently, $\mathfrak{D}_A(B) = B$ if and only if $A \longrightarrow B$ is unramified in codimension one.*

PROOF. For the first statement, take $M = A$ and $N = \mathfrak{C}_A(B)$ in Proposition B.15. In the second statement, necessity follows from the containment $\mathfrak{H}_A(B) \subseteq \mathfrak{D}_A(B)$ and Theorem B.11. Conversely, suppose $\mathfrak{D}_A(B) = B$. Let $\mathfrak{q}$ be a height-one prime of B and set $\mathfrak{p} = A \cap \mathfrak{q}$. Then $A_\mathfrak{p}$ is a DVR and $B_\mathfrak{p}$ is a finitely generated torsion-free $A_\mathfrak{p}$-module, whence free. Thus $\mathfrak{H}_{A_\mathfrak{p}}(B_\mathfrak{p}) = \mathfrak{D}_{A_\mathfrak{p}}(B_\mathfrak{p}) = (\mathfrak{D}_A(B))_\mathfrak{p} = B_\mathfrak{p}$. By Theorem B.11 $B_\mathfrak{p}$ is unramified over $A_\mathfrak{p}$, so in particular $\mathfrak{q}$ is unramified over $\mathfrak{p}$. □

B.17. COROLLARY. *If, in the setup of Corollary B.16, B is projective as an A-module, then $A \longrightarrow B$ is unramified if and only if it is unramified in codimension one.*

Now we turn to Auslander's proof of the theorem on the purity of the branch locus. The strategy is to reduce the general case to the situation of Corollary B.17 by proving a purely module-theoretic statement.

B.18. PROPOSITION. *Let $A \longrightarrow B$ be a module-finite extension of normal domains which is unramified in codimension one. Assume that A has the following property: If M is a finitely generated reflexive A-module such that $\mathrm{Hom}_A(M, M)$ is isomorphic to a direct sum of copies of M, then M is free. Then $A \longrightarrow B$ is unramified.*

PROOF. Let $K \longrightarrow L$ be the extension of quotient fields induced by $A \longrightarrow B$. Then L is a finite separable extension of K. By [**Aus62**, Prop. 1.1], we may assume in fact that $K \longrightarrow L$ is a Galois extension. (The proof of this result is somewhat technical, so we omit it.)

We are therefore in the situation of Theorem 5.12! Thus $\mathrm{End}_A(B)$ is isomorphic as a ring to the skew group ring $B\#G$, where $G = \mathrm{Gal}(L/K)$. As a B-module, and hence as an A-module, $B\#G$ is isomorphic to a direct sum of copies of B. By hypothesis, then, B is a free A-module. Corollary B.17 now says that $A \longrightarrow B$ is unramified. □

Auslander's argument that regular local rings satisfy the condition of Proposition B.18 seems to be unique in the field; we know of nothing else quite like it. We being with three preliminary results.

B.19. Lemma. *Let A be a Noetherian normal domain and let M be a finitely generated torsion-free A-module. Then*
$$\mathrm{Hom}_A(M,M)^* \cong \mathrm{Hom}_A(M^*,M^*).$$

Proof. We have the natural map $\rho\colon M^* \otimes_A M \longrightarrow \mathrm{Hom}_A(M,M)$ defined by $\rho(f \otimes y)(x) = f(x)y$, which is an isomorphism if and only if M is free; see Exercise 12.25. Dualizing yields $\rho^*\colon \mathrm{Hom}_A(M,M)^* \longrightarrow (M^* \otimes_A M)^* \cong \mathrm{Hom}_A(M^*,M^*)$ by Hom-tensor adjointness. Now ρ^* is a homomorphism between reflexive A-modules, which is an isomorphism in codimension one since A is normal and M is torsion-free. By Lemma 5.11, ρ^* is an isomorphism. $\square$

B.20. Lemma. *Let $(A,\mathfrak{m})$ be a local ring and $f\colon M \longrightarrow N$ a homomorphism of finitely generated A-modules. Assume that $f_\mathfrak{p}\colon M_\mathfrak{p} \longrightarrow N_\mathfrak{p}$ is an isomorphism for every non-maximal prime $\mathfrak{p}$ of A. Then $\mathrm{Ext}^i_A(f,A)\colon \mathrm{Ext}^i_A(N,A) \longrightarrow \mathrm{Ext}^i_A(M,A)$ is an isomorphism for each $i = 0,\ldots,\mathrm{depth}\,A - 2$.*

Proof. The kernel and cokernel of f both have finite length, so $\mathrm{Ext}^i_A(\ker f, A) = \mathrm{Ext}^i_A(\mathrm{cok}\,f, A) = 0$ for $i = 0,\ldots,\mathrm{depth}\,A-1$ [**Mat89**, Theorem 16.6]. The long exact sequence of Ext now gives the conclusion. $\square$

B.21. Proposition. *Let $(A,\mathfrak{m})$ be a local ring of depth at least 3 and let M be a reflexive A-module such that*

(i) M is locally free on the punctured spectrum of A; and
(ii) $\mathrm{pd}_A M \leqslant 1$.

If M is not free, then
$$\ell\left(\mathrm{Ext}^1_A(\mathrm{Hom}_A(M,M),A)\right) > (\mathrm{rank}_A M)\ \ell\left(\mathrm{Ext}^1_A(M,A)\right).$$

Proof. Assume that M is not free. We have the natural homomorphism $\rho_M \colon M^* \otimes_A M \longrightarrow \mathrm{Hom}_A(M,M)$, defined by $\rho_M(f \otimes x)(y) = f(y)x$, which is an isomorphism if and only if M is free; see Remark 12.2. In particular, ρ_M is locally an isomorphism on the punctured spectrum of A, so by Lemma B.20, we have
$$\mathrm{Ext}^1_A(M^* \otimes_A M, A) \cong \mathrm{Ext}^1_A(\mathrm{Hom}_A(M,M), A).$$

Next we claim that there is an injection $\mathrm{Ext}^1_A(M,M) \hookrightarrow \mathrm{Ext}^1_A(M^* \otimes_A M, A)$. Let

(B.2) $$0 \longrightarrow F_1 \longrightarrow F_0 \longrightarrow M \longrightarrow 0$$

be a free resolution. Dualizing gives an exact sequence

(B.3) $$0 \longrightarrow M^* \longrightarrow F_0^* \longrightarrow F_1^* \longrightarrow \mathrm{Ext}^1_A(M,A) \longrightarrow 0,$$

so that $\operatorname{Tor}_{i-2}^A(M^*, M) = \operatorname{Tor}_i^A(\operatorname{Ext}_A^1(M,A), M) = 0$ for all $i \geqslant 3$. In particular, applying $M^* \otimes_A -$ to (B.2) results in an exact sequence

$$0 \longrightarrow M^* \otimes_A F_1 \longrightarrow M^* \otimes_A F_0 \longrightarrow M^* \otimes_A M \longrightarrow 0.$$

Dualizing this yields an exact sequence

$$\operatorname{Hom}_A(M^* \otimes_A F_0, A) \xrightarrow{\eta} \operatorname{Hom}_A(M^* \otimes_A F_1, A) \longrightarrow \operatorname{Ext}_A^1(M^* \otimes_A M, A).$$

But the homomorphism η is naturally isomorphic to the homomorphism $\operatorname{Hom}_A(F_0, M^{**}) \longrightarrow \operatorname{Hom}_A(F_1, M^{**})$. Since M is reflexive, this implies that the cokernel of η is isomorphic to $\operatorname{Ext}_A^1(M, M)$, whence $\operatorname{Ext}_A^1(M, M) \hookrightarrow \operatorname{Ext}_A^1(M^* \otimes_A M, A)$, as claimed.

Next we claim that $\operatorname{Ext}_A^1(M, M) \cong \operatorname{Ext}_A^1(M, A) \otimes_A M$. This follows immediately from the commutative exact diagram

$$\begin{array}{ccccccc}
F_0^* \otimes_A M & \longrightarrow & F_1^* \otimes_A M & \longrightarrow & \operatorname{Ext}_A^1(M, A) \otimes_A M & \longrightarrow & 0 \\
\downarrow \rho_{F_0}^M & & \downarrow \rho_{F_1}^M & & \downarrow & & \\
\operatorname{Hom}_A(F_0, M) & \longrightarrow & \operatorname{Hom}_A(F_1, M) & \longrightarrow & \operatorname{Ext}_A^1(M, M) & \longrightarrow & 0
\end{array}$$

in which the rows are the result of applying $- \otimes_A M$ to (B.3) and $\operatorname{Hom}_A(-, M)$ to (B.2), respectively, the two vertical arrows $\rho_{F_i}^M$ are isomorphisms since each F_i is free, and the third vertical arrow is induced by the other two.

Putting the pieces together so far, we have

$$\ell\left(\operatorname{Hom}_A(\operatorname{Ext}_A^1(M, M), A)\right) = \ell\left(\operatorname{Ext}_A^1(M^* \otimes_A M, A)\right)$$
$$\geqslant \ell\left(\operatorname{Ext}_A^1(M, M)\right)$$
$$= \ell\left(\operatorname{Ext}_A^1(M, A) \otimes_A M\right)$$

Set $T = \operatorname{Ext}_A^1(M, A)$. Then $T \neq 0$, since $T = 0$ implies that M^* is free by (B.3), whence M is free as well, a contradiction. Then we have an exact sequence

$$0 \longrightarrow \operatorname{Tor}_1^A(T, M) \longrightarrow T \otimes_A F_1 \longrightarrow T \otimes_A F_0 \longrightarrow T \otimes_A M \longrightarrow 0.$$

The rank of M is equal to $\operatorname{rank}_A F_0 - \operatorname{rank}_A F_1$ by (B.2), so counting lengths shows that

$$\ell(T \otimes_A M) = (\operatorname{rank}_A M)\,\ell(T) + \ell\left(\operatorname{Tor}_1^A(T, M)\right).$$

But T is a non-zero module of finite length, so $\operatorname{Tor}_1^A(T, M) \neq 0$, and the proof is complete. $\square$

The next proposition will serve as a road map for Auslander's proof of the theorem on the purity of the branch locus.

§2. PURITY OF THE BRANCH LOCUS

B.22. PROPOSITION. *Let $\mathcal{C}$ be a set of pairs (A, M) where A is a local ring and M is a finitely generated reflexive A-module. Assume that*

(i) $(A, M) \in \mathcal{C}$ implies $(A_\mathfrak{p}, M_\mathfrak{p}) \in \mathcal{C}$ for every $\mathfrak{p} \in \operatorname{Spec} A$;
(ii) $(A, M) \in \mathcal{C}$ and $\operatorname{depth} A \leq 3$ imply that M is free; and
(iii) $(A, M) \in \mathcal{C}$, $\operatorname{depth} A > 3$, and M locally free on the punctured spectrum imply that there exists a non-zerodivisor x in the maximal ideal of A such that $(A/(x), (M/xM)^{**}) \in \mathcal{C}$.

Then M is free over A for every (A, M) in $\mathcal{C}$.

PROOF. If the statement fails, choose a witness $(A, M) \in \mathcal{C}$ with M not A-free and $\dim A$ minimal. By (ii), $\operatorname{depth} A > 3$, so that by (iii) we can find a non-zerodivisor x in the maximal ideal of A such that $(\overline{A}, \overline{M}^{**}) \in \mathcal{C}$, where overlines denote passage modulo x and the duals are taken over $\overline{A}$. Since both $\dim \overline{A}$ and $\dim A_\mathfrak{p}$, for $\mathfrak{p}$ a non-maximal prime, are less than $\dim A$, minimality implies that $\overline{M}^{**}$ is $\overline{A}$-free and $M_\mathfrak{p}$ is $A_\mathfrak{p}$-free for every non-maximal $\mathfrak{p}$. In particular, $\overline{M}_{\overline{\mathfrak{p}}}$ is $\overline{A}_{\overline{\mathfrak{p}}}$-free for every non-maximal prime $\overline{\mathfrak{p}}$ of $\overline{A}$. Thus the natural homomorphism of $\overline{A}$-modules $\overline{M} \longrightarrow \overline{M}^{**}$ is locally an isomorphism on the punctured spectrum of $\overline{A}$. Lemma B.20 then implies

(B.4) $$\operatorname{Ext}^i_{\overline{A}}(\overline{M}^{**}, \overline{A}) \cong \operatorname{Ext}^i_{\overline{A}}(\overline{M}, \overline{A})$$

for $i = 0, \ldots, \operatorname{depth} \overline{A} - 2$. In particular, (B.4) holds for $i = 0$ and $i = 1$ since $\operatorname{depth} \overline{A} - 2 = \operatorname{depth} A - 3 > 0$. In particular the case $i = 1$ says $\operatorname{Ext}^1_{\overline{A}}(\overline{M}, \overline{A}) = 0$ since $\overline{M}^{**}$ is free.

Now, since M is reflexive, the element x is also a non-zerodivisor on M, so standard index-shifting ([**Mat89**], p. 140]) gives $\operatorname{Ext}^1_A(M, \overline{A}) = \operatorname{Ext}^1_{\overline{A}}(\overline{M}, \overline{A}) = 0$. The short exact sequence $0 \longrightarrow A \xrightarrow{x} A \longrightarrow \overline{A} \longrightarrow 0$ induces the long exact sequence containing

$$\operatorname{Ext}^1_A(M, A) \xrightarrow{x} \operatorname{Ext}^1_A(M, A) \longrightarrow \operatorname{Ext}^1_A(M, \overline{A}) = 0$$

so that $\operatorname{Ext}^1_A(M, A) = 0$ by NAK. In particular $\operatorname{Hom}_A(M, \overline{A}) \cong \overline{M^*}$ from the rest of the long exact sequence. But $\operatorname{Hom}_A(M, \overline{A}) = \operatorname{Hom}_{\overline{A}}(\overline{M}, \overline{A})$ as well, so $\overline{M^*} \cong (\overline{M})^*$. Since $\overline{M}$ is $\overline{A}$-free, this shows that $\overline{M^*}$ is free over $\overline{A}$, and since x is a non-zerodivisor on M^* it follows that M^* is A-free. Thus M is A-free, which contradicts the choice of (A, M) and finishes the proof. □

B.23. PROPOSITION. *Let $\mathcal{C}$ be the set of all pairs (A, M) in which $(A, \mathfrak{m}_A)$ is a regular local ring and M is a reflexive A-module with $\operatorname{Hom}_A(M, M) \cong M^{(n)}$ for some n. Then $\mathcal{C}$ satisfies the conditions of Proposition B.22. Thus M is free over A for every such (A, M).*

PROOF. The fact that $\mathcal{C}$ satisfies (i) follows from the isomorphism $\operatorname{Hom}_{R_{\mathfrak{p}}}(M_{\mathfrak{p}}, M_{\mathfrak{p}}) \cong \operatorname{Hom}_R(M, M)_{\mathfrak{p}}$ and the fact that regularity localizes.

For (ii), we note that reflexive modules over a regular local ring of dimension $\leqslant 2$ are automatically free. Therefore M is locally free on the punctured spectrum; also, we may assume that $\dim A = 3$. Finally, the Auslander-Buchsbaum formula gives $\operatorname{pd}_A M \leqslant 1$; we want to show $\operatorname{pd}_A M = 0$. Observe that $n = \operatorname{rank}_A(M)$ (by passing to the quotient field of A), so $\operatorname{Ext}^1_A(\operatorname{Hom}_A(M, M), A) \cong \operatorname{Ext}^1_A(M^{(\operatorname{rank}_A M)}, A) \cong \operatorname{Ext}^1_A(M, A)^{(\operatorname{rank}_A M)}$. Thus by Proposition B.21 M is free.

As for (iii), let $(A, M) \in \mathcal{C}$ with $\dim A > 3$ and M locally free on the punctured spectrum. Let $x \in \mathfrak{m}_A \setminus \mathfrak{m}_A^2$ be a non-zerodivisor on A, hence on M as well since M is reflexive. Applying $\operatorname{Hom}_A(M, -)$ to the short exact sequence $0 \longrightarrow M \xrightarrow{x} M \longrightarrow \overline{M} \longrightarrow 0$ gives

$$\operatorname{Hom}_A(M, M) \xrightarrow{x} \operatorname{Hom}_A(M, M) \longrightarrow \operatorname{Hom}_A(M, \overline{M}) \longrightarrow \operatorname{Ext}^1_A(M, M).$$

As $\operatorname{Hom}_A(M, M) \cong M^{(n)}$, the cokernel of the map on $\operatorname{Hom}_A(M, M)$ defined by multiplication by x is $\overline{M}^{(n)}$. This gives an exact sequence

$$0 \longrightarrow \overline{M}^{(n)} \longrightarrow \operatorname{Hom}_A(M, \overline{M}) \longrightarrow \operatorname{Ext}^1_A(M, M).$$

The middle term is isomorphic to $\operatorname{Hom}_{\overline{A}}(\overline{M}, \overline{M})$, and the rightmost term has finite length as M is locally free. Apply the $i = 0$ version of Lemma B.20 to the $\overline{A}$-homomorphism $\overline{M}^{(n)} \longrightarrow \operatorname{Hom}_A(\overline{M}, \overline{M})$ to find that

$$\operatorname{Hom}_{\overline{A}}(\overline{M}, \overline{M})^* \cong \left(\overline{M}^*\right)^{(\operatorname{rank}_A M)},$$

whence

$$\operatorname{Hom}_{\overline{A}}(\overline{M}, \overline{M})^{**} \cong \left(\overline{M}^{**}\right)^{(\operatorname{rank}_A M)}.$$

Since A is regular and $x \notin \mathfrak{m}_A^2$, $\overline{A}$ is regular as well. In particular, $\overline{A}$ is a normal domain, so $\operatorname{Hom}_{\overline{A}}(\overline{M}, \overline{M})^{**} = \operatorname{Hom}_{\overline{A}}(\overline{M}^{**}, \overline{M}^{**})$. Thus $(\overline{A}, \overline{M}^{**}) \in \mathcal{C}$. □

B.24. REMARK. As Auslander observes, one can use the same strategy to prove that if A is a regular local ring and M is a reflexive A-module such that $\operatorname{Hom}_A(M, M)$ is a *free* A-module, then M is free. This is proved by other methods in [**AG60**], and has been extended to reflexive modules of finite projective dimension over arbitrary local rings by Braun [**Bra04**].

§3. Galois extensions

Let us now investigate ramification in Galois ring extensions. We will see that ramification in codimension one is attributable to the existence of pseudo-reflections in the Galois group, and prove the celebrated Chevalley-Shephard-Todd Theorem that finite groups generated by pseudo-reflections have polynomial rings of invariants. We also prove a result due to Prill, which roughly says that for the purposes of this book we may ignore the existence of pseudo-reflections.

B.25. DEFINITION. Let G be a group and V a finite-dimensional faithful representation of G over a field k. Say that $\sigma \in G$ is a *pseudo-reflection* if σ has finite order and the fixed subspace

$$V^\sigma = \{v \in V \mid \sigma v = v\}$$

has codimension one in V. This subspace is called the *reflecting hyperplane* of σ.

A *reflection* is a pseudo-reflection of order 2.

If the V-action of $\sigma \in G$ is diagonalizable, then to say σ is a pseudo-reflection is the same as saying $\sigma \sim \mathrm{diag}(1,\ldots,1,\lambda)$ where $\lambda \neq 1$ is a root of unity. In any case, the characteristic polynomial of a pseudo-reflection has 1 as a root of multiplicity at least $\dim V - 1$, hence splits into a product of linear factors $(t-1)^{n-1}(t-\lambda)$ with λ a root of unity. In fact, one can show (Exercise 5.38) that a pseudo-reflection with order prime to $\mathrm{char}(k)$ is necessarily diagonalizable.

B.26. NOTATION. Here is the notation we will use for the rest of the Appendix. In contrast to Chapter 5, where we consider the power series case, we will work in the graded polynomial situation, since it clarifies some of the arguments. We leave the translation between the two to the reader. Let k be a field and V an n-dimensional faithful k-representation of a finite group G, so that we may assume $G \subseteq \mathrm{GL}(V) \cong \mathrm{GL}(n,k)$. Set $S = k[V] \cong k[x_1,\ldots,x_n]$, viewed as the ring of polynomial functions on V. Then G acts on S by the rule $(\sigma f)(v) = f(\sigma^{-1}v)$, and we set $R = S^G$, the subring of polynomials fixed by this action. Then $R \longrightarrow S$ is a module-finite integral extension of Noetherian normal domains. Let K and L be the quotient fields of R and S, respectively; then L/K is a Galois extension with Galois group G, and S is the integral closure of R in L. Finally, let $\mathfrak{m}$ and $\mathfrak{n}$ denote the obvious homogeneous maximal ideals of R and S.

B.27. THEOREM (Chevalley-Shephard-Todd). *Consider the following conditions.*

(i) $R = S^G$ *is a polynomial ring.*

(ii) S *is free as an R-module.*
(iii) $\operatorname{Tor}_1^R(S, k) = 0$.
(iv) G *is generated by pseudo-reflections.*

We have (i) $\iff$ (ii) $\iff$ (iii) $\implies$ (iv), and all four conditions are equivalent if $|G|$ is invertible in k.

PROOF. (i) $\implies$ (ii): Note that S is always a MCM R-module, so if R is a polynomial ring then S is R-free by the Auslander-Buchsbaum formula.

(ii) $\implies$ (i): If S is free over R, then in particular it is flat. For any finitely generated R-module, then, we have
$$\operatorname{Tor}_i^R(M, k) \otimes_R S = \operatorname{Tor}_i^S(S \otimes_R M, S/\mathfrak{m}S).$$
Since S is regular of dimension n and $S/\mathfrak{m}S$ has finite length, the latter Tor vanishes for $i > n$, whence the former does as well. It follows that R is regular, hence a polynomial ring.

(ii) $\iff$ (iii): This is standard.

(i) $\implies$ (iv): Let $H \subseteq G$ be the subgroup of G generated by the pseudo-reflections. Then H is automatically normal. Localize the problem, setting $A = R_\mathfrak{m}$, a regular local ring by hypothesis, and $B = \left(k[V]^H\right)_{k[V]^H \cap \mathfrak{n}}$. Then $A \longrightarrow B$ is a module-finite extension of local normal domains, and $A = B^{G/H}$.

Consider as in Chapter 5 the skew group ring $B\#(G/H)$. There is, as in that chapter, a natural ring homomorphism $\delta \colon B\#(G/H) \longrightarrow \operatorname{Hom}_A(B, B)$, which considers an element $b\overline{\sigma} \in B\#(G/H)$ as the A-linear endomorphism $b' \mapsto b\overline{\sigma}(b')$ of B. We claim that δ is an isomorphism. Since source and target are reflexive over B, it suffices to check in codimension one. Let $\mathfrak{q}$ and $\mathfrak{p} = A \cap \mathfrak{q}$ be height-one primes of B and A respectively; then $B_\mathfrak{q}$ is a finitely generated free $A_\mathfrak{p}$-module and so $\delta_\mathfrak{q} \colon B_\mathfrak{q}\#(G/H) \longrightarrow \operatorname{Hom}_{A_\mathfrak{p}}(B_\mathfrak{q}, B_\mathfrak{q})$ is an isomorphism. This shows that δ is an isomorphism, and in particular $\operatorname{Hom}_A(B, B)$ is isomorphic as an A-module to a direct sum of copies of B. By Proposition B.23, B is free over A.

Since B is A-free, we have $\mathfrak{H}_A(B) = \mathfrak{D}_A(B)$ by Theorem B.14. But $\mathfrak{D}_A(B) = B$ since no non-identity element of G/H fixes a codimension-one subspace of V, i.e. a height-one prime of B. This implies $\mathfrak{H}_A(B) = B$ so that the branch locus is empty. However, if G/H is non-trivial then $A \longrightarrow B$ is ramified at the maximal ideal of B. Thus $G/H = 1$.

Finally, we prove (iv) $\implies$ (iii) under the assumption that $|G|$ is invertible in k. For an arbitrary finitely generated R-module M, set $T(M) = \operatorname{Tor}_1^R(M, k)$. We wish to show $T(S) = 0$. Note that G acts naturally on $T(S)$, which is a finitely generated graded S-module.

Let $\sigma \in G$ be a pseudo-reflection and set $W = V^\sigma$, a linear subspace of codimension one. Let $f \in S$ be a linear form vanishing on W. Then (f) is a prime ideal of S of height one, and σ acts trivially on the quotient $S/(f) \cong k[W]$. For each $g \in S$, then, there exists a unique element $h(g) \in S$ such that $\sigma(g) - g = h(g)f$. The function $g \mapsto h(g)$ is an R-linear endomorphism of S of degree -1, with $\sigma - 1_S = hf$ as functions on S. Applying the functor $T(-)$ gives $T(\sigma) - 1_{T(S)} = T(h)f_{T(S)}$ as functions on $T(S)$. It follows that $\sigma(s) \equiv x \bmod \mathfrak{n}T(S)$ for every $x \in T(S)$. Since G is generated by pseudo-reflections, we conclude that $\sigma(x) \equiv x \bmod \mathfrak{n}T(S)$ for every $\sigma \in G$ and every $x \in T(S)$.

Next we claim that $T(S)^G = 0$. Define $Q \colon S \longrightarrow S$ by

$$Q(f) = \frac{1}{|G|} \sum_{\sigma \in G} \sigma(f),$$

so that in particular $Q(S) = R$. Factor Q as $Q = iQ' \colon S \longrightarrow R \longrightarrow S$, so that $T(Q) = T(i)T(Q')$. Since $T(R) = 0$, $T(i)$ is the zero map, so $T(Q) = 0$ as well. Hence

$$0 = T(Q) = \frac{1}{|G|} \sum_{\sigma \in G} T(\sigma),$$

as R-linear maps $T(S) \longrightarrow T(S)$. But that R-linear map fixes the G-invariant elements of $T(S)$, so that $T(S)^G = 0$.

Finally, suppose $T(S) \neq 0$. Then there exists a homogeneous element $x \in T(S)$ of minimal positive degree. Since $\sigma(x) \equiv x \bmod \mathfrak{n}T(S)$ for every $\sigma \in G$, x is an invariant of $T(S)$. But then $x = 0$ as $T(S)^G = 0$. This completes the proof. $\square$

It is implicit in the proof of Theorem B.27 that pseudo-reflections are responsible for ramification. Let us now bring that out into the open. Briefly, the situation is this: let W be a codimension-one subspace of V, and $f \in S$ a linear form vanishing on W. Then (f) is a height-one prime of S, and (f) is ramified over R if and only if W is the reflecting hyperplane of a pseudo-reflection.

Keep the notation established so far, so that $R = k[V]^G \subseteq S = k[V]$ is a module-finite extension of normal domains inducing a Galois extension of quotient fields $K \longrightarrow L$. Since $R \longrightarrow S$ is integral, it follows from "going up" and "going down" that a prime ideal $\mathfrak{q}$ of S has height equal to the height of $R \cap \mathfrak{q}$. Furthermore, for a fixed $\mathfrak{p} \in \operatorname{Spec} R$, the primes $\mathfrak{q}$ lying over $\mathfrak{p}$ are all conjugate under the action of G. (If $\mathfrak{q}$ and $\mathfrak{q}'$ lying over $\mathfrak{p}$ are not conjugate, then by "lying over" no conjugate of $\mathfrak{q}$ contains $\mathfrak{q}'$. Use prime avoidance to find an element $s \in \mathfrak{q}'$ so that s avoids all conjugates of $\mathfrak{q}$. Then $\prod_{\sigma \in G} \sigma(s)$ is fixed by G, so in $R \cap \mathfrak{q} = \mathfrak{p}$, but not in $\mathfrak{q}'$.)

Assume now that $\mathfrak{p}$ is a fixed prime of R of height one, and let $\mathfrak{q} \subseteq S$ lie over $\mathfrak{p}$. Then $R_\mathfrak{p} \longrightarrow S_\mathfrak{q}$ is an extension of DVRs, so $\mathfrak{p} S_\mathfrak{q} = \mathfrak{q}^e S_\mathfrak{q}$ for some integer $e = e(\mathfrak{p})$, the *ramification index* of $\mathfrak{q}$ over $\mathfrak{p}$, which is independent of $\mathfrak{q}$ by the previous paragraph. Let $f = f(\mathfrak{p}, \mathfrak{q})$ be the *inertial degree* of $\mathfrak{q}$ over $\mathfrak{p}$, i.e. the degree of the field extension $R_\mathfrak{p}/\mathfrak{p} R_\mathfrak{p} \longrightarrow S_\mathfrak{q}/\mathfrak{q} S_\mathfrak{q}$. Then $S_\mathfrak{q}/\mathfrak{p} S_\mathfrak{q}$ is a free $R_\mathfrak{p}/\mathfrak{p} R_\mathfrak{p}$-module of rank ef, so $S_\mathfrak{q}$ is a free $R_\mathfrak{p}$-module of rank ef.

Let $\mathfrak{q}_1, \ldots, \mathfrak{q}_r$ be the distinct primes of S lying over $\mathfrak{p}$, and set $\mathfrak{q} = \mathfrak{q}_1$. Let $D(\mathfrak{q})$ be the *decomposition group* of $\mathfrak{q}$ over $\mathfrak{p}$,

$$D(\mathfrak{q}) = \{\sigma \in G \mid \sigma(\mathfrak{q}) = \mathfrak{q}\}\,.$$

By the orbit-stabilizer theorem, $D(\mathfrak{q})$ has index r in G. Furthermore, $S_\mathfrak{q}$ is an extension of $R_\mathfrak{p}$ of rank equal to $D(\mathfrak{q})$, which implies $|D(\mathfrak{q})| = ef$.

Notice that an element of $D(\mathfrak{q})$ induces an automorphism of $S/\mathfrak{q}$. We let $T(\mathfrak{q})$, the *inertia group* of $\mathfrak{q}$ over $\mathfrak{p}$, be the subgroup inducing the identity on $S/\mathfrak{q}$:

$$T(\mathfrak{q}) = \{\sigma \in G \mid \sigma(f) - f \in \mathfrak{q} \text{ for all } f \in S\}\,.$$

Then the quotient $D(\mathfrak{q})/T(\mathfrak{q})$ acts as Galois automorphisms of $S_\mathfrak{q}/\mathfrak{q} S_\mathfrak{q}$ fixing $R_\mathfrak{p}/\mathfrak{p} R_\mathfrak{p}$. It follows that $|D(\mathfrak{q})/T(\mathfrak{q})|$ divides the degree f of this field extension. Combining this with $|D(\mathfrak{q})| = ef$, we see that e divides $|T(\mathfrak{q})|$. In fact $e = |T(\mathfrak{q})|$ as long as $|G|$ is invertible in k:

B.28. PROPOSITION. *Let $\mathfrak{q}$ be a height one prime of S, set $\mathfrak{p} = R \cap \mathfrak{q}$, and suppose that $T(\mathfrak{q}) \neq 1$. Then $\mathfrak{q} = (f)$ for some linear form $f \in S$. If $W \subseteq V$ is the hyperplane on which f vanishes, then $T(\mathfrak{q})$ is the pointwise stabilizer of W, so every non-identity element of $T(\mathfrak{q})$ is a pseudo-reflection. Furthermore if $|G|$ is invertible in k then $e(\mathfrak{p}) = |T(\mathfrak{q})|$.*

PROOF. Since $\mathfrak{q}$ is a prime of height one in the UFD S, $\mathfrak{q} = (f)$ for some homogeneous element $f \in S$. If f has degree 2 or more, then every linear form of S survives in $S_\mathfrak{q}/\mathfrak{q} S_\mathfrak{q}$, so is acted upon trivially by $T(\mathfrak{q})$. Since $T(\mathfrak{q})$ is non-trivial, we must have $\deg f = 1$, so f is linear. The zero-set of f, $W = \operatorname{Spec} S/\mathfrak{q}$, is the subspace fixed pointwise by $T(\mathfrak{q})$.

For any $\sigma \in T(\mathfrak{q})$, $\sigma(f)$ vanishes on W, so $\sigma(f) = a_\sigma f$ for some scalar $a_\sigma \in k$. Define a linear character $\chi \colon T(\mathfrak{q}) \longrightarrow k^\times$ by $\chi(\sigma) = a_\sigma$. The image of χ is finite, so is cyclic of order prime to the characteristic of k. The kernel of χ consists of the non-diagonalizable pseudo-reflections in $T(\mathfrak{q})$ (see the discussion following Definition B.25). Since

$|G|$ is not divisible by p, the kernel of χ is trivial by Exercise 5.38, so that $T(\mathfrak{q})$ is cyclic.

Let $\sigma \in T(\mathfrak{q})$ be a generator, and let λ be the unique eigenvalue of σ different from 1. Then λ is an sth root of unity for some $s > 1$. We can find a basis $v_1, \ldots, v_n$ for V such that $v_1, \ldots, v_{n-1}$ span W, so are fixed by σ, and $\sigma v_n = \lambda v_n$. It follows that $k[V]^{T(\mathfrak{q})} \cong k[x_1, \ldots, x_{n-1}, x_n^s]$, and so $\mathfrak{p} = (x_n^s)$ and $e(\mathfrak{p}) = s = |T(\mathfrak{q})|$. □

Recall that we say the group G is *small* if it contains no pseudo-reflections.

B.29. THEOREM. *Let $G \subseteq \mathrm{GL}(V)$ be a finite group of linear automorphisms of a finite-dimensional vector space V over a field k. Set $S = k[V]$ and $R = S^G$. Assume that $|G|$ is invertible in k. Then a prime ideal $\mathfrak{q}$ of height one in S is ramified over R if and only if $T(\mathfrak{q}) = 1$. In particular, $R \longrightarrow S$ is unramified in codimension one if and only if G is small.*

PROOF. Let $e = e(\mathfrak{p})$ be the ramification index of $\mathfrak{p} = R \cap \mathfrak{q}$, and $f = f(\mathfrak{p}, \mathfrak{q})$ the degree of the field extension $R_{\mathfrak{p}}/\mathfrak{p} R_{\mathfrak{p}} \longrightarrow S_{\mathfrak{q}}/\mathfrak{q} S_{\mathfrak{q}}$. By the discussion before the Proposition, $ef = |D(\mathfrak{q})|$, where $D(\mathfrak{q})$ is the decomposition group of $\mathfrak{q}$ over $\mathfrak{p}$. Since the order of G is prime to the characteristic, we see that f is as well, so the field extension is separable. Therefore $\mathfrak{q}$ is ramified over R if and only if $e > 1$, which occurs if and only if $T(\mathfrak{q}) \neq 1$. □

To close the Appendix, we record a result due to Prill [**Pri67**].

B.30. PROPOSITION. *Let G be a finite subgroup of $\mathrm{GL}(V)$, where V is an n-dimensional vector space over a field k. Set $S = k[V]$ and $R = S^G$. Then there is an n-dimensional vector space V' and a small finite subgroup $G' \subseteq \mathrm{GL}(V')$ such that $R \cong k[V']^{G'}$.*

PROOF. Let H be the normal subgroup of G generated by pseudo-reflections. By the Chevalley-Shephard-Todd Theorem B.27, $S^H \cong k[f_1, \ldots, f_n]$ is a polynomial ring on algebraically independent elements $f_1, \ldots, f_n$. The quotient G/H acts naturally on S^H, with $(S^H)^{G/H} = S^G$, so it suffices to show that G/H acts on $V' = \mathrm{span}(f_1, \ldots, f_n)$ without pseudo-reflections. Fix $\sigma \in G \backslash H$ and let $\tau \in H$. Since $\sigma\tau \notin H$, the subspace $V^{\sigma\tau}$ fixed by $\sigma\tau$ has codimension at least two. The fixed locus of the action of the coset σH on V' is then the intersection of $V^{\sigma\tau}$ as τ runs over H, so also has codimension at least two. Therefore σH is not a pseudo-reflection. □

In fact the small subgroup G' of the Proposition is unique up to conjugacy in $\mathrm{GL}(n, k)$. We do not prove this; see [**Pri67**] for a proof in the complex-analytic situation, and [**DR69**] for a proof in an algebraic context.

Bibliography

[AB59] Maurice Auslander and David A. Buchsbaum, *On ramification theory in noetherian rings*, Amer. J. Math. **81** (1959), 749–765. MR0106929 [66, 322, 326]

[AB69] Maurice Auslander and Mark Bridger, *Stable module theory*, Memoirs of the American Mathematical Society, No. 94, American Mathematical Society, Providence, R.I., 1969. MR0269685 [187, 210]

[AB89] Maurice Auslander and Ragnar-Olaf Buchweitz, *The homological theory of maximal Cohen-Macaulay approximations*, Mém. Soc. Math. France (N.S.) (1989), no. 38, 5–37, Colloque en l'honneur de Pierre Samuel (Orsay, 1987). MR1044344 [175, 184]

[Abh90] Shreeram S. Abhyankar, *Algebraic geometry for scientists and engineers*, Mathematical Surveys and Monographs, vol. 35, American Mathematical Society, Providence, RI, 1990. MR1075991 [147]

[ADS93] Maurice Auslander, Songqing Ding, and Øyvind Solberg, *Liftings and weak liftings of modules*, J. Algebra **156** (1993), no. 2, 273–317. MR1216471 [43]

[AG60] Maurice Auslander and Oscar Goldman, *Maximal orders*, Trans. Amer. Math. Soc. **97** (1960), 1–24. MR0117252 [334]

[Aki37] Yasuo Akizuki, *Zur ldealtheorie der einartigen Ringbereiche mit dem Teilerkettensatz*, Proc. Imp. Acad. **13** (1937), no. 3, 53–55. MR1568451 [316]

[AM02] Luchezar L. Avramov and Alex Martsinkovsky, *Absolute, relative, and Tate cohomology of modules of finite Gorenstein dimension*, Proc. London Math. Soc. (3) **85** (2002), no. 2, 393–440. MR1912056 [209, 210]

[ANT44] Emil Artin, Cecil J. Nesbitt, and Robert M. Thrall, *Rings with Minimum Condition*, University of Michigan Publications in Mathematics, no. 1, University of Michigan Press, Ann Arbor, Mich., 1944. MR0010543 [267]

[AR89] Maurice Auslander and Idun Reiten, *The Cohen-Macaulay type of Cohen-Macaulay rings*, Adv. in Math. **73** (1989), no. 1, 1–23. MR979585 [287, 293]

[Arn81] Vladimir I. Arnol'd, *Singularity theory*, London Mathematical Society Lecture Note Series, vol. 53, Cambridge University Press, Cambridge, 1981, Selected papers, Translated from the Russian, With an introduction by C. T. C. Wall. MR631683 [260]

[Art66] Michael Artin, *On isolated rational singularities of surfaces*, Amer. J. Math. **88** (1966), 129–136. MR0199191 [97, 100]

[Art77] ———, *Coverings of the rational double points in characteristic p*, Complex analysis and algebraic geometry, Iwanami Shoten, Tokyo, 1977, pp. 11–22. MR0450263 [119, 160]

[Aus62] Maurice Auslander, *On the purity of the branch locus*, Amer. J. Math. **84** (1962), 116–125. MR0137733 [66, 326, 330]

[Aus67] _____, *Anneaux de Gorenstein, et torsion en algèbre commutative*, Séminaire d'Algèbre Commutative dirigé par Pierre Samuel, 1966/67., Secrétariat mathématique, Paris, 1967. MR0225844 (37 #1435) [187]

[Aus74] _____, *Representation theory of Artin algebras. I, II*, Comm. Algebra **1** (1974), 177–268; ibid. **1** (1974), 269–310. MR0349747 [267]

[Aus86a] _____, *Isolated singularities and existence of almost split sequences*, Representation theory, II (Ottawa, Ont., 1984), Lecture Notes in Math., vol. 1178, Springer, Berlin, 1986, pp. 194–242. MR842486 [113, 115]

[Aus86b] _____, *Rational singularities and almost split sequences*, Trans. Amer. Math. Soc. **293** (1986), no. 2, 511–531. MR816307 [117, 234]

[AV85] Michael Artin and Jean-Louis Verdier, *Reflexive modules over rational double points*, Math. Ann. **270** (1985), no. 1, 79–82. MR769609 [105, 160]

[Avr98] Luchezar L. Avramov, *Infinite free resolutions*, Six lectures on commutative algebra (Bellaterra, 1996), Progr. Math., vol. 166, Birkhäuser, Basel, 1998, pp. 1–118. MR1648664 [151, 210, 280]

[Azu48] Gorô Azumaya, *On generalized semi-primary rings and Krull-Remak-Schmidt's theorem*, Jap. J. Math. **19** (1948), 525–547. MR0032607 [3, 5]

[Azu50] _____, *Corrections and supplementaries to my paper concerning Krull-Remak-Schmidt's theorem*, Nagoya Math. J. **1** (1950), 117–124. MR0037832 [1]

[Bas62] Hyman Bass, *Torsion free and projective modules*, Trans. Amer. Math. Soc. **102** (1962), 319–327. MR0140542 [57]

[Bas63] _____, *On the ubiquity of Gorenstein rings*, Math. Z. **82** (1963), 8–28. MR0153708 [48, 54, 60, 176]

[Bas68] _____, *Algebraic K-theory*, W. A. Benjamin, Inc., New York-Amsterdam, 1968. MR0249491 [1]

[BD08] Igor Burban and Yuriy Drozd, *Maximal Cohen-Macaulay modules over surface singularities*, Trends in representation theory of algebras and related topics, EMS Ser. Congr. Rep., Eur. Math. Soc., Zürich, 2008, pp. 101–166. MR2484725 [63, 241]

[BD10] _____, *Cohen-Macaulay modules over non-isolated surface singularities*, http://arxiv.org/abs/1002.3042, 2010. [241, 245, 263]

[BGS87] Ragnar-Olaf Buchweitz, Gert-Martin Greuel, and Frank-Olaf Schreyer, *Cohen-Macaulay modules on hypersurface singularities. II*, Invent. Math. **88** (1987), no. 1, 165–182. MR877011 [141, 145, 241, 245, 284]

[BH93] Winfried Bruns and Jürgen Herzog, *Cohen-Macaulay rings*, Cambridge Studies in Advanced Mathematics, vol. 39, Cambridge University Press, Cambridge, 1993. MR1251956 [16, 17, 166, 168, 173, 176, 310]

[BHU87] Joseph P. Brennan, Jürgen Herzog, and Bernd Ulrich, *Maximally generated Cohen-Macaulay modules*, Math. Scand. **61** (1987), no. 2, 181–203. MR947472 [315]

[BL10] Nicholas R. Baeth and Melissa R. Luckas, *Monoids of torsion-free modules over rings with finite representation type*, to appear, 2010. [56, 57]

[Bon89] Klaus Bongartz, *A generalization of a theorem of M. Auslander*, Bull. London Math. Soc. **21** (1989), no. 3, 255–256. MR986367 [120]

[Bou87] Jean-François Boutot, *Singularités rationnelles et quotients par les groupes réductifs*, Invent. Math. **88** (1987), no. 1, 65–68. MR877006 [99]

[Bou98] Nicolas Bourbaki, *Commutative algebra. Chapters 1–7*, Elements of Mathematics (Berlin), Springer-Verlag, Berlin, 1998, Translated from the French, Reprint of the 1989 English translation. MR1727221 [23]

[Bou02] _____, *Lie groups and Lie algebras. Chapters 4–6*, Elements of Mathematics (Berlin), Springer-Verlag, Berlin, 2002, Translated from the 1968 French original by Andrew Pressley. MR1890629 [84]

[Bou06] _____, *Éléments de mathématique. Algèbre commutative. Chapitres 8 et 9*, Springer, Berlin, 2006, Reprint of the 1983 original. MR2284892 [170, 171]

[Bra41] Richard Brauer, *On the indecomposable representations of algebras*, Bull. A.M.S. **47** (1941), no. 9, 684. [267]

[Bra04] Amiram Braun, *On a question of M. Auslander*, J. Algebra **276** (2004), no. 2, 674–684. MR2058462 [334]

[Bri68] Egbert Brieskorn, *Rationale Singularitäten komplexer Flächen*, Invent. Math. **4** (1967/1968), 336–358. MR0222084 [99]

[Bru81] Winfried Bruns, *The Eisenbud-Evans generalized principal ideal theorem and determinantal ideals*, Proc. Amer. Math. Soc. **83** (1981), no. 1, 19–24. MR619972 [284]

[Buc81] Ragnar-Olaf Buchweitz, *Contributions à la théorie des singularités: Déformations de Diagrammes, Déploiements et Singularités très rigides, Liaison algébrique*, Ph.D. thesis, University of Paris VII, 1981, http://hdl.handle.net/1807/16684. [117]

[Buc86] _____, *Maximal Cohen-Macaulay modules and Tate-cohomology over Gorenstein rings*, unpublished manuscript available from http://hdl.handle.net/1807/16682, 1986. [187, 189, 203]

[Bur72] Lindsay Burch, *Codimension and analytic spread*, Proc. Cambridge Philos. Soc. **72** (1972), 369–373. MR0304377 [241]

[Bur74] Daniel M. Burns, *On rational singularities in dimensions > 2*, Math. Ann. **211** (1974), 237–244. MR0364672 [99]

[Car57] Henri Cartan, *Quotient d'un espace analytique par un groupe d'automorphismes*, Algebraic geometry and topology, Princeton University Press, Princeton, N. J., 1957, A symposium in honor of S. Lefschetz,, pp. 90–102. MR0084174 [62]

[CE99] Henri Cartan and Samuel Eilenberg, *Homological algebra*, Princeton Landmarks in Mathematics, Princeton University Press, Princeton, NJ, 1999, With an appendix by David A. Buchsbaum, Reprint of the 1956 original. MR1731415 [221]

[Çim94] Nuri Çimen, *One-dimensional rings of finite Cohen-Macaulay type*, Ph.D. thesis, University of Nebraska–Lincoln, Lincoln, NE, 1994, Thesis (Ph.D.)–The University of Nebraska - Lincoln, p. 105. MR2691778 [41, 45, 49, 50]

[Çim98] _____, *One-dimensional rings of finite Cohen-Macaulay type*, J. Pure Appl. Algebra **132** (1998), no. 3, 275–308. MR1642094 [41, 45, 49, 50]

[Coh76] Arjeh M. Cohen, *Finite complex reflection groups*, Ann. Sci. École Norm. Sup. (4) **9** (1976), no. 3, 379–436. MR0422448 [151]

[CPST08] Lars Winther Christensen, Greg Piepmeyer, Janet Striuli, and Ryo Takahashi, *Finite Gorenstein representation type implies simple singularity*, Adv. Math. **218** (2008), no. 4, 1012–1026. MR2419377 (2009b:13058) [203, 210, 211, 212, 214, 215]

[CS93] Steven Dale Cutkosky and Hema Srinivasan, *Local fundamental groups of surface singularities in characteristic p*, Comment. Math. Helv. **68** (1993), no. 2, 319–332. MR1214235 [117]

[ÇWW95] Nuri Çimen, Roger Wiegand, and Sylvia Wiegand, *One-dimensional rings of finite representation type*, Abelian groups and modules (Padova, 1994), Math. Appl., vol. 343, Kluwer Acad. Publ., Dordrecht, 1995, pp. 95–121. MR1378192 [34]

[Dad63] Everett C. Dade, *Some indecomposable group representations*, Ann. of Math. (2) **77** (1963), 406–412. MR0144981 [37]

[Dic13] Leonard Eugene Dickson, *Finiteness of the Odd Perfect and Primitive Abundant Numbers with n Distinct Prime Factors*, Amer. J. Math. **35** (1913), no. 4, 413–422. MR1506194 [28]

[Dic59] Leonard E. Dickson, *Algebraic theories*, Dover Publications Inc., New York, 1959. MR0105380 [87]

[Die46] Jean Dieudonné, *Sur la réduction canonique des couples de matrices*, Bull. Soc. Math. France **74** (1946), 130–146. MR0022826 [31]

[Die81] Ernst Dieterich, *Representation types of group rings over complete discrete valuation rings*, Integral representations and applications (Oberwolfach, 1980), Lecture Notes in Math., vol. 882, Springer, Berlin, 1981, pp. 369–389. MR646112 [268]

[Die87] _____, *Reduction of isolated singularities*, Comment. Math. Helv. **62** (1987), no. 4, 654–676. MR920064 [268, 276, 280, 284, 298]

[Din92] Songqing Ding, *Cohen-Macaulay approximation and multiplicity*, J. Algebra **153** (1992), no. 2, 271–288. MR1198202 [195]

[Din93] _____, *A note on the index of Cohen-Macaulay local rings*, Comm. Algebra **21** (1993), no. 1, 53–71. MR1194550 [195]

[Din94] _____, *The associated graded ring and the index of a Gorenstein local ring*, Proc. Amer. Math. Soc. **120** (1994), no. 4, 1029–1033. MR1181160 [192, 195]

[DR67] Yuriy A. Drozd and Andreĭ Vladimirovich Roĭter, *Commutative rings with a finite number of indecomposable integral representations*, Izv. Akad. Nauk SSSR Ser. Mat. **31** (1967), 783–798. MR0220716 [32, 34, 38, 50]

[DR69] Dieter Denneberg and Oswald Riemenschneider, *Verzweigung bei Galoiserweiterungen und Quotienten regulärer analytischer Raumkeime*, Invent. Math. **7** (1969), 111–119. MR0244254 [340]

[Dur79] Alan H. Durfee, *Fifteen characterizations of rational double points and simple critical points*, Enseign. Math. (2) **25** (1979), no. 1-2, 131–163. MR543555 [101]

[DV34] Patrick Du Val, *On isolated singularities of surfaces which do not affect the conditions of adjunction. I, II, III.*, Proc. Camb. Philos. Soc. **30** (1934), 453–459, 460–465, 483–491 (English). [97, 100]

[DV64] _____, *Homographies, quaternions and rotations*, Oxford Mathematical Monographs, Clarendon Press, Oxford, 1964. MR0169108 [87]

[EdlP98] David Eisenbud and José Antonio de la Peña, *Chains of maps between indecomposable modules*, J. Reine Angew. Math. **504** (1998), 29–35. MR1656826 [268, 270]

[EG85] E. Graham Evans, Jr. and Phillip A. Griffith, *Syzygies*, London Mathematical Society Lecture Note Series, vol. 106, Cambridge University Press, Cambridge, 1985. MR811636 [311]

[EH87] David Eisenbud and Joe Harris, *On varieties of minimal degree (a centennial account)*, Algebraic geometry, Bowdoin, 1985 (Brunswick, Maine, 1985), Proc. Sympos. Pure Math., vol. 46, Amer. Math. Soc., Providence, RI, 1987, pp. 3–13. MR927946 [295]

[EH88] David Eisenbud and Jürgen Herzog, *The classification of homogeneous Cohen-Macaulay rings of finite representation type*, Math. Ann. **280** (1988), no. 2, 347–352. MR929541 [119, 293]

[Eis80] David Eisenbud, *Homological algebra on a complete intersection, with an application to group representations*, Trans. Amer. Math. Soc. **260** (1980), no. 1, 35–64. MR570778 [123, 151]

[Eis95] _____, *Commutative algebra*, Graduate Texts in Mathematics, vol. 150, Springer-Verlag, New York, 1995, With a view toward algebraic geometry. MR1322960 [239, 321, 323, 324]

[EJ00] Edgar E. Enochs and Overtoun M. G. Jenda, *Relative homological algebra*, de Gruyter Expositions in Mathematics, vol. 30, Walter de Gruyter & Co., Berlin, 2000. MR1753146 (2001h:16013) [211]

[Elk73] Renée Elkik, *Solutions d'équations à coefficients dans un anneau hensélien*, Ann. Sci. École Norm. Sup. (4) **6** (1973), 553–603 (1974). MR0345966 [169]

[Esn85] Hélène Esnault, *Reflexive modules on quotient surface singularities*, J. Reine Angew. Math. **362** (1985), 63–71. MR809966 [117]

[Eva73] E. Graham Evans, Jr., *Krull-Schmidt and cancellation over local rings*, Pacific J. Math. **46** (1973), 115–121. MR0323815 [3, 7]

[Fac98] Alberto Facchini, *Module theory*, Progress in Mathematics, vol. 167, Birkhäuser Verlag, Basel, 1998, Endomorphism rings and direct sum decompositions in some classes of modules. MR1634015 [5, 11]

[Fer72] Daniel Ferrand, *Monomorphismes et morphismes absolument plats*, Bull. Soc. Math. France **100** (1972), 97–128. MR0318138 (47 #6687) [167, 168]

[FH91] William Fulton and Joe Harris, *Representation theory*, Graduate Texts in Mathematics, vol. 129, Springer-Verlag, New York, 1991, A first course, Readings in Mathematics. MR1153249 [73]

[Fle75] Hubert Flenner, *Reine lokale Ringe der Dimension zwei*, Math. Ann. **216** (1975), no. 3, 253–263. MR0382710 [117]

[Fle81] _____, *Rationale quasihomogene Singularitäten*, Arch. Math. (Basel) **36** (1981), no. 1, 35–44. MR612234 [98]

[Fog81] John Fogarty, *On the depth of local rings of invariants of cyclic groups*, Proc. Amer. Math. Soc. **83** (1981), no. 3, 448–452. MR627666 [64]

[Fox72] Hans-Bjørn Foxby, *Gorenstein modules and related modules*, Math. Scand. **31** (1972), 267–284 (1973). MR0327752 [177]

[FR70] Daniel Ferrand and Michel Raynaud, *Fibres formelles d'un anneau local noethérien*, Ann. Sci. École Norm. Sup. (4) **3** (1970), 295–311. MR0272779 [178]

[GAP08]   The GAP Group, *GAP – Groups, Algorithms, and Programming, Version 4.4.12*, 2008. [94]

[GD64]   Alexandre Grothendieck and Jean Dieudonné, *Éléments de géométrie algébrique. IV. Étude locale des schémas et des morphismes de schémas. I*, Inst. Hautes Études Sci. Publ. Math. (1964), no. 20, 259. MR0173675 [275]

[GD65]   _____, *Éléments de géométrie algébrique. IV. Étude locale des schémas et des morphismes de schémas. II*, Inst. Hautes Études Sci. Publ. Math. (1965), no. 24, 231. MR0199181 [310]

[GK85]   Gert-Martin Greuel and Horst Knörrer, *Einfache Kurvensingularitäten und torsionsfreie Moduln*, Math. Ann. **270** (1985), no. 3, 417–425. MR774367 [50, 51]

[GK90]   Gert-Martin Greuel and Heike Kröning, *Simple singularities in positive characteristic*, Math. Z. **203** (1990), 229–354. [159, 160]

[GR78]   Edward L. Green and Irving Reiner, *Integral representations and diagrams*, Michigan Math. J. **25** (1978), no. 1, 53–84. MR497882 [38, 41, 45, 49, 50, 172]

[Gre76]   Silvio Greco, *Two theorems on excellent rings*, Nagoya Math. J. **60** (1976), 139–149. MR0409452 [168]

[GSV81]   Gerardo González-Sprinberg and Jean-Louis Verdier, *Points doubles rationnels et représentations de groupes*, C. R. Acad. Sci. Paris Sér. I Math. **293** (1981), no. 2, 111–113. MR637103 [97, 152, 155]

[Gul80]   Tor Holtedahl Gulliksen, *On the deviations of a local ring*, Math. Scand. **47** (1980), no. 1, 5–20. MR600076 [151]

[Gur85]   Robert M. Guralnick, *Lifting homomorphisms of modules*, Illinois J. Math. **29** (1985), no. 1, 153–156. MR769764 [7]

[Gus82]   William H. Gustafson, *The history of algebras and their representations*, Representations of algebras (Puebla, 1980), Lecture Notes in Math., vol. 944, Springer, Berlin, 1982, pp. 1–28. MR672114 [267]

[Har77]   Robin Hartshorne, *Algebraic geometry*, Springer-Verlag, New York, 1977, Graduate Texts in Mathematics, No. 52. MR0463157 [25, 97, 98, 99, 100, 102]

[Hei93]   Raymond C. Heitmann, *Characterization of completions of unique factorization domains*, Trans. Amer. Math. Soc. **337** (1993), no. 1, 379–387. MR1102888 [26, 307]

[Her78a] Jürgen Herzog, *Ein Cohen-Macaulay-Kriterium mit Anwendungen auf den Konormalenmodul und den Differentialmodul*, Math. Z. **163** (1978), no. 2, 149–162. MR512469 [116]

[Her78b] _____, *Ringe mit nur endlich vielen Isomorphieklassen von maximalen, unzerlegbaren Cohen-Macaulay-Moduln*, Math. Ann. **233** (1978), no. 1, 21–34. MR0463155 [81, 141, 150]

[Her94]   _____, *On the index of a homogeneous Gorenstein ring*, Commutative algebra: syzygies, multiplicities, and birational algebra (South Hadley, MA, 1992), Contemp. Math., vol. 159, Amer. Math. Soc., Providence, RI, 1994, pp. 95–102. MR1266181 [195, 200]

[Hig54]   Donald Gordon Higman, *Indecomposable representations at characteristic p*, Duke Math. J. **21** (1954), 377–381. MR0067896 [31]

[Hig60] _____, *On representations of orders over Dedekind domains*, Canad. J. Math. **12** (1960), 107–125. MR0109175 [272]

[Hir95a] Friedrich Hirzebruch, *The topology of normal singularities of an algebraic surface (after D. Mumford)*, Séminaire Bourbaki, Vol. 8, Soc. Math. France, Paris, 1995, pp. Exp. No. 250, 129–137. MR1611536 [100]

[Hir95b] _____, *The topology of normal singularities of an algebraic surface (after D. Mumford)*, Séminaire Bourbaki, Vol. 8, Soc. Math. France, Paris, 1995, pp. Exp. No. 250, 129–137. MR1611536 [117]

[HK87] Jürgen Herzog and Michael Kühl, *Maximal Cohen-Macaulay modules over Gorenstein rings and Bourbaki-sequences*, Commutative algebra and combinatorics (Kyoto, 1985), Adv. Stud. Pure Math., vol. 11, North-Holland, Amsterdam, 1987, pp. 65–92. MR951197 [315]

[HL02] Craig Huneke and Graham J. Leuschke, *Two theorems about maximal Cohen-Macaulay modules*, Math. Ann. **324** (2002), no. 2, 391–404. MR1933863 [114]

[HM93] Jürgen Herzog and Alex Martsinkovsky, *Gluing Cohen-Macaulay modules with applications to quasihomogeneous complete intersections with isolated singularities*, Comment. Math. Helv. **68** (1993), no. 3, 365–384. MR1236760 [184, 194]

[Hoc73] Melvin Hochster, *Cohen-Macaulay modules*, Conference on Commutative Algebra (Univ. Kansas, Lawrence, Kan., 1972), Springer, Berlin, 1973, pp. 120–152. Lecture Notes in Math., Vol. 311. MR0340251 [170]

[HR61] Alex Heller and Irving Reiner, *Indecomposable representations*, Illinois J. Math. **5** (1961), 314–323. MR0122890 [31]

[HS88] Jürgen Herzog and Herbert Sanders, *Indecomposable syzygy-modules of high rank over hypersurface rings*, J. Pure Appl. Algebra **51** (1988), no. 1-2, 161–168. MR941897 [298]

[HS97] Mitsuyasu Hashimoto and Akira Shida, *Some remarks on index and generalized Loewy length of a Gorenstein local ring*, J. Algebra **187** (1997), no. 1, 150–162. MR1425563 [180, 182, 195]

[HS06] Craig Huneke and Irena Swanson, *Integral closure of ideals, rings, and modules*, London Mathematical Society Lecture Note Series, vol. 336, Cambridge University Press, Cambridge, 2006. MR2266432 [45]

[HU89] Craig Huneke and Bernd Ulrich, *Powers of licci ideals*, Commutative algebra (Berkeley, CA, 1987), Math. Sci. Res. Inst. Publ., vol. 15, Springer, New York, 1989, pp. 339–346. MR1015526 [117]

[Hum94] John F. Humphreys, *Character tables for the primitive finite unitary reflection groups*, Comm. Algebra **22** (1994), no. 14, 5777–5802. MR1298751 [94]

[HW09] Wolfgang Hassler and Roger Wiegand, *Extended modules*, J. Commut. Algebra **1** (2009), no. 3, 481–506. MR2524863 [26, 165]

[IN99] Yukari Ito and Iku Nakamura, *Hilbert schemes and simple singularities*, New trends in algebraic geometry (Warwick, 1996), London Math. Soc. Lecture Note Ser., vol. 264, Cambridge Univ. Press, Cambridge, 1999, pp. 151–233. MR1714824 [94]

[Isc69] Friedrich Ischebeck, *Eine Dualität zwischen den Funktoren Ext und Tor*, J. Algebra **11** (1969), 510–531. MR0237613 [176]

[Ive73] Birger Iversen, *Generic local structure of the morphisms in commutative algebra*, Lecture Notes in Mathematics, Vol. 310, Springer-Verlag, Berlin, 1973. MR0360575 [165, 319]

[Jac67] Heinz Jacobinski, *Sur les ordres commutatifs avec un nombre fini de réseaux indécomposables*, Acta Math. **118** (1967), 1–31. MR0212001 [50, 51]

[Jac75] Nathan Jacobson, *Lectures in abstract algebra*, Springer-Verlag, New York, 1975, Volume II: Linear algebra, Reprint of the 1953 edition [Van Nostrand, Toronto, Ont.], Graduate Texts in Mathematics, No. 31. MR0369381 [30]

[Jan57] James P. Jans, *On the indecomposable representations of algebras*, Ann. of Math. (2) **66** (1957), 418–429. MR0088485 [267]

[Jor77] Camille Jordan, *Sur une classe de groupes d'ordre fini contenus dans les groupes linéaires*, Bull. Soc. Math. France **5** (1877), 175–177. MR1503760 [84]

[Kat99] Kiriko Kato, *Cohen-Macaulay approximations from the viewpoint of triangulated categories*, Comm. Algebra **27** (1999), no. 3, 1103–1126. MR1669120 [189]

[Kat02] Karl Kattchee, *Monoids and direct-sum decompositions over local rings*, J. Algebra **256** (2002), no. 1, 51–65. MR1936878 [23]

[Kat07] Kiriko Kato, *Syzygies of modules with positive codimension*, J. Algebra **318** (2007), no. 1, 25–36. MR2363122 [188]

[Kaw96] Takesi Kawasaki, *Local cohomology modules of indecomposable surjective-Buchsbaum modules over Gorenstein local rings*, J. Math. Soc. Japan **48** (1996), no. 3, 551–566. MR1389995 [298]

[Kle93] Felix Klein, *Vorlesungen über das Ikosaeder und die Auflösung der Gleichungen vom fünften Grade (Lectures on the icosahedron and the solution of the 5th degree equations)*, Basel: Birkhäuser Verlag. Stuttgart: B. G. Teubner Verlagsgesellschaft. xxviii, 343 S. , 1993 (German). [81, 87]

[Knö87] Horst Knörrer, *Cohen-Macaulay modules on hypersurface singularities. I*, Invent. Math. **88** (1987), no. 1, 153–164. MR877010 [139, 141, 145]

[Kro74] Leopold Kronecker, *Über die congruenten Transformationen der bilinearen Formen*, Leopold Kroneckers Werke (K. Hensel, ed.), vol. I, Monatsberichte Königl. Preuß. Akad. Wiss. Berlin, 1874, pp. 423–483 (German). [31]

[Kru25] Wolfgang Krull, *Über verallgemeinerte endliche Abelsche Gruppen*, Math. Z. **23** (1925), no. 1, 161–196. MR1544736 [1]

[Kru30] _____, *Ein Satz über primäre Integritätsbereiche*, Math. Ann. **103** (1930), 450–465. [45]

[KS85] Karl-Heinz Kiyek and Günther Steinke, *Einfache Kurvensingularitäten in beliebiger Charakteristik*, Arch. Math. (Basel) **45** (1985), no. 6, 565–573. MR818299 [159, 160]

[Kun86] Ernst Kunz, *Kähler differentials*, Advanced Lectures in Mathematics, Friedr. Vieweg & Sohn, Braunschweig, 1986. MR864975 [116]

[KW09] Ryan Karr and Roger Wiegand, *Direct-sum behavior of modules over one-dimensional rings*, to appear in Springer volume: "Recent Developments in Commutative Algebra", 2009. [50]

[Lam86]   Klaus Lamotke, *Regular solids and isolated singularities*, Advanced Lectures in Mathematics, Friedr. Vieweg & Sohn, Braunschweig, 1986. MR845275 [87]

[Lam91]   T. Y. Lam, *A first course in noncommutative rings*, Graduate Texts in Mathematics, vol. 131, Springer-Verlag, New York, 1991. MR1125071 [5, 6]

[Lan02]   Serge Lang, *Algebra*, third ed., Graduate Texts in Mathematics, vol. 211, Springer-Verlag, New York, 2002. MR1878556 [63, 144]

[Lec86]   Christer Lech, *A method for constructing bad Noetherian local rings*, Algebra, algebraic topology and their interactions (Stockholm, 1983), Lecture Notes in Math., vol. 1183, Springer, Berlin, 1986, pp. 241–247. MR846452 [170]

[Lev65]   Gerson Levin, *Homology of local rings*, Ph.D. thesis, University of Chicago, 1965. [200]

[Lip69]   Joseph Lipman, *Rational singularities, with applications to algebraic surfaces and unique factorization*, Inst. Hautes Études Sci. Publ. Math. (1969), no. 36, 195–279. MR0276239 [99, 119]

[Lip71]   _____, *Stable ideals and Arf rings*, Amer. J. Math. **93** (1971), 649–685. MR0282969 [44]

[Lip78]   _____, *Desingularization of two-dimensional schemes*, Ann. Math. (2) **107** (1978), no. 1, 151–207. MR0491722 [98]

[LO96]   Lawrence S. Levy and Charles J. Odenthal, *Package deal theorems and splitting orders in dimension* 1, Trans. Amer. Math. Soc. **348** (1996), no. 9, 3457–3503. MR1351493 [17]

[LV68]   Gerson Levin and Wolmer V. Vasconcelos, *Homological dimensions and Macaulay rings*, Pacific J. Math. **25** (1968), 315–323. MR0230715 [200]

[LW00]   Graham J. Leuschke and Roger Wiegand, *Ascent of finite Cohen-Macaulay type*, J. Algebra **228** (2000), no. 2, 674–681. MR1764587 [170, 313]

[LW05]   _____, *Local rings of bounded Cohen-Macaulay type*, Algebr. Represent. Theory **8** (2005), no. 2, 225–238. MR2162283 [302]

[Mar53]   Jean-Marie Maranda, *On $\mathfrak{P}$-adic integral representations of finite groups*, Canadian J. Math. **5** (1953), 344–355. MR0056605 [272]

[Mar90]   Alex Martsinkovsky, *Almost split sequences and Zariski differentials*, Trans. Amer. Math. Soc. **319** (1990), no. 1, 285–307. MR955490 [192]

[Mar91]   _____, *Maximal Cohen-Macaulay modules and the quasihomogeneity of isolated Cohen-Macaulay singularities*, Proc. Amer. Math. Soc. **112** (1991), no. 1, 9–18. MR1042270 [192]

[Mat73]   Eben Matlis, *1-dimensional Cohen-Macaulay rings*, Lecture Notes in Mathematics, Vol. 327, Springer-Verlag, Berlin, 1973. MR0357391 [280]

[Mat89]   Hideyuki Matsumura, *Commutative ring theory*, second ed., Cambridge Studies in Advanced Mathematics, vol. 8, Cambridge University Press, Cambridge, 1989, Translated from the Japanese by M. Reid. MR1011461 [xvi, 45, 116, 118, 169, 170, 173, 244, 309, 310, 313, 314, 315, 321, 326, 331, 333]

[McK01]   John McKay, *A Rapid Introduction to ADE Theory*, http://math.ucr.edu/home/baez/ADE.html, 2001. [87]

[Mil08] James S. Milne, *Lectures on etale cohomology (v2.10)*, 2008, Available at www.jmilne.org/math/CourseNotes/lec.html, p. 196. [117]

[Miy67] Takehiko Miyata, *Note on direct summands of modules*, J. Math. Kyoto Univ. **7** (1967), 65–69. MR0214585 [109]

[ML95] Saunders Mac Lane, *Homology*, Classics in Mathematics, Springer-Verlag, Berlin, 1995, Reprint of the 1975 edition. MR1344215 [109]

[Mum61] David Mumford, *The topology of normal singularities of an algebraic surface and a criterion for simplicity*, Inst. Hautes Études Sci. Publ. Math. (1961), no. 9, 5–22. MR0153682 [99, 100, 117, 119]

[Nag58] Masayoshi Nagata, *A general theory of algebraic geometry over Dedekind domains. II. Separably generated extensions and regular local rings*, Amer. J. Math. **80** (1958), 382–420. MR0094344 [45]

[Noe15] Emmy Noether, *Der Endlichkeitssatz der Invarianten endlicher Gruppen*, Math. Ann. **77** (1915), no. 1, 89–92. MR1511848 [77]

[Noe50] _____, *Idealdifferentiation und Differente*, J. Reine Angew. Math. **188** (1950), 1–21. MR0038337 [328]

[NR73] Nazarova, Liudmila A. and Roĭter, Andreĭ Vladimirovich, *Kategornye matrichnye zadachi i problema Brauera-Trella*, Izdat. "Naukova Dumka", Kiev, 1973. MR0412233 [267]

[PR90] Dorin Popescu and Marko Roczen, *Indecomposable Cohen-Macaulay modules and irreducible maps*, Compositio Math. **76** (1990), no. 1-2, 277–294, Algebraic geometry (Berlin, 1988). MR1078867 [276, 280]

[PR91] _____, *The second Brauer-Thrall conjecture for isolated singularities of excellent hypersurfaces*, Manuscripta Math. **71** (1991), no. 4, 375–383. MR1104991 [280]

[Pri67] David Prill, *Local classification of quotients of complex manifolds by discontinuous groups*, Duke Math. J. **34** (1967), 375–386. MR0210944 [339]

[PS73] Christian Peskine and Lucien Szpiro, *Dimension projective finie et cohomologie locale. Applications à la démonstration de conjectures de M. Auslander, H. Bass et A. Grothendieck*, Inst. Hautes Études Sci. Publ. Math. (1973), no. 42, 47–119. MR0374130 [176, 215]

[Rei72] Idun Reiten, *The converse to a theorem of Sharp on Gorenstein modules*, Proc. Amer. Math. Soc. **32** (1972), 417–420. MR0296067 [177]

[Rei97] _____, *Dynkin diagrams and the representation theory of algebras*, Notices Amer. Math. Soc. **44** (1997), no. 5, 546–556. MR1444112 [91]

[Rem11] Robert Erich Remak, *Über die Zerlegung der endlichen Gruppen in direkte unzerlegbare Faktoren*, J. Reine Angew. Math. **139** (1911), 293–308. [1]

[Rie81] Oswald Riemenschneider, *Zweidimensionale Quotientensingularitäten: Gleichungen und Syzygien*, Arch. Math. (Basel) **37** (1981), no. 5, 406–417. MR643282 [263]

[Rin80] Claus Michael Ringel, *On algorithms for solving vector space problems. I. Report on the Brauer-Thrall conjectures: Rojter's theorem and the theorem of Nazarova and Rojter*, Representation theory, I (Proc. Workshop, Carleton Univ., Ottawa, Ont., 1979), Lecture Notes in Math., vol. 831, Springer, Berlin, 1980, pp. 104–136. MR607142 [267]

[Rob87]　Paul C. Roberts, *Le théorème d'intersection*, C. R. Acad. Sci. Paris Sér. I Math. **304** (1987), no. 7, 177–180. MR880574 [176]

[Roĭ68]　Andreĭ Vladimirovich Roĭter, *Unboundedness of the dimensions of the indecomposable representations of an algebra which has infinitely many indecomposable representations*, Izv. Akad. Nauk SSSR Ser. Mat. **32** (1968), 1275–1282. MR0238893 [267]

[RWW99]　Christel Rotthaus, Dana Weston, and Roger Wiegand, *Indecomposable Gorenstein modules of odd rank*, J. Algebra **214** (1999), no. 1, 122–127. MR1684896 [24]

[Sai71]　Kyoji Saito, *Quasihomogene isolierte Singularitäten von Hyperflächen*, Invent. Math. **14** (1971), 123–142. MR0294699 [192]

[Sal79]　Judith D. Sally, *Stretched Gorenstein rings*, J. London Math. Soc. (2) **20** (1979), no. 1, 19–26. MR545198 [294]

[Sch29]　Otto Schmidt, *Über unendliche Gruppen mit endlicher Kette*, Math. Z. **29** (1929), no. 1, 34–41. MR1544991 [1]

[Sch87]　Frank-Olaf Schreyer, *Finite and countable CM-representation type*, Singularities, representation of algebras, and vector bundles (Lambrecht, 1985), Lecture Notes in Math., vol. 1273, Springer, Berlin, 1987, pp. 9–34. MR915167 [163, 241, 260, 263]

[Sha70]　Rodney Y. Sharp, *Gorenstein modules*, Math. Z. **115** (1970), 117–139. MR0263801 [313]

[Sha71]　———, *On Gorenstein modules over a complete Cohen-Macaulay local ring*, Quart. J. Math. Oxford Ser. (2) **22** (1971), 425–434. MR0289504 [177]

[Slo83]　Peter Slodowy, *Platonic solids, Kleinian singularities, and Lie groups*, Algebraic geometry (Ann Arbor, Mich., 1981), Lecture Notes in Math., vol. 1008, Springer, Berlin, 1983, pp. 102–138. MR723712 [81]

[Sma80]　Sverre O. Smalø, *The inductive step of the second Brauer-Thrall conjecture*, Canad. J. Math. **32** (1980), no. 2, 342–349. MR571928 [280]

[Sol89]　Øyvind Solberg, *Hypersurface singularities of finite Cohen-Macaulay type*, Proc. London Math. Soc. (3) **58** (1989), no. 2, 258–280. MR977477 [160]

[SS02]　Anne-Marie Simon and Jan R. Strooker, *Reduced Bass numbers, Auslander's δ-invariant and certain homological conjectures*, J. Reine Angew. Math. **551** (2002), 173–218. MR1932178 [182, 201]

[ST54]　G. C. Shephard and John Arthur Todd, *Finite unitary reflection groups*, Canadian J. Math. **6** (1954), 274–304. MR0059914 [151, 155]

[Ste11]　Ernst Steinitz, *Rechteckige Systeme und Moduln in algebraischen Zahlkörpern. I*, Math. Ann. **71** (1911), no. 3, 328–354. MR1511661 [10]

[Str05]　Janet Striuli, *On extensions of modules*, J. Algebra **285** (2005), no. 1, 383–398. MR2119119 [110, 120]

[SV74]　Judith D. Sally and Wolmer V. Vasconcelos, *Stable rings*, J. Pure Appl. Algebra **4** (1974), 319–336. MR0409430 [44]

[SV85]　Rodney Y. Sharp and Peter Vámos, *Baire's category theorem and prime avoidance in complete local rings*, Arch. Math. (Basel) **44** (1985), no. 3, 243–248. MR784093 [241, 243]

[Tak04a]　Ryo Takahashi, *Modules of G-dimension zero over local rings of depth two*, Illinois J. Math. **48** (2004), no. 3, 945–952. MR2114261 [210]

[Tak04b] _____, *On the category of modules of Gorenstein dimension zero. II*, J. Algebra **278** (2004), no. 1, 402–410. MR2068085 [210]

[Tak05] _____, *On the category of modules of Gorenstein dimension zero*, Math. Z. **251** (2005), no. 2, 249–256. MR2191025 [210]

[Tat57] John Tate, *Homology of Noetherian rings and local rings*, Illinois J. Math. **1** (1957), 14–27. MR0086072 [151]

[Thr47] Robert M. Thrall, *On ahdir algebras*, Bull. A.M.S. **53** (1947), no. 1, 49. [267]

[Vie77] Eckart Viehweg, *Rational singularities of higher dimensional schemes*, Proc. Amer. Math. Soc. **63** (1977), no. 1, 6–8. MR0432637 [99]

[Wan94] Hsin-Ju Wang, *On the Fitting ideals in free resolutions*, Michigan Math. J. **41** (1994), no. 3, 587–608. MR1297711 [273, 275, 276]

[War70] Robert Breckenridge Warfield, Jr., *Decomposability of finitely presented modules*, Proc. Amer. Math. Soc. **25** (1970), 167–172. MR0254030 [31]

[Wat74] Keiichi Watanabe, *Certain invariant subrings are Gorenstein. I, II*, Osaka J. Math. **11** (1974), 1–8; ibid. 11 (1974), 379–388. MR0354646 [81, 84]

[Wat83] _____, *Rational singularities with k^*-action*, Commutative algebra (Trento, 1981), Lecture Notes in Pure and Appl. Math., vol. 84, Dekker, New York, 1983, pp. 339–351. MR686954 [98]

[Wed09] Joseph Henry Maclagan Wedderburn, *On the direct product in the theory of finite groups*, Ann. Math. **10** (1909), 173–176. [1]

[Wei68] Karl Weierstrass, *On the theory of bilinear and quadratic forms. (Zur Theorie der bilinearen und quadratischen Formen.)*, Monatsberichte Königl. Preuß. Akad. Wiss. Berlin, 1868 (German). [31]

[Wes88] Dana Weston, *On descent in dimension two and nonsplit Gorenstein modules*, J. Algebra **118** (1988), no. 2, 263–275. MR969672 [24]

[Wie88] Sylvia Wiegand, *Ranks of indecomposable modules over one-dimensional rings*, J. Pure Appl. Algebra **55** (1988), no. 3, 303–314. MR970697 [57]

[Wie89] Roger Wiegand, *Noetherian rings of bounded representation type*, Commutative algebra (Berkeley, CA, 1987), Math. Sci. Res. Inst. Publ., vol. 15, Springer, New York, 1989, pp. 497–516. MR1015536 [34, 38, 50, 283]

[Wie94] _____, *One-dimensional local rings with finite Cohen-Macaulay type*, Algebraic geometry and its applications (West Lafayette, IN, 1990), Springer, New York, 1994, pp. 381–389. MR1272043 [50, 53, 59, 60]

[Wie98] _____, *Local rings of finite Cohen-Macaulay type*, J. Algebra **203** (1998), no. 1, 156–168. MR1620725 [26, 163, 172]

[Wie99] _____, *Failure of Krull-Schmidt for direct sums of copies of a module*, Advances in commutative ring theory (Fez, 1997), Lecture Notes in Pure and Appl. Math., vol. 205, Dekker, New York, 1999, pp. 541–547. MR1767419 [15]

[Wie01] _____, *Direct-sum decompositions over local rings*, J. Algebra **240** (2001), no. 1, 83–97. MR1830544 [19, 20, 24, 28]

[WW94] Roger Wiegand and Sylvia Wiegand, *Bounds for one-dimensional rings of finite Cohen-Macaulay type*, J. Pure Appl. Algebra **93** (1994), no. 3, 311–342. MR1275969 [57]

[Xu96] Jinzhong Xu, *Flat covers of modules*, Lecture Notes in Mathematics, vol. 1634, Springer-Verlag, Berlin, 1996. MR1438789 (98b:16003) [211]

[YI00] Yuji Yoshino and Satoru Isogawa, *Linkage of Cohen-Macaulay modules over a Gorenstein ring*, J. Pure Appl. Algebra **149** (2000), no. 3, 305–318. MR1762771 [188]

[Yos87] Yuji Yoshino, *Brauer-Thrall type theorem for maximal Cohen-Macaulay modules*, J. Math. Soc. Japan **39** (1987), no. 4, 719–739. MR905636 [268, 276]

[Yos90] _____, *Cohen-Macaulay modules over Cohen-Macaulay rings*, London Mathematical Society Lecture Note Series, vol. 146, Cambridge University Press, Cambridge, 1990. MR1079937 [xii, xv, xvi, 146, 172, 276]

[Yos93] _____, *A note on minimal Cohen-Macaulay approximations*, Proceedings of the 4th Symposium on the Representation Theory of Algebras, unknown publisher, Izu, Japan, 1993, in Japanese, pp. 119–138. [180]

[Yos98] _____, *Auslander's work on Cohen-Macaulay modules and recent development*, Algebras and modules, I (Trondheim, 1996), CMS Conf. Proc., vol. 23, Amer. Math. Soc., Providence, RI, 1998, pp. 179–198. MR1648607 [200]

[ZS75] Oscar Zariski and Pierre Samuel, *Commutative algebra. Vol. II*, Springer-Verlag, New York, 1975, Reprint of the 1960 edition, Graduate Texts in Mathematics, Vol. 29. MR0389876 [25, 144, 300]

Index

$+(M)$, *13*
$M \cdot N$, *106*
$M \mid N$, *3*
M_{art}, 20, *42*
R^{h}, *165*
R_{art}, 20, *42*
$\overline{R}$, *41*
$\widehat{R}$, *7*
$\perp$, *211*
✗, *138*

a-invariant, 98
abelian category, *1*
Abhyankar, Shreeram, 101, 106
absolutely flat homomorphism, *167*, 168
abstract hypersurface, 141, *150*, 150–151, *298*, 298, 299
action by linear changes of variable, 61, 62, 65, 69–72, 81, 83, 86, 118, 151, 152, 262
acyclic complex, 208, 216
$\mathrm{add}_R(M)$, *13*, 15, 26, 106, 163, 165, 230
additive category, *1*, 2, 9, 10, 32
  in which idempotents split, 1, 3–5
additive function, 91, 97
ADE Coxeter-Dynkin diagram, 81, 91–97, 101, 102, 104, *see also* extended ADE Coxeter-Dynkin diagram
ADE hypersurface singularity, 42, 50–52, 81, 83, 87, 97, 133, 141, 146–161, 172–173, 229–232, 234–237, 295
  (A_1), 102, 103, 229, 291, 294, 295
  (A_2), 113

(A_∞), 78, 241, 244, 245, 252–256, 259, 260, 262–264, 297, 301
(A_n), 77, 87, 90, 93, 119, 133, 146, 153, 236
(D_4), 88, 103, 149, 294
(D_∞), 78, 241, 244, 245, 252, 253, 256–260, 262, 263, 265, 297, 300, 301, 304
(D_n), 87, 90, 93, 149, 154, 236, 262
(E_6), 88, 94, 148, 155, 234, 237
(E_7), 88, 95, 149, 237
(E_8), 89, 96, 148, 157, 237
adjoint pair, 71
adjunction formula, 101
affine diagram, *see* extended ADE Coxeter-Dynkin diagram
Akizuki, Yasuo, 316
algebra retraction, 77
algebraic closure, 117, 118, 173
algebraic duality, 203
algebraic field extension, 167, 174
algebraic fundamental group, *see* étale fundamental group
algebraic number field, 50
algebraically closed field, 51, 73, 90, 92, 98, 101, 118, 128, 133, 134, 137, 138, 141, 142, 144–147, 151, 159, 160, 163, 172, 192, 225, 234, 253, 259, 262, 268, 276, 277, 279–283, 285, 287, 288, 293, 299, 321, 323
almost split sequence, *see* Auslander-Reiten sequence
analytic branch, 19, 146–149
analytic local ring, 241
analytically
  normal, *23*, 24

ramified, 52, 268, 280
  unramified, 19, 41, *45*, 46, 48, 49, 53–57, 172, 268, 283, 316
Ann, *242*
anti-diagonal block matrix, 159, 235
antichain, 27, 47
apparently split, 109
approximation, *211*, 212, 215
Approximation Theorem, 187
AR, *see* Auslander-Reiten
Arnol'd, Vladimir I., 260
Artin, Michael, 97, 100, 104, 119, 160
Artin-Rees Lemma, 8
Artinian localization, 17
Artinian pair, 21, 28, 29, *32*, 32–39, 42–50, 246, 248, 252, 257, 283
Artinian ring, 5–6, 27, 29, 30, 32, 39, 120, 166, 173, 200, 238, 267, 281, 294
ascent
  of bounded CM type, 279, 298
  of excellence, 167
  of finite CM type, 48, 133, 138, 164, 165, 168, 169, 171
associated graded ring, 143, 160, 200
associativity formula, 315
asymmetry, 57, 224
atom, *13*, 22
Auslander transpose Tr, *204*, 203–223, 238, 239, 281
Auslander, Maurice, 43, 66, 67, 73, 109, 113–115, 117–119, 175, 184, 187, 189, 200, 208, 234, 241, 267, 276, 287, 291, 293, 326, 328, 330, 332, 334
Auslander-Buchsbaum formula, 123, 124, 284, *310*, 334, 336
Auslander-Reiten quiver, 217, *225*, 224–240, 269, 276–283, 290, 293, 298
Auslander-Reiten sequence, 114, 160, *217*, 217–232, 234, 238, 239, 267, 277, 281, 287–293, 295
  existence, 218
  uniqueness, 218
Auslander-Reiten translate τ, *222*, 222–232, 235–237, 281, 289, 292, 296

Azumaya, Gorô, 1, 3, 5

Baeth, Nicholas, 56
Bass numbers, 183, 201
Bass' Conjecture, 176
Bass, Hyman, 48, 54, 57, 176
BD, *see* Burban-Drozd
Betti numbers, 151, 184, 280, 298
biduality, 205, 264
binary polyhedral group, 84, *86*, 86–90, 92–97, 151, 263
biproduct, 1, 2, 10, 32
blowup, 102, 103
bounded CM type, 30, 150, 163, *268*, 268, 276, 279, 280, 297–307
bounded free resolution, 126
bounded representation type, 30, 267
Bourbaki, 23, 84, 170
Bourbaki sequence, *24*, 24
branch locus, 326, 336
branch of a curve, 19, 146–149
Brauer, Richard, 267
Brauer-Thrall Conjectures, 267–280, 285
  I, 268, 276–287
  II, 49, 268, 280–285
Braun, Amiram, 334
Bravais, Auguste, 84
Bruns, Winfried, 284
Bruns-Herzog, 310
Buchsbaum, David, 66, 326, 328
Buchweitz, Ragnar-Olaf, 141, 145, 175, 184, 187, 189, 203, 208, 241, 245, 252, 284
Burban, Igor, 241, 244, 245, 248, 263
Burban-Drozd triples BD, *248*, 244–259, 264–266

$\mathcal{C}$, *86*
$\mathfrak{c}$, *42*
cancellative semigroup, 14
canonical module ω, 116, *176*, 175–181, 183–185, 187–189, 193, 194, 196–202, 217, 218, 221–224, 228, 233, 234, 243, 263, 264, 281, 287, 288, 292, 313
canonical rank ω-rank, 190, 193, 195, 201
Cartan invariants, 267

Cartan, Henri, 62
case analysis, dreary, 33
Cauchy sequence, 6, 244
Cauchy's Theorem, 86
Caviglia, Giulio, 120
Cayley-Hamilton Theorem, 318
character, 73, 92–97, 153
character table, 94–96
characteristic
  bad, 64, 141, 159–160
  five, 159
  good, 51, 52, 83, 89–91, 133, 141, 144, 145, 151, 163, 172, 173, 234
  three, 38, 49, 50, 59, 159
  two, 36, 37, 41, 50, 128, 132–134, 137, 138, 140, 141, 145, 159, 160, 172, 253, 256, 259, 268, 280, 285, 288, 292, 297, 298
  zero, 50, 98, 99, 109, 116–119, 151, 192, 293
characteristic polynomial, 30, 318, 335
Chern class c_1, 105
Chevalley-Shephard-Todd Theorem, 69, 335, 339
Chinese Remainder Theorem, 318
Christensen, Lars Winther, 203, 210
Çimen, Nuri, 41, 45, 49, 50
Cl(R), *23*
closed under AR sequences, 287–293
closed under extensions, *211*, 212, 214, 215
clutter, *27*, 47
codepth, *177*, 178, 184, 185, 187–189, 193, 196, 208
coefficient field, 299
Cohen's Structure Theorems, 177, 275
Cohen-Macaulay type, *116*, 193
coherent sheaf, 98, 104, 106
cohomology, 64, 98, 105
complete intersection, 116
complete local ring
  is Henselian, 6
  satisfies Krull-Remak-Schmidt, 7
complete resolution, 207–209

complex numbers $\mathbb{C}$, 24, 81, 83–87, 89, 90, 93–97, 107, 119, 151–158, 241, 284
conductor $\mathfrak{c}$, *42*, 47, 48, 59, 60, 245–252, 254–257, 264, 283
conductor square, 20, *42*, 47, 48, 51, 53, 58, 59, 245, 283
conormal module I/I^2, 116
constant rank, 19, 23, 43–45, 49–51, 57, 184, 188–189, 268, 284, 299–306, *316*, 317
  of a module over an Artinian pair, 32
countable CM type, 132, 133, 141, 241–266, 268, 284, 302
countable prime avoidance, 241–243
countably simple singularity, *141*, 142–144, 245, 253, 259, 260, 297
covariants, 72, 152
cover, *211*, 212

$\mathcal{D}$, *86*
$\mathfrak{D}_A(B)$, *328*
Dade's construction, 21, 28, 37
Dade, Everett C., 37
de la Peña, José Antonio, 268, 270
decomposition group, 338–339
Dedekind different, 328–330, 336
Dedekind domain, 2, 10, 14
defects of an exact sequence, 239
deformation theory, 160
Del Pezzo-Bertini Theorem, 295
delta-invariant, 192
δ-invariant, 175, *190*, 189–200
depth, *309*, 309–310
Depth Lemma, 65, 66, 166, 178, 188, 246, 250, *310*, 312, 329
derivation, 323
derivative, 20, 59
  partial, 89, 192, 275
descent
  from the completion, 17
  of bounded CM type, 297, 298, 303, 306, 307
  of direct summands, 9
  of finite CM type, 48, 133, 138, 163, 171, 279, 280
  of finite representation type, 38
  of isomorphism, 9

desingularization graph, 81, 97, *101*, 101, 102, 104
determinant, 74
diagonal map, 58, 66–68, 152, 165, 274, *322*, 323–329
diagonalizable, 69, 79, 335, 338
diagonalization, 4, 265
Dickson's Lemma, 15, 16, 28, 47
Dieterich, Ernst, 268, 276, 280, 284, 285, 298
Dieudonné, Jean, 31
Ding, Songqing, 43, 192, 195, 196, 199, 200
Diophantine monoid, *see* Krull monoid
direct image, 98
direct limit, 164, 165, 167, 168
direct-sum cancellation, 3, 27
direct-sum decomposition, 1, 7, 10, 13–15, 19, 27, 73, 75, 106, 164, 180, 181, 189, 224, 230, 231, 263, 278, 298, 306
discrete valuation ring, DVR, 24, 46, 48, 56, 59, 126, 142, 143, 146, 249–251, 254, 257, 272, 303, 327, 330, 338
divisor, 100, 105
divisor class group Cl(R), 14, 23–26, 99, 289, 306, 307
divisor homomorphism, *14*, 14, 16, 26, 27
divisorial ideal, 24, 26, 307
double branched cover, *128*, 128–142, 145, 146, 160, 234, 245, 297, 299
double sharp ✖, *138*, 138–139
Drozd, Yuriy, 32, 38, 50, 241, 244, 245, 248, 263
Drozd-Roĭter conditions, 29, 32–39, 41–44, 48–52, 54, 56, 59, 60, 170, 172, 174, 283, 299
  necessity of, 42–45
  sufficiency of, 45–50, 52
Du Val singularity, *see* Kleinian singularity
Du Val, Patrick, 87, 97, 100
dual, 223, 234
duality, canonical, *177*, 222, 250, 261, 281

$e(R), e(M), e(I, M)$, *313*
EGA, 310
eigenvalue, 339
Eisenbud, David, 119, 126, 268, 270, 287, 293, 295
Elkik, Renée, 169
embedding dimension, 100
End, *203*
endomorphism ring, 1–3, 5, 11, 32, 44, 48, 55, 56, 59–61, 65, 67–70, 72, 83, 111, 177, 181, 196, 198, 203, 207, 225, 237, 239, 261, 262, 264, 302–304
  nc-local, 1, 3–6, 10, 134, 203, 218, 223, 225, 226, 238, 272, 273
enveloping algebra, 323
equivalence of matrix factorizations, 125, 126, 132, 134
Esnault, Hélène, 117, 118
essentially of finite type, 22, 66, 116, 164, 173, 176, 319, 321–326
étale covering, 117
étale fundamental group $\pi_1^{et}(X)$, 117–119
étale homomorphism, 117, *164*, 164, 167, 168, 173, 321–328
Euclid, 84
Euclidean diagram, *see* extended ADE Coxeter-Dynkin diagram
Euler characteristic, 98
evaluation homomorphism, 198
Evans, E. Graham, 3, 7, 311
excellent ring, 98, 118, 163, *167*, 167, 169, 171, 173, 242, 268, 276, 279, 280, 304
exceptional fiber, 99–105
expanded subsemigroup, 14, 16, 22, 23
Ext-minimal FID hull, 183
Ext-minimal MCM approximation, 183, 190, 192
extended ADE Coxeter-Dynkin diagram, 91–97, 232, 236–238, *see also* ADE Coxeter-Dynkin diagram
extended module, *17*, 18, 23, 24, 26, 43, 164, 169, 305, 306

INDEX

extension of modules, 109–111, 115, 120, 127, 179, 187, 223, 290, 291

exterior power $\bigwedge^p$, 65, 73, 74, 231, 292

faithful flatness, 17, 18, 26, 27, 163, 314

faithful ideal, 53, 55, 56, 283

faithful module, 305

faithful system of parameters, *271*, 271–280, 285

faithfully flat descent, 15, 24, 25, 215

Ferrand, Daniel, 178

fiber product, 323

FID hull, *see* hull of finite injective dimension

field
  algebraically closed, 51, 73, 90, 92, 98, 101, 118, 128, 133, 134, 137, 138, 141, 142, 144–147, 151, 159, 160, 163, 172, 192, 225, 234, 253, 259, 262, 268, 276, 277, 279–283, 285, 287, 288, 293, 299, 321, 323
  characteristic p, *see* characteristic
  characteristic zero, 109, 116
  coefficient, 299
  finite, 38
  imperfect, 275
  infinite, 33, 34, 39, 43, 44, 49–51, 142, 144, 171, 195, 202, 267, 268, 281, 285, 291, 297, 301, 302, 304, 314
  perfect, 52, 141, 267, 268, 275, 276, 279
  uncountable, 142, 144, 241, 242, 284, 285

field extension
  algebraic, 167, 174
  Galois, 63, 117–119, 330, 335, 337
  inseparable, 36, 49, 163, 172
  non-Galois, 38
  separable, 36, 38, 49–51, 66, 117, 163, 164, 170–172, 321, 324–330, 339

finite birational extension, *42*, 43, 44, 49, 51, 52, 172, 237, 295, 300–303, 307, 316, 317

finite CM type, *29*, 41, 45, 46, 48–54, 56, 57, 75, 81–83, 91, 115, 117–120, 132, 133, 141, 145, 146, 149–151, 159, 160, 163–166, 168–174, 199, 203, 210, 237, 243, 245, 268, 276, 279, 285, 287–296, 299, 304, 306
  $\implies$ isolated singularity, 109, 113, 115, 165, 169, 199, 276
  ascent of, 53, 54

finite field, 38

finite injective dimension, 175, 176, 178, 193, 201, 221

finite length module, 10, 17, 18, 24, 45, 49, 55, 60, 99, 106, 110, 112, 114, 115, 120, 121, 175, 183, 184, 206, 221, 222, 246, 247, 264, 267–270, 276, 278, 279, 296, 310, 313, 316, 332, 334, 336

finite representation type, 29, 30, 32, 38, 47, 49, 116, 212, 267
  of Artinian pairs, 29, *32*, 46
  TR modules, 209

finite TR type, *209*, 210

finite-dimensional algebra, 33, 267, 268, 317, 318

Fitting's Lemma, 10, 269

Five Lemma, 261

flatness, 164, 321

flatting, *128*, 130, 133, 137, 234, 237

Flenner, Hubert, 98, 117, 118

formally equidimensional, 170

Foxby, Hans-Bjørn, 177

fractional ideal, 10, 24, 292, 296, 328

fractional linear transformations, 85

free monoid, *13*, 16

free rank f-rank, 189, 191, 192, 194–196, 199

free resolution, 206–208
  bounded, 126, 151
  periodic, 126

free semigroup, 22

full subsemigroup, 14, 16, 25, 26

fundamental divisor Z_f, 100–102, 104

fundamental group π_1, 117

fundamental sequence, 233, 234

Gabriel quiver, 61, *73*, 74, *see also* McKay-Gabriel quiver
Galois field extension, 63, 117–119, 330, 335, 337
Galois ring extension, 335–340
γ-invariant, 189, *190*, 193–194
general linear group $\mathrm{GL}(n,k)$, 61, 62, 71, 78, 117, 335
generating invariants, 87–90, 152
generically Gorenstein, 177, 178, 184, 188, 189
genus, 25, 98
geometric McKay correspondence, 97–106
geometrically regular, *167*
global dimension, 64, 78
gluing construction, 184, 194
gluing map, 248, 254, 257
going down, 17, 168, 326, 337
going up, 326, 337
golden ratio, 89, 96
Goldie dimension, 60
gonflement, 44, *170*, 171–174, 281, 306, 314, 315
Gonzalez-Sprinberg, Gerardo, 97
Gorenstein
    generically, 177, 178, 184, 188, 189
    in codimension one, 82, 150, 220, 262, 264, 312
    on the punctured spectrum, 82, 116, 166, 196, 199, 206, 220, 262, 264, 313
Gorenstein dimension, 209
Gorenstein locus, 199
Gorenstein module, 179, *313*
great gross 1728, 89
Green, Edward, 38, 41, 45, 49, 50
Greuel, Gert-Martin, 42, 50–52, 141, 145, 159, 160, 241, 245, 252, 284
Griffith, Phillip A., 311
group
    binary polyhedral, 86–90, 92–97, 151, 263
    étale fundamental, 117–119
    fundamental, 117
    Galois, 63, 119, 335
    general linear, 61, 62, 71, 78, 117, 335
    special linear, 81, 83, 86, 92, 151, 229
group algebra kG, 71, 268
Gulliksen, Tor, 151
Guralnick, Robert, 7, 199, 271

$\mathfrak{H}_T(R)$, *273*
Harada-Sai Lemma, 268, 270, 275, 277, 278, 282
Harada-Sai sequence, *269*, 269, 270, 276
Hashimoto, Mitsuyasu, 180, 182
Heitmann's amazing theorem, 26, 307
Heitmann, Raymond C., 26
Heller, Alex, 31
Hensel's Lemma, 6, 145, 146, 148
Henselian local ring, *6*, 6, 7, 10, 48, 114, 118, 165, 181, 189, 210, 212, 217, 223–225, 228, 249, 272, 280, 317, 319
    classical definition, 6, 145, 147, 317
    complete local ring is, 6
    Krull-Remak-Schmidt holds for, 6
Henselization, 163, 164, *165*, 168, 169
Herzog, Jürgen, 81, 82, 90, 109, 116, 117, 119, 141, 150, 184, 189, 194, 195, 200, 262, 287, 293, 295, 298, 299
Hessel, Johann F. C., 84
Hessian, 89
higher direct image, 98
Higman, Donald G., 31, 272
Hilbert function, 58, 101, 160, 313
Hilbert polynomial, 58, 313
Hilbert-Burch Theorem, 288
Hom-tensor adjointness, 93, 221, 331
homogeneous coordinate ring, 25
homogeneous ring, 195, 287, *293*, 294, 295
homological different, *274*, 274, 285, 326–330, 336
homomorphism of matrix factorizations, *124*, 127, 134
hosohedron, 84
hull of finite injective dimension, 175, *179*, 179–194
    Ext-minimal, 183
    left minimal, 183

INDEX

minimal, 183, 186, 189
Huneke, Craig, 109, 114
hypersurface singularity, 50, 52, 83, 89, 90, 99, 101, 106, 120, 123–141, 163, 192, 203, 216, 241, 244, 245, 252–260, 280, 284, 285, 297–299, 301–303, 315

$\mathcal{I}$, 86
icosahedron, 84, 89
ideal
　divisorial, 24, 26
　faithful, 53, 55, 56, 283
　fractional, 10, 24, 292, 296, 328
　of pure height one, 177, 178, 188, 326, 327, 329, 330
　reflexive, 328
　stable, 44, 56
idempotent, 1–3, 9, 32, 70, 165, 181, 254, 273, 300, 318, 324
　lifting, 5, 6, 72, 317, 318
　split, 1, 2, 4, 9, 32
imperfect field, 172, 275
index, 195, 199, 200, 202
inertia group, 338
inertial degree, 338
infinite CM type, 19, 43, 51, 53, 170, 172, 253, 260, 267, 268, 280, 285, 291, 295, 301–304, 306
infinite field, 33, 34, 39, 43, 44, 49–51, 142, 144, 171, 195, 202, 267, 268, 281, 285, 291, 297, 301–303, 314
infinite syzygy, 209
injective hull of the residue field, 175, 221
inner product, 84, 85
inseparable field extension, 36, 49, 163, 172
integral closure, 41, 42, 45–59, 117–119, 245, 248, 256, 301, 316, 335
integral extension, 52, 63, 98, 323, 326, 328
integrally closed, see normal
intersection matrix, 100, 102
intersection multiplicity, 99
Intersection Theorem, 176
intersection theory, 99

intumescence, 170
invariant ring, 81, 82, 86, 97, 109, 117–119, 233, 262, 263, 291–293
invariant theory, 61
inverse determinant representation, 74
Irr(M, N) and irr(M, N), 225
irreducible homomorphism, 224, 224–229, 235, 276–278, 281–283, 287, 288
irreducible representation, 71–73, 77, 83, 91, 94, 97, 104, 152, 292
Ischebeck, Friedrich, 176
Isogawa, Satoru, 188
isolated singularity, 42, 53, 98, 101, 109, 113–116, 121, 192, 199, 219, 223–225, 228, 230, 243, 248, 268, 275–285, 287, 288, 290, 291, 295, 298, 304

$\mathcal{J}(-)$, 3
Jacobian
　criterion, 275, 288
　determinant, 89
　ideal, 192, 274, 275, 285
　matrix, 274
Jacobinski, Heinz, 50, 51
Jacobson radical $\mathcal{J}(-)$, 3, 225, 317, 318
Jans, James P., 267
Jordan block, 28, 29, 31, 34, 37
Jordan, Camille, 84

Kähler differentials Ω, 116, 323
Kaplansky, Irving, 200
Karoubian, see idempotent, split
Karr, Ryan, 50
Kato, Kiriko, 188, 189
Kattchee, Karl, 23
Kawasaki, Takesi, 298, 299
Kiyek, Karl-Heinz, 159, 160
Klein, Felix, 81, 87, 89
Kleinian singularity, 81, 83, 90, 83–97, 99, 102, 104, 133, 141, 151–159, 229–232
Knörrer's periodicity, 133, 253
Knörrer, Horst, 42, 50–52, 134, 138, 139, 141, 145, 252, 297

Koszul complex, 65, 73, 74, 228, 230, 292, 296
Koszul relations, 289
Kronecker, Leopold, 31
Kröning, Heike, 159, 160
KRS, *see* Krull-Remak-Schmidt Theorem
Krull Intersection Theorem, 77, 115, 219
Krull monoid, 14, *16*, 16, 21–23
Krull, Wolfgang, 1
Krull-Remak-Schmidt Theorem, 1–7, 10, 19, 27, 47, 48, 114, 131, 137, 210, 212, 298
    failure of, 3, 13, 19, 22
    for Artinian pairs, 59
    for skew group rings, 70, 72
    in an additive category, 4
    over Artinian pairs, 32
    over Artinian rings, 5, 27
    over complete local rings, 7, 9, 13, 15, 82, 83
    over Henselian local rings, 6

$\ell_R(M)$, *313*
left minimal FID hull, 183
Lemme d'Acyclicité, 206, 215
length sequence, *269*, 270
Leray spectral sequence, 98
Leuschke, Graham J., 109, 114
Levin, Gerson, 200
Levy, Lawrence S., 17
lifting
    Drozd-Roĭter conditions, 43
    factorizations, 317
    field extensions, 170, 171
    homomorphisms, 7, 8
    idempotents, 5, 6, 72, 317, 318
    modules from Artinian pairs, 43
    simple roots, 145, 319
lifting number, 7, 8, 9
lifting property, *210*, 211–215, 217–219, 223, 226–228, 238
    of FID hulls, 179, 180
    of MCM approximations, 179, 180
Lipman, Joseph, 119
$\ell_R(M)$, 195
local cohomology, 99, 106, 183, 201
local duality, 202

local fundamental group, 117
local-global theorem, 57, 66
locally finite, AR quiver is, 228, 279, 282
locally free
    in codimension one, 264
    on the punctured spectrum, 114, 116, 121, 169, 219, 220, 222, 223, 331, 333, 334
locally free sheaf, 105, 106
Loewy length, 195, 196
Luckas, Melissa, 56
lying over, 326, 337

Maclaurin series, 134
mapping cone, 127
Maranda's Theorem, 272
Maranda, Jean-Marie, 272
Martsinkovsky, Alex, 184, 189, 192, 194
Maschke's Theorem, 79
Matlis duality, 223
matrix decompositions, 38, 49, 50, 172
matrix factorization, *123*, 123–141, 151–159, 234–238, 245, 253, 255, 256, 258–260
maximally generated, *315*
McKay correspondence, 71–77, 81, 91–97, 229–233
    geometric, 97–106
McKay quiver, 61, *73*, 73–75, *see also* McKay-Gabriel quiver
McKay, John, 81, 91, 104
McKay-Gabriel quiver, 71, *75*, 76, 77, 81, 91–97, 104, 229, 230
MCM approximation, 175, *179*, 179–202, 208, 243
    Ext-minimal, 183, 190, 192
    minimal, *180*, 180, 182–184, 186, 188–192, 196, 200, 231, 234, 278, 282
    right minimal, *180*, 180, 182, 183
MCM module, *29*, 57
    over the skew group ring, 135
mess, 34, 36
minimal FID hull, 186, 189
    uniqueness, 183

minimal MCM approximation, *180*,
   180, 182–184, 186, 188–192, 196,
   200, 201, 228, 231, 234, 278, 282
  uniqueness, 182
minimal multiplicity, *101*, 106, 120,
   151, 202, 294, 295
minimal polynomial, 30, 38
minimal prime ideal, 13, 17, 19, 20,
   39, 45, 56, 58, 60, 127, 170, 177,
   178, 305, 306, 310, 312, 316, 318
minimal reduction, 44, 101, 171, 301,
   314
minimal resolution of singularities,
   98–101, 104
Miyata's Theorem, 109, 112–114, 197
Miyata, Takehiko, 109
modified Burban-Drozd triples BD',
   see Burban-Drozd triples BD
module
  canonical, 116, 175, 180, 181,
     183–185, 187–189, 193, 194,
     196–202, 217, 218, 221–224, 228,
     233, 234, 243, 263, 264, 281,
     287, 288, 292, 313
  conormal I/I^2, 116
  extended, 17, 18, 23, 24, 26, 43,
     164, 169, 305, 306
  faithful, 305
  Gorenstein, 179, 313
  maximal Cohen-Macaulay, *29*
  MCM, *29*
  n-torsionless, 205, 206
  of covariants, 72, 152
  of finite length, 10, 17, 18, 24, 45,
     49, 55, 60, 99, 106, 110, 112,
     114, 115, 120, 121, 175, 183,
     184, 206, 221, 222, 246, 247,
     264, 267–270, 276, 278, 279,
     296, 310, 313, 316, 332, 334, 336
  projective, 20, 32, 43, 44, 46, 47,
     56, 61, 64, 65, 70–73, 78, 83,
     177, 230, 238, 249, 250, 306,
     311, 323, 325, 328–330
  reflexive, 14, 24, 25, 81, *82*, 82, 83,
     104, 106, 119, 150, 205, 206,
     220, 221, 246, 250, 252, 264,
     311–313, 327, 329–334, 336
  simple, 73, 270

  stable, 126, 129, 131, 132, 136,
     137, 150, 190, 192, 194, 195, 284
  torsion, 178, 189, 250, 252, 329
  torsion-free, 14, 20, 21, 24, 41, 42,
     52, 55–58, 69, 166, 178, 206,
     247, 250, 251, 264, 307, 309,
     327, 330, 331
  torsionless, 205, 206
  totally reflexive, *209*, 209–215
  trivial, 65, 73
  Ulrich, *315*
  weakly extended, *43*, 47, 59,
     165–167, 171
  weakly liftable, 43
monoid, see semigroup
$\mu_R(M)$, *315*
multiplicity, 41, 44, 54–56, 100, 101,
   106, 116, 160, 267, 276,
   280–283, 294, 297, 299, 301,
   302, 307, *313*, 314–316
multiplicity two, 42, 54, 57, 60, 101,
   106, 142, 143, 145, 146, 268,
   295, 298, 299, 303, 304
Mumford, David, 100, 117–119

n-torsionless module, 205, 206
Nagata, Masayoshi, 275
NAK, see Nakayama's Lemma
Nakayama's Lemma, 6, 9, 20, 31, 44,
   47, 72, 181, 215, 272, 279, 282,
   318, 321, 333
Nazarova, Liudmila A., 267
nc-local, 3, 5, 10
  endomorphism ring, 1, 3–6, 10,
     134, 203, 218, 223, 225, 226,
     238, 272, 273
negative definite matrix, 100, 102,
   117
nilpotent Jordan block, 29
nilpotent Jordan block, 28, 31, 34, 37
nilradical, 53, 54, 301, 302, 305
nodal cubic curve, 46, 58, 148
Noether, Emmy, 77, 328
non-constant ranks, 57
non-derogatory matrix, 30
non-free locus, 284
non-normal locus, 246
non-singular ring, 116
norm, 78

normal bundle, 100, 101
normal domain, 23–25, 63, 67, 81, 82, 98, 101, 104, 106, 117–119, 229, 230, 233, 234, 245, 246, 288, 291, 295, 306, 307, 311, 313, 325–331, 334, 336
normal forms, 89, 90, 146, 159–160
normal scheme, 117
number of generators, 297

$\mathcal{O}$, 86
octahedron, 84
Odenthal, Charles, 17
ω_R, 176
one-one correspondence, 61, 70–73, 83, 97, 105
orbit-stabilizer theorem, 338
orthonormal basis, 85
overring, 299

perfect field, 52, 141, 173, 267, 268, 275, 276, 279
periodic free resolution, 126
Peskine, Christian, 176
Picard group Pic, 25, 105
Piepmeyer, Greg, 203, 210
pitchfork construction, 184, 208
Platonic solids, 84, 86
point at infinity, 85
pointed étale neighborhood, 164, 165, 319
Popescu, Dorin, 280
poset, 27
positive characteristic, 90, 99, 119, 146, 176
positive normal affine semigroup, *see* Krull monoid
Prill, David, 69, 335, 339
prime avoidance, 285, 337
  countable, 241, 242
primelike, 3
primitive element theorem, 38, 39, 283
principal ideal domain, 41, 42, 45, 47
principal ideal ring, 29, 30, 38, 39, 46, 55, 56, 294, 317
projective line, 85, 99, 102, 103
projective module, 20, 32, 43, 44, 46, 47, 56, 61, 64, 65, 70–73, 78, 83, 177, 230, 238, 249, 250, 306, 311, 323, 325, 328–330
projective plane, 102
projective space, 143
pseudo-reflection, *69*, 69, 70, 72, 78, 79, 83, 107, 230, 335
pullback, 20, 42, 43, 48, 110, 111, 196, 197, 223, 234, 245, 246, 251, 252, 255, 257, 258
pure
  homomorphism, 110, 239
  submodule, 110, 120, 239
pure height one, 177, 178, 188, 326, 327, 329, 330
pure local ring, 118
purely transcendental extension, 170
purity of the branch locus, 118, 326–334
pushout, 110–114, 127, 185–187, 219, 238

quasi-homogeneous polynomial, *89*, 99, 192
quotient field, 63, 77, 98, 117, 118

rad and rad^2, *224*, 229, 239
ramification, 19, 66, 69, 321, 335, 337
ramification index, 338, 339
rank, 13, *20*, 20, 22, *56*, 69, 126, 176, 177, 179, 184, 189, 190, 234, 249, 268, 295, 296, 305, 313, *316*, 332, 334
rational curve, 99
rational double point, *see* Kleinian singularity
rational singularity, 98–101, 104, 119, 120, 151
Raynaud, Michel, 178
real numbers $\mathbb{R}$, 79, 128
reduced matrix factorization, 125, 126, 131, 141
reduced ring, 38, 41, 45, 53, 172, 245, 246, 248, 259, 264, 268, 279, 299, 302, 305, 310, 316, 324, 325
reduced semigroup, 14
reduced syzygy redsyz, 219–223, 235, 289, 292, 296, *311*
reduction, 101, 314
reflecting hyperplane, 335, 337

reflection, *79*, 151, 153, 335
reflection group, 155
reflexive module, 14, 24, 25, 81, *82*, 82, 83, 104, 106, 119, 150, 205–207, 220, 221, 246, 250, 252, 264, 311–313, 327, 329–334, 336
reflexive product, *106*, 234, 246, 264, 327
reflexive subcategory, *211*, 212, 215
regular homomorphism, *167*, 169, 170
regular in codimension one, 245, 246, 311
regular local ring, 53, 69, 78, 106, 115, 116, 118, 119, 123, 124, 128, 149, 151, 160, 169, 194, 195, 200, 242, 243, 246, 249, 257, 273–275, 285, 311, 315, 330, 333, 334, 336
Reiner, Irving, 31, 38, 41, 45, 49, 50
Reiten, Idun, 91, 177, 287, 291, 293
Remak, Robert E., 1
representation theory
  of Artin algebras, 217
  of finite groups, 61, 71, 72, 81, 97, 151
  of finite-dimensional algebras, 267
  of lattices over orders, 50
resolution of singularities, 97–104
retraction, 77
Reynolds operator, *62*, 63, 65, 90
Riemann-Roch Theorem, 101
right minimal homomorphism, 180, 211–213, 226
right minimal MCM approximation, *180*, 180, 182, 183
Roberts, Paul C., 176
Roczen, Marko, 280
Roĭter, Andreĭ V., 32, 38, 50, 267

$S\#G$, *64*
(S_1), 66, 67, 310
(S_2), 66, 67, 82, 245, 264, 266, 311, 312, 326, 329
Saito, Kyoji, 192
Sally, Judith D., 44
Sanders, Herbert, 298
Schanuel's Lemma, 166, 180, 311
Schmidt, Otto, 1

Schreyer, Frank-Olaf, 141, 145, 163, 241, 245, 252, 260, 262, 263, 284
Schur's Lemma, 93
scroll, *291*, 291, 294, 295
self-intersection, 100, 101
semigroup, 13, 14
  cancellative, 14
  finitely generated, 13
  reduced, 14
seminormal, *49*, 50, 59
seminormalization, 49
separable algebra, 324, 325
separable field extension, 36, 38, 49–51, 66, 117, 163, 164, 170–172, 321, 324–330, 339
series of singularities, 260, 262
Serre's conditions
  (S_1), 66, 67, 310
  (S_2), 66, 67, 82, 245, 264, 266, 311, 312, 326, 329
  (S_n), 309, *310*, 311, 312, 325
Serre's criterion for normality, 245, 295, 311, 325
Sharp, Rodney, 177
sharping, *128*, 130, 133, 137
sheafification, 106
shenanigans, 137
Shida, Akira, 180, 182
Simon, Anne-Marie, 182
simple module, 73, 270
simple singularity, 51, *141*, 142, 143, 145, 146, 149, 160, 173, 253, 259, 297
singular locus, 167, 241, 242, 263, 275, 284
skew group ring $S\#G$, 61, *64*, 65, 67, 69–74, 78, 83, 135, 230, 292, 330, 336
Smalø, Sverre, 268, 280–283
small subgroup, 69, 72, 339
(S_n), *310*
Snake Lemma, 109, 188, 218, 226, 247, 251, 252
socle, 101, 201, 223
Solberg, Øyvind, 43, 160
special linear group $SL(n,k)$, 81, 83, 84, 86, 90, 92, 93, 104, 151, 152, 229

special orthogonal group SO(3), 84, 85
special unitary group SU(n), 84, 85
spectral sequence, 221
splitting number spl(R), 22, 23
square root of -1, 86, 133, 137, 154, 253
stable
  Hom module Hom, 203–207, 216
  category, 139
  endomorphism ring End, 203, 207, 222, 223, 290
  equivalence, 204
  MCM trace, 201
  module, 126, 129, 131, 132, 136, 137, *150*, 190, 192, 194, 195, 284
stable ideal, *44*, 56
Steinitz, Ernst, 10
Steinke, Günther, 159, 160
stereographic projection, 85
stretched ring, 294, 295
strict transform, 104
strictly positive vector, *21*, 25
striped matrix, 29
Striuli, Janet, 110, 111, 203, 210
strongly unbounded representation type, 267, 283
Strooker, Jan R., 182
stubbornness, 175
sub-additive function, 91, 102
subsemigroup
  expanded, 14, 16, 22, 23
  full, 14, 16, 25, 26
surface singularity, 98–101, 104, 119, 245
Swan, Richard G., 3
system of parameters, faithful, *271*, 271–280, 285
syzygy, 112, 113, 116, 126, 128, 134, 136–138, 150, 166, 185, 189, 194, 196, 197, 200, 205, 206, 219, 220, 242, 280, 284, 289, 309, *311*, 312, 313
  infinite, 209
  reduced, *see* reduced sygyzy
Szpiro, Lucien, 176

$\mathcal{T}$, *86*
Takahashi, Ryo, 203, 210

Tangent Lemma, 147
tangent line of an analytic curve, 146–149
Tate, John, 151
τ, *222*
tedium, 172
tetrahedron, 84
Thrall, Robert M., 267
Threlfall, William, 86
topology, 117
torsion, *309*
torsion module, 178, 189, 250, 252, *309*, 329
torsion submodule, 41, 166, 246, 247, *310*
torsion-free module, 14, 20, 21, 24, 41, 42, 52, 55–58, 69, 166, 178, 206, 247, 250, 251, 264, 307, *309*, 327, 330, 331
torsionless module, 205, 206
total quotient ring, 17, 41, 42, 44, 45, 52, 59, 60, 177, 245, 246, 248, 249, 254, 306, 309, 316
totally acyclic complex, *209*
totally reflexive module, 203, *209*, 209–215
Tr, *204*
trace, 63, 93, 328
trace ideal, *198*, 198, 199, 201
trivial matrix factorization, *124*, 125, 132
trivial module, 65, 73
trivial representation, 73, 74, 104, 292
truncated diagonal, 28
twisted group ring, *see* skew group ring
type, *116*, 193

ubiquity, 54, 83
Ulrich, Bernd, 313
unbounded CM type, 31, 297, 301
unbounded representation type, 267
uncountable CM type, 263, 268, 280, 284, 285
uncountable field, 142, 144, 241, 242, 284, 285
unique factorization, 3, 300, 317

unique factorization domain, 14, 25, 26, 307, 338
unitary group U(n), 84, 151
unitary transformation, 84
universal bound, 57
universally catenary, 167, 170
unramified
   homomorphism, 321–328, 330
   local homomorphism, *66*, 67, *164*, 164, 321–322, 325
   prime ideal, 322, 325, 326, 330, 336, 337, 339
unramified in codimension one, 66, 67, 69, 118, 327, 328, 330, 339

V(R), *13*
valued tree, 117
Vandermonde matrix, 244
Vasconcelos, Wolmer V., 44
Verdier, Jean-Louis, 97, 104, 160

Wakamatsu's Lemma, 211, 213
Warfield, Robert B., 31
Watanabe, Kei-ichi, 81, 84, 98
weakly extended module, *43*, 47, 59, 165–167, 171
weakly liftable module, 43
Wedderburn, J. H. M., 1
Wedderburn-Artin Theorem, 5, 6
Weierstrass Preparation Theorem, 142, 144, 300
Weierstrass, Karl, 31
Wiegand, Roger, 38, 50, 57
Wiegand, Sylvia, 57

Yoneda correspondence, 109, 110
Yoshino, Yuji, 114, 146, 188, 200, 268, 275, 276

Zariski and Samuel, 144
Zariski topology, 35, 102, 103
Zariski's Main Theorem, 99

Selected Titles in This Series

181 **Graham J. Leuschke and Roger Wiegand,** Cohen-Macaulay Representations, 2012
180 **Martin W. Liebeck and Gary M. Seitz,** Unipotent and Nilpotent Classes in Simple Algebraic Groups and Lie Algebras, 2012
179 **Stephen D Smith,** Subgroup Complexes, 2011
178 **Helmut Brass and Knut Petras,** Quadrature Theory, 2011
177 **Alexei Myasnikov, Vladimir Shpilrain, and Alexander Ushakov,** Non-commutative Cryptography and Complexity of Group-theoretic Problems, 2011
176 **Peter E. Kloeden and Martin Rasmussen,** Nonautonomous Dynamical Systems, 2011
175 **Warwick de Launey and Dane Flannery,** Algebraic Design Theory, 2011
174 **Lawrence S. Levy and J. Chris Robson,** Hereditary Noetherian Prime Rings and Idealizers, 2011
173 **Sariel Har-Peled,** Geometric Approximation Algorithms, 2011
172 **Michael Aschbacher, Richard Lyons, Stephen D. Smith, and Ronald Solomon,** The Classification of Finite Simple Groups, 2011
171 **Leonid Pastur and Mariya Shcherbina,** Eigenvalue Distribution of Large Random Matrices, 2011
170 **Kevin Costello,** Renormalization and Effective Field Theory, 2011
169 **Robert R. Bruner and J. P. C. Greenlees,** Connective Real K-Theory of Finite Groups, 2010
168 **Michiel Hazewinkel, Nadiya Gubareni, and V. V. Kirichenko,** Algebras, Rings and Modules, 2010
167 **Michael Gekhtman, Michael Shapiro, and Alek Vainshtein,** Cluster Algebras and Poisson Geometry, 2010
166 **Kyung Bai Lee and Frank Raymond,** Seifert Fiberings, 2010
165 **Fuensanta Andreu-Vaillo, José M. Mazón, Julio D. Rossi, and J. Julián Toledo-Melero,** Nonlocal Diffusion Problems, 2010
164 **Vladimir I. Bogachev,** Differentiable Measures and the Malliavin Calculus, 2010
163 **Bennett Chow, Sun-Chin Chu, David Glickenstein, Christine Guenther, James Isenberg, Tom Ivey, Dan Knopf, Peng Lu, Feng Luo, and Lei Ni,** The Ricci Flow: Techniques and Applications, 2010
162 **Vladimir Maz'ya and Jürgen Rossmann,** Elliptic Equations in Polyhedral Domains, 2010
161 **Kanishka Perera, Ravi P. Agarwal, and Donal O'Regan,** Morse Theoretic Aspects of p-Laplacian Type Operators, 2010
160 **Alexander S. Kechris,** Global Aspects of Ergodic Group Actions, 2010
159 **Matthew Baker and Robert Rumely,** Potential Theory and Dynamics on the Berkovich Projective Line, 2010
158 **D. R. Yafaev,** Mathematical Scattering Theory, 2010
157 **Xia Chen,** Random Walk Intersections, 2010
156 **Jaime Angulo Pava,** Nonlinear Dispersive Equations, 2009
155 **Yiannis N. Moschovakis and Yiannis N. Moschovakis,** Descriptive Set Theory, 2009
154 **Andreas Čap, Andreas Čap, and Jan Slovák,** Parabolic Geometries I, 2009
153 **Habib Ammari, Hyeonbae Kang, and Hyundae Lee,** Layer Potential Techniques in Spectral Analysis, 2009
152 **János Pach and Micha Sharir,** Combinatorial Geometry and Its Algorithmic Applications, 2009
151 **Ernst Binz and Sonja Pods,** The Geometry of Heisenberg Groups, 2008
150 **Bangming Deng, Jie Du, Brian Parshall, and Jianpan Wang,** Finite Dimensional Algebras and Quantum Groups, 2008
149 **Gerald B. Folland,** Quantum Field Theory, 2008
148 **Patrick Dehornoy, Patrick Dehornoy, Ivan Dynnikov, Dale Rolfsen, and Bert Wiest,** Ordering Braids, 2008

SELECTED TITLES IN THIS SERIES

147 **David J. Benson and Stephen D. Smith,** Classifying Spaces of Sporadic Groups, 2008
146 **Murray Marshall,** Positive Polynomials and Sums of Squares, 2008
145 **Tuna Altınel, Alexandre V. Borovik, and Gregory Cherlin,** Simple Groups of Finite Morley Rank, 2008
144 **Bennett Chow, Bennett Chow, Sun-Chin Chu, David Glickenstein, Christine Guenther, James Isenberg, Tom Ivey, Dan Knopf, Peng Lu, Feng Luo, and Lei Ni,** The Ricci Flow: Techniques and Applications, 2008
143 **Alexander Molev,** Yangians and Classical Lie Algebras, 2007
142 **Joseph A. Wolf,** Harmonic Analysis on Commutative Spaces, 2007
141 **Vladimir Maz'ya, Vladimir Maz'ya, Vladimir Maz'ya, and Gunther Schmidt,** Approximate Approximations, 2007
140 **Elisabetta Barletta, Sorin Dragomir, and Krishan L. Duggal,** Foliations in Cauchy-Riemann Geometry, 2007
139 **Michael Tsfasman, Michael Tsfasman, Michael Tsfasman, Serge Vlăduţ, Serge Vlăduţ, and Dmitry Nogin,** Algebraic Geometric Codes: Basic Notions, 2007
138 **Kehe Zhu,** Operator Theory in Function Spaces, 2007
137 **Mikhail G. Katz,** Systolic Geometry and Topology, 2007
136 **Jean-Michel Coron,** Control and Nonlinearity, 2007
135 **Bennett Chow, Bennett Chow, Sun-Chin Chu, David Glickenstein, Christine Guenther, James Isenberg, Tom Ivey, Dan Knopf, Peng Lu, Feng Luo, and Lei Ni,** The Ricci Flow: Techniques and Applications, 2007
134 **Dana P. Williams,** Crossed Products of C*-Algebras, 2007
133 **Andrew Knightly and Charles Li,** Traces of Hecke Operators, 2006
132 **J. P. May and J. Sigurdsson,** Parametrized Homotopy Theory, 2006
131 **Jin Feng and Thomas G. Kurtz,** Large Deviations for Stochastic Processes, 2006
130 **Qing Han and Jia-Xing Hong,** Isometric Embedding of Riemannian Manifolds in Euclidean Spaces, 2006
129 **William M. Singer,** Steenrod Squares in Spectral Sequences, 2006
128 **Athanassios S. Fokas, Alexander R. Its, Andrei A. Kapaev, and Victor Yu. Novokshenov,** Painlevé Transcendents, 2006
127 **Nikolai Chernov and Roberto Markarian,** Chaotic Billiards, 2006
126 **Sen-Zhong Huang,** Gradient Inequalities, 2006
125 **Joseph A. Cima, Alec L. Matheson, and William T. Ross,** The Cauchy Transform, 2006
124 **Ido Efrat,** Valuations, Orderings, and Milnor K-Theory, 2006
123 **Barbara Fantechi, Lothar Göttsche, Luc Illusie, Steven L. Kleiman, Nitin Nitsure, and Angelo Vistoli,** Fundamental Algebraic Geometry, 2005
122 **Antonio Giambruno and Mikhail Zaicev,** Polynomial Identities and Asymptotic Methods, 2005
121 **Anton Zettl,** Sturm-Liouville Theory, 2005
120 **Barry Simon,** Trace Ideals and Their Applications, 2005
119 **Tian Ma and Shouhong Wang,** Geometric Theory of Incompressible Flows with Applications to Fluid Dynamics, 2005
118 **Alexandru Buium,** Arithmetic Differential Equations, 2005
117 **Volodymyr Nekrashevych and Volodymyr Nekrashevych,** Self-Similar Groups, 2005
116 **Alexander Koldobsky,** Fourier Analysis in Convex Geometry, 2005

For a complete list of titles in this series, visit the
AMS Bookstore at **www.ams.org/bookstore/**.